AF566796

IET POWER AND ENERGY SERIES 76

Power System Stability

Other volumes in this series:

Volume 1	**Power Circuit Breaker Theory and Design** C.H. Flurscheim (Editor)
Volume 4	**Industrial Microwave Heating** A.C. Metaxas and R.J. Meredith
Volume 7	**Insulators for High Voltages** J.S.T. Looms
Volume 8	**Variable Frequency AC Motor Drive Systems** D. Finney
Volume 10	**SF_6 Switchgear** H.M. Ryan and G.R. Jones
Volume 11	**Conduction and Induction Heating** E.J. Davies
Volume 13	**Statistical Techniques for High Voltage Engineering** W. Hauschild and W. Mosch
Volume 14	**Uninterruptible Power Supplies** J. Platts and J.D. St Aubyn (Editors)
Volume 15	**Digital Protection for Power Systems** A.T. Johns and S.K. Salman
Volume 16	**Electricity Economics and Planning** T.W. Berrie
Volume 18	**Vacuum Switchgear** A. Greenwood
Volume 19	**Electrical Safety: A guide to causes and prevention of hazards** J. Maxwell Adams
Volume 21	**Electricity Distribution Network Design, 2nd Edition** E. Lakervi and E.J. Holmes
Volume 22	**Artificial Intelligence Techniques in Power Systems** K. Warwick, A.O. Ekwue and R. Aggarwal (Editors)
Volume 24	**Power System Commissioning and Maintenance Practice** K. Harker
Volume 25	**Engineers' Handbook of Industrial Microwave Heating** R.J. Meredith
Volume 26	**Small Electric Motors** H. Moczala *et al.*
Volume 27	**AC-DC Power System Analysis** J. Arrillaga and B.C. Smith
Volume 29	**High Voltage Direct Current Transmission, 2nd Edition** J. Arrillaga
Volume 30	**Flexible AC Transmission Systems (FACTS)** Y-H. Song (Editor)
Volume 31	**Embedded Generation** N. Jenkins *et al.*
Volume 32	**High Voltage Engineering and Testing, 2nd Edition** H.M. Ryan (Editor)
Volume 33	**Overvoltage Protection of Low-Voltage Systems, Revised Edition** P. Hasse
Volume 36	**Voltage Quality in Electrical Power Systems** J. Schlabbach *et al.*
Volume 37	**Electrical Steels for Rotating Machines** P. Beckley
Volume 38	**The Electric Car: Development and future of battery, hybrid and fuel-cell cars** M. Westbrook
Volume 39	**Power Systems Electromagnetic Transients Simulation** J. Arrillaga and N. Watson
Volume 40	**Advances in High Voltage Engineering** M. Haddad and D. Warne
Volume 41	**Electrical Operation of Electrostatic Precipitators** K. Parker
Volume 43	**Thermal Power Plant Simulation and Control** D. Flynn
Volume 44	**Economic Evaluation of Projects in the Electricity Supply Industry** H. Khatib
Volume 45	**Propulsion Systems for Hybrid Vehicles** J. Miller
Volume 46	**Distribution Switchgear** S. Stewart
Volume 47	**Protection of Electricity Distribution Networks, 2nd Edition** J. Gers and E. Holmes
Volume 48	**Wood Pole Overhead Lines** B. Wareing
Volume 49	**Electric Fuses, 3rd Edition** A. Wright and G. Newbery
Volume 50	**Wind Power Integration: Connection and system operational aspects** B. Fox *et al.*
Volume 51	**Short Circuit Currents** J. Schlabbach
Volume 52	**Nuclear Power** J. Wood
Volume 53	**Condition Assessment of High Voltage Insulation in Power System Equipment** R.E. James and Q. Su
Volume 55	**Local Energy: Distributed generation of heat and power** J. Wood
Volume 56	**Condition Monitoring of Rotating Electrical Machines** P. Tavner, L. Ran, J. Penman and H. Sedding
Volume 57	**The Control Techniques Drives and Controls Handbook, 2nd Edition** B. Drury
Volume 58	**Lightning Protection** V. Cooray (Editor)
Volume 59	**Ultracapacitor Applications** J.M. Miller
Volume 62	**Lightning Electromagnetics** V. Cooray
Volume 63	**Energy Storage for Power Systems, 2nd Edition** A. Ter-Gazarian
Volume 65	**Protection of Electricity Distribution Networks, 3rd Edition** J. Gers
Volume 66	**High Voltage Engineering Testing, 3rd Edition** H. Ryan (Editor)
Volume 67	**Multicore Simulation of Power System Transients** F.M. Uriate
Volume 68	**Distribution System Analysis and Automation** J. Gers
Volume 69	**The Lightening Flash, 2nd Edition** V. Cooray (Editor)
Volume 70	**Economic Evaluation of Projects in the Electricity Supply Industry, 3rd Edition** H. Khatib
Volume 76	**Power System Stability: Modelling, analysis and control** Abdelhay A. Sallam and Om P. Malik
Volume 78	**Numerical Analysis of Power System Transients and Dynamics** A. Ametani (Editor)
Volume 79	**Vehicle-to-Grid: Linking electric vehicles to the smart grid** J. Lu and J. Hossain (Editors)
Volume 905	**Power System Protection, 4 volumes**

Power System Stability

Modelling, analysis and control

Abdelhay A. Sallam and Om P. Malik

The Institution of Engineering and Technology

Published by The Institution of Engineering and Technology, London, United Kingdom

The Institution of Engineering and Technology is registered as a Charity in England & Wales (no. 211014) and Scotland (no. SC038698).

First published 2015

The Institution of Engineering and Technology
Michael Faraday House
Six Hills Way, Stevenage
Herts, SG1 2AY, United Kingdom

www.theiet.org

British Library Cataloguing in Publication Data
A catalogue record for this product is available from the British Library

ISBN 978-1-84919-944-5 (hardback)
ISBN 978-1-84919-945-2 (PDF)

Typeset in India by MPS Limited
Printed in the UK by CPI Group (UK) Ltd, Croydon

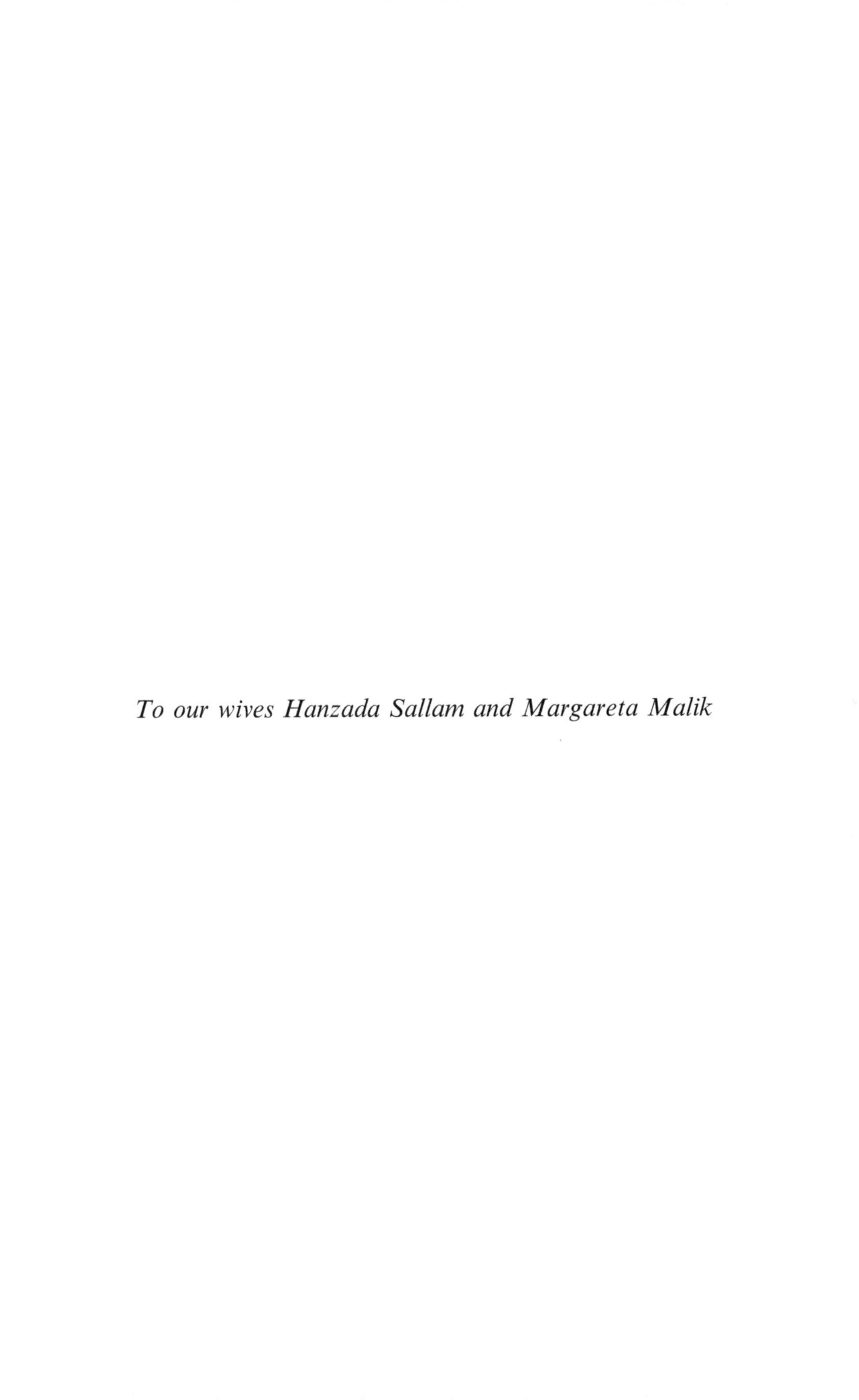

To our wives Hanzada Sallam and Margareta Malik

Contents

Preface

Modern day large power systems are essentially dynamic systems with stringent requirements of high reliability for the continuous availability of electricity. Reliability is contingent on the power system retaining stable operation during steady-state operation and also following disturbances. The subject of power system stability has been studied for many decades. With new developments, and there have been many over the past couple of decades, new concerns and problems arise that need to be studied and analysed. The objective of this book is a step in that direction though not ignoring the conventional and well-established approaches.

To ensure stable operation of the power system, it is necessary to analyse the power system performance under various operating conditions. Analysis includes studies such as power flow and both steady-state and transient stabilities. To perform such studies requires knowledge about the models to represent the various components that constitute an integrated power system. In situations where there is a risk of loss of stability, it is necessary to apply controls that can ensure stable and uninterrupted supply of electricity following a disturbance.

The subject of stability thus encompasses modelling, computation of load flow in the transmission grid, stability analysis under both steady-state and disturbed conditions and appropriate controls to enhance stability. All these topics are covered in this book in that order to provide a fairly comprehensive treatment of the overall subject of stability of power systems. The subject matter is covered at a level that is suitable for students, scientists and engineers involved in the study, design, analysis and control of power systems.

The stage for a study of power system stability is set in Chapter 1 where an overview of the problem of power system stability is provided. The following part of the book is divided into four parts, each part dedicated to a specific topic, i.e. Modelling, Power Flow, Stability Analysis, and Stability Enhancement and Control.

Part I Modelling consists of three chapters. A comprehensive description of modelling synchronous machines is provided, which is followed by the models for transformers, transmission lines and loads.

Part II Power Flow consists of two chapters. Description of the general concept of power flow is provided, followed by a description of the various commonly used techniques for load flow and optimal load flow.

Part III Stability Analysis consists of three chapters. Small signal stability and conventional methods of transient stability assessment are covered in Chapters 7 and 8, respectively. Description of transient stability calculation using transient energy function methods is discussed in Chapter 9. These methods are useful for online assessment of transient stability of large power systems. They can provide continuous assessment of the state of the power system so that measures can be taken in advance, in case a gradual degradation in the power system secure state operation is noted.

Part IV Stability Enhancement and Control consists of six chapters. Various measures for stability enhancement are described in this part. Not only the conventional techniques but also the newly emerging techniques for power system stability enhancement and control are described. Power system stabilisers, initially developed in the 1950s, are the most common devices used in the power systems to provide damping following disturbances. Advances making use of adaptive control and artificial intelligence (AI) techniques have taken place in the development of new algorithms for the power system stabiliser. A brief introduction to AI techniques is given in Chapter 10. The conventional power system stabiliser as well as the developments in adaptive and AI-based power system stabiliser is described in Chapter 11. Use of power electronics-based compensation, series and shunt compensation, is described in Chapters 12 and 13, respectively. These devices, commonly known as FACTS devices, are described in Chapter 14.

With the deployment of satellites and the commensurate developments in communication technologies, such as GPS, investigations in many additional directions are taking place to take advantage of these new technologies in improving power system stability and reliability. In addition, the significant move towards generation of electricity from renewable sources has resulted in many new developments, in particular energy storage. These technologies are in the initial state of development, and a brief introduction to the newly developing technologies is given in Chapter 15.

Certain supporting material is described in Appendices I–IV.

The subject matter of this book covers a broad spectrum of topics. The material covered in the book includes only a very limited part of the authors' own research over the last 40 years. Similarly, because of the broad scope, it is possible to cover certain topics only briefly. However, a good set of references is included as resource material so that a reader interested in pursuing a specific topic in more detail can do so by going to these references.

These days no man is unto self. It is the cooperative effort of many. Even though not individually named, the authors wish to acknowledge the work and help of innumerable graduate students, colleagues and others whose association over the many years has helped them to compile this book.

To conclude, the authors hope that the readers will derive benefit from reading the book and wish them all success in their endeavours.

Chapter 1
Power system stability overview

1.1 General

A dynamic system, in general, would necessarily entail a detailed study of some concepts interrelated to each other. Particularly, in system planning these concepts are system reliability, security and stability. Definitions of these concepts may help in understanding the relationships and differences between them [1, 2].

System reliability is defined as the probability of a system's ability to provide a desired function under specific operating conditions during its lifetime.

System security refers to the degree of risk in its ability to withstand contingencies without interrupting the system function. It pertains to system robustness to contingencies. Thus, it depends on the system operating condition and probability of contingency occurrence.

System stability is defined as the ability of the system to continue its intact operation and remain stable following a disturbance. Consequently, it depends on the operating condition and the nature of the physical disturbance.

A power system is similar to any dynamic system. Its function is to provide electricity to loads at a desired quality with as few interruptions as possible. The power system is commonly subjected to disturbances during operation. According to the three concepts defined above reliability can be seen as the primary objective in power system design and operation. To attain system reliability the system must be secure most of the time, during and post fault periods. This necessitates that the system must be stable. Therefore, the aspects of security and stability are time-varying attributes that can be judged by analysing the power system performance under a specified set of conditions. On the other hand, reliability is determined in terms of the time average performance of the power system and can be judged by studying the behaviour of the system over a period of time.

1.2 Understanding power system stability

Synchronous generators in an interconnected power system are the main source of producing electrical power. A necessary condition for the transmission and power exchange is that all generators must rotate in synchronism, that is, the average electrical speed of all generators must remain the same anywhere in the system. Each generator is driven by a prime mover. The prime mover applies mechanical power to the generator that in turn delivers electrical power into the connected system. In a steady-state operation, the input mechanical power to the generator balances the output electrical power. Both the input mechanical power and the output electrical power produce mechanical and electrical torques, respectively, when applied to the shaft. The mechanical torque is in the direction of rotation, whereas the electrical torque is in a direction opposite of rotation.

If a fault occurs in the system, the output electrical power changes rapidly at a rate faster than the input mechanical power. This is because the excitation system of the generator has a fast response whereas the prime mover controller has a relatively slow response. Accordingly, a temporary imbalance of power exists causing a difference in torque applied to the shaft. This results in a change of rotor speed (increase or decrease) and, thus, the relative rotor angle changes. The rotor angle δ (also called torque or power angle) is the angle between the rotor mmf and the resultant of the rotor and stator mmfs (Figure 1.1) [3].

If the change of rotor speed continues perhaps beyond the limits of generator synchronous operation, the protective relaying system operates to isolate the generator from the rest of the system. Then the remaining system is disturbed due to the loss of generation. This disturbance may result in additional units tripping offline, and potentially a cascading outage.

Therefore, the concept of power system stability relates to the ability of generators on a system to maintain synchronism and the tendency to return to a steady-state operation point following a system disturbance [4].

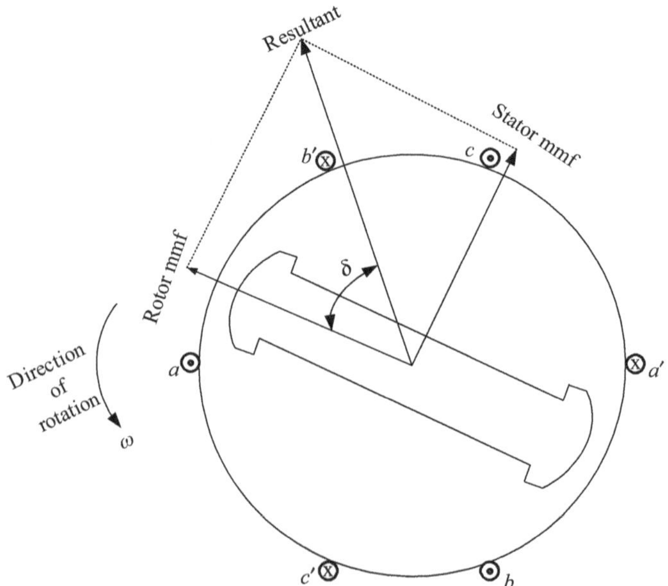

Figure 1.1 Rotor angle and resultant of stator and rotor mmfs

1.3 Classification of power system stability

Classification of power system stability is based on the type of disturbance. Disturbances can be divided into two types: small and large. A small disturbance affects the system by small changes in its behaviour, e.g. small changes in the load and tripping a line carrying insignificant power. The system dynamics can be analysed using linearised equations, known as small signal analysis. A large disturbance results in a sudden big change in some of the system parameters. The system dynamics is studied using non-linear equations. For instance, a sudden change in the load, loss of generation, switching out overloaded transmission lines, symmetrical and unsymmetrical faults and lightning strokes can be considered large disturbances.

Accordingly, the system stability for the purpose of analysis can be classified into two classes: small signal stability and transient stability.

1.3.1 Small signal stability

The synchronous machine in an interconnected power system can be simply represented by an internal voltage source, E_g, behind the generator reactance, X_g, which is equal to synchronous reactance, X_d, for steady-state analysis. More explanation of generator reactance and its variation is given in Chapter 2, Part I. The output electrical power, P_e, on a steady-state basis can approximately be expressed as

$$P_e = P_{\max} \sin \delta = \frac{E_g E_t}{X_g} \sin \delta \tag{1.1}$$

where E_t is the machine terminal voltage and δ is the power angle (angle between machine terminal voltage and machine internal voltage). $P_{\max} = (E_g E_t / X_g)$, called steady-state stability limit, equals the output power at $\delta = 90°$. Equation (1.1) is typically plotted as shown in Figure 1.2.

The synchronous generator is assumed to be in steady-state operation at point ◙ as shown in Figure 1.2, where the input mechanical power, P_m, equals the output electrical power, P_e, at power angle δ_o. When a small temporary disturbance occurs, e.g. a small reduction in the load, resulting in a decrease of the output power to P_{e1} (point #1) for a short period, the rotor will accelerate as P_m is greater than P_e. Thus, the rotor speed initially increases to absorb the excess energy in the rotor inertia and the increase of angle continues until point #2. From point ◙ to point #2, P_m is less than P_e and the rotor decelerates and tries to overcome the effect of inertia that vanishes at point #2. Then, due to excess output P_{e2} than input P_m the rotor decelerates, the power angle decreases and the operating point moves again towards the point ◙. The oscillation of operating point between points #1 and #2 continues. It is to be noted that the rate of power angle change at points #1 and #2 is zero. If the amplitude of oscillation decays with time (damped oscillation) and the operating point stands at point ◙ or at a neighbouring equilibrium point, system stability is attained. Conversely, in the case of increasing oscillation excursions

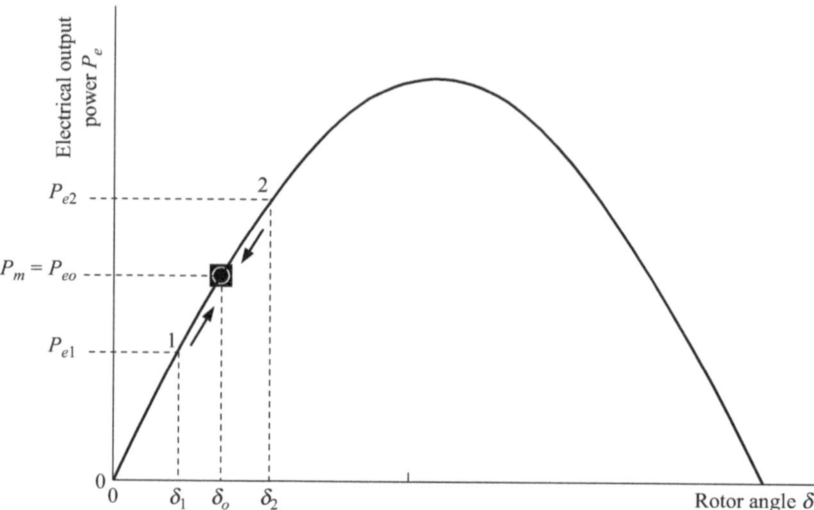

Figure 1.2 Power–angle curve illustrating machine oscillations

(un-damped oscillation) a balanced operating point cannot be achieved and the system is unstable. The capability of the power system to damp the oscillations is affected by a number of factors such as the generator design, the strength of the machine interconnection to the network and the setting of the excitation system. Power systems generally have efficient damping of oscillations at normal operating conditions. Under special circumstances they may experience a significant reduction of damping capability following disturbances, and in the worst circumstance the damping may become negative. Thus, the oscillations grow and eventually the synchronism is lost. This form of instability is referred to as small signal instability. Therefore, small signal stability is defined as the ability of the power system to remain stable in the presence of small disturbances.

Loss of small signal stability results in one or more of the types of oscillations that have been experienced with large interconnected power system involving rotor swings. Rotor swing may grow without bound or take a long time to dampen. Three main types of oscillations are dealt with for small signal stability analysis. First, *local mode oscillations* that involve one or more synchronous generators at a power plant swinging together against a comparatively large power system or load centre. Their frequency is in the range of 0.7–2 Hz. Second, *inter-unit oscillations* that involve two or more synchronous generators at a power plant or nearby power plants swinging against each other with a frequency range of 1.5–3 Hz. Third, *inter-area oscillations* that usually involve a group of generators on one part of a power system swinging against another group in another part of that system. The frequency of this type of oscillations normally is in the range of less than 0.7 Hz.

Damping of generator oscillations has a prominent role in small signal stability analysis. The power system contains inherent damping effects that tend to damp out dynamic oscillations. The natural damping of the system is represented by the positive term D in the swing (1.2). It is generally sufficient to prevent any sustained oscillations unless a source of negative damping is introduced.

$$\frac{2H}{\omega_s}\frac{\mathrm{d}^2\delta}{\mathrm{d}t^2}+\frac{D}{\omega_s}\frac{\mathrm{d}\delta}{\mathrm{d}t}+K\Delta\delta=0 \tag{1.2}$$

where

$H \triangleq$ rotor inertia constant (MW.s/MVA)
$\omega_s \triangleq$ synchronous speed (elec.rad/s)
$D \triangleq$ damping coefficient (pu power/pu freq. change)
$K \triangleq$ synchronising coefficient (pu ΔP/rad) = the slope of power – angle curve at the particular steady-state operating point
$\Delta\delta \triangleq$ rotor angle deviation from the steady-state operating point (rad)

A major source of negative damping is the high gain voltage regulator with fast excitation system. The main function of the voltage regulator is to continually adjust the generator excitation level in response to changes in generator terminal voltage. It acts to accurately maintain a desired generator voltage and change the excitation level in response to disturbances on the system. It is found that increasing forcing capability and decreasing response time of the excitation system provide tremendous benefits to transient stability (as explained in Section 1.3.2). On the contrary, it can contribute a significant amount of negative damping to oscillations because it can reduce damping torque. Thus, an excitation system has the potential to contribute to small signal instability of power systems. On the other hand, it is recognised that the normal feedback control actions of voltage regulators and speed governors on generating units have the potential of contributing negative damping that can cause un-damped modes of dynamic oscillations.

Further understanding of both positive and negative effects of high-performance voltage regulator-excitation systems can be described by the phase relationship of the rotor torque components.

The output electrical power of a synchronous generator, P_e, is the product of electrical torque, T_e, and the angular speed, ω. Following a disturbance the change in electrical torque, ΔT_e, can be expressed as a sum of two components: synchronising component ($K_s\Delta\delta$) in phase with the rotor angle change and damping component ($K_D\Delta\omega$) in phase with the speed change.

$$\Delta T_e = K_s\Delta\delta + K_D\Delta\omega \tag{1.3}$$

where

$K_s \triangleq$ synchronising coefficient (pu ΔT/rad)
$\Delta\delta \triangleq$ change of rotor angle (rad)
$\Delta\omega \triangleq$ change of rotor angular speed (elec.rad/s)
$K_D \triangleq$ damping coefficient (pu ΔT.s/rad)

It can be found that for a positive value of K_s the synchronising torque component opposes changes in rotor angle from the equilibrium point. This means that an

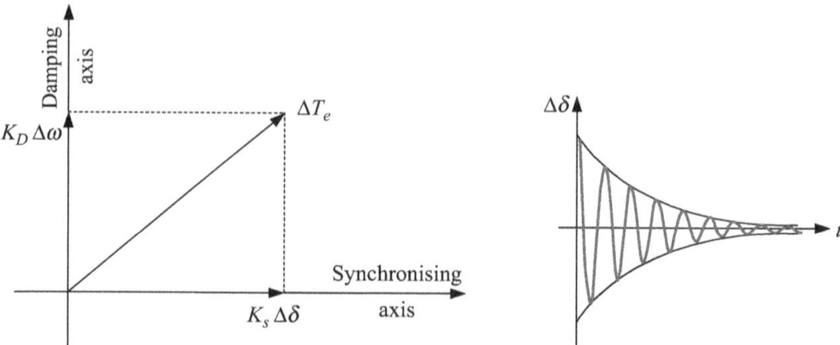

Figure 1.3 Two torque components positive with damped oscillations 'stable state'

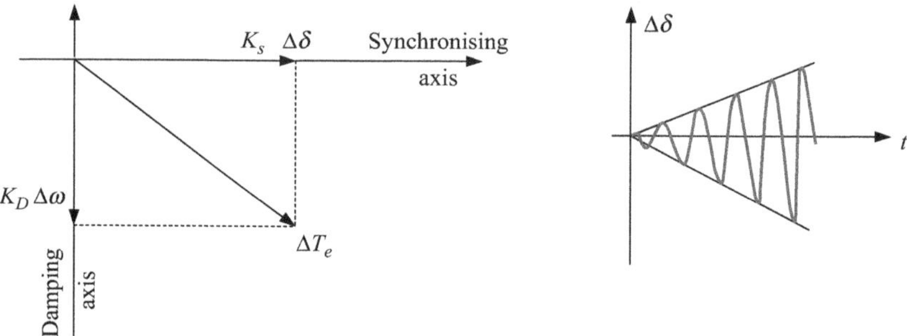

Figure 1.4 Positive synchronising torque and negative damping torque components with un-damped oscillations 'unstable state'

increase in rotor angle leads to a net decelerating torque resulting in the unit to slow down relative to the power system. The slowdown continues until the rotor angle is restored to its equilibrium point and the change of rotor angle vanishes. Similarly, when K_D has positive values the damping torque component opposes the change in the rotor speed from the initial steady operating point. Therefore, the generator remains in a stable state when sufficient positive synchronising and damping torques are acting on the rotor for all operating conditions. Figures 1.3 and 1.4 depict the relation between the two torque components and the corresponding state of the power system. The calculation methods of small signal stability are explained in Chapter 7, Part III.

1.3.2 Transient stability

Transient stability is concerned with the generator stability for the first swing when a large disturbance occurs in the system, e.g. transmission line faults. As in (1.3) the change of electrical torque is resolved into two components acting on each generator in the system: the synchronising torque and the damping torque. If the synchronising torque is insufficient to oppose the change of rotor angle, the generator may lose its synchronism. This can be treated by developing sufficient magnetic flux that can be provided by an excitation system having fast response and sufficient positive and negative forcing capability to resist acceleration or deceleration of the rotor. When the mechanical torque is higher than the electrical torque, the rotor accelerates with respect to the stator flux and its angle increases. The exciter system must increase excitation by applying as quickly as possible a high positive voltage to the generator field. On the other hand, when the mechanical torque is less than the electrical torque the rotor decelerates and its angle decreases. Thus, the excitation system must rapidly apply a high negative voltage to the generator field circuit.

Referring to Figure 1.5, a generator connected with a grid is supposed to operate steadily at the operating point ◙ at which P_m equals P_{eo}. A large disturbance (transient disturbance) in the transmission network very close to the generator will result in a reduction of output electrical power from P_{eo} to zero. This reduction leads to the rotor accelerating with respect to the system and increasing the power angle from δ_o to δ_1 at which the fault is cleared. The electrical power is restored to a level corresponding to the appropriate point on the power–angle curve after the fault (point #1). This curve is lower than that before the fault as the system may become weaker, high impedance transmission network, due to the isolation of faulted line. After clearing the fault the electrical power is higher than the mechanical power causing the generator to decelerate reducing the momentum the rotor had gained during the fault period (point #3). If a sufficient retarding torque exists, the generator moves back towards its operating point, and on the first swing it will be transiently stable. If the retarding torque is insufficient, the power angle continues to increase until the generator loses synchronism with the power system. Determination of generator stability in this case depends to a large extent

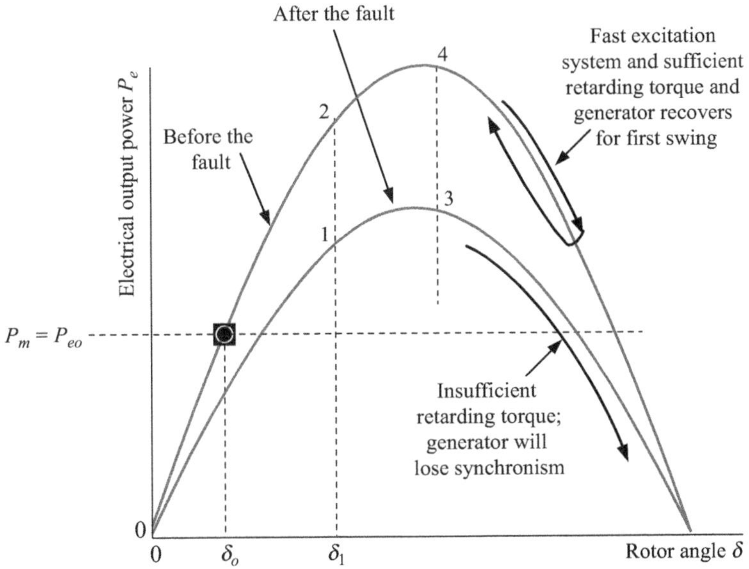

Figure 1.5 Power–angle curve for transient disturbance

on the time of fault clearance. If the fault is cleared earlier, the probability of generator stability becomes higher.

With the effect of excitation system, it is found that maintaining system stability depends also on excitation behaviour and response speed. Thus, increasing the forcing capability and decreasing the response time help in restoring the power–angle curve to that before the fault, i.e. substituting the weakness of transmission network. In this case points #2 and #4 correspond to points #1 and #3, respectively, and generator stability on the first swing is much improved. More details are presented in Chapters 8 and 9, Part III.

1.4 Need for modelling

As described in Section 1.1, assessment of both security and stability of power system is essentially required to avoid catastrophic consequences of system disturbances such as blackouts. Accurate assessment of security and stability is based on accurate modelling of power system components. This is a great challenge and more attention must be paid to it as the power systems today are becoming more complicated along with the increased complexity of operation and control. Models of power system components are the basis of methods to analyse. They are composed of mathematical relations established from the physical behaviour of components. For the steady-state analysis, the models mainly are network structure and distribution of generators and loads; while for the dynamic calculations the models also include the parameters of generators and the dynamic characteristics of loads besides their static parameters.

For instance, load modelling is difficult due to the random behaviour of the load and the accumulation of large volumes of measurement data. It involves two major issues: modelling and parameter identification. The measurement data are used for parameter identification of the composite load on the load bus. The dynamic part of the composite load is sometimes represented by an induction machine. Power system stability is affected by the sum of dynamic motor loads connected to the system and the line loading level, so load behaviour differs under different system conditions. This indicates that choosing an accurate induction machine model is important for the accuracy of system stability analysis [5].

Methods of power system analysis and simulation are based on a proper design of adequate models of system components for the purpose of a study. They, generally, are in time domain and include the calculation of system behaviour in the past, present and future [6, 7]. The system analysis for the past time focuses on the analysis of the historical data, summarising the experience, recognising the disturbances occurred and studying the intrinsic characteristics that help improve the operating conditions. The present time system analysis means real-time calculations [8, 9] that include state estimation to eliminate bad data from the measurement system in addition to estimating the un-measured information [10, 11]. Also, they include power flow analysis to calculate the real-time power flow distribution of the system. The analysis for future time analyses the supposed system by simulation to provide decision for system development plan, system operation and emergency control strategy. All these calculations are important, in general, for power system study and, in particular, for power system stability. Their accuracy can be measured by the degree of conformation between the results of calculations and the real system. Selecting appropriate models of system components will contribute to a large extent to improving the accuracy.

1.5 Stability margin increase

Operation of power systems at operating conditions close to stability boundary has increased the importance of increasing stability margin to maintain system stability, small signal stability and transient stability. Small signal stability can be enhanced by increasing the system damping to damp the oscillations that may happen due to small disturbances or following severe disturbances. Unfortunately, the voltage regulator is a major source of negative damping but its use is inevitable. Removing voltage regulators from service is not a realistic solution of the problem because of the need of their beneficial features. The problem of negative damping effect of the voltage regulator, fortunately, has been solved by providing supplementary controls to contribute positive damping for oscillatory angle stabilisation. These controls are known as power system stabilisers (PSSs) (Figure 1.6). More details are given in Chapter 11, Part IV.

Increasing the stability limit improves the system stability and increases the stability margin as well. Referring to (1.1) the stability limit (P_{max}) can be increased by modifying the bus voltage or modifying the line reactance. Shunt compensation can be used for modifying the voltage at the compensator bus while series compensation is required to modify the line reactance.

Thyristor switched capacitor, thyristor-controlled reactor, static var compensator and static synchronous compensator can be used for shunt compensation. Also, fixed series capacitor, thyristor-protected series capacitor, static synchronous series

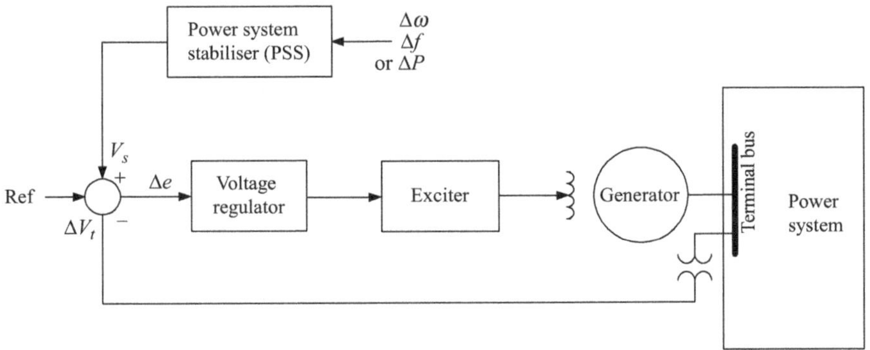

Figure 1.6 A synchronous generator with exciter, voltage regulator and PSS (V_s = output voltage of PSS; $\Delta\omega$ = change of shaft speed; Δf = change of generator; electrical frequency; ΔP = change of electrical power; ΔV_t = change of terminal voltage, Δe = voltage error)

compensator, unified power flow controller, interline power flow controller and interphase power controller can be used as series compensators. Part IV deals with more details about the compensation in power systems related to system stability, and in addition, the flexible AC transmission system devices are described.

References

1. Kundur P., Paserba J., Viter S. (eds.). 'Overview on definition and classification of power system stability'. *Quality and Security of Electric Power Delivery Systems 2003. CIGRE/PES 2003. CIGRE/IEEE PES International Symposium*; Montreal, Canada, Oct 2003. pp. 1–4
2. Kundur P., Paserba J., Ajjarapu V., Andersson G. 'Definition and classification of power system stability'. *IEEE Transactions on Power Systems*. 2004; **19**(3):1387–401
3. Basler M.J., Schaefer R.C. 'Understanding power system stability'. *IEEE Transactions on Industry Applications*. 2008;**44**(2):463–74
4. IEEE Task Force on Power System Stabilizers (eds.). 'Overview of power system stability concepts'. *Proceedings of IEEE PES General Meeting*; Toronto, Canada, Jul 2003, vol. 3. pp. 1–7
5. Dahal S., Attaviriyanupap P., Kataoka Y., Saha T. (eds.). 'Effects of induction machines dynamics on power system stability'. *Power Engineering Conference, AUPEC 2009, Australasian Universities*; Adelaide, SA, Sept 2009. pp. 1–6
6. Anjia M., Zhizhong G. (eds.). 'The influence of model mismatch to power system calculation, Part II: On the stability calculation'. *Proceedings of Power Engineering Conference, IPEC 2005, The 7th International*; Singapore, Nov/Dec 2005, vol. 2. pp. 1127–32
7. Avramenko V.N. (eds.). 'Power system stability assessment for current states of the system'. *Power Tech Conference, IEEE*; St. Petersburg, Russia, Jun 2005. pp. 1–6
8. Fishov A.G., Toutoundaeva D.V. (eds.). 'Power system stability standardization under present-day conditions'. *Strategic Technology, IFOST 2007, International Forum on*; Ulaanbaatar, Mongolia, Oct 2007. pp. 411–5
9. Shirai Y., Nitta T. (eds.). 'On-line evaluation of power system stability by use of SMES'. *Proceedings of IEEE Power Engineering Society Winter Meeting*; New York, USA, Jan 2002, vol. 2. pp. 900–5
10. Youfang X. (ed.). 'Measures to ensure the security and stability of the central China power system'. *Power System Technology, 1998. Proceedings. POWERCON '98. 1998 International Conference on*; Beijing, China, Aug 1998, vol. 2. pp. 1374–7
11. Dai Y., Zhao T., Tian Y., Gao L. (eds.). 'Research on the influence of primary frequency control distribution on power system security and stability'. *Industrial Electronics and Applications, ICIEA 2007, 2nd IEEE Conference on*; Harbin, China, May 2007. pp. 222–6

Part I

Modelling

Chapter 2
Modelling of the synchronous machine

2.1 Introduction

The synchronous machine, as one of the very important power system components, must be modelled mathematically in an adequate manner for dynamic and stability studies. Two models based on the state space formulation of the machine equations have been developed depending on using either the currents or the flux linkages as state variables [1, 2].

The synchronous machine considered in this chapter has six magnetically coupled windings: three stator 'armature' windings and three rotor windings, 'one for the field circuit and two for the damper circuits' as demonstrated in Figure 2.1. The field circuit and one of the two damper circuits are located on the same axis called the direct or d-axis. The second damper circuit is located on an axis lagging the d-axis by 90° elec. and is called the quadrature or q-axis. The d-axis defines the rotor position in space at some instant of time to be at angle θ elec. with respect to a fixed reference position. In the case of a larger number of damper windings, the

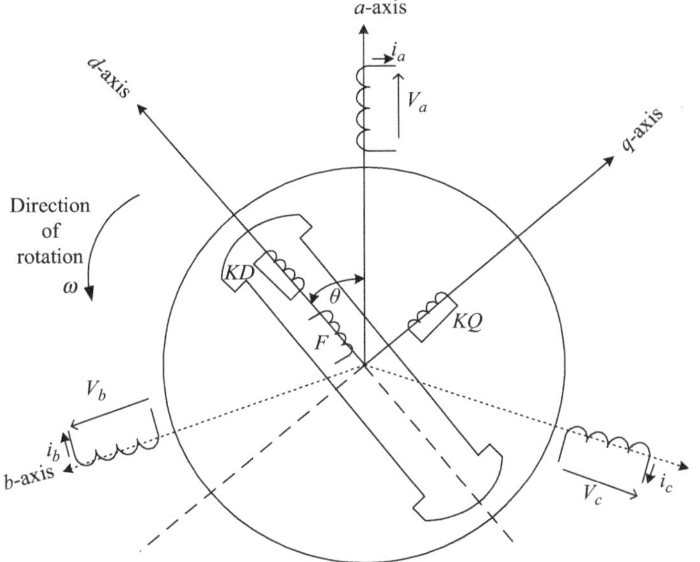

Figure 2.1 Schematic representation of a synchronous machine

same methodology of derivation, as derived here for one damper winding on each of the two axes, can be applied to model the synchronous machine. The modelling process is based on considering a uniformly distributed sinusoidal mmf in the air gap and without harmonics. It commences, for simplicity, with neglecting the magnetic saturation that will be represented later.

2.2 Synchronous machine equations

2.2.1 Flux linkage equations

As depicted in Figure 2.1, the synchronous machine consists of three-phase stator windings a, b and c and three rotor windings: F denotes the field winding and KD and KQ denote the damper windings. Vectors and matrices are designated by bold, italic symbols. The symbols for the stator are subscripted by 's' and for the rotor by 'r'. Equation for the flux linkage, Ψ, can be written in matrix form as

$$\begin{bmatrix} \boldsymbol{\Psi}_s \\ \boldsymbol{\Psi}_r \end{bmatrix} = \begin{bmatrix} \boldsymbol{L}_{ss} & \boldsymbol{L}_{sr} \\ \boldsymbol{L}_{rs} & \boldsymbol{L}_{rr} \end{bmatrix} \begin{bmatrix} \boldsymbol{i}_s \\ \boldsymbol{i}_r \end{bmatrix} \tag{2.1}$$

where

$$\boldsymbol{\Psi}_s = \begin{bmatrix} \Psi_a \\ \Psi_b \\ \Psi_c \end{bmatrix} \quad \boldsymbol{\Psi}_r = \begin{bmatrix} \Psi_f \\ \Psi_{kd} \\ \Psi_{kq} \end{bmatrix} \quad \boldsymbol{i}_s = \begin{bmatrix} i_a \\ i_b \\ i_c \end{bmatrix} \quad \boldsymbol{i}_r = \begin{bmatrix} i_f \\ i_{kd} \\ i_{kq} \end{bmatrix} \quad \boldsymbol{L}_{rs} = \boldsymbol{L}_{sr}^t$$

$$\boldsymbol{L}_{ss} = \begin{bmatrix} L_{aa} & L_{ab} & L_{ac} \\ L_{ba} & L_{bb} & L_{bc} \\ L_{ca} & L_{cb} & L_{cc} \end{bmatrix}$$

$L_{jk} = L_{kj} \triangleq$ mutual inductance between the stator windings if $j \neq k$ and is defined as the self-inductance 'L_{jj}' of the jth winding. On the other hand, $\boldsymbol{L}_{ss}$ representing the self and mutual inductances of stator phase windings can be expressed as

$$\boldsymbol{L}_{ss} = \begin{bmatrix} L_s & M_s & M_s \\ M_s & L_s & M_s \\ M_s & M_s & L_s \end{bmatrix} + L_m \times \begin{bmatrix} \cos 2\theta & \cos\left(2\theta - \frac{2\pi}{3}\right) & \cos\left(2\theta + \frac{2\pi}{3}\right) \\ \cos\left(2\theta - \frac{2\pi}{3}\right) & \cos\left(2\theta + \frac{2\pi}{3}\right) & \cos 2\theta \\ \cos\left(2\theta + \frac{2\pi}{3}\right) & \cos 2\theta & \cos\left(2\theta - \frac{2\pi}{3}\right) \end{bmatrix} \tag{2.2}$$

L_m, L_s and M_s are constants.

$\boldsymbol{L}_{sr}$ = matrix of mutual inductances of stator-to-rotor windings

$$= \begin{bmatrix} M_{af} & M_{akd} & M_{akq} \\ M_{bf} & M_{bkd} & M_{bkq} \\ M_{cf} & M_{ckd} & M_{ckq} \end{bmatrix}$$

$$= \begin{bmatrix} M_f \cos\theta & M_{kd}\cos\theta & M_{kq}\sin\theta \\ M_f \cos\left(\theta - \frac{2\pi}{3}\right) & M_{kd}\cos\left(\theta - \frac{2\pi}{3}\right) & M_{kq}\sin\left(\theta - \frac{2\pi}{3}\right) \\ M_f \cos\left(\theta + \frac{2\pi}{3}\right) & M_{kd}\cos\left(\theta + \frac{2\pi}{3}\right) & M_{kq}\sin\left(\theta + \frac{2\pi}{3}\right) \end{bmatrix} \tag{2.3}$$

$$\boldsymbol{L}_{rr} = \begin{bmatrix} L_f & L_{fkd} & 0 \\ L_{fkd} & L_{kd} & 0 \\ 0 & 0 & L_{kq} \end{bmatrix} = \text{constant matrix} \tag{2.4}$$

where L_f, L_{kd} and L_{kq} are the self-inductances of the field winding, F; damper circuit in d-axis, KD; and damper circuit in q-axis, KQ, respectively. L_{fkd} is the mutual inductance between windings F and KD.

It can be observed from (2.2) and (2.3) that both $\boldsymbol{L}_{ss}$ (if $L_m \neq 0$) and $\boldsymbol{L}_{sr}$ are time varying and functions of the rotor position angle θ. In (2.4) the matrix elements representing the mutual inductance between the winding KQ and the winding F or KD are 0 as the angle between them is 90° elec.

2.2.2 Voltage equations

The voltage equations of stator and rotor windings, considering the generator convention as positive, are given by

$$\begin{bmatrix} \boldsymbol{v}_s \\ \boldsymbol{v}_r \end{bmatrix} = -\begin{bmatrix} \dot{\boldsymbol{\Psi}}_s \\ \dot{\boldsymbol{\Psi}}_r \end{bmatrix} - \begin{bmatrix} \boldsymbol{R}_s & 0 \\ 0 & \boldsymbol{R}_r \end{bmatrix}\begin{bmatrix} \boldsymbol{i}_s \\ \boldsymbol{i}_r \end{bmatrix} = -\begin{bmatrix} \dot{\boldsymbol{\Psi}}_s \\ \dot{\boldsymbol{\Psi}}_r \end{bmatrix} - \boldsymbol{R}\begin{bmatrix} \boldsymbol{i}_s \\ \boldsymbol{i}_r \end{bmatrix} \tag{2.5}$$

where

$$\boldsymbol{v}_s = \begin{bmatrix} v_a \\ v_b \\ v_c \end{bmatrix}, \quad \boldsymbol{v}_r = \begin{bmatrix} -v_f \\ 0 \\ 0 \end{bmatrix},$$

$$\boldsymbol{R} = \begin{bmatrix} \boldsymbol{R}_s & 0 \\ 0 & \boldsymbol{R}_r \end{bmatrix}, \quad \boldsymbol{R}_s = \begin{bmatrix} R_a & 0 & 0 \\ 0 & R_b & 0 \\ 0 & 0 & R_c \end{bmatrix}, \quad \boldsymbol{R}_r = \begin{bmatrix} R_f & 0 & 0 \\ 0 & R_{kd} & 0 \\ 0 & 0 & R_{kq} \end{bmatrix}$$

and a dot on a symbol represents the differential.

Usually $R_a = R_b = R_c$. Hence, $\boldsymbol{R}_s = R_a\mathbf{U}_3$, where $\mathbf{U}_3$ is a 3×3 unit matrix.

In the case of defining the neutral voltage contribution to the stator voltages v_a, v_b and v_c, the voltage equations become

$$\begin{bmatrix} \boldsymbol{v}_s \\ \boldsymbol{v}_r \end{bmatrix} = -\begin{bmatrix} \dot{\boldsymbol{\Psi}}_s \\ \dot{\boldsymbol{\Psi}}_r \end{bmatrix} - [\boldsymbol{R}]\begin{bmatrix} \boldsymbol{i}_s \\ \boldsymbol{i}_r \end{bmatrix} + \begin{bmatrix} \boldsymbol{v}_n \\ 0 \end{bmatrix} \tag{2.6}$$

where

$$\begin{aligned} \boldsymbol{v}_n &= -R_n\begin{bmatrix} 1 & 1 & 1 \\ 1 & 1 & 1 \\ 1 & 1 & 1 \end{bmatrix}\begin{bmatrix} i_a \\ i_b \\ i_c \end{bmatrix} - L_n\begin{bmatrix} 1 & 1 & 1 \\ 1 & 1 & 1 \\ 1 & 1 & 1 \end{bmatrix}\begin{bmatrix} \mathrm{d}i_a/\mathrm{d}t \\ \mathrm{d}i_b/\mathrm{d}t \\ \mathrm{d}i_c/\mathrm{d}t \end{bmatrix} \\ &= -\boldsymbol{R}_n\boldsymbol{i}_s - L_n p\boldsymbol{i}_s \end{aligned} \tag{2.7}$$

where p is the operator d/dt. R_n and L_n are the resistance and inductance of the neutral to earth, respectively.

It is to be noted that the positive convention is stator currents flowing out of the machine terminals, i.e. the machine operating as a generator.

2.2.3 Torque equation

The rotor equation of motion is given by $J\ddot{\theta}_m + D\dot{\theta}_m = T_m - T_e$ or

$$\frac{2}{p}\left(J\ddot{\theta} + D\dot{\theta}\right) = T_m - T_e = T_a \tag{2.8}$$

where

$p \triangleq$ the number of poles
$D \triangleq$ the damping coefficient
$\Theta \triangleq$ the rotor angle in electrical radians
$J \triangleq$ the moment of inertia in $\mathrm{kg \cdot m^2}$
$\Theta_m \triangleq$ the rotor angle in mechanical radians with respect to a fixed frame = $(2/p)\Theta$
$T_m \triangleq$ the mechanical torque in the direction of rotation in $\mathrm{N \cdot m}$
$T_e \triangleq$ the electrical torque opposing the mechanical torque in $\mathrm{N \cdot m}$
$T_a \triangleq$ the accelerating torque in $\mathrm{N \cdot m}$

$$T_e = -\frac{\partial W}{\partial \theta_m} = -\frac{p}{2}\frac{\partial W}{\partial \theta} = \frac{p}{2}T'_e \tag{2.9}$$

$T'_e = -\frac{\partial W}{\partial \theta} \triangleq$ the electrical torque of the equivalent two pole machine and W is the co-energy expressed as

$$W = \frac{1}{2}\begin{bmatrix} \boldsymbol{i}_s^t & \boldsymbol{i}_r^t \end{bmatrix}\begin{bmatrix} \boldsymbol{L}_{ss} & \boldsymbol{L}_{sr} \\ \boldsymbol{L}_{rs} & \boldsymbol{L}_{rr} \end{bmatrix}\begin{bmatrix} \boldsymbol{i}_s \\ \boldsymbol{i}_r \end{bmatrix} \tag{2.10}$$

$$\text{Therefore, } T'_e = -\frac{1}{2}\left[\boldsymbol{i}_s^t\left(\frac{\partial \boldsymbol{L}_{ss}}{\partial \theta}\right)\boldsymbol{i}_s + 2\boldsymbol{i}_s^t\left(\frac{\partial \boldsymbol{L}_{sr}}{\partial \theta}\right)\boldsymbol{i}_r\right] \tag{2.11}$$

From (2.8) and (2.9), the equation of motion can be rewritten as

$$\left(\frac{2}{p}\right)^2 (J\ddot{\theta} + D\dot{\theta}) = \frac{2}{p}T_m - T'_e \tag{2.12}$$

It is observed that (2.12) represents the transformation of a p-pole machine to a two-pole machine. In expressing the equations in per unit 'pu' see Appendix I. It is found that there is no loss of generality in assuming that the machine has two poles as explained in Section 2.6.

2.3 Park's transformation

Equation (2.5) can be written as

$$\dot{\Psi} = -[R]\boldsymbol{i} - \boldsymbol{v} \tag{2.13}$$

where

$$\dot{\Psi} = \begin{bmatrix} \dot{\Psi}_s \\ \dot{\Psi}_r \end{bmatrix}, \quad \boldsymbol{i} = \begin{bmatrix} \boldsymbol{i}_s \\ \boldsymbol{i}_r \end{bmatrix} = \boldsymbol{L}^{-1}\Psi, \quad \boldsymbol{v} = \begin{bmatrix} \boldsymbol{v}_s \\ \boldsymbol{v}_r \end{bmatrix}$$

Also

$$\dot{\Psi} = \frac{\mathrm{d}}{\mathrm{d}t}(\boldsymbol{L}\boldsymbol{i}) = \boldsymbol{L}\frac{\mathrm{d}\boldsymbol{i}}{\mathrm{d}t} + \boldsymbol{i}\frac{\partial \boldsymbol{L}}{\partial \theta}\frac{\mathrm{d}\theta}{\mathrm{d}t} \tag{2.14}$$

Equating the RHS of (2.13) and (2.14) gives

$$\frac{\mathrm{d}\boldsymbol{i}}{\mathrm{d}t} = \boldsymbol{L}^{-1}\left\{-\boldsymbol{R}\boldsymbol{i} - \boldsymbol{i}\frac{\partial \boldsymbol{L}}{\partial \theta}\frac{\mathrm{d}\theta}{\mathrm{d}t} - \boldsymbol{v}\right\} \tag{2.15}$$

It is difficult to solve (2.15) as the inductances are time varying. Using Park's transformation [3], the equations are simplified by referring all quantities to a rotor frame of reference constituting time invariant rather than time variant equations. Consequently, this attains the simplification of both steady-state and transient calculations.

Defining f_{abc} as the voltage or current or flux linkage of the stator windings in *a-b-c* frame of reference and f_{dqo} as the same quantities in the *d-q-o* frame of reference, Park's transformation '**P**' is expressed as [3]

$$f_{abc} = \mathbf{P} f_{dqo} \tag{2.16}$$

where

$$\boldsymbol{f}_{abc} = \begin{bmatrix} f_a \\ f_b \\ f_c \end{bmatrix}, \quad \boldsymbol{f}_{dqo} = \begin{bmatrix} f_d \\ f_q \\ f_o \end{bmatrix}$$

$$\mathbf{P} = \sqrt{\frac{2}{3}} \begin{bmatrix} \cos\theta & \sin\theta & \frac{1}{\sqrt{2}} \\ \cos\left(\theta - \frac{2\pi}{3}\right) & \sin\left(\theta - \frac{2\pi}{3}\right) & \frac{1}{\sqrt{2}} \\ \cos\left(\theta + \frac{2\pi}{3}\right) & \sin\left(\theta + \frac{2\pi}{3}\right) & \frac{1}{\sqrt{2}} \end{bmatrix}$$

Thus,

$$\boldsymbol{f}_{dqo} = \mathbf{P}^{-1}\boldsymbol{f}_{abc} \tag{2.17}$$

where

$$\mathbf{P}^{-1} = \sqrt{\frac{2}{3}} \begin{bmatrix} \cos\theta & \cos\left(\theta - \frac{2\pi}{3}\right) & \cos\left(\theta + \frac{2\pi}{3}\right) \\ \sin\theta & \sin\left(\theta - \frac{2\pi}{3}\right) & \sin\left(\theta + \frac{2\pi}{3}\right) \\ \frac{1}{\sqrt{2}} & \frac{1}{\sqrt{2}} & \frac{1}{\sqrt{2}} \end{bmatrix}$$

2.4 Transformation of synchronous machine equations

2.4.1 Transformation of flux linkage equations

Applying Park's transformation as in (2.16) to the flux linkages gives

$$\begin{bmatrix} \Psi_s \\ \Psi_r \end{bmatrix} = \begin{bmatrix} \mathbf{P} & 0 \\ 0 & \mathbf{U}_3 \end{bmatrix} \begin{bmatrix} \Psi_{dqo} \\ \Psi_r \end{bmatrix} \tag{2.18}$$

Equation (2.18) can be transformed to become

$$\begin{bmatrix} \Psi_s \\ \Psi_r \end{bmatrix} = \begin{bmatrix} \boldsymbol{L}_{ss} & \boldsymbol{L}_{sr} \\ \boldsymbol{L}_{rs} & \boldsymbol{L}_{rr} \end{bmatrix} \begin{bmatrix} \mathbf{P} & 0 \\ 0 & \mathbf{U}_3 \end{bmatrix} \begin{bmatrix} \boldsymbol{i}_{dqo} \\ \boldsymbol{i}_r \end{bmatrix} \tag{2.19}$$

Equating the RHS of (2.18) and (2.19) gives

$$\begin{bmatrix} \Psi_{dqo} \\ \Psi_r \end{bmatrix} = \begin{bmatrix} \boldsymbol{L}'_{ss} & \boldsymbol{L}'_{sr} \\ \boldsymbol{L}'_{rs} & \boldsymbol{L}_{rr} \end{bmatrix} \begin{bmatrix} \boldsymbol{i}_{dqo} \\ \boldsymbol{i}_r \end{bmatrix} \tag{2.20}$$

where

$$\boldsymbol{L}'_{ss} = \mathbf{P}^{-1}\boldsymbol{L}_{ss}\mathbf{P} = \begin{bmatrix} L_d & 0 & 0 \\ 0 & L_q & 0 \\ 0 & 0 & L_o \end{bmatrix} \tag{2.21}$$

$$L_d = L_s - M_s + \frac{3}{2}L_m, \quad L_q = L_s - M_s - \frac{3}{2}L_m, \quad L_o = L_s + 2M_s$$

$$\boldsymbol{L}'_{sr} = \mathbf{P}^{-1}\boldsymbol{L}_{sr} = \sqrt{\frac{3}{2}}\begin{bmatrix} M_f & M_{kd} & 0 \\ 0 & 0 & M_{kq} \\ 0 & 0 & 0 \end{bmatrix} \tag{2.22}$$

$$\boldsymbol{L}'_{rs} = \boldsymbol{L}_{rs}\mathbf{P} = \sqrt{\frac{3}{2}}\begin{bmatrix} M_f & 0 & 0 \\ M_{kd} & 0 & 0 \\ 0 & M_{kq} & 0 \end{bmatrix} \tag{2.23}$$

Equations (2.22) and (2.23) show that $\boldsymbol{L}'_{sr} = \boldsymbol{L}'^{t}_{rs}$.

The matrix $\boldsymbol{L}_{rr}$ is a constant matrix as in (2.4) representing the inductance of rotor windings as there is no transformation of rotor currents and flux linkages.

From (2.20) it can be concluded that:

- Stator windings *a*, *b* and *c* are replaced by virtual windings *d*, *q* and *o* by applying the Park's transformation.
- The '*o*' winding can be neglected under balanced conditions as there is no coupling with rotor windings and no flow of zero sequence current, i_o.
- *d* and *q* windings rotate at the same speed as the rotor because the transformed mutual inductance terms between them are constants.
- Mutual inductance between the *d*-winding and the rotor windings on the *q*-axis, and the mutual inductance between the *q*-winding and the rotor windings on the *d*-axis are zero. Consequently, the *d*-winding is aligned with the *d*-axis and *q*-winding is aligned with the *q*-axis.
- The time-varying coefficients are removed from the machine equations.

Therefore, the synchronous machine can be represented as shown in Figure 2.2.

2.4.2 Transformation of stator voltage equations

The stator voltage (2.6) can be rewritten, by applying Park's transformation, as below:

$$\mathbf{P}\boldsymbol{v}_{dqo} = -\frac{\mathrm{d}}{\mathrm{d}t}\left(\mathbf{P}\boldsymbol{\Psi}_{dqo}\right) - \boldsymbol{R}_s\mathbf{P}\boldsymbol{i}_{dqo} + \mathbf{P}\boldsymbol{v}_{n(dqo)} \tag{2.24}$$

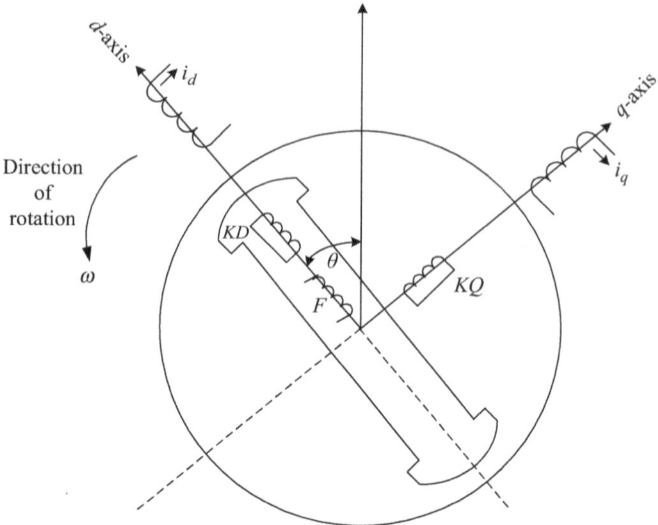

Figure 2.2 Synchronous machine representation by transformed windings

The first term on the RHS is derived as

$$-\frac{\mathrm{d}}{\mathrm{d}t}(\mathbf{P}\Psi_{dqo}) = -\dot{\theta}\frac{\mathrm{d}\mathbf{P}}{\mathrm{d}\theta}\Psi_{dqo} - \mathbf{P}\frac{\mathrm{d}\Psi_{dqo}}{\mathrm{d}t} \tag{2.25}$$

where

$$\frac{\mathrm{d}\mathbf{P}}{\mathrm{d}\theta} = \sqrt{\frac{2}{3}}\begin{bmatrix} -\sin\theta & \cos\theta & 0 \\ -\sin\left(\theta - \frac{2\pi}{3}\right) & \cos\left(\theta - \frac{2\pi}{3}\right) & 0 \\ -\sin\left(\theta + \frac{2\pi}{3}\right) & \cos\left(\theta + \frac{2\pi}{3}\right) & 0 \end{bmatrix} = \mathbf{P}\mathbf{P}_1 \tag{2.26}$$

and

$$\mathbf{P}_1 = \begin{bmatrix} 0 & 1 & 0 \\ -1 & 0 & 0 \\ 0 & 0 & 0 \end{bmatrix}$$

Substituting (2.25) and (2.26) into (2.24) gives the following:

$$\mathbf{P}\boldsymbol{v}_{dqo} = -\omega\mathbf{P}\mathbf{P}_1\Psi_{dqo} - \mathbf{P}\frac{\mathrm{d}\Psi_{dqo}}{\mathrm{d}t} - \boldsymbol{R}_s\mathbf{P}\boldsymbol{i}_{dqo} + \mathbf{P}\boldsymbol{v}_{n(dqo)}$$

Thus,

$$\boldsymbol{v}_{dqo} = -\omega\mathbf{P}_1\Psi_{dqo} - \frac{\mathrm{d}\Psi_{dqo}}{\mathrm{d}t} - \mathbf{P}^{-1}\boldsymbol{R}_s\mathbf{P}\boldsymbol{i}_{dqo} + \boldsymbol{v}_{n(dqo)} \tag{2.27}$$

where $\omega = \frac{\mathrm{d}\theta}{\mathrm{d}t}$ is the rotor angular speed in elec. rad/s.

The term $\boldsymbol{v}_{n(dqo)}$ can be derived as below.

Applying Park's transformation to (2.7):

$$\boldsymbol{v}_{n(dqo)} = -\mathbf{P}^{-1}\boldsymbol{R}_n\mathbf{P}\boldsymbol{i}_{dqo} - \mathbf{P}^{-1}L_n\mathbf{P}\frac{\mathrm{d}\boldsymbol{i}_{dqo}}{\mathrm{d}t} = -\begin{bmatrix} 0 \\ 0 \\ 3R_n i_o \end{bmatrix} - \begin{bmatrix} 0 \\ 0 \\ 3L_n i_o \end{bmatrix} \tag{2.28}$$

The rotor voltage equations are unchanged. Then, the combined voltage equations of stator and rotor can be expressed as

$$\begin{bmatrix} \boldsymbol{v}_{dqo} \\ \boldsymbol{v}_r \end{bmatrix} = -\begin{bmatrix} \dot{\boldsymbol{\Psi}}_{dqo} \\ \dot{\boldsymbol{\Psi}}_r \end{bmatrix} - \begin{bmatrix} \omega\mathbf{P}_1\boldsymbol{\Psi}_{dqo} \\ \mathbf{0} \end{bmatrix} - \begin{bmatrix} \boldsymbol{R}_s & \mathbf{0} \\ \mathbf{0} & \boldsymbol{R}_r \end{bmatrix}\begin{bmatrix} \boldsymbol{i}_{dqo} \\ \boldsymbol{i}_r \end{bmatrix} + \begin{bmatrix} \boldsymbol{v}_{n(dqo)} \\ \mathbf{0} \end{bmatrix} \tag{2.29}$$

where

$$\boldsymbol{R}_s = R_a\mathbf{U}_3.$$

Equation (2.29) determines the synchronous machine voltages, for both stator and rotor, in terms of flux linkages and currents as state variables in the *d-q-o* frame of reference. Using the relation between flux linkages and currents given by (2.20), the machine voltages can be expressed in terms of either currents only or flux linkages only. The formulation of machine voltages in terms of currents as state variables, $\boldsymbol{x}^t = [i_d, i_q, i_o, i_f, i_{kd}, i_{kq}]$ can be derived as below.

Substituting (2.28) into (2.29), the following set of equations in the expanded form is obtained – it is noted that $\boldsymbol{v}_r$ is unchanged due to Park's transformation:

$$\left.\begin{aligned} v_d &= -\dot{\Psi}_d - \omega\Psi_q - R_a i_d \\ v_q &= -\dot{\Psi}_q + \omega\Psi_d - R_a i_q \\ v_o &= -\dot{\Psi}_o - 3L_n p i_o - 3R_n i_o \\ v_f &= \dot{\Psi}_f + R_f i_f \\ v_{kd} &= 0 = \dot{\Psi}_{kd} + R_{kd} i_{kd} \\ v_{kq} &= 0 = \dot{\Psi}_{kq} + R_{kq} i_{kq} \end{aligned}\right\} \tag{2.30}$$

Equation (2.20) can be written in expanded form to give a set of equations as

$$\left.\begin{aligned} \Psi_d &= L_d i_d + kM_f i_f + kM_{kd} i_{kd} \\ \Psi_q &= L_q i_q + kM_{kq} i_{kq} \\ \Psi_o &= L_o i_o \\ \Psi_f &= kM_f i_d + L_f i_f + L_{fkd} i_{kd} \\ \Psi_{kd} &= kM_{kd} i_d + L_{kdf} i_f + L_{kd} i_{kd} \\ \Psi_{kq} &= kM_{kq} i_q + L_{kq} i_{kq} \end{aligned}\right\} \tag{2.31}$$

where $k = \sqrt{\frac{3}{2}}$

From the two sets of (2.30) and (2.31), the machine voltages in terms of currents are given by

$$\begin{bmatrix} v_d \\ v_q \\ v_o \\ -v_f \\ v_{kd}=0 \\ v_{kq}=0 \end{bmatrix} = -\left[\begin{array}{ccc:ccc} R_a & \omega L_q & 0 & 0 & 0 & \omega k M_{kq} \\ -\omega L_d & R_a & 0 & -\omega k M_f & \omega k M_{kd} & 0 \\ 0 & 0 & R_a+3R_n & 0 & 0 & 0 \\ \hdashline 0 & 0 & 0 & R_f & 0 & 0 \\ 0 & 0 & 0 & 0 & R_{kd} & 0 \\ 0 & 0 & 0 & 0 & 0 & R_{kq} \end{array}\right] \begin{bmatrix} i_d \\ i_q \\ i_o \\ i_f \\ i_{kd} \\ i_{kq} \end{bmatrix}$$

$$-\left[\begin{array}{ccc:ccc} L_d & 0 & 0 & kM_f & kM_{kd} & 0 \\ 0 & L_q & 0 & 0 & 0 & kM_{kq} \\ 0 & 0 & L_o+3L_n & 0 & 0 & 0 \\ \hdashline kM_f & 0 & 0 & L_f & L_{fkd} & 0 \\ kM_{kd} & 0 & 0 & L_{fkd} & L_{kd} & 0 \\ 0 & kM_{kq} & 0 & 0 & 0 & L_{kq} \end{array}\right] \begin{bmatrix} \mathrm{p}i_d \\ \mathrm{p}i_q \\ \mathrm{p}i_o \\ \mathrm{p}i_f \\ \mathrm{p}i_{kd} \\ \mathrm{p}i_{kq} \end{bmatrix} \quad (2.32)$$

Interesting points observed from the set of (2.32) are:

- Stator, i.e. armature, equations except that of v_o include the speed voltage terms, ωL_i and ωM_i, the speed voltage providing a difference from those of a passive network.
- The speed voltage terms in the d-axis equation are only due to q-axis currents and the q-axis speed voltages are due to d-axis currents.
  - The zero-sequence voltage v_o depends only on i_o and its first derivative, i.e. its equation can be solved separately by knowing the initial conditions on i_o.
  - Under balanced condition, $v_o = 0$, i.e. the row of v_o and the corresponding column are omitted. Thus, the set of equations can be depicted schematically by the equivalent circuits shown in Figure 2.3.
- It is noted that the self-inductances of stator windings in d and q axes are L_d and L_q, respectively. The mutual inductance between the stator winding and any of rotor windings, e.g. ith winding, is denoted by the symbol kM_i. The self-inductance of the ith rotor winding and the mutual inductance between the ith and jth rotor windings are denoted by L_i and L_{ij}, respectively. The equivalent circuits in d and q axes include the speed voltage terms as controlled sources.
- Regardless of the angular velocity, ω, all other terms in the coefficient matrices are constants, i.e. not time varying unlike the coefficients in (2.5) in a-b-c frame of reference. Under the situation of the variation of ω with time, (2.32) becomes non-linear and the formulation of state space equations is in the form:

$$\dot{x} = f(x, u, t) \quad (2.33)$$

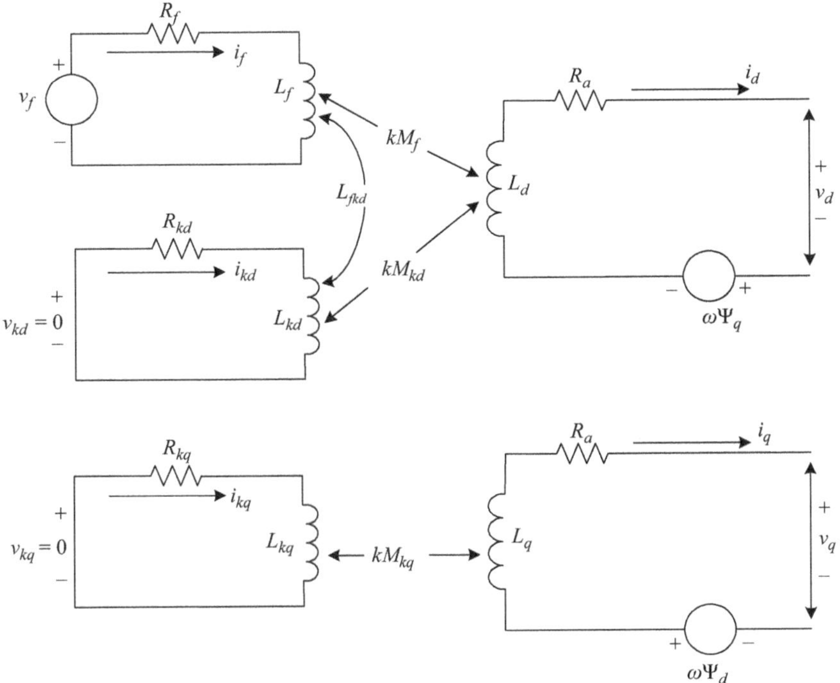

Figure 2.3 d–q equivalent circuit of synchronous generator

where $\boldsymbol{f}$ is a set of non-linear functions, $\boldsymbol{x}$ a vector of the state variables and $\boldsymbol{u}$ the system-driving functions.

On the other hand, if ω is assumed to be a constant 'it is an accepted approximation in the steady state'; (2.32) is a linear time invariant and the formulation of state space equations as a set of first-order differential equations is in the form

$$\dot{\boldsymbol{x}} = \boldsymbol{A}\boldsymbol{x} + \boldsymbol{B}\boldsymbol{u} \tag{2.34}$$

In the steady state, if the synchronous generator is *unloaded* all currents (i_d, i_q, i_{kd}, i_{kq}) are zero except the field current $i_{f(NL)} = v_{f(NL)}/R_f$, 'the subscript "$_{NL}$" denotes the value at unloaded conditions'. Therefore, from (2.31) the flux linkages are

$$\begin{aligned} &\Psi_{d(NL)} = kM_f v_{f(NL)}/R_f, \quad \Psi_{q(NL)} = \Psi_{o(NL)} = 0, \\ &\Psi_{f(NL)} = L_f v_{f(NL)}/R_f, \quad \Psi_{kd(NL)} = L_{kdf} v_{f(NL)}/R_f, \quad \Psi_{kq(NL)} = 0 \end{aligned} \tag{2.35}$$

Similarly, from (2.32), the stator voltages in *d-q* frame of reference are

$$v_{d(NL)} = 0 \quad \text{and} \quad v_{q(NL)} = \omega_o kM_f v_{f(NL)}/R_f \tag{2.36}$$

Consequently, the machine terminal voltages at no-load conditions equal the induced voltages in stator windings that are given from (2.16), considering $v_o = i_o = 0$ and $\theta = \omega_o t + \delta$ 'ω_o is the rated angular speed' as

$$\left.\begin{aligned} v_a &= \sqrt{\frac{2}{3}}\left[v_{d(NL)}\cos(\omega_o t + \delta) + v_{q(NL)}\sin(\omega_o t + \delta)\right] \\ &= \sqrt{\frac{2}{3}}\, v_{q(NL)}\sin(\omega_o t + \delta) \\ v_b &= \sqrt{\frac{2}{3}}\left[v_{d(NL)}\cos\left(\omega_o t + \delta - \frac{2\pi}{3}\right) + v_{q(NL)}\sin\left(\omega_o t + \delta - \frac{2\pi}{3}\right)\right] \\ &= \sqrt{\frac{2}{3}}\, v_{q(NL)}\sin\left(\omega_o t + \delta - \frac{2\pi}{3}\right) \\ v_c &= \sqrt{\frac{2}{3}}\left[v_{d(NL)}\cos\left(\omega_o t + \delta + \frac{2\pi}{3}\right) + v_{q(NL)}\sin\left(\omega_o t + \delta + \frac{2\pi}{3}\right)\right] \\ &= \sqrt{\frac{2}{3}}\, v_{q(NL)}\sin\left(\omega_o t + \delta + \frac{2\pi}{3}\right) \end{aligned}\right\} \tag{2.37}$$

Under no-load conditions '$\delta = 0$'. Therefore, δ in the machine terminal voltages (2.37) should be equal to zero.

For *loaded* generator, the synchronous machine delivers electric power determined by the prime mover output. Thus, the currents and flux linkages in the machine are functions of the mechanical torque 'T_m' and the rms value of the sinusoidal line-to-line terminal voltage V.

Therefore, the voltage sources at generator terminals can be assumed as

$$\left.\begin{aligned} v_a &= \sqrt{\frac{2}{3}}\, V\sin(\omega_o t) \\ v_b &= \sqrt{\frac{2}{3}}\, V\sin\left(\omega_o t - \frac{2\pi}{3}\right) \\ v_c &= \sqrt{\frac{2}{3}}\, V\sin\left(\omega_o t + \frac{2\pi}{3}\right) \end{aligned}\right\} \tag{2.38}$$

and the *d-q* axis components of terminal voltages can be calculated as

$$v_{do} = -V\sin\delta, \quad v_{qo} = V\cos\delta \tag{2.39}$$

The subscript '*o*' denotes the rated value. Neglecting the armature resistance and considering no currents flowing in the damper windings in the steady state, it is observed from (2.30) that

$$v_{do} = -\omega_o\Psi_{qo}, \quad v_{qo} = \omega_o\Psi_{do} \tag{2.40}$$

Under the same conditions, from (2.32), (2.39) and (2.40), it is found that

$$\begin{aligned} v_{do} &= -\omega_o L_q i_{qo} && \text{i.e. } i_{qo} = -v_{do}/(\omega_o L_q) \quad \text{and} \\ v_{qo} &= \omega_o L_d i_{do} + \omega_o k M_f i_{fo} && \text{i.e. } i_{do} = (1/\omega_o L_d)\left[v_{qo} - \omega_o k M_f v_{fo}/R_f\right] \end{aligned} \tag{2.41}$$

2.4.3 Transformation of the torque equation

Applying Park's transformation to (2.11), the transformed electrical torque is

$$T'_e = -\frac{1}{2}\left[\boldsymbol{i}^t_{dqo}\mathbf{P}^t\left(\frac{\partial \boldsymbol{L}_{ss}}{\partial\theta}\right)\mathbf{P}\boldsymbol{i}_{dqo} + 2\boldsymbol{i}^t_{dqo}\mathbf{P}^t\left(\frac{\partial \boldsymbol{L}_{sr}}{\partial\theta}\right)\boldsymbol{i}_r\right] \tag{2.42}$$

where

$$\frac{\partial \boldsymbol{L}_{ss}}{\partial\theta} = -2L_m\begin{bmatrix} \sin 2\theta & \sin\left(2\theta - \frac{2\pi}{3}\right) & \sin\left(2\theta + \frac{2\pi}{3}\right) \\ \sin\left(2\theta - \frac{2\pi}{3}\right) & \sin\left(2\theta + \frac{2\pi}{3}\right) & \sin 2\theta \\ \sin\left(2\theta + \frac{2\pi}{3}\right) & \sin 2\theta & \sin\left(2\theta - \frac{2\pi}{3}\right) \end{bmatrix} \tag{2.43}$$

$$\frac{\partial \boldsymbol{L}_{sr}}{\partial\theta} = \begin{bmatrix} -M_f \sin\theta & -M_{kd}\sin\theta & M_{kq}\cos\theta \\ -M_f\sin\left(\theta - \frac{2\pi}{3}\right) & -M_{kd}\sin\left(\theta - \frac{2\pi}{3}\right) & M_{kq}\cos\left(\theta - \frac{2\pi}{3}\right) \\ -M_f\sin\left(2\theta + \frac{2\pi}{3}\right) & -M_{kd}\sin\left(2\theta + \frac{2\pi}{3}\right) & M_{kq}\cos\left(\theta + \frac{2\pi}{3}\right) \end{bmatrix} \tag{2.44}$$

Assuming

$$\mathbf{P}_2 = \begin{bmatrix} 0 & 1 & 0 \\ 1 & 0 & 0 \\ 0 & 0 & 0 \end{bmatrix} \tag{2.45}$$

and substituting (2.43), (2.44) and (2.45) into (2.42), it can be proved that the electrical torque is given by

$$T'_e = \sqrt{\frac{3}{2}}i_q\left(M_f i_f + M_{kd}i_{kd} + \sqrt{\frac{3}{2}}L_m i_d\right) - \sqrt{\frac{3}{2}}i_d\left(M_{kq}i_{kq} - \sqrt{\frac{3}{2}}L_m i_q\right) \tag{2.46}$$

From (2.20), it can be observed that

$$\Psi_d = L_d i_d + \sqrt{\frac{3}{2}} M_f i_f + \sqrt{\frac{3}{2}} M_{kd} i_{kd} \quad \text{and} \quad \Psi_q = L_q i_q + \sqrt{\frac{3}{2}} M_{kq} i_{kq} \tag{2.47}$$

From the definition of L_d and L_q in (2.21), it is found that

$$L_d - \frac{3}{2} L_m = L_q + \frac{3}{2} L_m = L_s - M_s \tag{2.48}$$

From (2.46), (2.47) and (2.48), the electrical torque can be expressed as

$$T'_e = i_q \Psi_d - i_d \Psi_q \tag{2.49}$$

In the steady-state and unloaded generator, the electrical torque is zero as the currents are zero. For loaded generator, the electrical torque in (2.49) can be written as

$$T'_{eo} = i_{qo} \Psi_{do} - i_{do} \Psi_{qo} \tag{2.50}$$

Substituting (2.39), (2.40) and (2.41) into (2.50), the electrical torque becomes

$$T'_{eo} = \frac{E_{fdo}}{\omega_o x_d} V \sin\delta + \frac{(x_d - x_q)}{2\omega_o x_d x_q} V^2 \sin 2\delta \tag{2.51}$$

where

$$x_d = \omega_o L_d, \quad x_q = \omega_o L_q, \quad E_{fdo} = x_{fdo} v_{fo} / R_f, \quad x_{dfo} = \omega_o k M_f$$

2.5 Machine parameters in per unit values

All machine variables in the preceding sections, such as voltage, current, power, flux linkage, inductance, are mainly based on three quantities: volt (V), ampere (A) and time 't' (s). The difficulty that the power engineers meet is that the quantities in physical units pertaining to the machine stator are in a much higher range than those for the machine rotor, e.g. the stator voltage may be in kilovolts and the field voltage of a much smaller value. Consequently, their magnitudes are very different involving inadequacy for engineering use.

Therefore, the equations used to calculate the machine variables are normalised by applying a convenient base to obviate this problem and the quantities are then expressed in percent of the base (per unit 'pu' values) [4].

Fixed base quantities must be chosen in such a way that all three quantities V, A and t are involved. Further information on per unit/normalised form is given in Appendix I.

Example 2.1 A two-pole, three-phase synchronous generator has the data given below. Find the per unit values of the machine parameters that can be calculated from these data.

Frequency = 60 Hz, line-to-line voltage = 24 kV, rating = 555 MV, power factor = 0.9, $L_s = 3.2758$ mH, $L_m = 0.0458$ mH, $M_s = -1.6379$, $M_f = 32.653$ mH

The stator leakage inductances $\ell_d = \ell_q \triangleq \ell_a = 0.4129$ mH

$L_f = 576.92$ mH

Solution:

From (2.20), L_d and L_q are defined as

$$L_d = L_s - M_s + \frac{3}{2}L_m, \quad L_q = L_s - M_s - \frac{3}{2}L_m$$

Hence,

$$L_d = 3.2758 + 1.6379 + \frac{3}{2} \times 0.0458 = 4.9824 \text{ mH}$$

$$L_q = 3.2758 + 1.6379 - \frac{3}{2} \times 0.0458 = 4.845 \text{ mH}$$

As in (I.7) and (I.9) of Appendix I, L_{md} and L_{mq} can be calculated by

$$L_{md} = L_d - \ell_a = 4.9824 - 0.4129 = 4.5695 \text{ mH}$$

$$L_{mq} = L_q - \ell_a = 4.845 - 0.4129 = 4.4321 \text{ mH}$$

$$kM_f = \sqrt{\frac{3}{2}} \times 32.653 = 40.0 \text{ mH}$$

Base quantities:

i. For the stator:

S_B = three-phase rating = 555 MVA
V_B = line-to-line voltage = 24 kV
$\omega_B = 120\pi = 377$ elec. rad/s
$t_B = (1/\omega_B) = 2.65258 \times 10^{-3}$ s

$$I_B = \frac{555}{24} \times 10^3 = 23.125 \text{ kA}$$

$$Z_B = \frac{V_B}{I_B} = \frac{24}{23.125} = 1.0378\ \Omega$$

$$\Psi_B = \frac{V_B}{\omega_B} = \frac{24 \times 10^3}{377} = 63.66 \text{ Wb turn}$$

$$L_B = \frac{\Psi_B}{I_B} = \frac{Z_B}{\omega_B} = \frac{1.0378}{377} = 2.75 \text{ mH}$$

ii. For the rotor:

As in (I.6) and (I.7) of Appendix I, it can be found that

$$I_{fB} = \frac{L_{md}}{kM_f} I_B = \frac{4.5695}{40.0} \times 23.125 = 2.64\,\text{kA}$$

$$M_{fB} = \frac{kM_f}{L_{md}} L_B = \frac{40.0}{4.5695} \times 2.75 \times 10^{-3} = 24.07\,\text{mH}$$

$$V_{fB} = \frac{S_B}{I_{fB}} = \frac{555 \times 10^6}{2.64 \times 10^3} = 210.23\,\text{kV}$$

$$Z_{fB} = \frac{210.23}{2.64} = 79.6325\,\Omega$$

$$L_{fB} = \frac{Z_{fB}}{\omega_B} = \frac{79.6325}{377} \times 10^3 = 211.227\,\text{mH}$$

Per unit values of machine parameters:

Applying the rule: per unit value $= \frac{\text{actual value}}{\text{base value}}$; the parameters in pu are obtained as below:

$L_d = 4.9824/2.75 = 1.81$
$L_f = 576.92/211.227 = 2.73$
$\ell_d = \ell_q = \ell_a = 0.4129/2.75 = 0.15$
$L_q = 4.845/2.75 = 1.76$
$L_{md} = 4.5695/2.75 = 1.66$
$L_{mq} = 4.4321/2.75 = 1.61$
$kM_f = kM_{kd} = L_d - \ell_d = 1.81 - 0.15 = 1.66$
$R_a = 0.0031/1.0378 = 2.99 \times 10^{-3}$
$R_f = 0.0715/79.6325 = 0.898 \times 10^{-3}$

Example 2.2 For the machine in Example 2.1 considering the following parameters 'in pu values' calculate the coefficient matrices to obtain the vector $d\boldsymbol{i}/dt$ in terms of the vectors $\boldsymbol{i}$ and $\boldsymbol{v}$.

$kM_{kq} = 1.59$, $L_{kd} = 0.1713$, $L_{kq} = 0.7252$, $R_{kd} = 0.0284$, $R_{kq} = 0.00619$ and assuming $kM_f = L_{fkd} = 1.66$

Solution:

Equation I.23 derived in Appendix I is written below:

$$\begin{bmatrix} v_d \\ -v_f \\ 0 \\ v_q \\ 0 \end{bmatrix} = -\begin{bmatrix} R_a & 0 & 0 & \omega L_q & \omega k M_{kq} \\ 0 & R_f & 0 & 0 & 0 \\ 0 & 0 & R_{kd} & 0 & 0 \\ -\omega L_d & -\omega k M_f & -\omega k M_{kd} & R_a & 0 \\ 0 & 0 & 0 & 0 & R_{kq} \end{bmatrix} \begin{bmatrix} i_d \\ i_f \\ i_{kd} \\ i_q \\ i_{kq} \end{bmatrix} - \begin{bmatrix} L_d & kM_f & kM_{kd} & 0 & 0 \\ kM_f & L_f & L_{fkd} & 0 & 0 \\ kM_{kd} & L_{fkd} & L_{kd} & 0 & 0 \\ 0 & 0 & 0 & L_q & kM_{kq} \\ 0 & 0 & 0 & kM_{kq} & L_{kq} \end{bmatrix} \begin{bmatrix} \mathrm{p}i_d \\ \mathrm{p}i_f \\ \mathrm{p}i_{kd} \\ \mathrm{p}i_q \\ \mathrm{p}i_{kq} \end{bmatrix} \tag{2.52}$$

Therefore, substituting the pu machine parameters calculated in Example 2.1 as well as the parameters given above, (2.52) becomes

$$\begin{bmatrix} v_d \\ -v_f \\ 0 \\ v_q \\ 0 \end{bmatrix} = -\begin{bmatrix} 0.00299 & 0 & 0 & 1.7618\omega & 1.59\omega \\ 0 & 0.00898 & 0 & 0 & 0 \\ 0 & 0 & 0.0284 & 0 & 0 \\ -1.812\omega & -1.66\omega & -1.66\omega & 0.00299 & 0 \\ 0 & 0 & 0 & 0 & 0.00619 \end{bmatrix} \begin{bmatrix} i_d \\ i_f \\ i_{kd} \\ i_q \\ i_{kq} \end{bmatrix} - \begin{bmatrix} 1.812 & 1.66 & 1.66 & 0 & 0 \\ 1.66 & 2.73 & 1.66 & 0 & 0 \\ 1.66 & 1.66 & 0.1713 & 0 & 0 \\ 0 & 0 & 0 & 1.7618 & 1.59 \\ 0 & 0 & 0 & 1.59 & 0.7252 \end{bmatrix} \begin{bmatrix} \mathrm{p}i_d \\ \mathrm{p}i_f \\ \mathrm{p}i_{kd} \\ \mathrm{p}i_q \\ \mathrm{p}i_{kq} \end{bmatrix}$$

Thus, the vector $\mathrm{d}\boldsymbol{i}/\mathrm{d}t$ can be written as

$$\frac{\mathrm{d}\boldsymbol{i}}{\mathrm{d}t} = -\mathbf{B}_2^{-1}\mathbf{B}_1\boldsymbol{i} - \mathbf{B}_2^{-1}\mathbf{v}$$

where

$$\mathbf{B}_1 = \begin{bmatrix} 0.00299 & 0 & 0 & 1.7618\omega & 1.59\omega \\ 0 & 0.00898 & 0 & 0 & 0 \\ 0 & 0 & 0.0284 & 0 & 0 \\ -1.812\omega & -1.66\omega & -1.66\omega & 0.00299 & 0 \\ 0 & 0 & 0 & 0 & 0.00619 \end{bmatrix}$$

$$\mathbf{B}_2 = \begin{bmatrix} 1.812 & 1.66 & 1.66 & 0 & 0 \\ 1.66 & 2.73 & 1.66 & 0 & 0 \\ 1.66 & 1.66 & 0.1713 & 0 & 0 \\ 0 & 0 & 0 & 1.7618 & 1.59 \\ 0 & 0 & 0 & 1.59 & 0.7252 \end{bmatrix}$$

Hence,

$$\mathbf{B}_2^{-1} = \begin{bmatrix} 0.763 & -0.825 & 0.594 & 0 & 0 \\ -0.825 & 0.817 & 0.084 & 0 & 0 \\ 0.594 & 0.084 & -0.732 & 0 & 0 \\ 0 & 0 & 0 & -0.581 & 1.272 \\ 0 & 0 & 0 & 1.274 & -1.41 \end{bmatrix}$$

and

$$\mathbf{B}_2^{-1}\mathbf{B}_1 = 10^{-3}\begin{bmatrix} 2.289 & 7.425 & 16.632 & 1344.4\omega & 1487.8\omega \\ 2.475 & 7.353 & 2.352 & -1453.6\omega & -1311.7\omega \\ 1.782 & 0.765 & -0.020 & 1046.6\omega & 944.5\omega \\ 1052.8\omega & 964.5\omega & 964.5\omega & -1.743 & 7.632 \\ -2308.5\omega & -2114.8\omega & -2114.8\omega & 3.822 & -8.460 \end{bmatrix}$$

Based on per unit system and normalising voltage equations as explained in Appendix I, the torque, power and swing equations can be derived as below.

2.5.1 Torque and power equations

According to (2.8), $J\ddot{\theta}_m = T_a$ when the damping term is neglected. It is convenient to express θ_m as $\theta_m = (\omega_o t + \alpha) + \delta_m$ as the angular reference may be chosen relative to a synchronously rotating reference frame moving with constant velocity ω_o, and α 'a constant' expresses the angle between the rotor position and the angular reference frame, while δ_m is the mechanical torque angle in radians. The electrical (torque) angle $\delta = (p/2)\delta_m$. Thus, this equation may be written as

$$J\ddot{\delta}_m = J\dot{\omega}_m = T_a \quad \text{or} \quad (2J/p)\ddot{\delta} = (2J/p)\dot{\omega} = T_a \quad \text{or} \quad (2/p)M\dot{\omega} = P_a \tag{2.53}$$

where M is the angular momentum $= J\omega$

The base quantity of the torque 'T_B' is equal to the rated torque at rated speed, i.e.

$$T_B = S_B/\omega_{mo} = 60S_B/2\pi n_o \tag{2.54}$$

where S_B is the three-phase stator-rated power (VA rms) and n_o is the rated shaft speed in rpm.

Thus, dividing (I.24) by (I.25) and substituting $p = 120f_o/n_o$ gives the pu quantity of torque, T_{au}, as

$$T_a/T_B = \frac{J\pi^2 n_o^2}{900S_B\omega_o}\dot{\omega} = T_{au} \quad \text{pu} \tag{2.55}$$

The quantity H 'called the inertia constant' is defined as the ratio of kinetic energy in megajoules to the rating in MVA, i.e. $H = \frac{\text{kinetic energy in megajoules}}{\text{Rating in MVA}} = (J\omega_m^2)/(2S_B)$ and has the dimension of time in seconds. Consequently,

$$T_a/T_B = \frac{J\dot{\omega}_m}{\dfrac{S_B}{\omega_{mo}}} = \frac{2H}{\omega_{mo}}\dot{\omega}_m = (2H/\omega_B)\dot{\omega} = T_{au} \quad \text{pu} \tag{2.56}$$

where ω is commonly given in the units of electrical rad/s as it is the angular velocity of revolving magnetic field in the air gap and, therefore, pertains directly to the network voltages and currents. The value of rated angular speed 'ω_o' is taken as the base quantity 'ω_B'. The form of (2.56) is called the swing equation and has been adapted for machines with any number of poles as all machines are incorporated in the same system and synchronised to the same rated angular velocity.

The swing equation may be written in another approximate form that is convenient for use with the classical model of the synchronous machine. This form is based on considering that the angular speed is nearly constant, which in turn yields that the numerical value of the pu accelerating torque is nearly equal to the accelerating power P_a. Thus, the approximated form becomes

$$(2H/\omega_B)\dot{\omega} \approx P_{au} \quad \text{pu} \tag{2.57}$$

Referring to the definition of basic pu quantities $t = t_u t_B$, $\omega = \omega_u\omega_B$ and $\omega_B = 1/t_B$, the following relations may be written as

$$\frac{1}{\mathrm{d}t} = \omega_B \frac{1}{\mathrm{d}t_u} \quad \text{and} \quad \mathrm{d}\omega = \omega_B \mathrm{d}\omega_u \tag{2.58}$$

Based on which of the terms in (2.56) are given in pu, different forms of swing equation are obtained by incorporating (2.58) as below:

- *T in pu, t in second, ω in electrical rad/s*

$$(2H/\omega_B)\frac{\mathrm{d}\omega}{\mathrm{d}t} = T_{au} \tag{2.59}$$

If ω is given in electrical degree/s, it should be multiplied by $(\pi/180)$ and the swing equation becomes

$$\frac{H}{180 f_B}\frac{d\omega}{dt} = T_{au} \tag{2.60}$$

- *T and t are in pu and ω is in electrical rad/s*

$$2H\frac{d\omega}{dt_u} = T_{au} \tag{2.61}$$

If ω is given in electrical degree/s, then the swing equation is in the form

$$\frac{\pi H}{90}\frac{d\omega}{dt_u} = T_{au} \tag{2.62}$$

- *T, t and ω are in pu*

$$(2H\omega_B)\frac{d\omega_u}{dt_u} = T_{au} \tag{2.63}$$

2.6 Synchronous machine equivalent circuits

The synchronous machine can be represented by two equivalent circuits: one corresponding to the d-axis and the other corresponding to the q-axis. The normalised flux linkages under balanced conditions 'Ψ_o is omitted' given in (2.20) can be rewritten as

$$\left.\begin{aligned} \Psi_d &= [(L_d - \ell_a) + \ell_a] i_d + kM_f i_f + kM_{kd} i_{kd} \\ \Psi_f &= kM_f i_d + \left[(L_f - \ell_f) + \ell_f\right] i_f + L_{fkd} i_{kd} \\ \Psi_{kd} &= kM_{kd} i_d + L_{fkd} i_f + [(L_{kd} - \ell_{kd}) + \ell_{kd}] i_{kd} \end{aligned}\right\} \tag{2.64}$$

where ℓ_a, ℓ_f and ℓ_{kd} are the leakage inductances of the coupled circuits on the d-axis, armature d-circuit, field circuit 'f' and damper circuit 'KD', respectively.

Also,

$$\left.\begin{aligned} \Psi_q &= \left[(L_q - \ell_a) + \ell_a\right] i_q + kM_{kq} i_{kq} \\ \Psi_{kq} &= kM_{kq} i_{kq} + \left[(L_{kq} - \ell_{kq}) + \ell_{kq}\right] i_{kq} \end{aligned}\right\} \tag{2.65}$$

where ℓ_a and ℓ_{kq} are the leakage inductances of the coupled circuits on the q-axis, armature q-circuit and damper circuit 'KQ', respectively.

For the d-axis: If $i_f = i_{kd} = 0$, the d-axis flux linkage mutually coupled to the other circuits is $(L_d - \ell_a) i_d$ or $L_{md} i_d$. In this case, the flux linkages in the f and KD windings are given by $\Psi_f = kM_f i_d$ and $\Psi_{kd} = kM_{kd} i_d$, respectively.

As the choice of the base rotor current is based on giving equal mutual flux, the pu values of $L_{md}i_d$, Ψ_f and Ψ_{kd} must be equal. Then,

$$L_d - \ell_a = kM_f = kM_{kd} = L_{md} \quad \text{pu} \tag{2.66}$$

It can be proved that

$$L_d - \ell_a = L_f - \ell_f = L_{kd} - \ell_{kd} = kM_f = kM_{kd} \triangleq L_{md} \quad \text{pu} \tag{2.67}$$

Subtracting the pu leakage flux linkage in each circuit results in the equality of the remaining flux linkages for all other coupled circuits. Therefore,

$$\Psi_d - \ell_a i_d = \Psi_f - \ell_f i_f = \Psi_{kd} - \ell_{kd} i_{kd} \triangleq \Psi_{Ad} \quad \text{pu} \tag{2.68}$$

and the pu d-axis mutual flux linkage, Ψ_{Ad}, is given by

$$\Psi_{Ad} = (L_d - \ell_a)i_d + kM_f i_f + kM_{kd} i_{kd} = L_{md}\left(i_d + i_f + i_{kd}\right) \tag{2.69}$$

From (2.30) and (2.64), the voltage equations are

$$\begin{aligned} v_d &= -\dot{\Psi}_d - \omega\Psi_q - R_a i_d \\ &= -\ell_a \mathrm{p} i_d - \left\{L_{md}\mathrm{p} i_d + kM_f \mathrm{p} i_f + kM_{kd}\mathrm{p} i_{kd}\right\} - R_a i_a - \omega\Psi_q \end{aligned}$$

Thus,

$$v_d = -\ell_a \mathrm{p} i_d - L_{md}\left(\mathrm{p} i_d + \mathrm{p} i_f + \mathrm{p} i_{kd}\right) - R_a i_a - \omega\Psi_q \tag{2.70}$$

Similarly,

$$-v_f = -\ell_f \mathrm{p} i_f - L_{md}(\mathrm{p} i_d + \mathrm{p} i_f + \mathrm{p} i_{kd}) - R_f i_f \tag{2.71}$$

$$v_{kd} = 0 = -\ell_{kd}\mathrm{p} i_{kd} - L_{md}(\mathrm{p} i_d + \mathrm{p} i_f + \mathrm{p} i_{kd}) - R_{kd} i_{kd} \tag{2.72}$$

Applying the procedure used in developing the equivalent circuit of transformers to represent the relations given in (2.66) through (2.69) and to satisfy the voltage equations (2.70) through (2.72) the equivalent circuit for the d-axis is shown in Figure 2.4. The d-axis circuits (D, F and KD) are coupled through the

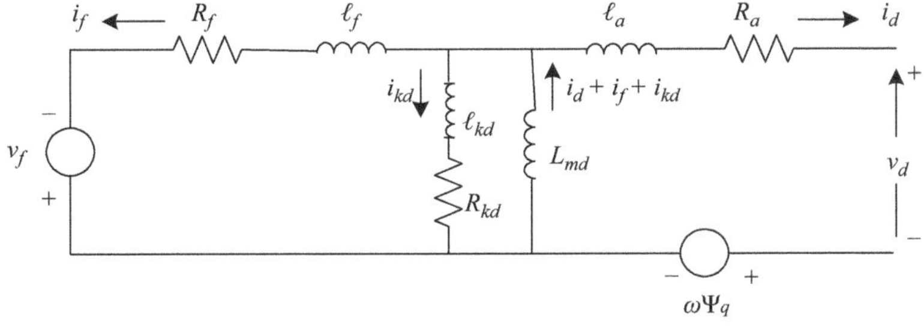

Figure 2.4 Equivalent circuit for the d-axis

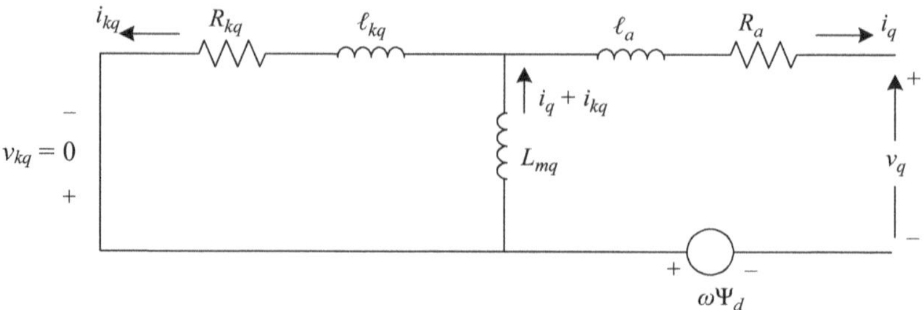

Figure 2.5 Equivalent circuit for the q-axis

common magnetising inductance $L_{md} = (L_d - \ell_a)$ that carries the sum of currents i_d, i_f and i_{kd}. The equivalent circuit contains a controlled voltage source $\omega\Psi_q$.

For the q-axis, following the same procedure as used above, the pu q-axis mutual flux linkage, Ψ_{aq}, and voltage equations are given by

$$\Psi_{Aq} = L_{mq} i_q + kM_{kq} i_{kq} = L_{mq}\left(i_q + i_{kq}\right) \tag{2.73}$$

$$v_q = -\ell_a \mathrm{p} i_q - L_{mq}\left(\mathrm{p} i_q + \mathrm{p} i_{kq}\right) - R_a i_q + \omega\Psi_d \tag{2.74}$$

$$v_{kq} = 0 = -\ell_{kq} \mathrm{p} i_{kq} - L_{mq}\left(\mathrm{p} i_q + \mathrm{p} i_{kq}\right) - R_{kq} i_{kq} \tag{2.75}$$

where L_{mq} is defined as

$$L_{mq} = L_q - \ell_a = L_{kq} - \ell_{kq} = kM_{kq} \quad \text{pu}$$

The equivalent circuit for the q-axis satisfying these relations is shown in Figure 2.5. It is to be noted that it contains a controlled voltage source $\omega\Psi_d$.

2.7 Flux linkage state space model

2.7.1 Modelling without saturation

The relations in Section 2.6 can be used to develop an alternative state space model based on choosing Ψ_d, Ψ_f, Ψ_{kd}, Ψ_{kq} and Ψ_q as state variables. From (2.68) the d-axis currents are given by

$$i_d = \frac{1}{\ell_a}(\Psi_d - \Psi_{Ad}), \quad i_f = \frac{1}{\ell_f}(\Psi_f - \Psi_{Ad}), \quad i_{kd} = \frac{1}{\ell_{kd}}(\Psi_{kd} - \Psi_{Ad}) \tag{2.76}$$

Incorporating (2.69), where $\Psi_{Ad} = L_{md}(i_d + i_f + i_{kd})$ into (2.76) gives

$$\Psi_{Ad}\left(\frac{1}{L_{md}} + \frac{1}{\ell_a} + \frac{1}{\ell_f} + \frac{1}{\ell_{kd}}\right) = \frac{\Psi_d}{\ell_a} + \frac{\Psi_f}{\ell_f} + \frac{\Psi_{kd}}{\ell_{kd}} \tag{2.77}$$

If L_{Md} is defined as

$$\frac{1}{L_{Md}} \triangleq \frac{1}{L_{md}} + \frac{1}{\ell_a} + \frac{1}{\ell_f} + \frac{1}{\ell_{kd}}$$

then

$$\Psi_{Ad} = \frac{L_{Md}}{\ell_a}\Psi_d + \frac{L_{Md}}{\ell_f}\Psi_f + \frac{L_{Md}}{\ell_{kd}}\Psi_{kd} \tag{2.78}$$

Similarly, it can be found that

$$\Psi_{Aq} = \frac{L_{Mq}}{\ell_a}\Psi_q + \frac{L_{Mq}}{\ell_{kq}}\Psi_{kq} \tag{2.79}$$

where L_{Mq} is defined as

$$\frac{1}{L_{Mq}} \triangleq \frac{1}{L_{mq}} + \frac{1}{\ell_a} + \frac{1}{\ell_{kq}} \tag{2.80}$$

and the q-axis currents are given by

$$i_q = \frac{1}{\ell_a}(\Psi_q - \Psi_{Aq}), \quad i_{kq} = \frac{1}{\ell_{kq}}(\Psi_{kq} - \Psi_{Aq}) \tag{2.81}$$

Equations (2.76) and (2.81) can be rewritten in the matrix form as

$$\begin{bmatrix} i_d \\ i_f \\ i_{kd} \\ \hline i_q \\ i_{kq} \end{bmatrix} = \left[\begin{array}{cccc|ccc} \frac{1}{\ell_a} & 0 & 0 & -\frac{1}{\ell_d} & 0 & 0 & 0 \\ 0 & \frac{1}{\ell_f} & 0 & -\frac{1}{\ell_f} & 0 & 0 & 0 \\ 0 & 0 & \frac{1}{\ell_{kd}} & -\frac{1}{\ell_{kd}} & 0 & 0 & 0 \\ \hline 0 & 0 & 0 & 0 & \frac{1}{\ell_a} & 0 & -\frac{1}{\ell_q} \\ 0 & 0 & 0 & 0 & 0 & \frac{1}{\ell_{kq}} & -\frac{1}{\ell_{kq}} \end{array}\right] \begin{bmatrix} \Psi_d \\ \Psi_f \\ \Psi_{kd} \\ \Psi_{Ad} \\ \hline \Psi_q \\ \Psi_{kq} \\ \Psi_{Aq} \end{bmatrix} \tag{2.82}$$

Incorporate the currents in (2.82) into voltage equations (2.30) to get the derivatives of flux linkages as

$$\left.\begin{aligned}
\dot{\Psi}_d &= -\frac{R_a}{\ell_a}\Psi_d + \frac{R_a}{\ell_a}\Psi_{Ad} - \omega\Psi_q - v_d \\
\dot{\Psi}_f &= -\frac{R_f}{\ell_f}\Psi_f + \frac{R_f}{\ell_f}\Psi_{Ad} - (-v_f) \\
\dot{\Psi}_{kd} &= -\frac{R_{kd}}{\ell_{kd}}\Psi_{kd} + \frac{R_{kd}}{\ell_{kd}}\Psi_{Ad} \\
\dot{\Psi}_q &= -\frac{R_a}{\ell_a}\Psi_q + \frac{R_a}{\ell_a}\Psi_{Aq} - \omega\Psi_d - v_q \\
\dot{\Psi}_{kq} &= -\frac{R_{kq}}{\ell_{kq}}\Psi_g + \frac{R_{kq}}{\ell_{kq}}\Psi_{Aq}
\end{aligned}\right\} \tag{2.83}$$

Also, substituting the currents in (2.82) in (2.49), where the electrical torque $T_e = i_q\Psi_d - i_d\Psi_q$, gives

$$T_e = -\Psi_q\left(\frac{\Psi_d - \Psi_{Ad}}{\ell_a}\right) + \Psi_d\left(\frac{\Psi_q - \Psi_{Aq}}{\ell_a}\right)$$

Thus,

$$T_e = -\frac{1}{\ell_a}\Psi_d\Psi_{Aq} + \frac{1}{\ell_a}\Psi_q\Psi_{Ad} + \left(\frac{1}{\ell_a} - \frac{1}{\ell_a}\right)\Psi_d\Psi_q \tag{2.84}$$

Considering the electrical torque, time and angular speed in pu values, the swing equation, Section 2.5.2, is

$$(2H\omega_B)\frac{d\omega_u}{dt_u} = T_{au} \tag{2.85}$$

Substituting (2.73) in (2.74) and ignoring the damping coefficient to give $\dot{\omega}$ as

$$\dot{\omega} = \frac{1}{2H\omega_B}\left[T_m - \frac{\Psi_{Ad}}{\ell_a}\Psi_q + \frac{\Psi_{Aq}}{\ell_a}\Psi_d\right] \tag{2.86}$$

If the damping is considered, the term $(-D/2H\omega_B)\omega$ is added to the RHS.
The equation of electrical torque angle in pu is given by

$$\dot{\delta} = \omega - 1 \tag{2.87}$$

Therefore, (2.83), (2.86) and (2.87) are in state space form: $\dot{\boldsymbol{x}} = \boldsymbol{f}(\boldsymbol{x}, \boldsymbol{u}, t)$, where $\boldsymbol{x} \triangleq$ the state variable $= \left[\Psi_d, \Psi_f, \Psi_{kd}, \Psi_q, \Psi_{kq}, \omega, \delta\right]$ and $\boldsymbol{u} \triangleq$ the forcing function is $[v_d, v_q, v_f]$, and T_m. Equations (2.67) and (2.68) are used to calculate Ψ_{Ad} and Ψ_{Aq}. It is to be noted that this form of equations is adequate for the analysis when saturation is considered, as the terms Ψ_{Ad} and Ψ_{Aq} are affected by saturation.

If the saturation is neglected, the terms L_{md} and L_{mq} are constant. This implies that L_{Md} and L_{Mq} are constant as well. Consequently, the relationships of the magnetising flux linkages Ψ_{Ad} and Ψ_{Aq} to the state variables (2.78) and (2.79) are constant and can be omitted from the machine equations. The d-axis currents in (2.76) can be rewritten by substituting the value of Ψ_{Ad} given in (2.78) as

$$\left.\begin{aligned} i_d &= \left(1-\frac{L_{Md}}{\ell_a}\right)\frac{\Psi_d}{\ell_a}-\frac{L_{Md}}{\ell_a}\frac{\Psi_f}{\ell_f}-\frac{L_{Md}}{\ell_a}\frac{\Psi_{kd}}{\ell_{kd}} \\ i_f &= -\frac{L_{Md}}{\ell_f}\frac{\Psi_d}{\ell_a}+\left(1-\frac{L_{Md}}{\ell_f}\right)\frac{\Psi_f}{\ell_f}-\frac{L_{Md}}{\ell_f}\frac{\Psi_{kd}}{\ell_{kd}} \\ i_{kd} &= -\frac{L_{Md}}{\ell_{kd}}\frac{\Psi_d}{\ell_a}-\frac{L_{Md}}{\ell_{kd}}\frac{\Psi_f}{\ell_f}+\left(1-\frac{L_{Md}}{\ell_{kd}}\right)\frac{\Psi_{kd}}{\ell_{kd}} \end{aligned}\right\} \tag{2.88}$$

Applying the same procedure using (2.79) and (2.85) to obtain the q-axis currents and then substituting all currents in d and q axes in voltage equation (2.30) gives

$$\left.\begin{aligned} \dot{\Psi}_d &= -R_a\left(1-\frac{L_{Md}}{\ell_a}\right)\frac{\Psi_d}{\ell_a}+R_a\frac{L_{Md}}{\ell_a}\frac{\Psi_f}{\ell_f}+R_a\frac{L_{Md}}{\ell_a}\frac{\Psi_{kd}}{\ell_{kd}}-\omega\Psi_q-v_d \\ \dot{\Psi}_f &= R_f\frac{L_{Md}}{\ell_f}\frac{\Psi_d}{\ell_a}-R_f\left(1-\frac{L_{Md}}{\ell_f}\right)\frac{\Psi_f}{\ell_f}+R_f\frac{L_{Md}}{\ell_f}\frac{\Psi_{kd}}{\ell_{kd}}+v_f \\ \dot{\Psi}_{kd} &= R_{kd}\frac{L_{Md}}{\ell_{kd}}\frac{\Psi_d}{\ell_a}+R_{kd}\frac{L_{Md}}{\ell_{kd}}\frac{\Psi_f}{\ell_f}R_{kd}\left(1-\frac{L_{Md}}{\ell_{kd}}\right)\frac{\Psi_{kd}}{\ell_{kd}} \\ \dot{\Psi}_q &= -R_a\left(1-\frac{L_{Mq}}{\ell_a}\right)\frac{\Psi_q}{\ell_a}+R_a\frac{L_{Mq}}{\ell_a}\frac{\Psi_{kq}}{\ell_{kq}}+\omega\Psi_d-v_q \\ \dot{\Psi}_{kq} &= R_{kq}\frac{L_{Mq}}{\ell_{kq}}\frac{\Psi_q}{\ell_a}-R_{kq}\left(1-\frac{L_{Mq}}{\ell_{kq}}\right)\frac{\Psi_{kq}}{\ell_{kq}} \end{aligned}\right\} \tag{2.89}$$

Using (2.78), (2.79) and (2.84) gives the electrical torque as

$$T_e = \Psi_d\Psi_q\left(\frac{L_{Md}-L_{Mq}}{\ell_a^2}\right)-\Psi_d\Psi_{kq}\frac{L_{Mq}}{\ell_a\ell_{kq}}+\Psi_q\Psi_f\frac{L_{Md}}{\ell_a\ell_f}+\Psi_q\Psi_{kd}\frac{L_{Md}}{\ell_a\ell_{kd}} \tag{2.90}$$

Incorporating (2.90) into (2.85):

$$\dot{\omega} = \frac{1}{2H\omega_B}\left\{-\left[\Psi_d\Psi_q\left(\frac{L_{Md}-L_{mq}}{\ell_a^2}\right)-\Psi_d\Psi_{kq}\frac{L_{Mq}}{\ell_a\ell_{kq}}+\Psi_q\Psi_f\frac{L_{Md}}{\ell_a\ell_f}+\Psi_a\Psi_{kd}\frac{L_{Md}}{\ell_a\ell_{kd}}\right]\right\} \tag{2.91}$$

Again, the equation of electrical torque angle in pu is given by

$$\dot{\delta} = \omega - 1 \tag{2.92}$$

Therefore, (2.89), (2.91) and (2.92) form the state space model describing the system in terms of v_d and v_q that are functions of the currents demanded by the external loads. In matrix form the state space model can be written as

$$\begin{bmatrix} \dot{\Psi}_d \\ \dot{\Psi}_f \\ \dot{\Psi}_{kd} \\ \dot{\Psi}_q \\ \dot{\Psi}_{kq} \\ \dot{\omega} \\ \dot{\delta} \end{bmatrix} = \begin{bmatrix} -\frac{R_a}{\ell_a}\left(1-\frac{L_{Md}}{\ell_a}\right) & \frac{R_a}{\ell_a}\frac{L_{Md}}{\ell_f} & \frac{R_a}{\ell_a}\frac{L_{Md}}{\ell_{kd}} & -\omega & 0 & 0 & 0 \\ \frac{R_f}{\ell_f}\frac{L_{Md}}{\ell_a} & -\frac{R_f}{\ell_f}\left(1-\frac{L_{Md}}{\ell_f}\right) & \frac{R_f}{\ell_f}\frac{L_{Md}}{\ell_{kd}} & 0 & 0 & 0 & 0 \\ \frac{R_{kd}}{\ell_{kd}}\frac{L_{Md}}{\ell_a} & \frac{R_{kd}}{\ell_{kd}}\frac{L_{Md}}{\ell_f} & -\frac{R_{kd}}{\ell_{kd}}\left(1-\frac{L_{Md}}{\ell_{kd}}\right) & 0 & 0 & 0 & 0 \\ \omega & 0 & 0 & -\frac{R_a}{\ell_a}\left(1-\frac{L_{Mq}}{\ell_a}\right) & \frac{R_a}{\ell_q}\frac{L_{Mq}}{\ell_{kq}} & 0 & 0 \\ 0 & 0 & 0 & \frac{R_{kq}}{\ell_{kq}}\frac{L_{Mq}}{\ell_a} & -\frac{R_{kq}}{\ell_{kq}}\left(1-\frac{L_{Mq}}{\ell_{kq}}\right) & 0 & 0 \\ -\frac{L_{Md}}{2H\omega_B\ell_a^2}\Psi_q & -\frac{L_{Md}}{2H\omega_B\ell_a\ell_f}\Psi_q & -\frac{L_{Md}}{2H\omega_B\ell_a\ell_{kd}}\Psi_q & \frac{L_{Mq}}{2H\omega_B\ell_a^2}\Psi_d & \frac{L_{Mq}}{2H\omega_B\ell_a\ell_{kq}}\Psi_d & -Đ & 0 \\ 0 & 0 & 0 & 0 & 0 & 1 & 0 \end{bmatrix}$$

$$\times \begin{bmatrix} \Psi_d \\ \Psi_f \\ \Psi_{kd} \\ \Psi_q \\ \Psi_{kq} \\ \omega \\ \delta \end{bmatrix} + \begin{bmatrix} -v_d \\ v_f \\ 0 \\ -v_q \\ 0 \\ \frac{T_m}{2H\omega_B} \\ -1 \end{bmatrix} \tag{2.93}$$

where $Đ = D/(2H\omega_B)$

Example 2.3 Write the coefficient matrix of the flux linkage model (2.93) for the machine data given in Examples 2.1 and 2.2 considering the inertia constant $H = 3.5$ s, $\omega_o = 3600$ rpm and the damping coefficient is neglected.

Solution:

$$\ell_f = L_f - kM_f = 2.73 - 1.66 = 1.07, \quad \ell_{kd} = L_d - kM_{kd} = 1.81 - 1.66 = 0.15$$

$$L_{md} = kM_{kd} = kM_f = 1.66, \quad \ell_{kq} = L_q - kM_{kq} = 1.76 - 1.59 = 0.17$$

$$L_{mq} = L_q - \ell_a = 1.76 - 0.15 = 1.61$$

$$\frac{1}{L_{Md}} = \frac{1}{L_{md}} + \frac{1}{\ell_a} + \frac{1}{\ell_f} + \frac{1}{\ell_{kd}} = \frac{1}{1.66} + \frac{1}{0.15} + \frac{1}{1.07} + \frac{1}{0.15} = 14.87$$

$$L_{Md} = 0.067$$

$$\frac{1}{L_{Mq}} = \frac{1}{L_{mq}} + \frac{1}{\ell_a} + \frac{1}{\ell_{kq}} = \frac{1}{1.61} + \frac{1}{0.15} + \frac{1}{0.17} = 13.17$$

$$L_{Mq} = 0.076$$

$$\frac{R_a}{\ell_d}\left(1 - \frac{L_{Md}}{\ell_a}\right) = \frac{2.99}{0.15} \times 10^{-3}\left(1 - \frac{0.067}{0.15}\right) = 10.96 \times 10^{-3}$$

$$\frac{R_a}{\ell_a}\frac{L_{Md}}{\ell_f} = 19.93 \times 10^{-3} \times 0.0626 = 1.25 \times 10^{-3}$$

$$\frac{R_a}{\ell_a}\frac{L_{Md}}{\ell_{kd}} = 19.93 \times 10^{-3} \times 0.45 = 8.90 \times 10^{-3}$$

$$\frac{R_f}{\ell_f}\frac{L_{Md}}{\ell_a} = (0.898 \times 10^{-3} \times 0.067)/(1.07 \times 0.15) = 0.375 \times 10^{-3}$$

$$\frac{R_f}{\ell_f}\left(1 - \frac{L_{Md}}{\ell_f}\right) = 0.786 \times 10^{-3}$$

$$\frac{R_f}{\ell_f}\frac{L_{Md}}{\ell_{kd}} = 0.375 \times 10^{-3}$$

$$\frac{R_{kd}}{\ell_{kd}}\frac{L_{Md}}{\ell_a} = (28.4 \times 10^{-3} \times 0.067)/(0.15 \times 0.15) = 84.57 \times 10^{-3}$$

$$\frac{R_{kd}}{\ell_{kd}}\frac{L_{Md}}{\ell_f} = (28.4 \times 10^{-3} \times 0.067)/(0.15 \times 1.07) = 11.86 \times 10^{-3}$$

$$\frac{R_{kd}}{\ell_{kd}}\left(1 - \frac{L_{Md}}{\ell_{kd}}\right) = 104.13 \times 10^{-3}$$

$$\frac{R_a}{\ell_a}\left(1 - \frac{L_{Mq}}{\ell_a}\right) = 9.827 \times 10^{-3}$$

$$\frac{R_a}{\ell_a}\frac{L_{Mq}}{\ell_{kq}} = 8.911 \times 10^{-3}, \qquad \frac{R_{kq}}{\ell_{kq}}\frac{L_{Mq}}{\ell_a} = 18.45 \times 10^{-3}$$

$$\frac{R_{kq}}{\ell_{kq}}\left(1 - \frac{L_{Mq}}{\ell_{kq}}\right) = 20.135 \times 10^{-3}$$

$$\frac{L_{Md}}{2H\omega_B\ell_a^2} = 1.13 \times 10^{-3} \qquad \frac{L_{Md}}{2H\omega_B\ell_a\ell_f} = 0.158 \times 10^{-3}$$

$$\frac{L_{Md}}{2H\omega_B\ell_a\ell_{kd}} = 1.132 \times 10^{-3} \qquad \frac{L_{Mq}}{2H\omega_B\ell_a^2} = 1.283 \times 10^{-3}$$

$$\frac{L_{Mq}}{2H\omega_B\ell_a\ell_{kq}} = 1.132 \times 10^{-3}$$

$\omega = 2\pi\dfrac{P}{2}\dfrac{\text{rpm}}{60} = 120\pi = \omega_B$ and then the per unit value of ω equals 1.

Thus, the coefficient matrix is

$$\begin{bmatrix} -10.96 & 1.25 & 8.90 & -1 & 0 & 0 & 0 \\ 0.375 & -0.786 & 0.375 & 0 & 0 & 0 & 0 \\ 84.57 & 11.86 & -104.13 & 0 & 0 & 0 & 0 \\ 1 & 0 & 0 & -9.827 & 8.911 & 0 & 0 \\ 0 & 0 & 0 & 18.45 & -20.135 & 0 & 0 \\ -1.13\Psi_q & -0.158\Psi_q & -1.132\Psi_q & 1.283\Psi_d & 1.132\Psi_d & 0 & 0 \\ 0 & 0 & 0 & 0 & 0 & 1 & 0 \end{bmatrix} \times 10^{-3}$$

2.7.2 Modelling with saturation

The stator and rotor of the synchronous machine exhibit saturation because they contain magnetic materials. Equations (2.83) through (2.87) form the flux linkage state space model that can be used in the case of considering the effect of saturation. This implies that the machine mutual inductances L_{md} and L_{mq} are not constant and depend on the levels of magnetising flux linkages Ψ_{Ad} and Ψ_{Aq} that have non-linear characteristics. So, the terms L_{md} and L_{mq} should be corrected for saturation. The exact analysis of saturation may present some difficulty in parameter identification [5–7] and may require finite element analysis [8].

Several approaches have been proposed for modelling the saturated machine such as (i) representing the saturation function as an arctangent function in terms of the initial constant slope and the final constant slope of saturation characteristics [9, 10], (ii) by a polynomial series or (iii) incorporating the saturation characteristics as a lookup table function. Other approaches introduce machine models that replace the rotor equivalent circuits by arbitrary linear networks to allow for elimination of parameter identification procedure of the equivalent circuit [11–13] or by determining an intermediate axis saturation characteristic [14]. A comparative study of saturation models used in stability analysis has been presented in [15]. It concludes that the complication of modifying the inductances at every solution step may not be essential particularly for large-scale system studies.

The predominant method proposed in the literature mostly considers that the mutual inductances L_{md} and L_{mq} in a machine are affected by saturation and should be corrected to be L_{mds} and L_{mqs} 'the subscript s denotes saturation' given by

$$L_{mds} = S_d L_{md} \quad \text{and} \quad L_{mqs} = S_q L_{mq} \tag{2.94}$$

where S_d and S_q are non-linear factors that depend on the flux levels.

Experimentally, the q-axis saturation is neglected and this is sufficiently accurate for calculating power system transient stability involving salient pole machines. Therefore, for salient pole machines, $S_q = 1$ and S_d is a function of Ψ_{Ad} while for round rotor machines, $S_d = S_q$ and is a function of linkages Ψ_{Ad} and Ψ_{Aq}. Thus, for salient pole machines

$$L_{mds} = S_d L_{md}, \quad L_{mqs} = L_{mq} \quad \text{and} \quad S_d = f(\Psi_{Ad}) \tag{2.95}$$

and for round rotor machines

$$L_{Ads} = S_d L_{Ad}, \quad L_{Adqs} = S_q L_{Aq}, \quad S_d = S_q = f(\Psi) \quad \text{and} \quad \Psi = \sqrt{\Psi_{Ad}^2 + \Psi_{Aq}^2} \tag{2.96}$$

The factor S_d is normally derived from the saturation curve of the machine (Figure 2.6), indicating the relation between the flux linkage Ψ_{Ad} and the magnetising current, i_{sum}, the sum of currents $i_d + i_f + i_{kd}$ as in (2.69).

This relation is linear for both unsaturated and highly saturated conditions, although the slopes and intercepts of the two regions are different. The slope of this characteristic is, therefore, initially constant, undergoes a transition and finally becomes constant again.

For a given value of Ψ_{Ad} the unsaturated magnetising current is $i_{(sum)o}$ corresponding to L_{mdo}, while the saturated value is $i_{(sum)s}$. The saturation factor S_d is a function of this magnetising current, i.e. is a function of Ψ_{Ad} as well. Thus, the value of unsaturated inductance L_{md} is given by

$$L_{md} = \frac{A_3A_4}{oA_4} \text{ and the saturated inductance } L_{mds} = \frac{A_2A_4}{oA_4}$$

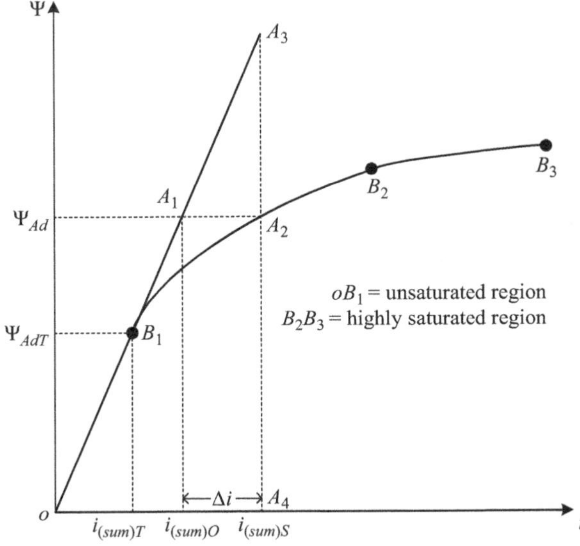

Figure 2.6 Magnetic saturation curve

Then

$$L_{mds} = L_{md}\frac{A_2A_4}{A_3A_4} = S_d L_{md}$$

where

$$S_d = \frac{A_2A_4}{A_3A_4} = \frac{i_{(sum)o}}{i_{(sum)s}} \tag{2.97}$$

The value $i_{(sum)s}$ can be calculated by adding the current increment Δi to the unsaturated magnetising current $i_{(sum)o}$. So, Δi should be calculated first when the flux linkage lies in the saturation region ($\mathbf{B}_1\mathbf{B}_2$ in Figure 2.6). This can be obtained by applying the approximate relation:

$$\Delta i = A_s \exp[B_s(\Psi_{Ad} - \Psi_{AdT})] \quad \text{and} \quad \Psi_{Ad} > \Psi_{AdT} \tag{2.98}$$

where A_s and B_s are constants and determined from the saturation curve of the machine. Ψ_{AdT} is the flux linkage at the transition point from the initial constant slope region to the saturated region. Then, $i_{(sum)s}$ is calculated for a given Ψ_{Ad} as

$$i_{(sum)s} = i_{(sum)o} + \Delta i \tag{2.99}$$

Accordingly, S_d is determined by using (2.97). The solution is achieved through an iterative process that is terminated when $\Psi_{Ad}S_d = L_{Ado}i_{(sum)s}$ is satisfied.

2.8 The current state space model

In matrix notation, (2.52) is written as

$$v = -\mathbf{B}_1\boldsymbol{i} - \mathbf{B}_2\frac{\mathrm{d}\boldsymbol{i}}{\mathrm{d}t} \tag{2.100}$$

where the two matrices $\mathbf{B}_1$ and $\mathbf{B}_2$ are

$$\mathbf{B}_1 = \begin{bmatrix} R_a & 0 & 0 & \omega L_q & \omega kM_{kq} \\ 0 & R_f & 0 & 0 & 0 \\ 0 & 0 & R_h & 0 & 0 \\ -\omega L_d & -\omega kM_f & -\omega kM_{kd} & R_a & 0 \\ 0 & 0 & 0 & 0 & R_{kq} \end{bmatrix}$$

$$\mathbf{B}_2 = \begin{bmatrix} L_d & kM_f & kM_{kd} & 0 & 0 \\ kM_f & L_f & L_{fkd} & 0 & 0 \\ kM_{kd} & L_{fkd} & L_{kd} & 0 & 0 \\ 0 & 0 & 0 & L_q & kM_{kq} \\ 0 & 0 & 0 & kM_{kq} & l_{kq} \end{bmatrix}$$

Thus,

$$\frac{\mathrm{d}\boldsymbol{i}}{\mathrm{d}t} = -\mathbf{B}_2^{-1}\mathbf{B}_1\boldsymbol{i} - \mathbf{B}_2^{-1}\mathbf{v} \tag{2.101}$$

From (2.85) and taking the damping term 'T_d' into account, it is observed that

$$(2H\omega_B)\dot{\omega} = T_a = T_m - T_e - T_d$$

where $T_e = i_q\Psi_d - i_d\Psi_q$ as given by (2.49)

$\Psi_d = L_d i_d + kM_f i_f + kM_{kd}i_{kd}, \Psi_q = L_q i_q + kM_{kq}i_{kq}$ as given by (2.20) and $T_d = D\omega$

Thus,

$$\begin{aligned}\dot{\omega} &= \frac{1}{2H\omega_B}(T_m - D\omega) \\ &\quad - \frac{1}{2H\omega_B}\left[i_q\left(L_d i_d + kM_f i_f + kM_{kd}i_{kd}\right) - i_d\left(L_q i_q + kM_{kq}i_{kq}\right)\right]\end{aligned} \tag{2.102}$$

and

$$\dot{\delta} = \omega - 1 \tag{2.103}$$

Incorporating (2.101) through (2.103), the current state space model can be written as

$$\begin{bmatrix} pi_d \\ pi_f \\ pi_{kd} \\ pi_q \\ pi_{kq} \\ \dot{\omega} \\ \dot{\delta} \end{bmatrix} = \left[\begin{array}{ccccc|cc} \multicolumn{5}{c|}{-\mathbf{B}_2^{-1}\mathbf{B}_1} & \multicolumn{2}{c}{\mathbf{0}} \\ \hline -\frac{L_d i_q}{2H\omega_B} & -\frac{kM_f i_q}{2H\omega_B} & -\frac{kM_{kd} i_q}{2H\omega_B} & \frac{L_q i_d}{2H\omega_B} & \frac{kM_{kq} i_d}{2H\omega_B} & -\frac{D}{2H\omega_B} & 0 \\ 0 & 0 & 0 & 0 & 0 & 1 & 0 \end{array}\right] \begin{bmatrix} i_d \\ i_f \\ i_{kd} \\ i_q \\ i_{kq} \\ \omega \\ \delta \end{bmatrix} + \begin{bmatrix} \mathbf{B}_2^{-1}\mathbf{v} \\ \frac{T_m}{2H\omega_B} \\ -1 \end{bmatrix} \tag{2.104}$$

Example 2.4 Referring to Example 2.2, find the current state space model.

Solution:

The current state space model is expressed by (2.104). The matrices $\mathbf{B}_1$, $\mathbf{B}_2$, $\mathbf{B}_2^{-1}$ and $\mathbf{B}_2^{-1}\mathbf{B}_1$ have been calculated in Example 2.2. Therefore, the last two rows in (2.104) should be calculated to form the current state space model as below:

$$\frac{L_d i_q}{2H\omega_B} = 0.688 \times 10^{-3} i_q \qquad \frac{kM_f i_q}{2H\omega_B} = 0.631 \times 10^{-3} i_q$$

$$\frac{kM_{kd} i_q}{2H\omega_B} = 0.631 \times 10^{-3} i_q \qquad \frac{L_q i_d}{2H\omega_B} = 0.669 \times 10^{-3} i_d$$

$$\frac{kM_{kq} i_d}{2H\omega_B} = 0.604 \times 10^{-3} i_d$$

Thus, the current state space model is

$$\begin{bmatrix} \mathrm{p}i_d \\ \mathrm{p}i_f \\ \mathrm{p}i_{kd} \\ \mathrm{p}i_q \\ \mathrm{p}i_{kq} \\ \hline \dot{\omega} \\ \dot{\delta} \end{bmatrix} = 10^{-3} \left[\begin{array}{cccccc|c} & & & & & & \\ & & & & & & \\ & & & -\mathbf{B}_2^{-1}\mathbf{B}_1 & & & \mathbf{0} \\ & & & & & & \\ & & & & & & \\ \hline -0.688 i_q & -0.631 i_q & -0.631 i_q & 0.669 i_d & 0.604 i_d & 0 & 0 \\ 0 & 0 & 0 & 0 & 0 & 1 & 0 \end{array}\right] \begin{bmatrix} i_d \\ i_f \\ i_{kd} \\ i_q \\ i_{kq} \\ \hline \omega \\ \delta \end{bmatrix}$$

$$+ \begin{bmatrix} \\ \mathbf{B}_2^{-1}\mathbf{v} \\ \\ 0.38 \times 10^{-3} T_m \\ -1 \end{bmatrix}$$

References

1. Anderson P.M., Fouad A.A. *Power System Control and Stability*. 2nd edn. United States: IEEE – John Wiley & Sons, Inc.; 2003
2. Kundur P. *Power System Stability and Control*. United States: McGraw-Hill, Inc.; 1994
3. Park R.H. ‘Two reaction theory of synchronous machine, generalized method of analysis – Part I’. *Transactions on AIEE*. 1929;**48**(3):716–30

4. Padiar K.R. *Power System Dynamics Stability and Control*. 2nd edn. India: BS Publications; 2008
5. Keyhani A., Tsai H. 'Identification of high-order synchronous generator models from SSFR test data'. *IEEE Transactions on Energy Conversion*. 1994;**9**(3):593–603
6. Sanchez Gasca J.J., Bridenbaugh C.J., Bowler C.E.J., Edmonds J.S. 'Trajectory sensitivity based identification of synchronous generator and excitation system parameters'. *IEEE Transactions on Power Systems*. 1998;**3**(4): 1814–22
7. Martinez J.A., Johnson B., Grande-Moran C. 'Parameter determination for modeling system transients-Part IV: Rotating machines'. *IEEE Transactions on Power Delivery*. 2005;**20**(3):2063–72
8. Minnich S.H., Schulz R.P., Baker D.H., Sharma D.K., Farmer R.G., Fish J.H. 'Saturation functions for synchronous generators from finite elements'. *IEEE Trans. Energy Conversion*. 1987;**2**(4):680–92
9. Corzine K.A., Kuhn B.T., Sudhoff S.D., Hegner H.J. 'An improved method for incorporating saturation in the Q-D synchronous machine model'. *IEEE Transactions on Energy Conversion*. 1998;**13**(3):270–5
10. Pekarek S.D., Walters E.A., Kuhn B.T. 'An efficient and accurate method of representing magnetic saturation in physical-variable models of synchronous machines'. *IEEE Transactions on Energy Conversion*. 1999;**14**(1):72–9
11. Aliprantis D.C., Sudhoff S.D., Kuhn B.T. 'A synchronous machine model and arbitrary rotor network representation'. *IEEE Transactions on Energy Conversion*. 2005;**20**(3):584–94
12. Aliprantis D.C., Sudhoff S.D., Kuhn B.T. 'Experimental characterization procedure for a synchronous machine model with saturation and arbitrary rotor network representation'. *IEEE Transactions on Energy Conversion*. 2005;**20**(3):595–603
13. Aliprantis D.C., Wasynczuk O., Valdez C.D.R. 'A voltage-behind-reactance synchronous machine model with saturation and arbitrary rotor network representation'. *IEEE Transactions on Energy Conversion*. 2008;**23**(2): 499–508
14. El-Serafi A.M., Kar N.C. 'Methods for determining the intermediate-axis saturation characteristics of salient-pole synchronous machines from the measured D-axis characteristics'. *IEEE Transactions on Energy Conversion*. 2005;**20**(1):88–97
15. Harley R.J., Limebeer D.J., Chirricozzi E. 'Comparative study of saturation methods in synchronous machine models'. *IEE Proceedings*. 1980;**127**(1) Pt B:1–7

Chapter 3
Synchronous machine connected to a power system

As stated in Section 2.7.1, (2.93) formulates the flux linkage state space model in a general form: $\dot{x} = f(x, u, t)$, where $x \triangleq$ the vector of state variables $= [\Psi_d, \Psi_f, \Psi_{kd}, \Psi_q, \Psi_{kq}, \omega, \delta]$ and $u \triangleq$ the forcing functions v_d, v_q, v_f and T_m. The same is applied to the current state space model developed in Section 2.8 and represented by (2.104) where $x \triangleq$ the vector of state variables $= [i_d, i_f, i_{kd}, i_q, i_{kq}, \omega, \delta]$. To describe the machine completely, the forcing functions must be known. In the case of v_f and T_m *being* known, the two functions of v_d, v_q must be identified by relations added to the model equations. This necessitates identifying the machine terminal conditions by describing and modelling its load. The load modelling is explained in the forthcoming chapters. However, the load is connected to the machine through a network that is either simple, one machine-infinite bus, or integrated, multi-machine system.

Moreover, each synchronous machine in the power system is equipped with an excitation control system and its prime mover is controlled by a governor control system. These controllers decide the values of v_f and T_m and may be involved in the machine equations that are written in this chapter in pu using base phase quantities (the alternative per unit/normalising system is given in Appendix I, Section I.3). These features are presented in this chapter.

3.1 Synchronous machine connected to an infinite bus

Figure 3.1 depicts a simple power system, one machine connected to an infinite bus through a transmission line and its single-line equivalent circuit where the transmission line is represented by an external impedance, resistance R_e and inductance L_e. Ignoring the mutual coupling between phases a, b and c, and considering the generation as positive convention, the three phase to neutral voltages can be expressed by (3.1):

$$\left.\begin{aligned} v_{ta} &= v_{a\infty} + R_e i_a + L_e \mathrm{p} i_a \\ v_{tb} &= v_{b\infty} + R_e i_b + L_e \mathrm{p} i_b \\ v_{tc} &= v_{c\infty} + R_e i_c + L_e \mathrm{p} i_c \end{aligned}\right\} \tag{3.1}$$

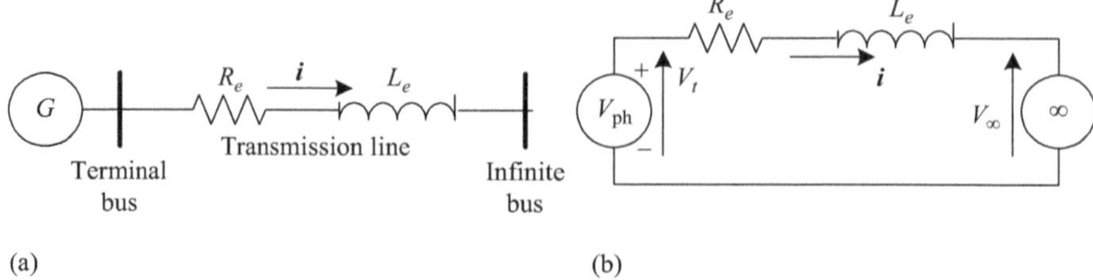

Figure 3.1 One machine–infinite bus system: (a) one-line diagram and (b) equivalent circuit

In matrix form, (3.1) can be rewritten as

$$\begin{bmatrix} v_{ta} \\ v_{tb} \\ v_{tc} \end{bmatrix} = \begin{bmatrix} v_{a\infty} \\ v_{b\infty} \\ v_{c\infty} \end{bmatrix} + R_e\mathbf{U}\begin{bmatrix} i_a \\ i_b \\ i_c \end{bmatrix} + L_e\mathbf{U}\begin{bmatrix} \mathrm{p}i_a \\ \mathrm{p}i_b \\ \mathrm{p}i_c \end{bmatrix} \tag{3.2}$$

or

$$\boldsymbol{v}_{abc} = \boldsymbol{v}_{abc\infty} + R_e\mathbf{U}\boldsymbol{i}_{abc} + L_e\mathbf{U}\mathrm{p}\boldsymbol{i}_{abc} \tag{3.3}$$

where $\mathbf{U}$ is a unit matrix.

Applying Park's transformation to obtain the voltages in *dqo* frame of reference gives

$$\boldsymbol{v}_{dqo} = \mathbf{P}^{-1}\boldsymbol{v}_{abc} = \boldsymbol{v}_{dqo\infty} + R_e\boldsymbol{i}_{dqo} + L_e\mathbf{P}^{-1}\mathrm{p}\boldsymbol{i}_{abc} \tag{3.4}$$

$\boldsymbol{v}_{dqo\infty} = \mathbf{P}^{-1}\boldsymbol{v}_{abc\infty}$, $\boldsymbol{i}_{dqo} = \mathbf{P}^{-1}\boldsymbol{i}_{abc}$ and
$\boldsymbol{v}_{abc\infty} \triangleq$ a set of balanced three-phase voltages. The last term in the RHS of (3.4) can be determined as below:

$\boldsymbol{i}_{dqo} = \mathbf{P}^{-1}\boldsymbol{i}_{abc}$ and by taking the derivative of both sides gives

$$\frac{\mathrm{d}\boldsymbol{i}_{dqo}}{\mathrm{d}t} = \mathbf{P}^{-1}\mathrm{p}\boldsymbol{i}_{abc} + \frac{\mathrm{d}\mathbf{P}^{-1}}{\mathrm{d}t}\boldsymbol{i}_{abc}$$

Then,

$$\mathbf{P}^{-1}\mathrm{p}\boldsymbol{i}_{abc} = \mathrm{p}\boldsymbol{i}_{dqo} - \frac{\mathrm{d}\mathbf{P}^{-1}}{\mathrm{d}t}\mathbf{P}\boldsymbol{i}_{dqo} \tag{3.5}$$

By the definition of **P** in Section 2.3 (Chapter 2), and assuming $\theta = \omega_o t + \delta$, it is found that

$$\frac{\mathrm{d}\mathbf{P}^{-1}}{\mathrm{d}t}\mathbf{P} = \omega\begin{bmatrix} 0 & -1 & 0 \\ 1 & 0 & 0 \\ 0 & 0 & 0 \end{bmatrix} \tag{3.6}$$

Incorporating (3.6) and (3.5) into (3.4) gives

$$\boldsymbol{v}_{dqo} = \boldsymbol{v}_{dqo\infty} + R_e\boldsymbol{i}_{dqo} + L_e\mathrm{p}\boldsymbol{i}_{dqo} - \omega L_e\begin{bmatrix} -i_q \\ i_d \\ 0 \end{bmatrix} \tag{3.7}$$

where $\boldsymbol{v}_{dqo\infty}$ can be determined in terms of $\boldsymbol{v}_{abc\infty}$ as below:

$$\boldsymbol{v}_{dqo\infty} = \mathbf{P}^{-1}\sqrt{2}V_\infty\begin{bmatrix} \cos(\omega t + \alpha) \\ \cos(\omega t + \alpha - 120) \\ \cos(\omega t + \alpha + 120) \end{bmatrix} \quad \text{and} \quad V_\infty \text{ is the rms phase voltage} \tag{3.8}$$

Substituting (3.8) into (3.7) gives

$$\boldsymbol{v}_{dqo} = \sqrt{3}V_\infty\begin{bmatrix} -\sin(\delta - \alpha) \\ \cos(\delta - \alpha) \\ 0 \end{bmatrix} + R_e\boldsymbol{i}_{dqo} + L_e\frac{\mathrm{d}\boldsymbol{i}_{dqo}}{\mathrm{d}t} - \omega L_e\begin{bmatrix} -i_q \\ i_d \\ 0 \end{bmatrix} \tag{3.9}$$

The set of (3.9) gives the two non-linear relations of v_d and v_q to be added to the machine equations in order to completely describe the machine by either flux linkage state space model or current state space model. For readers more interested in machine equations, more details considering mutual effects, cross magnetising, saturation, etc., are discussed in [1–5].

3.1.1 Flux linkage state space model

From (2.79) and (2.81) i_q in terms of flux linkages is given by

$$i_q = \frac{1}{\ell_q}\left(1 - \frac{L_{Mq}}{\ell_q}\right)\Psi_q - \frac{L_{Mq}}{\ell_q\ell_{kq}}\Psi_{kq} \tag{3.10}$$

and i_d is given by (2.88). Substitute the currents i_d and i_q into the two relations of v_d and v_q (3.9) to obtain

$$\begin{aligned} v_d = {} & -\sqrt{3}V_\infty \sin(\delta - \alpha) + \frac{R_e}{\ell_d}\left(1 - \frac{L_{Md}}{\ell_a}\right)\Psi_d - \frac{R_e L_{Md}}{\ell_a \ell_f}\Psi_f - \frac{R_e L_{Md}}{\ell_a \ell_{kd}}\Psi_{kd} \\ & + \frac{\omega L_e}{\ell_a}\left(1 - \frac{L_{Mq}}{\ell_a}\right)\Psi_q - \frac{\omega L_e L_{Mq}}{\ell_a \ell_{kq}}\Psi_{kq} + \frac{L_e}{\ell_a}\left(1 - \frac{L_{Md}}{\ell_a}\right)\dot{\Psi}_d \\ & - \frac{L_e L_{Md}}{\ell_a \ell_f}\dot{\Psi}_f - \frac{L_e L_{Md}}{\ell_a \ell_{kd}}\dot{\Psi}_{kd} \end{aligned} \tag{3.11}$$

$$\begin{aligned} v_q = {} & -\sqrt{3}V_\infty \cos(\delta - \alpha) + \frac{R_e}{\ell_a}\left(1 - \frac{L_{Mq}}{\ell_a}\right)\Psi_q - \frac{R_e L_{Mq}}{\ell_a \ell_{kq}}\Psi_{kq} \\ & - \frac{\omega L_e}{\ell_a}\left(1 - \frac{L_{Md}}{\ell_a}\right)\Psi_d + \frac{\omega L_e}{\ell_a \ell_f}\Psi_f + \frac{\omega L_e}{\ell_a \ell_{kd}}\Psi_{kd} \\ & + \frac{L_e}{\ell_a}\left(1 - \frac{L_{Mq}}{\ell_a}\right)\dot{\Psi}_q - \frac{L_e L_{Mq}}{\ell_a \ell_{kq}}\dot{\Psi}_{kq} \end{aligned} \tag{3.12}$$

Substituting v_d and v_q as in (3.11) and (3.12) into (2.93) to compose the flux linkage state space model in the form

$$\mathbf{A}\dot{x} = \mathbf{B}x + \mathbf{C} \tag{3.13}$$

Equation (3.13) can be expressed in the general form $\dot{x} = \boldsymbol{f}(\boldsymbol{x}, \boldsymbol{u}, t)$ as

$$\dot{x} = \mathbf{A}^{-1}\mathbf{B}x + \mathbf{A}^{-1}\mathbf{C} \tag{3.14}$$

where $\boldsymbol{x}^t = [\Psi_d, \Psi_f, \Psi_{kd}, \Psi_q, \Psi_{kq}, \omega, \delta]$

$$\mathbf{A} = \left[\begin{array}{ccc:cc|cc} 1 + \frac{L_e}{\ell_a}\left(1 - \frac{L_{Md}}{\ell_a}\right) & -\frac{L_e L_{Md}}{\ell_a \ell_f} & -\frac{L_e L_{Md}}{\ell_a \ell_{kd}} & 0 & 0 & 0 & 0 \\ 0 & 1 & 0 & 0 & 0 & 0 & 0 \\ 0 & 0 & 1 & 0 & 0 & 0 & 0 \\ \hdashline 0 & 0 & 0 & 1 + \frac{L_e}{\ell_a}\left(1 - \frac{L_{Mq}}{\ell_q}\right) & -\frac{L_e L_{Mq}}{\ell_a \ell_{kq}} & 0 & 0 \\ 0 & 0 & 0 & 0 & 1 & 0 & 0 \\ \hdashline 0 & 0 & 0 & 0 & 0 & 1 & 0 \\ 0 & 0 & 0 & 0 & 0 & 0 & 1 \end{array}\right]$$

$$
\mathbf{B} = \left[\begin{array}{ccc|cc|cc}
\frac{\bar{R}}{\ell_a}\left(1 - \frac{L_{Md}}{\ell_a}\right) & \frac{\bar{R}L_{Md}}{\ell_a \ell_f} & \frac{\bar{R}L_{Md}}{\ell_a \ell_{kd}} & -\omega\left[1 + \frac{L_e}{\ell_a}\left(1 - \frac{L_{Mq}}{\ell_a}\right)\right] & \frac{\omega L_e L_{Mq}}{\ell_a \ell_{kq}} & 0 & 0 \\
\frac{R_f L_{Md}}{\ell_f \ell_a} & \frac{R_f}{\ell_f}\left(1 - \frac{L_{Md}}{\ell_f}\right) & \frac{R_f L_{Md}}{\ell_f \ell_{kd}} & 0 & 0 & 0 & 0 \\
\frac{R_{kd} L_{Md}}{\ell_{kd} \ell_a} & \frac{R_{kd} L_{Md}}{\ell_{kd} \ell_f} & \frac{R_{kd}}{\ell_{kd}}\left(1 - \frac{L_{Md}}{\ell_{kd}}\right) & 0 & 0 & 0 & 0 \\
\hline
-\omega\left[1 + \frac{L_e}{\ell_a}\left(1 - \frac{L_{Md}}{\ell_a}\right)\right] & -\frac{\omega L_e L_{Md}}{\ell_a \ell_f} & -\frac{\omega L_e L_{Md}}{\ell_a \ell_{kd}} & \frac{\bar{R}}{\ell_a}\left(1 - \frac{L_{Mq}}{\ell_a}\right) & \frac{\bar{R}L_{Mq}}{\ell_a \ell_{kq}} & 0 & 0 \\
0 & 0 & 0 & \frac{R_{kq} L_{Mq}}{\ell_a \ell_{kq}} & \frac{R_{kq}}{\ell_{kq}}\left(1 - \frac{L_{Mq}}{\ell_{kq}}\right) & 0 & 0 \\
\hline
-\frac{L_{Md}\Psi_q}{2H\omega_B \ell_a^2} & -\frac{L_{Md}\Psi_q}{2H\omega_B \ell_a \ell_f} & \frac{L_{Md}\Psi_q}{2H\omega_B \ell_a \ell_{kd}} & \frac{L_{Mq}\Psi_d}{2H\omega_B \ell_a^2} & \frac{L_{Mq}\Psi_d}{2H\omega_B \ell_a \ell_{kq}} & -\frac{D}{2H\omega_B} & 0 \\
0 & 0 & 0 & 0 & 0 & 1 & 0
\end{array}\right]
$$

$$\mathbf{C} = \begin{bmatrix} \sqrt{3}V_\infty \sin(\delta - \alpha) \\ v_f \\ 0 \\ -\sqrt{3}V_\infty \cos(\delta - \alpha) \\ 0 \\ \dfrac{T_m}{2H\omega_B} \\ -1 \end{bmatrix} \quad \text{and} \quad \bar{R} = R_a + R_e$$

Example 3.1 Use the synchronous generator data given in Examples 2.1–2.4 to compute the flux linkage state space model when the machine is connected to an infinite bus through a transmission line represented by a resistance $R_e = 0.05$ pu and an inductance $L_e = 0.35$ pu.

Solution (all values are in pu, otherwise will be cited):

$$\bar{R} = 0.00299 + 0.05 = 0.053$$

L_d and L_q should be modified to be $\overline{L}_d = L_d + L_e$ and $\overline{L}_q = L_q + L_e$, respectively. Hence,

$$\overline{L}_d = 1.81 + 0.35 = 2.16 \quad \text{and} \quad \overline{L}_q = 1.76 + 0.35 = 2.11$$

$$L_{md} = \overline{L}_d - \ell_a = 2.16 - 0.15 = 2.01 \quad \text{and} \quad L_{mq} = \overline{L}_q - \ell_a = 2.11 - 0.15 = 1.96$$

$$\frac{1}{L_{Md}} = \frac{1}{L_{md}} + \frac{1}{\ell_a} + \frac{1}{\ell_f} + \frac{1}{\ell_{kd}} = 14.765$$

$$L_{Md} = 0.068$$

$$\frac{1}{L_{Mq}} = \frac{1}{L_{mq}} + \frac{1}{\ell_a} + \frac{1}{\ell_{kq}} = 13.023$$

$$L_{Mq} = 0.077$$

$$\mathbf{A} = \left[\begin{array}{ccc:cc:cc} 2.276 & -0.148 & -1.057 & 0 & 0 & 0 & 0 \\ 0 & 1 & 0 & 0 & 0 & 0 & 0 \\ 0 & 0 & 1 & 0 & 0 & 0 & 0 \\ \hdashline 0 & 0 & 0 & 2.135 & -1.057 & 0 & 0 \\ 0 & 0 & 0 & 0 & 1 & 0 & 0 \\ \hdashline 0 & 0 & 0 & 0 & 0 & 1 & 0 \\ 0 & 0 & 0 & 0 & 0 & 0 & 1 \end{array}\right]$$

$$\mathbf{B} = \left[\begin{array}{ccc:cc:cc} 0.193 & 0.022 & 0.160 & -2.720 & 1.060 & 0 & 0 \\ 0.0004 & 0.0008 & 0.0004 & 0 & 0 & 0 & 0 \\ 0.086 & 0.012 & 0.103 & 0 & 0 & 0 & 0 \\ \hdashline -2.275 & -0.148 & -1.058 & 0.172 & 0.160 & 0 & 0 \\ 0 & 0 & 0 & 0.018 & 0.019 & 0 & 0 \\ \hdashline -0.00115\Psi_q & -0.00016\Psi_q & 0.00115\Psi_q & 0.0013\Psi_d & 0.0011\Psi_d & 0 & 0 \\ 0 & 0 & 0 & 0 & 0 & 1 & 0 \end{array}\right]$$

$$\mathbf{C} = \begin{bmatrix} \sqrt{3}V_\infty \sin(\delta - \alpha) \\ v_f \\ 0 \\ -\sqrt{3}V_\infty \cos(\delta - \alpha) \\ 0 \\ 0.00038T_m \\ -1 \end{bmatrix}$$

Hence,

$$\mathbf{A}^{-1} = \left[\begin{array}{ccc:cc:cc} 0.439 & 0.065 & 0.464 & 0 & 0 & 0 & 0 \\ 0 & 1 & 0 & 0 & 0 & 0 & 0 \\ 0 & 0 & 1 & 0 & 0 & 0 & 0 \\ \hdashline 0 & 0 & 0 & 0.468 & 0.495 & 0 & 0 \\ 0 & 0 & 0 & 0 & 1 & 0 & 0 \\ \hdashline 0 & 0 & 0 & 0 & 0 & 1 & 0 \\ 0 & 0 & 0 & 0 & 0 & 0 & 1 \end{array}\right]$$

$$\mathbf{A}^{-1}\mathbf{B} = \left[\begin{array}{ccc|cc|cc} 0.125 & 0.065 & 0.118 & -1.194 & 0.465 & 0 & 0 \\ 0.0004 & 0.0008 & 0.0004 & 0 & 0 & 0 & 0 \\ 0.086 & 0.012 & 0.103 & 0 & 0 & 0 & 0 \\ \hline -1.065 & -0.069 & -0.495 & 0.089 & 0.084 & 0 & 0 \\ 0 & 0 & 0 & 0.018 & 0.019 & 0 & 0 \\ \hline -0.00115\Psi_q & 0.00016\Psi_q & 0.00115\Psi_q & 0.0013\Psi_d & 0.00114\Psi_d & 0 & 0 \\ 0 & 0 & 0 & 0 & 0 & 1 & 0 \end{array}\right]$$

$$\mathbf{A}^{-1}\mathbf{C} = \begin{bmatrix} 0.76V_\infty \sin(\delta - \alpha) + 0.065v_f \\ v_f \\ 0 \\ -0.81V_\infty \cos(\delta - \alpha) \\ 0 \\ 0.00038T_m \\ -1 \end{bmatrix}$$

Substituting into (3.14) gives the flux state space model as

$$\begin{bmatrix} \dot{\Psi}_d \\ \dot{\Psi}_f \\ \dot{\Psi}_{kd} \\ \dot{\Psi}_q \\ \dot{\Psi}_{kq} \\ \dot{\omega} \\ \dot{\delta} \end{bmatrix} = \left[\begin{array}{ccc|cc|cc} 0.125 & 0.065 & 0.118 & -1.194 & 0.465 & 0 & 0 \\ 0.0004 & 0.0008 & 0.0004 & 0 & 0 & 0 & 0 \\ 0.086 & 0.012 & 0.103 & 0 & 0 & 0 & 0 \\ \hline -1.065 & -0.069 & -0.495 & 0.089 & 0.084 & 0 & 0 \\ 0 & 0 & 0 & 0.018 & 0.019 & 0 & 0 \\ \hline -0.00115\Psi_q & -0.00016\Psi_q & 0.00115\Psi_q & 0.0013\Psi_d & 0.00114\Psi_d & 0 & 0 \\ 0 & 0 & 0 & 0 & 0 & 1 & 0 \end{array}\right] \begin{bmatrix} \Psi_d \\ \Psi_f \\ \Psi_{kd} \\ \Psi_q \\ \Psi_{kq} \\ \omega \\ \delta \end{bmatrix}$$

$$+ \begin{bmatrix} 0.76V_\infty \sin(\delta - \alpha) + 0.065v_f \\ v_f \\ 0 \\ -0.81V_\infty \cos(\delta - \alpha) \\ 0 \\ 0.00038T_m \\ -1 \end{bmatrix}$$

3.1.2 Current state space model

Equation (2.100) can be rewritten as $-\mathbf{B}_2 \frac{di}{dt} = \mathbf{B}_1 \boldsymbol{i} + \boldsymbol{v}$. Substituting the value of $\boldsymbol{v}$ from (3.9) gives

$$-\mathbf{B}_2 \frac{d\boldsymbol{i}}{dt} = \mathbf{B}_1 \boldsymbol{i} + \begin{bmatrix} \sqrt{3}V_\infty \sin(\delta - \alpha) + R_e i_d + L_e \dfrac{di_d}{dt} + \omega L_e i_q \\ -v_f \\ 0 \\ \sqrt{3}V_\infty \cos(\delta - \alpha) + R_e i_q + L_e \dfrac{di_q}{dt} + \omega L_e i_d \\ 0 \end{bmatrix} \tag{3.15}$$

Using $\overline{R} = R_a + R_e$, $\overline{L}_d = L_d + L_e$ and $\overline{L}_q = L_q + L_e$ in (3.15) gives the corresponding matrices $\overline{\mathbf{B}}_1$ and $\overline{\mathbf{B}}_2$. Then adding the relations of $\dot{\omega}$ and $\dot{\delta}$ gives the current state space model as

$$\begin{bmatrix} pi_d \\ pi_f \\ pi_{kd} \\ pi_q \\ pi_{kq} \\ \dot{\omega} \\ \dot{\delta} \end{bmatrix} = \left[\begin{array}{ccccc|cc} \multicolumn{5}{c|}{-\overline{\mathbf{B}}_2^{-1}\overline{\mathbf{B}}_1} & \multicolumn{2}{c}{0} \\ \hline -\dfrac{\overline{L}_d i_q}{2H\omega_B} & -\dfrac{kM_f i_q}{2H\omega_B} & -\dfrac{kM_{kd} i_{kd}}{2H\omega_B} & \dfrac{\overline{L}_q i_d}{2H\omega_B} & \dfrac{kM_{kq} i_d}{2H\omega_B} & -\dfrac{D}{2H\omega_B} & 0 \\ 0 & 0 & 0 & 0 & 0 & 1 & 0 \end{array}\right] \begin{bmatrix} i_d \\ i_f \\ i_{kd} \\ i_q \\ i_{kq} \\ \omega \\ \delta \end{bmatrix}$$

$$+ \left[\begin{array}{c|cc} -\overline{\mathbf{B}}_2^{-1} & \multicolumn{2}{c}{0} \\ \hline 0 & 1 & 0 \\ & 0 & 1 \end{array}\right] \begin{bmatrix} -\sqrt{3}V_\infty \sin(\delta - \alpha) \\ -v_f \\ 0 \\ \sqrt{3}V_\infty \cos(\delta - \alpha) \\ 0 \\ \dfrac{T_m}{2H\omega_B} \\ -1 \end{bmatrix} \tag{3.16}$$

Example 3.2 Repeat Example 3.1 for current state space model.

Solution:

Referring to (2.100) the coefficient matrices are

$$\overline{\mathbf{B}}_1 = \begin{bmatrix} \overline{R} & 0 & 0 & \omega\overline{L}_q & \omega kM_{kq} \\ 0 & R_f & 0 & 0 & 0 \\ 0 & 0 & R_{kd} & 0 & 0 \\ -\omega\overline{L}_d & -\omega kM_f & -\omega kM_{kd} & \overline{R} & 0 \\ 0 & 0 & 0 & 0 & R_{kq} \end{bmatrix}$$

$$= \begin{bmatrix} 0.05299 & 0 & 0 & 2.11\omega & 1.59\omega \\ 0 & 0.00898 & 0 & 0 & 0 \\ 0 & 0 & 0.0284 & 0 & 0 \\ -2.16\omega & -1.66\omega & -1.66\omega & 0.05299 & 0 \\ 0 & 0 & 0 & 0 & 0.00619 \end{bmatrix}$$

$$\overline{\mathbf{B}}_2 = \begin{bmatrix} \overline{L}_d & kM_f & kM_{kd} & 0 & 0 \\ kM_f & L_f & L_{fkd} & 0 & 0 \\ kM_{kd} & L_{fh} & L_{kd} & 0 & 0 \\ 0 & 0 & 0 & \overline{L}_q & kM_{kq} \\ 0 & 0 & 0 & kM_{kq} & L_{kq} \end{bmatrix}$$

$$= \begin{bmatrix} 2.16 & 1.66 & 1.66 & 0 & 0 \\ 1.66 & 2.73 & 1.66 & 0 & 0 \\ 1.66 & 1.66 & 0.1713 & 0 & 0 \\ 0 & 0 & 0 & 2.11 & 1.59 \\ 0 & 0 & 0 & 1.59 & 0.7252 \end{bmatrix}$$

$$\overline{\mathbf{B}}_2^{-1} = \begin{bmatrix} 0.604 & -0.652 & 0.467 & 0 & 0 \\ -0.652 & 0.629 & 0.219 & 0 & 0 \\ 0.468 & 0.219 & -0.829 & 0 & 0 \\ 0 & 0 & 0 & 1.674 & -1.596 \\ 0 & 0 & 0 & -1.597 & 2.119 \end{bmatrix}$$

$$\overline{\mathbf{B}}_2^{-1}\overline{\mathbf{B}}_1 = \begin{bmatrix} 0.032 & -0.006 & 0.013 & 1.270\omega & 0.960\omega \\ -0.034 & 0.006 & 0.006 & -1.376\omega & -1.034\omega \\ 0.025 & 0.002 & -0.023 & 0.987\omega & 0.744\omega \\ -3.616\omega & -2.779\omega & -2.779\omega & 0.089 & 0.010 \\ 3.449\omega & 2.651\omega & 2.651\omega & -0.085 & -0.009 \end{bmatrix}$$

Thus, the current state space model can be expressed as

$$\begin{bmatrix} pi_d \\ pi_f \\ pi_{kd} \\ pi_q \\ pi_{kq} \\ \dot{\omega} \\ \dot{\delta} \end{bmatrix} = \left[\begin{array}{ccccc:cc} \multicolumn{5}{c:}{-\overline{\mathbf{B}}_2^{-1}\overline{\mathbf{B}}_1} & \multicolumn{2}{c}{0} \\ \hdashline -0.0008i_q & -0.0006i_q & -0.0006i_{kd} & 0.0008i_d & 0.0006i_d & 0 & 0 \\ 0 & 0 & 0 & 0 & 0 & 1 & 0 \end{array}\right] \begin{bmatrix} i_d \\ i_f \\ i_{kd} \\ i_q \\ i_{kq} \\ \omega \\ \delta \end{bmatrix}$$

$$+ \left[\begin{array}{c:cc} -\overline{\mathbf{B}}_2^{-1} & \multicolumn{2}{c}{0} \\ \hdashline 0 & 1 & 0 \\ & 0 & 1 \end{array}\right] \begin{bmatrix} -\sqrt{3}V_\infty \sin(\delta - \alpha) \\ -v_f \\ 0 \\ \sqrt{3}V_\infty \cos(\delta - \alpha) \\ 0 \\ \dfrac{T_m}{2H\omega_B} \\ -1 \end{bmatrix}$$

3.2 Synchronous machine connected to an integrated power system

Terminal or load conditions of the synchronous machine determined by v_d and v_q can be achieved by direct relations when the machine is connected to an infinite bus. In the case of multi-machine system, each machine is connected to the power system that comprises numerous static elements, e.g. transmission lines, static loads, shunt capacitors, transformers and dynamic elements, e.g. generators and their control systems and dynamic loads. Therefore, the determination of v_d and v_q representing the machine terminal conditions is more complicated because of the system non-linearity, the complexity of modelling, the dynamic interaction between some of system components, the large and complex size of the system and the operating mode (normal or contingencies). Load flow techniques, more details in Part III, are commonly used to determine the terminal conditions of each machine in the system and to know the values v_d and v_q. Then, these values are substituted into (2.89) or (2.100) to construct the flux or current state space model. Load flow results depend on the representation of the system components according to the operating mode. So, to accurately model the synchronous machine, its parameters, inductances and time constants must be known in different operating modes (steady state, transient state and sub-transient state).

3.3 Synchronous machine parameters in different operating modes

In sub-transient, transient and steady-state operating conditions, both inductances and time constants must be identified (the values of inductances are numerically equal to the corresponding reactances in pu with the synchronous speed as base, i.e. $\omega_B = 2\pi f$ rad/s). Different testing and measuring methods such as standstill frequency response and rotating time domain response can be applied to determine the machine parameters and synthesise their models from test data [6–10]. When balanced three-phase voltages are suddenly applied to the stator terminals while the rotor circuits are short circuited, the flux linkage in the d-axis frame of reference, Ψ_d, depends initially on the sub-transient inductances and then on the transient inductances after a few cycles. Assuming the three-phase balanced voltages, v_{abc}, suddenly applied to the stator terminals are expressed as

$$\begin{bmatrix} v_a \\ v_b \\ v_c \end{bmatrix} = \sqrt{2}V_{rms}\begin{bmatrix} \cos\theta \\ \cos(\theta - 120) \\ \cos(\theta + 120) \end{bmatrix} u(t) \tag{3.17}$$

where V_{rms} = rms phase voltage and $u(t)$ is a unit step function.

Applying Park's transformation gives v_{dqo} as

$$\begin{bmatrix} v_d \\ v_q \\ v_o \end{bmatrix} = \begin{bmatrix} \sqrt{3}V_{rms}\, u(t) \\ 0 \\ 0 \end{bmatrix} \tag{3.18}$$

The flux linking the field circuit, Ψ_f, and the damper winding, Ψ_{kd}, remain zero at the instant voltage is applied, as flux cannot change instantaneously. Thus, at that instant t_o^+ and using (2.31) it is seen that

$$\left.\begin{aligned} \Psi_f &= 0 = kM_f i_d + L_f i_f + L_{fkd} i_{kd} \\ \Psi_{kd} &= 0 = kM_{kd} i_d + L_{kd} i_{kd} + L_{fkd} i_f \end{aligned}\right\} \tag{3.19}$$

Thus,

$$\left.\begin{aligned} i_f &= -\frac{kM_f L_{kd} - kM_{kd} L_{fkd}}{L_f L_{kd} - L_{fkd}^2} i_d \\ i_{kd} &= -\frac{kM_{kd} L_f - kM_f L_{fkd}}{L_f L_{kd} - L_{fkd}^2} i_d \end{aligned}\right\} \tag{3.20}$$

Using (3.20) and substituting in (2.20) Ψ_d can be written as a function of i_d as

$$\Psi_d = \left(L_d - \frac{k^2 M_f^2 L_{kd} + L_f k^2 M_{kd}^2 - 2kM_f kM_{kd} L_{fkd}}{L_f L_{kd} - L_{fkd}^2} \right) i_d \tag{3.21}$$

$$\text{By definition: } \Psi_d \triangleq L_d'' i_d \tag{3.22}$$

where L_d'' is the d-axis sub-transient inductance.

Comparing (3.21) and (3.22) it is found that

$$L_d'' = \left(L_d - \frac{k^2 M_f^2 L_{kd} + L_f k^2 M_{kd}^2 - 2kM_f kM_{kd} L_{fkd}}{L_f L_{kd} - L_{fkd}^2} \right)$$

and by using the definition of L_{md} as in (2.67) L_d'' can be written as

$$L_d'' = L_d - \frac{L_f + L_{kd} - 2L_{md}}{\left(\dfrac{L_f L_{kd}}{L_{md}^2} \right) - 1} \tag{3.23}$$

The d-axis transient inductance, L_d', is obtained when a balanced three-phase voltage is suddenly applied to a machine without damper windings. Applying the same procedure gives

$$i_f = -\frac{kM_f}{L_f} i_d \tag{3.24}$$

$$\Psi_d = \left[L_d - \frac{(kM_f)^2}{L_f} \right] i_d \triangleq L_d' i_d \tag{3.25}$$

Therefore,

$$L_d' = L_d - \frac{(kM_f)^2}{L_f} = L_d - \frac{L_{md}^2}{L_f} \tag{3.26}$$

It is to be noted that after a few cycles from the start of transients in a machine with damper windings, the damper winding current decays rapidly to zero and the stator inductance is the transient inductance.

To calculate the inductances in the q-axis frame of reference, the suddenly applied three-phase voltages are shifted by 90° to be expressed as

$$\begin{bmatrix} v_a \\ v_b \\ v_c \end{bmatrix} = \sqrt{2} V_{rms} \begin{bmatrix} \sin\theta \\ \sin(\theta - 120) \\ \sin(\theta + 120) \end{bmatrix} u(t) \tag{3.27}$$

Applying Park's transformation gives v_{dqo} as

$$\begin{bmatrix} v_d \\ v_q \\ v_o \end{bmatrix} = \begin{bmatrix} 0 \\ \sqrt{3} V_{rms}\, u(t) \\ 0 \end{bmatrix} \tag{3.28}$$

For salient pole machines with damper windings when the initial sub-transient decays to zero, the stator flux linkage is determined by the same circuit of the steady-state q-axis flux linkage. Therefore, the q-axis transient inductance can be considered the same as the q-axis steady-state inductance, $L'_q = L_q$. The same procedure above is applied for q-axis circuits to determine L''_q as below:

$$\Psi_{kd} = kM_{kq}i_q + L_{kq}i_{kq} = 0$$

Thus,

$$i_{kq} = -\frac{kM_{kq}}{L_{kq}} \tag{3.29}$$

Substituting (3.29) in the relation $\Psi_q = L_q i_q + kM_{kq}i_{kq}$ gives

$$\Psi_q = \left[L_q - \frac{(kM_{kq})^2}{L_{kq}}\right] i_q \triangleq L''_q i_q \tag{3.30}$$

Hence,

$$L'_q = \left[L_q - \frac{(kM_{kq})^2}{L_{kq}}\right] = L_q - \frac{L^2_{mq}}{L_{kq}} \tag{3.31}$$

For round rotor machines, multiple paths of eddy currents are provided by the solid iron rotor and act as equivalent circuits during the sub-transient and transient periods. Therefore, the q-axis sub-transient and transient inductances are determined by the q-axis rotor circuits resulting in q-axis transient inductance much smaller than q-axis steady-state inductance, $L''_q \ll L'_q \ll L_q$. This can be verified by considering two q-axis rotor damper circuits [11].

On the other hand, to determine the *time constants of a salient pole machine*, as in (3.30) the voltage equations of v_f and v_{kd}, when the stator circuits are open circuited and a step voltage $V_f u(t)$ is applied to the field circuit, become

$$V_f u(t) = \dot{\Psi}_f + R_f i_f \quad v_{kd} = 0 = \dot{\Psi}_{kd} + R_{kd} i_{kd} \tag{3.32}$$

From (2.31) the flux linkages Ψ_f and Ψ_{kd} immediately after the sub-transient can be rewritten as

$$\Psi_f = L_f i_f + L_{fkd} i_{kd} \qquad \Psi_{kd} = L_{fkd} i_f + L_{kd} i_{kd} = 0 \tag{3.33}$$

where $i_d = 0$ as the stator circuits are open.

$$\text{Hence, } i_f = -\frac{L_{kd}}{L_{fkd}} i_{kd} \tag{3.34}$$

From (3.32) and using (3.33) the relations below can be written as

$$\left.\begin{aligned} \frac{V_f}{L_f} &= \frac{R_f}{L_f} i_f + \mathrm{p} i_f + \frac{L_{fkd}}{L_f} \mathrm{p} i_{kd} \\ 0 &= \frac{R_{kd}}{L_{fkd}} i_{kd} + \mathrm{p} i_f + \frac{L_{kd}}{L_{fkd}} \mathrm{p} i_{kd} \end{aligned}\right\} \tag{3.35}$$

Solving the relations (3.35) using (3.34) gives a relation of i_{kd} as

$$\mathrm{p}i_{kd} + \frac{R_{kd}L_f + R_f L_{kd}}{L_f L_{kd} - L_{fkd}^2} i_{kd} = -V_f \frac{L_{fkd}}{l_f L_{kd} - L_{fkd}^2} \tag{3.36}$$

This relation can be approximated by considering that $R_{kd} \gg R_f$ and L_f and L_{kd} in per unit are almost equal in magnitude, as

$$\mathrm{p}i_{kd} + \frac{R_{kd}}{L_{kd} - L_{fkd}^2/L_f} i_{kd} = -V_f \frac{L_{fkd}/L_f}{L_{kd} - L_{fkd}^2/L_f} \tag{3.37}$$

It is, therefore, seen that i_{kd} decays with a time constant T''_{do}, where

$$\mathrm{T}''_{do} = \frac{L_{kd} - (L_{fkd}^2/L_f)}{R_{kd}} \triangleq d\text{-axis open circuit sub-transient time constant} \tag{3.38}$$

After the sub-transient current decays, i.e. in the transient period, the field current is only affected by the field circuit parameters. Then, the relation below can be written as

$$R_f i_f + L_f \mathrm{p} i_f = V_f u(t) \tag{3.39}$$

Therefore, the time constant of (3.39) denoted by T'_{do} is called d-axis transient open circuit time constant and is given by

$$\mathrm{T}'_{do} = \frac{L_f}{R_f} \tag{3.40}$$

The time constants when the stator is short circuited can be calculated using the following approximate relations [12].

In the d-axis:
$$\left.\begin{aligned} \mathrm{T}''_d &= \mathrm{T}''_{do} L''_d / L'_d \\ \mathrm{T}'_d &= \mathrm{T}'_{do} L'_d / L_d \end{aligned}\right\} \tag{3.41}$$

Similarly, in the q-axis:
$$\left.\begin{aligned} \mathrm{T}''_{qo} &= L_{kq}/R_{kq} \\ \mathrm{T}''_q &= \mathrm{T}''_{qo} L''_q / L_q \end{aligned}\right\} \tag{3.42}$$

For the *time constants of round rotor machines*, a time constant, T_c, is added to the sub-transient and transient time constants. It is associated with the rate of change of stator direct current or with the alternating currents enveloped in the field windings in the case of exposing the machine to a three-phase short circuit. T_c is given by [12]

$$T_c = \frac{L_2}{R_a} \quad \text{and} \quad L_2 = \frac{L'_d + L_q}{2} \triangleq \text{the negative-sequence inductance} \tag{3.43}$$

As explained in this chapter the synchronous machine is modelled by either current or flux linkage state space model. Both include five currents or flux linkage state variables (may be increased as the rotor circuits increase) plus two state variables of angular speed and rotor angle. The state variables are presented by non-linear first-order differential equations. Electric power systems are composed of a large number of components, equipment and control devices, interacting with each other and exhibiting non-linear dynamic behaviour with a wide range of timescales. For instance, equations describing network constraints, loads, excitation system and prime movers must be included in the mathematical model to study the response of a large number of synchronous machines to a given disturbance. Therefore, a complete mathematical description of the system yields more complexity and more computation time. So, some simplifications are needed such as simplified machine models that are described in Section 3.4. The choice between those models depends on the type of study and the extent of its adequacy to that study.

Example 3.3 Calculate the transient and sub-transient inductances of a cylindrical rotor generator having the parameters in per unit values as below:

$L_d = 1.81, L_q = 1.76, L_{md} = 1.66, L_{mq} = 1.61, L_f = 2.73, L_{kd} = 1.601, L_{kq} = 1.72,$
$L_{fkd} = 1.66, R_f = 0.898 \times 10^{-3}, R_{kq} = 6.19 \times 10^{-3}, R_{kd} = 0.0284$

Solution:

The direct and quadrature axis sub-transient and transient inductances are

$$\text{By (3.23)}, L''_d = 1.81 - \frac{2.73 + 1.67 - 2 \times 1.66}{\left(\frac{2.73 \times 1.6}{1.66 \times 1.66}\right) - 1} = 0.125$$

$$\text{From (3.26)}, L'_d = 1.81 - \frac{1.66 \times 1.66}{2.73} = 0.8006$$

$$\text{Referring to (3.31)}, L''_q = 1.76 - \frac{1.61 \times 1.61}{1.72} = 0.253$$

$$L'_q = L_q = 1.76$$

3.4 Synchronous machine-simplified models

3.4.1 The classical model

This model is relevant to determine the system stability in the first swing of the rotor angle in the order of one second or less. The machine is represented by a constant voltage 'E' at initial angle δ, behind a direct axis transient reactance, X'_d. Whether the machine is connected to an infinite bus or a multi-machine system, the model is based on the following assumptions:

- constant input mechanical power
- damping is negligible

- synchronous machine is represented by a constant voltage behind transient reactance
- mechanical rotor angle of the machine coincides with the angle of the voltage behind the transient reactance
- loads at the machine terminals are represented by a constant impedance or admittance

3.4.1.1 Classical model for one machine connected to an infinite bus

The general configuration of the system consisting of one machine connected to the infinite bus through a transmission line with impedance Z_{TL} and a shunt impedance Z_s connected to the machine terminals (it may represent a load) is shown in Figure 3.2(a). The terminal voltage of the machine is denoted by V_t and the voltage of the infinite bus is $V_\infty \angle 0°$ (it is used as reference). Its equivalent circuit is depicted in Figure 3.2(b) where the node representing the machine terminal in Figure 3.2(a) can be eliminated by a Y–Δ transformation.

The transient stability of the machine is determined by solving the swing equation, (2.85), and therefore the power delivered by the machine, P_1, is needed to calculate the accelerating power. As in the equivalent circuit (Figure 3.2(b)), it can be found by network theory that

$$P_1 = \mathcal{R}e\left(\boldsymbol{E}\boldsymbol{I}_1^*\right) = E^2 Y_{11} \cos\theta_{11} + EV_\infty Y_{12} \cos(\theta_{12} - \delta)$$

where

$Y_{11} \angle \theta_{11} = \boldsymbol{y}_{12} + \boldsymbol{y}_{1o}$ and $Y_{12} \angle \theta_{12} = -y_{12}$. It is to be noted that y_{2o} is not needed.

By defining $G_{11} \triangleq Y_{11} \cos\theta_{11}$ and $\gamma = \theta_{12} - \pi/2$, the power P_1 can be expressed as

$$P_1 = E^2 G_{11} + EV_\infty Y_{12} \sin(\delta - \gamma) \tag{3.44}$$

3.4.1.2 Classical model for multi-machine system

The multi-machine system (Figure 3.3) comprises a number of generators (n) feeding different loads (m) through a transmission network. Each machine is represented by a constant internal voltage source behind its direct axis transient reactance and the loads are represented by constant impedances. The electric output

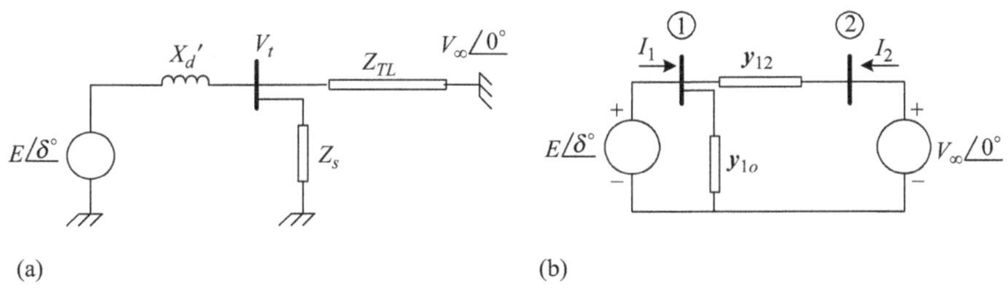

Figure 3.2 One machine to an infinite bus system

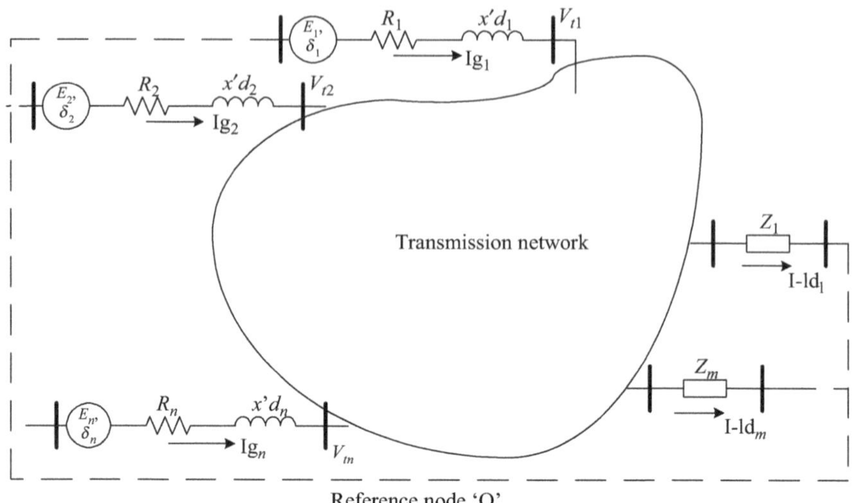

Figure 3.3 Multi-machine power system

power of each generator can be obtained by solving the set of non-linear algebraic equations that represent the relations between currents and voltages, and load flow techniques to obtain the initial values of system parameters that are needed to solve the equations of motion of the generators. Then, the system stability is determined. Detailed explanation for load flow techniques is provided in Part II.

3.4.2 The E'_q model

In this model the effect of the damper circuits in the d-axis is neglected and consequently i_{kd} is omitted from the current state space model (Section 2.8) or Ψ_{kd} is omitted from the flux linkages state space model (Section 2.7.1). The effect of damper circuits in the q-axis also can be neglected, but, in particular, for solid round rotor machines and in the absence of damper circuits, the rotor acts as a q-axis damper winding. However, this effect is small enough to be neglected or can be included by increasing the damping coefficient D in the torque equation. Thus, i_{kq} in the current state space model is omitted and Ψ_{kq} is omitted from the flux linkages state space model.

Under the condition of neglecting the effect of damper circuits, an alternative machine model containing the well-known machine parameters can be deduced as below.

Using (2.76), (2.78), (2.79) and (2.81) with the KD and KQ circuits omitted gives

$$\begin{bmatrix} i_d \\ i_f \\ i_q \end{bmatrix} = \begin{bmatrix} \dfrac{(\ell_d - L_{Md})}{\ell_d^2} & \dfrac{-L_{Md}}{\ell_d \ell_f} & 0 \\ \dfrac{-L_{Md}}{\ell_d \ell_f} & \dfrac{(\ell_f - L_{Md})}{\ell_f^2} & 0 \\ 0 & 0 & \dfrac{1}{L_q} \end{bmatrix} \begin{bmatrix} \Psi_d \\ \Psi_f \\ \Psi_q \end{bmatrix} \tag{3.45}$$

The elements of the matrix in (3.45) can be written in terms of L'_d, L_{md} and L_f as

$$\begin{bmatrix} i_d \\ i_f \\ i_q \end{bmatrix} = \begin{bmatrix} \frac{1}{L'_d} & \frac{-L_{md}}{L'_d L_f} & 0 \\ \frac{-L_{md}}{L'_d L_f} & \frac{L_d}{L'_d L_f} & 0 \\ 0 & 0 & \frac{1}{L_q} \end{bmatrix} \begin{bmatrix} \Psi_d \\ \Psi_f \\ \Psi_q \end{bmatrix} \tag{3.46}$$

From the set of (2.30) the first equation is rewritten as $\dot{\Psi}_d = -R_a i_d - \omega \Psi_q - v_d$ and using (3.46) gives

$$\dot{\Psi}_d = -\left(\frac{R_a}{L'_d}\right)\Psi_d + \left(\frac{R_a L_{md}}{L'_d L_f}\right)\Psi_f - R_a i_d - \omega \Psi_q - v_d \tag{3.47}$$

Equation (I.11), Appendix I: $E'_q = \omega_o k M_f \Psi_f / L_f$ can be converted into pu values, it gives

$$E'_q = L_{md}\Psi_f / L_f \tag{3.48}$$

Substitute (3.48) in (3.47) to obtain

$$\dot{\Psi}_d = -\left(\frac{R_a}{L'_d}\right)\Psi_d + \left(\frac{R_a}{L'_d}\right)E'_q - \omega \Psi_q - v_d \tag{3.49}$$

Similarly, apply the same procedure for $\dot{\Psi}_q$ as in (2.30) to write
$\dot{\Psi}_q = -(R_a/L_q)\Psi_q + \omega \Psi_d - v_q$ and $v_f = \dot{\Psi}_f + R_f i_f$ and use (3.45) to obtain

$$v_f = R_f\left[-\left(\frac{L_{md}}{L'_d L_f}\right)\Psi_d + \left(\frac{L_d}{L'_d L_f}\right)\Psi_f\right] + \dot{\Psi}_f \tag{3.50}$$

The definition given in (I.12), Appendix I, states that $E_{fd} = (v_f/R_f)\omega_o k M_f$. It can be converted into pu values as

$$E_{fd} = \frac{L_{md} v_f}{R_f} \tag{3.51}$$

Combining (3.48), (3.50) and (3.51) gives

$$\frac{R_f}{L_{md}} E_{fd} = -\frac{L_{md}}{L'_d}\frac{R_f}{l_f}\Psi_d + \frac{L_d}{L'_d}\frac{R_f}{L_{md}}E'_q + \frac{L_f}{L_{md}}\dot{E}'_q \tag{3.52}$$

Substituting $\frac{L^2_{md}}{L_f} = L_d - L'_d$ and $T'_{do} = \frac{L_f}{R_f}$

$$\dot{E}'_q = \frac{1}{T'_{do}}\left(E_{fd} - \frac{L_d}{L'_d}E'_q + \frac{L_d - L'_d}{L'_d}\Psi_d\right) \tag{3.53}$$

It is to be noted that the above equations include all variables in pu as well as the voltages v_d, v_q, E_{fd} and E'_q are considered as line-to-line pu values.

The variables i_d and i_q in (3.46) are substituted in the torque equation $T_e = i_q\Psi_d - i_d\Psi_q$ and (3.48) is used as well to express the torque as

$$T_e = \frac{1}{L'_d}E'_q\Psi_q - \left(\frac{1}{L'_d} - \frac{1}{L_q}\right)\Psi_d\Psi_q \tag{3.54}$$

The swing equation including the damping term is

$$\left.\begin{aligned} \dot{\omega} &= \frac{1}{2H\omega_B}[T_m - T_e - D\omega] \\ \dot{\delta} &= \omega - 1 \end{aligned}\right\} \tag{3.55}$$

Thus, the E'_q model, described by (3.49), (3.50), (3.53) and (3.55) in time-domain, is a fifth-order system and can be represented in the *s* domain by the block diagram shown in Figure 3.4.

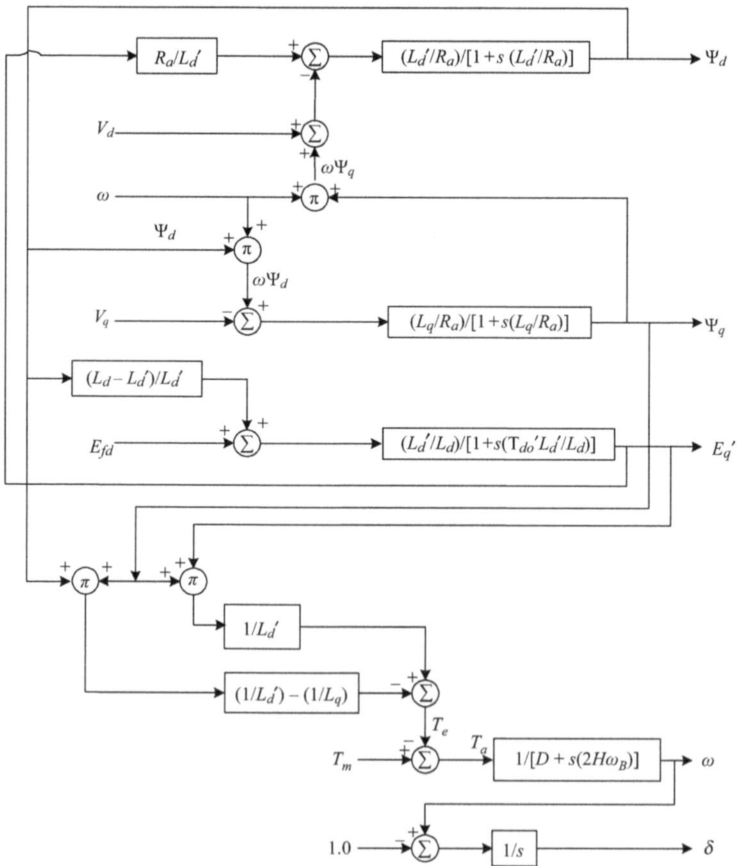

Figure 3.4 Block diagram representation of E'_q model in s domain

Taking the saturated operating conditions into consideration necessitates adding additional field current, Δi, to substitute the saturation effect as explained in Section 2.7.2. Equation (3.46) implies that $i_d = \frac{1}{L'_d}\Psi_d - \frac{L_{Ad}}{L'_d l_f}\Psi_f$. Substituting this value of i_d in equation $\Psi_d = L_d i_d + L_{md} i_f$ gives

$$L_{md} i_f = \left[\left(1 - \left(\frac{L_d}{L'_d}\right)\right)\Psi_d\right] + \left[\left(\frac{L_d L_{Ad}}{(L'_d L_f)}\right)\Psi_f\right] \tag{3.56}$$

It is given from '(I.10)', Appendix I, that $\omega_o k M_f i_f = L_{md} i_f = E_I$. It is incorporated with (3.48) and (3.56) to write

$$E_I = \frac{L_d}{L'_d}E'_q - \frac{L_d - L'_d}{L'_d}\Psi_d \tag{3.57}$$

Equations (3.53) and (3.57) show that

$$\dot{E}'_q = \frac{1}{T'_{do}}(E_{fd} - E_I) \tag{3.58}$$

Assuming ΔE is a component that must be added to (3.57) corresponding to Δi to obtain the same EMF on the no-load saturation curve, it can be written as

$$E_I = \frac{L_d}{L'_d}E'_q - \frac{L_d - L'_d}{L'_d}\Psi_d + \Delta E \tag{3.59}$$

Equations (3.58) and (3.59) should be represented by a block diagram to be added to that depicted in Figure 3.4 when considering the effect of machine saturation, thus giving the block diagram shown in Figure 3.5.

Each generating unit is individually provided with an excitation control system and a prime mover control system. Therefore, modelling of both the controllers should be considered as a supplementary part of the synchronous machine model, in particular, when studying the stability of a large-scale power system.

3.5 Excitation system

The regulation of synchronous machine terminal voltage is the main function of the excitation system that controls the field current. As the field circuit has 'to some extent' a high time constant in order of a few seconds, field forcing is required for fast control of the field current. Consequently, the exciter should have a high ceiling voltage to be able to operate transiently with voltage levels of the order of three to four times the normal in addition to changing the voltage at a fast rate.

The excitation system must be modelled in such a way that the model is relevant for use in large-scale power system stability studies. The order of the model should be chosen in a condition of adequacy to the aim of the study, avoiding the complexity of analysis and keeping the accuracy of results. Therefore,

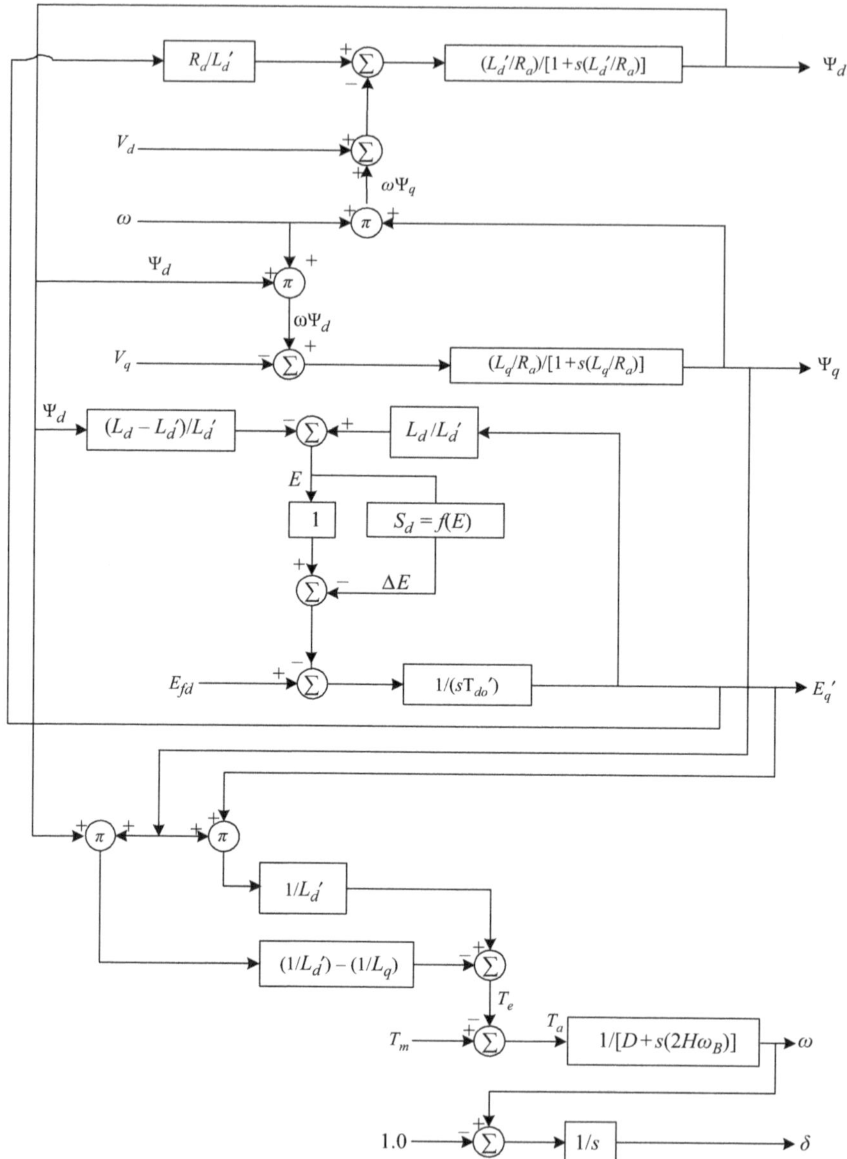

Figure 3.5 Block diagram representation of E_q' model with saturation in s domain

based on these requirements, the model can be a reduced order model and may not represent the system in more detail than that required for stability studies.

3.5.1 Excitation system modelling

Regardless of the type of excitation system, the main parts that comprise the excitation system, as shown in Figure 3.6, are voltage transducer and load

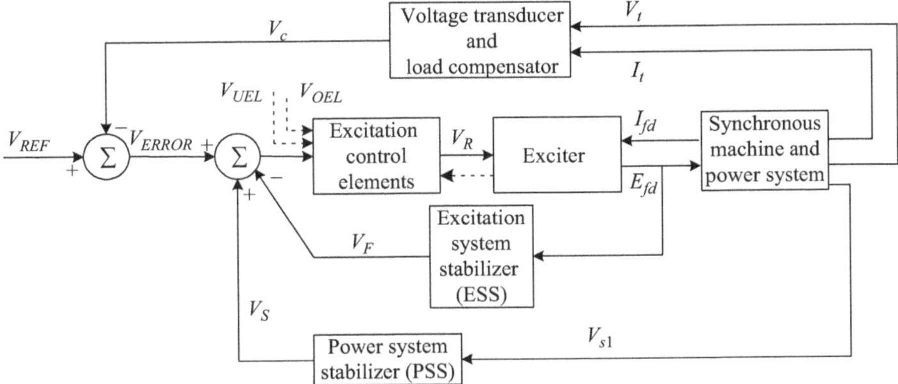

Figure 3.6 Basic functional block diagram of excitation control system

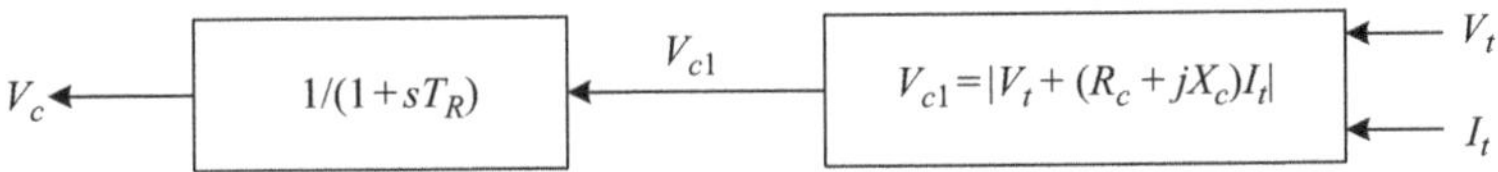

Figure 3.7 Elements of terminal voltage transducer and load compensator

compensator, excitation control elements, an exciter, excitation system stabiliser (ESS) and commonly a power system stabiliser (PSS).

(I) Terminal voltage transducer and load compensator

The objective of the load compensation is to provide an output voltage, V_c, equal to the machine terminal voltage plus the voltage drop in an impedance ($R_c + jX_c$). The impedance or range of adjustment should be specified. Both voltage and current phasors at the machine terminal are sensed and used to compute V_c, which is then compared with a reference voltage representing the desired terminal voltage setting. Without load compensation, both R_c and X_c are zero; the excitation system attempts to maintain a terminal voltage determined by the reference signal within its regulation characteristics. The function of the load compensator is to regulate the voltage at some point other than the machine terminals in two ways: First, to regulate the voltage at a point internal to the generator by the sharing of reactive power among units connected to the same bus with zero impedance in between. R_c and X_c are in this case positive values. Second, to regulate the voltage at a point beyond the generator terminals when the generating units are operating in parallel through unit transformers, a compensation of a portion of transformer impedance is required. In this case, R_c and X_c are negative values. Generally, in practice, the resistive component of compensation is neglected when generators are synchronised to a large grid over high-voltage interconnections. Thus, R_c is assumed to be zero to simplify the analysis. The terminal voltage transducer and load compensator can be represented by the block diagram shown in Figure 3.7 where a single time constant, T_R, is used for the combined voltage sensing and compensation signal.

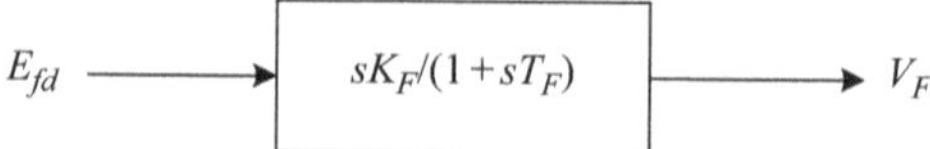

Figure 3.8 Transfer function of excitation system stabiliser

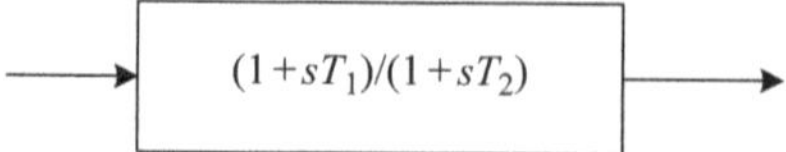

Figure 3.9 Transfer function of transient gain reduction

(II) Excitation control elements

They include the functions of both regulating and stabilising the excitation. The terms ESS and *transient gain reduction* (TGR) are used for increasing the stable region of operation of the excitation system and permit higher regulator gains. It is to be noted that feedback control systems (of which the excitation system is an example) often require lead/lag compensation or derivative, rate, feedback. The feedback transfer function for ESS is shown in Figure 3.8.

A typical value of the time constant is taken as one second. The feedback compensation for ESS can be replaced by using a series-connected lead/lag circuit as shown in Figure 3.9 where T_1 is commonly less than T_2. Consequently, this means of stabilisation is termed as TGR. The main goal of TGR is to reduce the transient gain or gain at higher frequencies, thereby minimising the negative contribution of the regulator to system damping and accordingly the system damping is enhanced. Therefore, if PSS is specifically used to enhance system damping, the TGR may not be required. A typical value of the TGR factor (T_2/T_1) is 10.

Recently, modelling of field current limiters has become increasingly important, resulting in the addition to this over-excitation and under-excitation limiters, OELs and UELs, respectively. Output of the UEL may be received as an input to the excitation system (V_{UEL}) at various locations, either as a summing input or as a gated input, but for any one application of the model, only one of these inputs would be used. For the OEL some models provide a gate through which the output of the over-excitation limiter or terminal voltage limiter (V_{OEL}) could enter the regulator loop.

(III) Power system stabiliser

The stabilisation provided by PSS differs from that provided by ESS. ESS provides effective voltage regulation under open- or short-circuit conditions while the function of PSS is to provide damping of the rotor oscillations at the occurrence of transient disturbances. The damping of these oscillations can be impaired by the provision of high gain AVR, particularly at high loading conditions when a generator is connected through high external impedance. Detailed discussion of PSS is presented in Chapter 11. It is to be noted that the input signal for PSS is derived from speed/frequency, accelerating power or a combination of these signals.

Several rotor oscillation frequencies must be considered when designing the PSS in a multi-machine system. However, the stabiliser is designed to have zero output in steady state. Also the output is limited in order not to adversely affect the voltage control. The stabiliser output 'V_s' is added to the terminal voltage error signal as shown in Figure 3.6.

(IV) Types of excitation system

Excitation systems are classified into three types based on the source of excitation power: DC excitation systems, AC excitation systems and static ST excitation systems. Modelling of excitation system is essential for stability studies [13–16] and identification of model parameters has a prominent role, in particular, identifying non-linearity such as limits and saturation as discussed in [17–21]. Commonly, the excitation systems have under-excitation and over-excitation limiters, thus their modelling must be considered as well [22, 23]. In this section, each type is presented through an example indicating the model in *s* domain associated with sample data. It is to be noted that for DC and AC types, the exciter saturation and loading effects should be accounted for as below.

The increase in excitation requirements due to saturation is represented by an exciter saturation function $S_E(E_{fd})$ defined as a multiplier of pu exciter voltage. This function at a given exciter output voltage, representing the load saturation as depicted in Figure 3.10, can be calculated by (3.60):

$$S_E(E_{fd}) = \frac{A - B}{B} \tag{3.60}$$

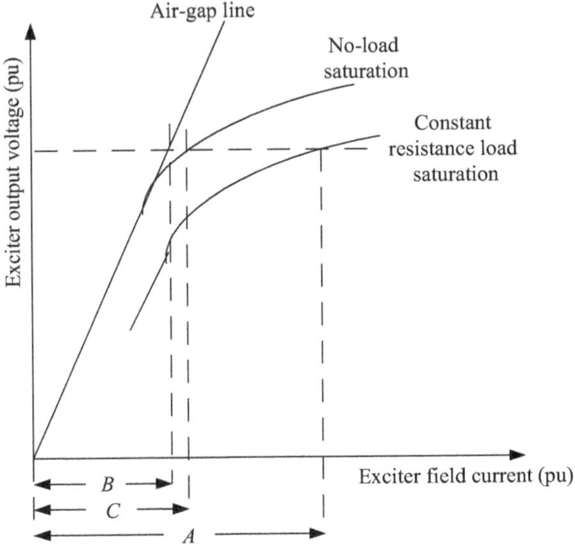

Figure 3.10 Exciter saturation characteristics

where

$A \triangleq$ the excitation required to produce the exciter output voltage on the constant-load–saturation curve

$B \triangleq$ the excitation required to produce the exciter output voltage on the air-gap line

$C \triangleq$ the excitation required to produce the exciter output voltage on the no-load saturation curve

For some alternator-rectifier exciters, the no-load saturation curve is used in defining $S_E(V_E)$ as

$$S_E(V_E) = \frac{C - B}{B} \tag{3.61}$$

because the exciter regulation effects are accounted for by inclusion of both a demagnetising factor and commutating reactance voltage drop in the model.

In general, the saturation function can be specified adequately by two points: the first is at the voltage E_1 near the exciter ceiling voltage for DC types or V_1 near the exciter open circuit ceiling voltage for the AC types. The second is at the voltage E_2 at a lower value, commonly near 75 per cent of E_1 for DC types or near 75 per cent of V_1 for AC types. Computer programs based on these higher and lower voltages along with the corresponding saturation data as inputs have been designed to represent the exciter saturation characteristics with different mathematical expressions.

(a) Type DC excitation systems

For this type, a direct current generator with a commutator is used as the source of excitation power. The block diagram of a typical model is shown in Figure 3.11 and sample data is given in Table 3.1 [24]. It includes a proportional, integral and differential generator voltage regulator (AVR). An alternative feedback loop (K_F, T_F) is designed for stabilisation if the derivative term is not included in the AVR.

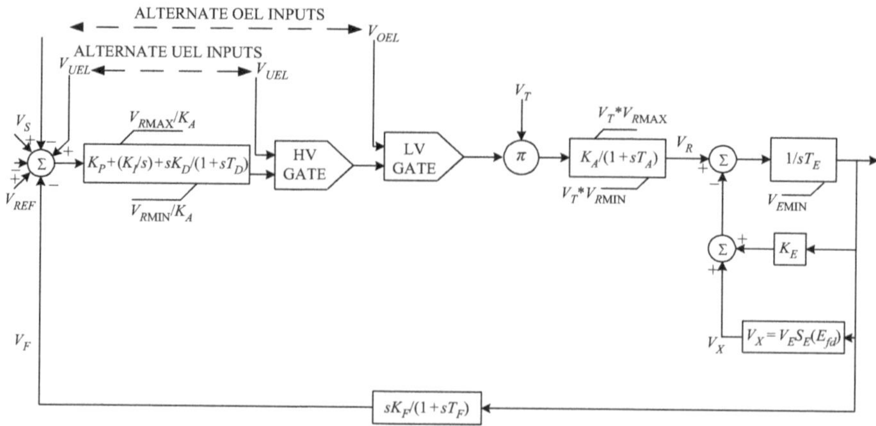

Figure 3.11 A typical example of DC excitation systems (Type DC4B)

Table 3.1 Sample data for excitation system (Type DC4B)

Description	Parameter	Value	Units
Regulator proportional gain	K_P	80.0	pu
Regulator integral gain	K_I	20.0	pu
Regulator derivative gain	K_D	20.0	pu
Regulator derivative filter time constant	T_D	0.01	s
Regulator output gain	K_A	1.0	pu
Regulator output time constant	T_A	0.2	S
Max controller output	V_{RMAX}	2.7	pu
Exciter field time constant	T_E	0.8	pu
Exciter field proportional constant	K_E	1.0	pu
Exciter minimum output voltage	V_{EMIN}	0.0	pu
Exciter flux at S_{E1}	E_1	1.75	pu
Saturation factor at E_1	S_{E1}	0.08	
Exciter flux at S_{E2}	E_2	2.33	pu
Saturation factor at E_2	S_{E2}	0.27	
Rate feedback gain	K_F	0.0	pu
Rate feedback time constant	T_F	0.0	s

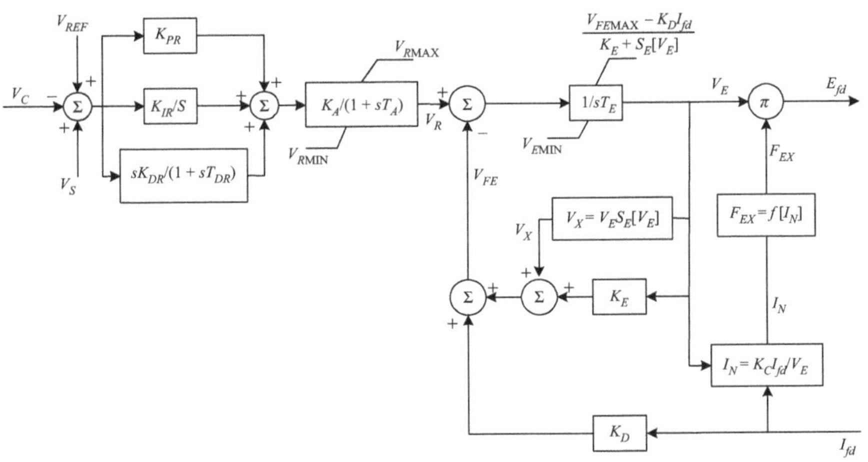

Figure 3.12 Type AC8B – Alternator-rectifier excitation system

(b) Type AC excitation systems

These excitation systems use an a.c. alternator and either stationary or rotating rectifiers to produce the d.c. field requirements. Loading effects on such exciters are significant, and the use of generator field current as an input to the models allows these effects to be represented accurately.

An example of this type is shown in Figure 3.12 and its sample data are summarised in Table 3.2.

Table 3.2 Sample data of excitation system-type AC8B

$K_{PR}=80$	$V_{RMAX}=35.0$	$S_E(E_1)=0.3$
$K_{IR}=5$	$V_{RMIN}=00.0$	$E_1=6.5$
$K_{DR}=10$	$K_E=1.0$	$S_E(E_2)=3.0$
$T_{DR}=0.1$	$T_E=1.2$	$E_2=9.0$
$V_{FEMAX}=6.0$	$K_C=0.55$	$K_D=1.1$

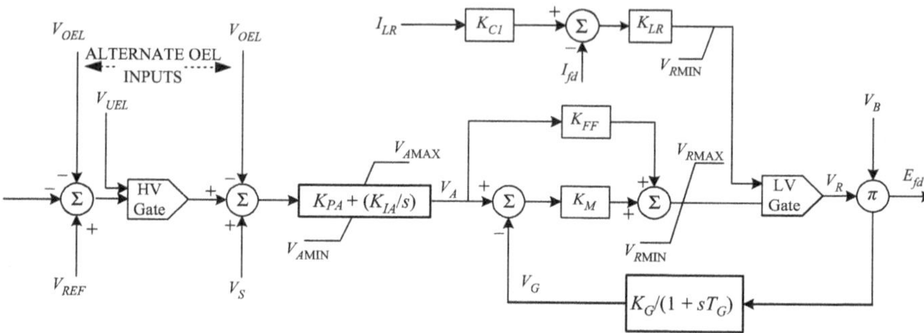

Figure 3.13 Type ST6B – Static potential-source excitation system with field current limiter

Table 3.3 Sample data of excitation system-type ST6B

$K_{PA}=18.038$	$T_G=0.02$ s	$V_{AMIN}=-3.85$
$K_{IA}=45.094\ s^{-1}$	$T_R=0.012$ s	$K_{LR}=17.33$
$K_{FF}=1.0$	$V_{AMAX}=4.81$	$I_{LR}=4.164$
$K_M=1.0$	$V_{AMIN}=-3.85$	$V_{RMAX}=4.81$
$K_G=1.0$		$V_{RMIN}=-3.85$

(c) Type ST excitation systems

In these excitation systems, voltage (and also current in compounded systems) is transformed to an appropriate level. Rectifiers, either controlled or non-controlled, provide the necessary direct current for the generator field. An example of this type is shown in Figure 3.13 and its sample data is summarised in Table 3.3.

3.6 Modelling of prime mover control system

The prime mover is responsible for providing the synchronous generator with the input mechanical power required. The power demand from the generator changes corresponding to load characteristics, and the operating conditions of the power system vary continuously. The input mechanical power to the generator must vary in response to these variations to maintain the frequency constant. So, to regulate the power system frequency, speed control of the prime mover using a governor is required. For stability studies, relevant models of the prime mover and the associated speed control system should be developed to verify accurate performance. The most popular type of prime mover used in power systems is the turbine, either hydraulic or steam, while the speed-governing systems are either mechanical-hydraulic or electro-hydraulic. In both types of speed-governing systems the valve position or gate controlling water or steam flow are determined by using hydraulic motors. The speed sensing for mechanical hydraulic governors is carried out by mechanical components, but for electro-hydraulic type it is implemented using electronic circuits. However, both types have similar dynamic performance. The following description of turbines and governing systems is based on the IEEE definitions and standards [25].

3.6.1 Hydraulic turbines

The hydraulic turbine is simply represented in a suitable manner for stability studies by the block diagram depicted in Figure 3.14.

The transfer function is expressed as

$$P_m = \frac{1 - sT_W}{1 + 0.5\, sT_W} P_{GV} \tag{3.62}$$

where

$T_W \triangleq$ water starting time constant $= (L \times V)/(H_T \times g)$ 'typical value around 1.0 s'
$L \triangleq$ length of penstock
$V \triangleq$ water velocity
$H_T \triangleq$ total head
$g \triangleq$ acceleration due to gravity
$P_{GV} \triangleq$ gate opening expressed in pu and provided by speed governor

Construction of a governing system for hydro-turbines is based on the main parts shown in Figure 3.15, and a typical non-linear model is given in Figure 3.16 as a block diagram with its data summarised in Table 3.4. A dashpot feedback is employed for stability performance due to the effect of water inertia [26].

For stability studies, the speed governing system is commonly simplified as represented by the block diagram shown in Figure 3.17. Its parameters can be calculated in terms of those defined in Figure 3.15 by using the relations below:

$$T_1, T_3 = \frac{T_B}{2} \pm \sqrt{\frac{T_B^2}{4} - T_A} \tag{3.63}$$

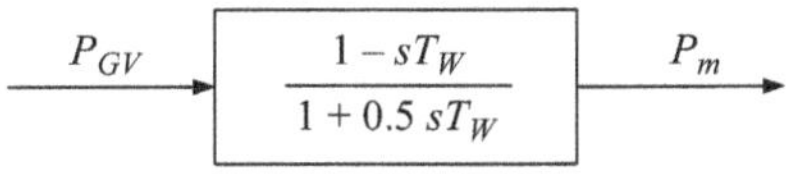

$$P_{GV} \longrightarrow \boxed{\frac{1 - sT_W}{1 + 0.5\, sT_W}} \longrightarrow P_m$$

Figure 3.14 Model of a hydro-turbine

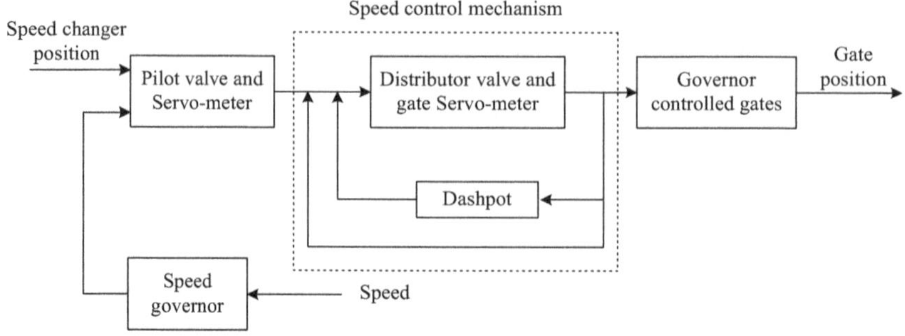

Figure 3.15 Functional block diagram for hydro-turbines governing system

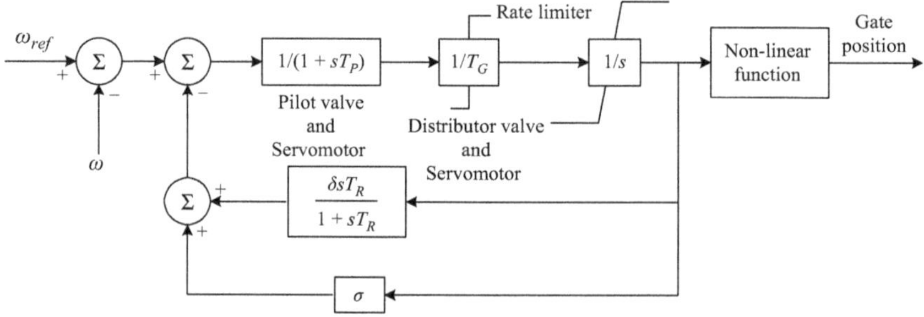

Figure 3.16 Model of speed governing system for hydro-turbines

Table 3.4 Data of the model in Figure 3.16

Parameter	Typical value	Range
T_R	5.0	2.5–25.0
T_G	0.2	0.2–0.4
T_P	0.04	0.03–0.05
δ	0.3	0.2–1.0
σ	0.05	0.03–0.06

$T_R = 5T_W$ and $\delta = 1.25T_W/H$, generator inertia constant.

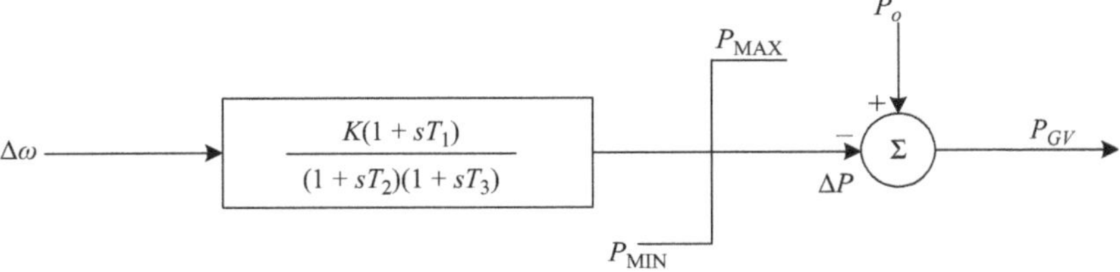

Figure 3.17 General simplified model of speed governing system for hydro-turbines

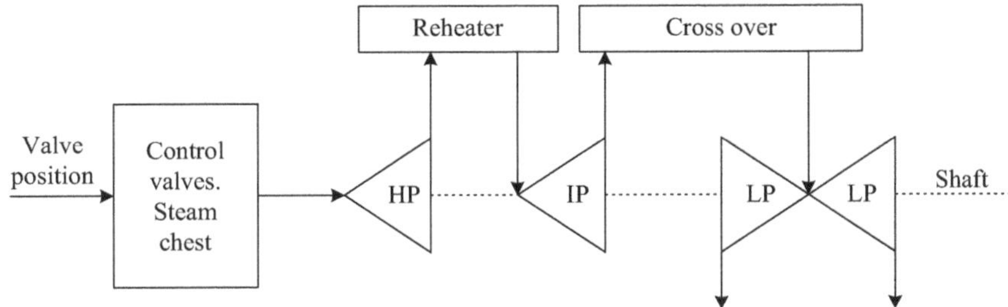

Figure 3.18 Configuration of steam system – tandem compound, single reheat

where

$$T_A = \left(\frac{1}{\sigma}\right) T_R T_G, \quad T_B = \left(\frac{1}{\sigma}\right)[(\sigma + \delta)T_R + T_G]$$

$$K = \left(\frac{1}{\sigma}\right) \quad \text{and} \quad T_2 = 0$$

$P_o \triangleq$ the initial power (load reference)

3.6.2 Steam turbines

As an example, the turbine system shown in Figure 3.18 (tandem compound-single reheat) is one of the various types of turbine systems, such as tandem compound-double reheat, cross compound-single reheat with one or two low-pressure (LP) turbines and cross compound-double reheat. It has one shaft on which all turbines; high pressure (HP), intermediate pressure (IP) and LP turbines are mounted. It can be represented by the block diagram in Figure 3.19 where governor control valves are used at the inlet to HP turbine to control the steam flow. The steam chest, re-heater and crossover piping introduce delays that are represented by time constants T_{CH}, T_{RH} and T_{CO}, respectively. The total power developed in HP, IP and LP turbines is represented by the fractions F_{HP}, F_{IP} and F_{LP}, respectively.

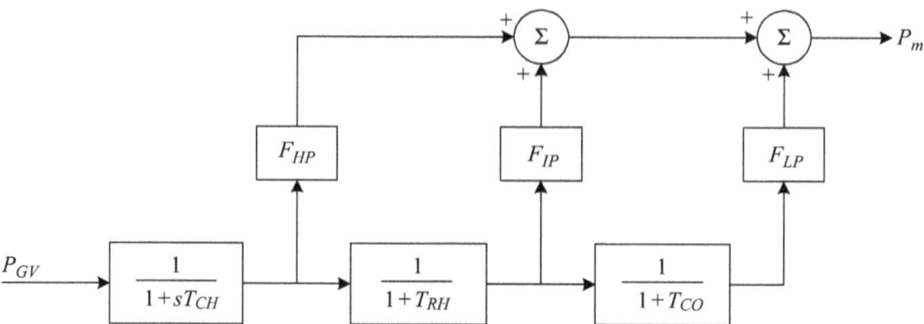

Figure 3.19 Block diagram of Tandem compound, single reheat (typical values: $T_{CH}=0.1–0.4$ s, $T_{RH}=4–11$ s, $T_{CO}=0.3–0.5$ s, $F_{HP}=0.3$, $F_{IP}=0.3$, $F_{LP}=0.4$)

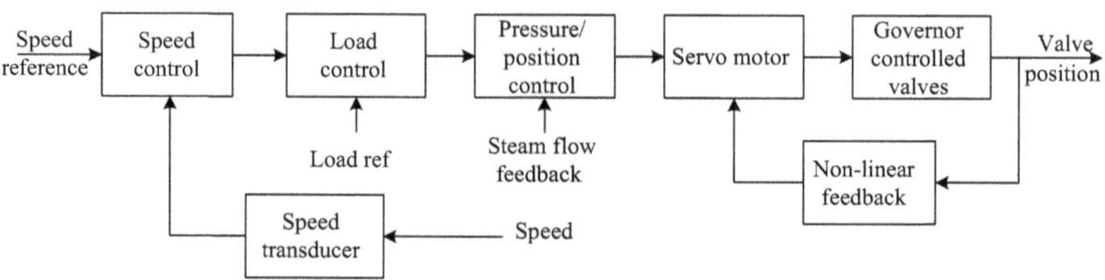

Figure 3.20 Functional block diagram of electro-hydraulic speed governing system for steam turbines

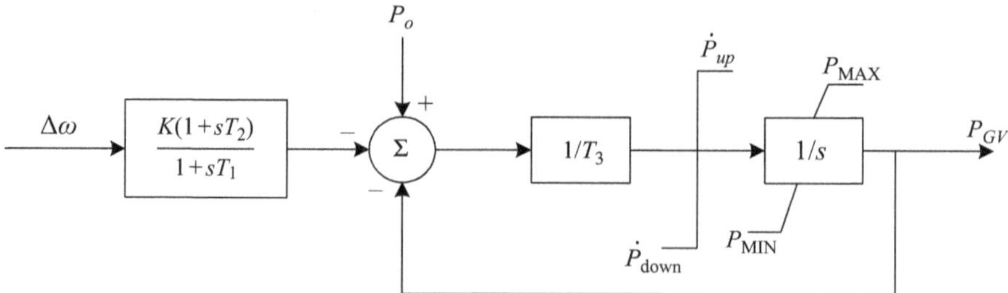

Figure 3.21 General simplified model for speed governor for steam turbine

The main construction of the governing system (Figure 3.20) includes a feedback from steam flow or pressure in the first stage turbine as well as a servo-motor feedback loop to improve linearity. A simplified model of governor control system for steam turbines is shown by the block diagram in Figure 3.21. Typical values of time constants in seconds are

Electro-hydraulic governor: $T_1 = T_2 = T_3 = 0.025 - 0.15$ s
Mechanical-hydraulic governor: $T_1 = 0.2 - 0.3$, $T_2 = 0$, $T_3 = 0.1$ s

References

1. De Oliveira S.E.M. 'Modeling of synchronous machines for dynamic studies with different mutual couplings between direct axis windings'. *IEEE Transactions on Energy Conversion.* 1989;**4**(4):591–9
2. El-Serafi A.M., Abdallah A.S. 'Effect of saturation on the steady-state stability of a synchronous machine connected to an infinite bus system'. *IEEE Transactions on Energy Conversion.* 1991;**6**(3):514–21
3. Xu W.W., Dommel H.W., Marti J.R. 'A synchronous machine model for three-phase harmonic analysis and EMTP initialization'. *IEEE Transactions on Power Systems.* 1991;**6**(4):1530–8
4. Wang L., Jatskevich J. 'A voltage-behind-reactance synchronous machine model for the EMTP-type solution'. *IEEE Transactions on Power Systems.* 2006;**21**(4):1539–49
5. Wang L., Jatskevich J., Domme H.W. 'Re-examination of synchronous machine modeling techniques for electromagnetic transient simulations'. *IEEE Transactions on Power Systems.* 2007;**22**(3):1221–30
6. Sriharan S., Hiong K.W. 'Synchronous machine modeling by standstill frequency response tests'. *IEEE Transactions on Energy Conversion.* 1987; **EC-2**(2):239–45
7. Rusche P.E.A., Brock G.J., Hannett L.N., Willis J.R. 'Test and simulation of network dynamic response using SSFR and RTDR derived synchronous machine models'. *IEEE Transactions on Energy Conversion.* 1990;**5**(1): 145–55
8. Verbeeck J., Pintelon R., Lataire P. 'Relationships between parameter sets of equivalent synchronous machine models'. *IEEE Transactions on Energy Conversion.* 1999;**14**(4):1075–80
9. Verbeeck J., Pintelon R., Lataire P. 'Influence of saturation on estimated synchronous machine parameters in standstill frequency response tests'. *IEEE Transactions on Energy Conversion.* 2000;**15**(3):277–83
10. Dedene N., Pintelon R., Lataire P. 'Estimation of a global synchronous machine model using a multiple-input multiple-output estimator'. *IEEE Transactions on Energy Conversion.* 2003;**18**(1):11–6
11. Jackson W.B., Winchester R.L. 'Direct and quadrature axis equivalent circuits for solid-rotor turbine generators'. *IEEE Transactions on Power Apparatus and Systems.* 1969;**PAS-88**(7):1121–36

12. Anderson P.M. *Analysis of Faulted Power Systems*. Ames, IA, US: Iowa State Univ. Press; 1973
13. IEEE working group on computer modelling of excitation systems 'Excitation system models for power system stability studies'. *IEEE Transactions on Power Apparatus and Systems*. 1981;**PAS-100**(2):494–509
14. Ruuskanen V., Niemelä M., Pyrhönen J., Kanerva S., Kaukonen J. 'Modelling the brushless excitation system for a synchronous machine'. *IET Electric Power Applications*. 2009;**3**(3):231–9
15. The Digital Excitation Task Force of the Equipment Working Group 'Computer models for representation of digital based excitation systems'. *IEEE Transactions on Energy Conversion*. 1996;**11**(3):607–15
16. IEEE Committee Report 'Computer representation of excitation systems'. *IEEE Transactions on Power Apparatus and Systems*. 1968;**PAS-87**(6):1460–4
17. Wang J.C., Chiang H.D., Haung C.T., Chen Y.T., Chang C.L., Huang C.Y. 'Identification of excitation system models based on on-line digital measurements'. *IEEE Transactions on Power Systems*. 1995;**10**(3):1286–93
18. Benchluch S.M., Chow J.H. 'A trajectory sensitivity method for the identification of nonlinear excitation system models'. *IEEE Transactions on Energy Conversion*. 1993;**8**(2):159–64
19. Puma J.Q., Colomé D.C. 'Parameters identification of excitation system models using genetic algorithms'. *IET Generation, Transmission & Distribution*. 2008;**2**(3):456–67
20. Liu C.S., Yuan-Yih H., Jeng L.H., Lin C.J., Huang C.T., Liu A.H., Li T.H. 'Identification of exciter constants using a coherence function based weighted least squares approach'. *IEEE Transactions on Energy Conversion*. 1993; **8**(3):460–7
21. IEEE Committee Report (eds.). 'Excitation system dynamic characteristics'. *Proceedings of Power and Energy Society PES Summer Meeting*; San-Francisco, CA, US, 1972. pp. 64–75
22. IEEE Task Force on Excitation Limiters. 'Under-excitation limiter models for power system stability studies'. *IEEE Transactions on Energy Conversion*. 1995;**10**(3):524–31
23. IEEE Task Force on Excitation Limiters. 'Recommended models for over-excitation limiting devices'. *IEEE Transactions on Energy Conversion*. 1995;**10**(4):706–13
24. IEEE Std 421.5^{TM}-2005. 'IEEE Recommended Practice for Excitation System Models for Power System Stability Studies'.
25. IEEE Std 421.1^{TM}-2007. 'IEEE Standard Definitions for Excitation Systems for Synchronous Machines'.
26. IEEE Working Group on Prime Mover and Energy Supply Models for System Dynamic Performance Studies. 'Hydraulic turbine and turbine control models for system dynamic studies'. *Transactions on Power Systems*. 1992;**7**(1):167–79
27. IEEE Committee Report, 'Dynamic models for steam and hydro turbine in power system studies', *IEEE Transactions on Power Apparatus and Systems*. Nov/Dec 1973, vol. PAS-92, pp. 1904–15.

Chapter 4
Modelling of transformers, transmission lines and loads

Generation, transmission and distribution are the main three parts that comprise a power system. Each part has its own function for electric power: generation part to generate the electric power commonly at medium voltage level by using synchronous generators, transmission part to transmit this power through high voltage or extra-high voltage transmission lines and finally distribution part to distribute the electric power to feed the consumers' loads at medium voltage or low voltage levels. The interconnections between these three parts that operate at different voltages necessitate use of transformers, e.g. step-up/step-down power transformers, distribution transformers, autotransformers. Accordingly, to start the stability studies it is essential to model all elements – generators, transformers, transmission lines and loads – in a manner that is convenient for this purpose. Modelling of synchronous generator has been explained in Chapters 2 and 3, and this chapter deals with the modelling of the rest of the elements.

4.1 Transformers

Transformers in power systems are used not only to enable power transfer from generator sending end to consumers at the receiving end but also to control the voltage and reactive power flow using taps on transformer windings that may be used to change the transformer turns ratio. The most commonly used is either a single-phase or a three-phase two-winding transformer. In some applications the transformer may have a third winding, called the tertiary winding. Autotransformers may be used to regulate the voltage when the transformation ratio is small. In some applications, such as controlling the power circulation and preventing overloading of lines, phase-shift transformers are used.

4.1.1 Modelling of two-winding transformers

The equivalent circuit of a two-winding transformer is shown in Figure 4.1. The saturation effect can be neglected as the magnetising reactance X_{m1} is large and the quantities are in physical units. The relations below can be deduced from the equivalent circuit. It is to be noted that the parameters written in bold font are

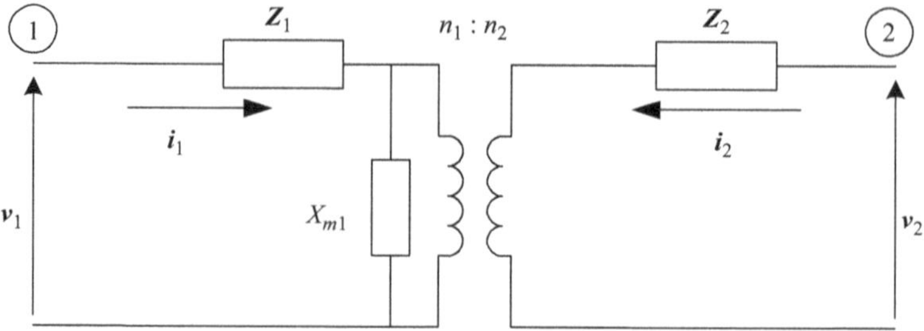

Figure 4.1 Basic equivalent circuit of a two-winding transformer

vectors, the subscript '1' denotes the primary side and subscript '2' denotes the secondary side of the transformer.

$$\left.\begin{aligned} v_1 &= Z_1 i_1 + \frac{n_1}{n_2} v_2 - \frac{n_1}{n_2} Z_2 i_2 \\ v_2 &= \frac{n_2}{n_1} v_1 - \frac{n_2}{n_1} Z_1 i_1 + Z_2 i_2 \end{aligned}\right\} \tag{4.1}$$

where

$Z_i = R_i + jX_i$ and $i = 1$ for primary winding, $i = 2$ for secondary winding.
R_i, X_i = resistance and leakage reactance of ith winding, respectively.
n_1, n_2 = number of turns of primary and secondary windings, respectively.

For power system stability analysis, (4.1) should be written in per unit values. This requires a proper choice of primary and secondary base quantities. They usually are chosen based on the nominal turns ratio. Assume that the nominal numbers of turns of primary and secondary sides are n_{1o} and n_{2o}, respectively. Accordingly, $Z_{io} = Z_i$ at the nominal ith side tap position. Therefore, in terms of these nominal values, and considering that the impedance is proportional to the square of the number of turns, this consideration is accepted as $R \ll X$ and $(n_i - n_{io})$ is not very large; (4.1) can be rewritten as

$$\left.\begin{aligned} v_1 &= \left(\frac{n_1}{n_{1o}}\right)^2 Z_{1o} i_1 + \frac{n_1}{n_2} v_2 - \frac{n_1}{n_2}\left(\frac{n_2}{n_{2o}}\right)^2 Z_{2o} i_2 \\ v_2 &= \frac{n_2}{n_1} v_1 - \frac{n_2}{n_1}\left(\frac{n_1}{n_{1o}}\right)^2 Z_{1o} i_1 + \left(\frac{n_2}{n_{2o}}\right)^2 Z_{2o} i_2 \end{aligned}\right\} \tag{4.2}$$

The transformer connection is assumed to be Y–Y. The nominal number of turns is related to the base voltages by

$$\frac{n_{1o}}{n_{2o}} = \frac{v_{1(base)}}{v_{2(base)}}, \quad v_{1(base)} = Z_{1(base)} i_{1(base)} \quad \text{and} \quad v_{2(base)} = Z_{2(base)} i_{2(base)} \tag{4.3}$$

Using (4.3), (4.2) in per unit form becomes

$$\left.\begin{aligned} \boldsymbol{v}_{1(\text{pu})} &= n_{1(\text{pu})}{}^2 \boldsymbol{Z}_{1o(\text{pu})} \boldsymbol{i}_{1(\text{pu})} + \frac{n_{1(\text{pu})}}{n_{2(\text{pu})}} \boldsymbol{v}_{2(\text{pu})} - \frac{n_{1(\text{pu})}}{n_{2(\text{pu})}} n_{2(\text{pu})}{}^2 \boldsymbol{Z}_{2o(\text{pu})} \boldsymbol{i}_{2(\text{pu})} \\ \boldsymbol{v}_{2(\text{pu})} &= \frac{n_{2(\text{pu})}}{n_{1(\text{pu})}} \boldsymbol{v}_{1(\text{pu})} - \frac{n_{2(\text{pu})}}{n_{1(\text{pu})}} n_{1(\text{pu})}{}^2 \boldsymbol{Z}_{1o(\text{pu})} \boldsymbol{i}_{1(\text{pu})} + n_{2(\text{pu})}{}^2 \boldsymbol{Z}_{2o(\text{pu})} \boldsymbol{i}_{2(\text{pu})} \end{aligned}\right\} \tag{4.4}$$

where

$$n_{1(\text{pu})} = \frac{n_1}{n_{1o}} \quad \text{and} \quad n_{2(\text{pu})} = \frac{n_2}{n_{2o}}$$

Thus, (4.4) can be represented by the per unit equivalent circuit shown in Figure 4.2.

The per unit turns ratio $n_{(\text{pu})}$ is expressed as

$$n_{(\text{pu})} = \frac{n_{1(\text{pu})}}{n_{2(\text{pu})}} = \frac{n_1 n_{2o}}{n_{1o} n_2} \triangleq \text{off-nominal ratio (ONR)} \tag{4.5}$$

and the equivalent impedance, $\boldsymbol{Z}_e$, is given by

$$\boldsymbol{Z}_{e(\text{pu})} = n_{2(\text{pu})}^2 \left(\boldsymbol{Z}_{1o(\text{pu})} + \boldsymbol{Z}_{2o(\text{pu})} \right) = \left(\frac{n_2}{n_{2o}} \right)^2 \left(\boldsymbol{Z}_{1o(\text{pu})} + \boldsymbol{Z}_{2o(\text{pu})} \right) \tag{4.6}$$

Applying (4.5) and (4.6) to the parameters of Figure 4.2 the standard form of the equivalent circuit in per unit values for a two-winding transformer can be drawn as in Figure 4.3. This form is valid to be used for representing transformers equipped with tap changers (off-load or on-load tap changers). It needs only to calculate the corresponding values of ONR and $\boldsymbol{Z}_e$. On the other hand, this is not adequate for stability and load flow studies, and hence, it is converted to an equivalent π circuit. Assuming that the transformer is connected between bus, p, and bus, q, in the power network (Figure 4.4(a)), the π-equivalent circuit in general form is as shown in Figure 4.4(b). The deduction of parameters in π circuit form is as below [1].

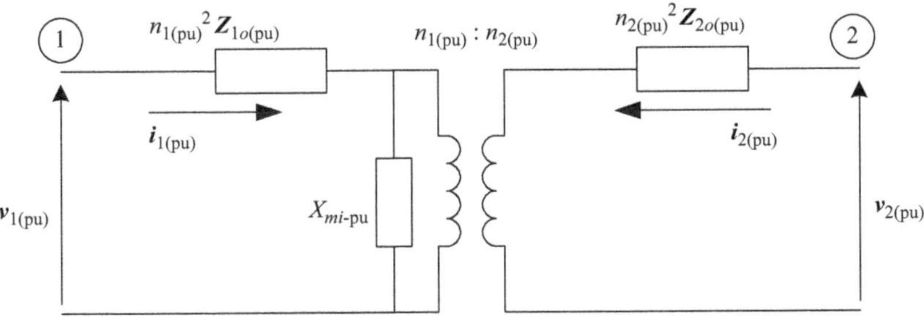

Figure 4.2 Equivalent circuit in per unit

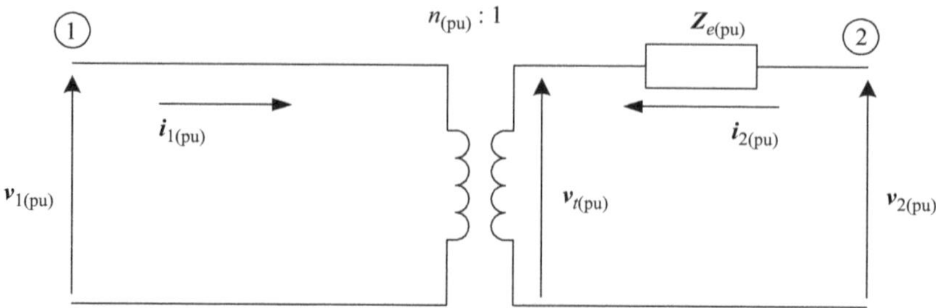

Figure 4.3 Standard per unit equivalent circuit for a two-winding ideal transformer

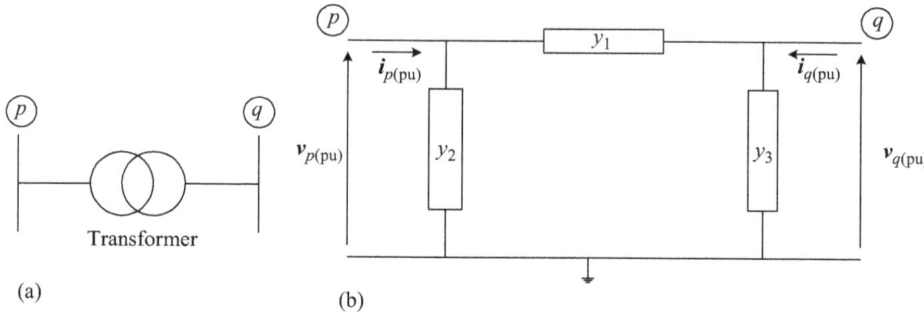

Figure 4.4 (a) Transformer connected in power network and (b) general form of π circuit

From Figure 4.3, replacing the subscripts 1 and 2 by p and q, respectively, and assuming $Y_{e(pu)} = 1/\boldsymbol{Z}_{e(pu)}$, the current at bus p, $\boldsymbol{i}_{p(pu)}$, is given by

$$\boldsymbol{i}_{p(pu)} = \left(\boldsymbol{v}_{t(pu)} - \boldsymbol{v}_{q(pu)}\right)\frac{Y_e}{n_{(pu)}} = \left(\frac{\boldsymbol{v}_{p(pu)}}{n_{(pu)}} - \boldsymbol{v}_{q(pu)}\right)\frac{Y_e}{n_{(pu)}}$$

$$= \left(\boldsymbol{v}_{p(pu)} - n_{(pu)}\boldsymbol{v}_{q(pu)}\right)\frac{Y_{e(pu)}}{n^2_{(pu)}} \tag{4.7}$$

The current at bus q can be obtained by the same procedure as

$$\boldsymbol{i}_{q(pu)} = \left(n_{(pu)}\boldsymbol{v}_{q(pu)} - \boldsymbol{v}_{p(pu)}\right)\frac{Y_{e(pu)}}{n_{(pu)}} \tag{4.8}$$

Meanwhile, the currents at buses p and q for the π circuit (Figure 4.3(b)) are calculated by

$$\boldsymbol{i}_{p(pu)} = y_1\left(\boldsymbol{v}_{p(pu)} - \boldsymbol{v}_{q(pu)}\right) + y_2\boldsymbol{v}_{p(pu)} \tag{4.9}$$

$$\boldsymbol{i}_{q(pu)} = y_1\left(\boldsymbol{v}_{q(pu)} - \boldsymbol{v}_{p(pu)}\right) + y_3\boldsymbol{v}_{q(pu)} \tag{4.10}$$

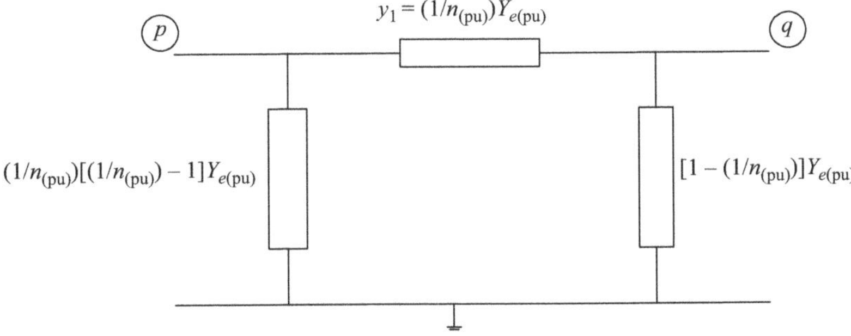

Figure 4.5 Transformer π-equivalent circuit

Equating the admittance terms in (4.7) and (4.9) and also in (4.8) and (4.10), the parameters of the π circuit are found to be

$$\left.\begin{aligned} y_1 &= \frac{1}{n_{(pu)}} Y_{e(pu)}, \quad y_2 = \frac{1}{n_{(pu)}}\left(\frac{1}{n_{(pu)}} - 1\right) Y_{e(pu)}, \\ y_3 &= \left(1 - \frac{1}{n_{(pu)}}\right) Y_{e(pu)} \end{aligned}\right\} \tag{4.11}$$

The π-equivalent circuit is shown in Figure 4.5 where its parameters are given in terms of ONR and the leakage impedance of the transformer.

It is to be noted that the standard equivalent circuit (Figure 4.3) represents the single-phase equivalent of a three-phase transformer. To be taken into consideration is the value of the nominal turns ratio (n_{1o}/n_{2o}) equal to the ratio of line-to-line base voltages on the primary and secondary sides regardless of winding connections, Y–Y or Δ–Δ. For Y–Δ connection, a factor of $\sqrt{3}$ should be accounted for but the phase shift of 30° provided by this connection can be neglected as it has no effect on the stability studies.

Example 4.1 Find the model parameters of a three-phase, 60-Hz, two-winding transformer with the data given below:

Transformer rating = 500 MVA
Nominal voltage of primary side and secondary side = 400 kV and 10.5 kV, respectively
Winding connection: Y/Y
Off-load tap changer on primary side: 4 steps, 2.5% kV each
On-load tap changer on secondary side: ±10% kV in 8 steps
$R_{10(pu)} + R_{20(pu)} = 0.003$ pu on rating/phase, $X_{10(pu)} + X_{20(pu)} = 0.12$ pu on rating/phase

Solution:

As initial operating condition, it is assumed that the secondary winding is at its nominal position and the primary winding is set one step above its nominal position (410 kV). The parameters of the equivalent circuit (Figure 4.3) calculated in pu of transformer rated values with the ONR on the secondary side are

$$\text{The initial ONR is } n_{(\text{pu})} = \frac{400}{410}\frac{10.5}{10.5} = 0.976 \text{ by (4.5)}$$

The pu equivalent impedance referred to secondary side $Z_{e(\text{pu})} = (1/0.976)^2 \times (0.003 + j0.12) = 0.003152 + j0.1261$

The step value of tap changer on the primary side is 10 kV and on the secondary side is 0.13125 kV.

$$\text{Maximum pu turns ratio, } n_{\max(\text{pu})} = \frac{400}{410}\frac{10.63125}{10.5} = 0.9878$$

$$\text{Minimum pu turns ratio, } n_{\min(\text{pu})} = \frac{400}{410}\frac{10.36875}{10.5} = 0.9634$$

$$\text{pu turns ratio step, } \Delta n_{(\text{pu})} = \frac{1.05}{8 \times 10.5}\frac{400}{410} = 0.01219$$

Per unit parameters can be recalculated according to the system voltage and MVA base values. For instance, assuming primary system voltage base = 410 kV, secondary system base voltage = 10.5 kV and system MVA base = 100 MVA; the corresponding pu parameters are

$$\text{the initial ONR, } n_{(\text{pu})} = 0.976\frac{410}{400}\frac{10.5}{10.5} = 0.9994$$

$$\text{pu equivalent impedance, } Z_{e(\text{pu})} = (0.003152 + j0.1261)\left(\frac{400}{410}\right)^2\frac{100}{500}$$

$$= 0.0006 + j0.024$$

$$\text{minimum pu turns ratio, } n_{\min(\text{pu})} = 0.9878\frac{410}{400}\frac{10.5}{10.5} = 1.0125$$

$$\text{minimum pu turns ratio, } n_{\min(\text{pu})} = 0.9634\frac{410}{400}\frac{10.5}{10.5} = 0.9875$$

$$\text{pu turns ratio step, } \Delta n_{(\text{pu})} = 0.01219\frac{410}{400}\frac{10.5}{10.5} = 0.01249$$

Referring to Figure 4.5 and (4.11) the parameters of transformer π-equivalent circuit representing the initial tap position are

$$y_1 = \frac{1}{n_{(\mathrm{pu})}} Y_{e(\mathrm{pu})} = \frac{1}{0.9994} \frac{1}{0.0006 + j0.024} = 1.04157 - j41.6627$$

$$y_2 = \frac{1}{n_{(\mathrm{pu})}} \left(\frac{1}{n_{(\mathrm{pu})}} - 1 \right) Y_{e(\mathrm{pu})} = \frac{1}{0.9994} \left(\frac{1}{0.9994} - 1 \right) \frac{1}{0.0006 + j0.024}$$

$$= 0.00062 - j0.00249$$

$$y_3 = \left(1 - \frac{1}{n_{(\mathrm{pu})}} \right) Y_{e(\mathrm{pu})} = \left(1 - \frac{1}{0.9994} \right) \frac{1}{0.0006 + j0.024}$$

$$= -0.00062 + j0.0249$$

Example 4.2 Find the model parameters for a three-phase, three-winding, 60-Hz transformer with the data below:

Rating = 500 MVA, high/low/tertiary nominal voltages = 400/240/10.5 kV
Winding connection (H/L/T): Y/Y/Δ

The positive-sequence impedances in pu on transformer MVA rating and nominal voltages tap position are given as

$$\mathbf{Z}_{ps} = 0.0016 + j0.1392, \quad \mathbf{Z}_{st} = 0 + j0.1633, \quad \mathbf{Z}_{pt} = 0 + j0.4741$$

The on-load tap changer at high-voltage side: 400 ± 40 kV in 20 steps.

Solution:

Modelling of three-winding transformer is based on measuring the impedances defined below from short-circuit tests and then obtaining the equivalent impedance of each of the three windings (primary, secondary and tertiary) in pu on the same MVA base. It should be taken into account that the primary, secondary and tertiary windings may or may not have the same MVA rating. Under balanced conditions and neglecting the effect of magnetising reactance, the single-phase equivalent circuit of three-winding transformer is represented by three impedances connected in a Y Figure 4.6. It is to be noted that the common star point is fictitious, i.e. it is not related to the system neutral. It is most convenient if the impedances are expressed in pu. The pu values must be calculated on the same MVA base even in case of having different ratings for the three windings. The same procedure of using ONR for two-winding transformers is applied to three-winding transformers as well to consider the difference between the actual turns ratio and base voltages.

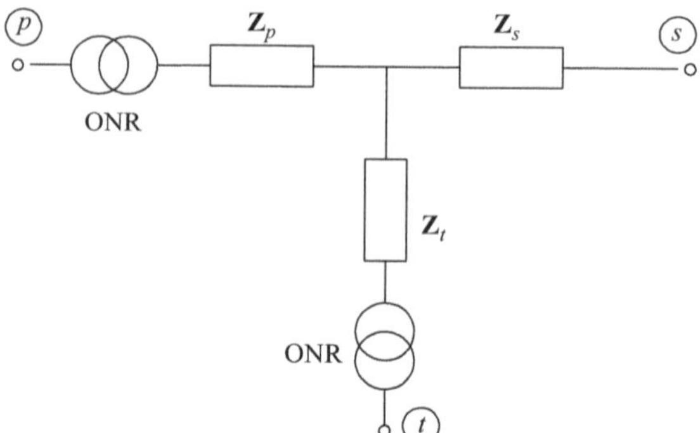

Figure 4.6 Three-winding transformer representation

If short-circuit tests are carried out on a three-winding transformer, the following impedances, composed of resistance and leakage reactance, may be measured:

$\mathbf{Z}_{ps}$ = impedance measured in the primary circuit with the secondary short-circuited and the tertiary open.

$\mathbf{Z}_{pt}$ = impedance measured in the primary circuit with the tertiary short-circuited and the secondary open.

$\mathbf{Z}_{st}$ = impedance measured in the secondary circuit with the tertiary short-circuited and the primary open.

If the above ohmic impedances are referred to the same voltage base and rating, then their values in terms of the equivalent impedances of the three separate windings $\mathbf{Z}_p$, $\mathbf{Z}_s$, $\mathbf{Z}_t$ are given by

$$\left.\begin{aligned}\mathbf{Z}_{ps} &= \mathbf{Z}_p + \mathbf{Z}_s\\ \mathbf{Z}_{pt} &= \mathbf{Z}_p + \mathbf{Z}_t\\ \mathbf{Z}_{st} &= \mathbf{Z}_s + \mathbf{Z}_t\end{aligned}\right\} \tag{4.12}$$

Thus,

$$\left.\begin{aligned}\mathbf{Z}_p &= \frac{1}{2}(\mathbf{Z}_{ps} + \mathbf{Z}_{pt} - \mathbf{Z}_{st})\\ \mathbf{Z}_s &= \frac{1}{2}(\mathbf{Z}_{ps} + \mathbf{Z}_{st} - \mathbf{Z}_{pt})\\ \mathbf{Z}_t &= \frac{1}{2}(\mathbf{Z}_{pt} + \mathbf{Z}_{st} - \mathbf{Z}_{ps})\end{aligned}\right\} \tag{4.13}$$

By substituting the data given above, it is found that

$$\mathbf{Z}_p = 0.0008 + j0.225, \quad \mathbf{Z}_s = 0.0008 - j0.0858, \quad \mathbf{Z}_t = -0.0008 + j0.2491$$

The equivalent circuit parameters, in Δ form, in pu on transformer MVA rating and nominal voltages are (the subscripts have been written in upper case letters to denote Δ form)

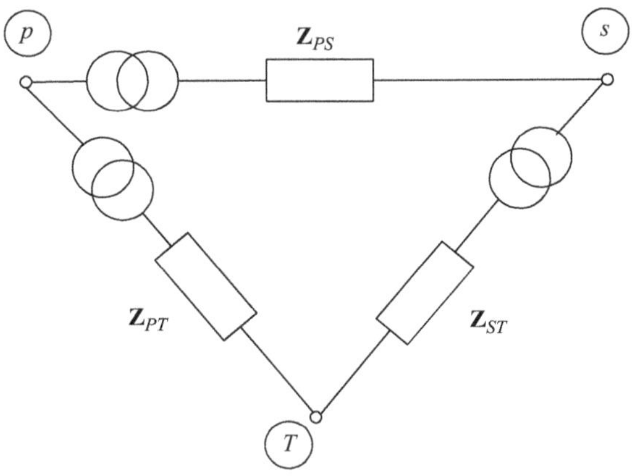

$$\mathbf{Z}_{PS} = \frac{\mathbf{Z}_p\mathbf{Z}_s + \mathbf{Z}_p\mathbf{Z}_t + \mathbf{Z}_s\mathbf{Z}_t}{\mathbf{Z}_t} = \frac{-0.01537 + j0.00035}{-0.0008 + j0.2491} = 0.0016 + j0.061$$

$$\mathbf{Z}_{ST} = \frac{\mathbf{Z}_p\mathbf{Z}_s + \mathbf{Z}_p\mathbf{Z}_t + \mathbf{Z}_s\mathbf{Z}_t}{\mathbf{Z}_p} = \frac{-0.01537 + j0.00035}{0.0008 + j0.225} = 0.0016 + j0.0684$$

$$\mathbf{Z}_{PT} = \frac{\mathbf{Z}_p\mathbf{Z}_s + \mathbf{Z}_p\mathbf{Z}_t + \mathbf{Z}_s\mathbf{Z}_t}{\mathbf{Z}_s} = \frac{-0.01537 + j0.00035}{0.0008 - j0.0858} = -0.0057 - j0.1793$$

The corresponding parameters of equivalent Δ circuit, superscripted by ′, in pu on system MVA base of 100 MVA and voltage bases, *P*/*S*/*T*, of 400/220/12.47 kV can be calculated as below:

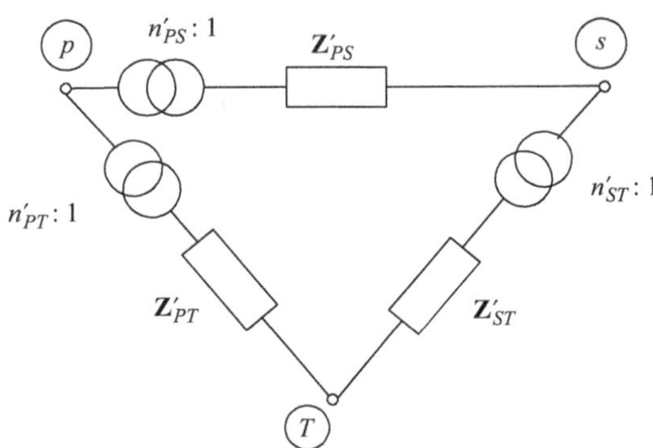

$$\mathbf{Z}'_{PS} = \mathbf{Z}_{PS}\left(\frac{240}{220}\right)^2 \frac{100}{500} = 0.0004 + j0.0145$$

$$\mathbf{Z}'_{ST} = \mathbf{Z}_{ST}\left(\frac{10.5}{12.47}\right)^2 \frac{100}{500} = 0.0002 + j0.0097$$

$$\mathbf{Z}'_{PT} = \mathbf{Z}_{PT}\left(\frac{10.5}{12.47}\right)^2 \frac{100}{500} = -0.0008 - j0.0254$$

$$n'_{PS} = \frac{400}{400}\frac{220}{240} = 0.9167$$

$$n'_{ST} = \frac{240}{220}\frac{12.47}{10.5} = 1.2956$$

$$n'_{PT} = \frac{400}{400}\frac{12.47}{10.5} = 1.1876$$

Data of on-load tap changer:

$$n'_{PS\max} = \frac{440}{400}\frac{400}{400}\frac{220}{240} = 1.0083$$

$$n'_{PS\min} = \frac{360}{400}\frac{400}{400}\frac{220}{240} = 0.825$$

$$\Delta n'_{PS} = \frac{1.0083 - 0.825}{20} = 0.0092$$

$$n'_{PT\max} = \frac{440}{400}\frac{400}{400}\frac{12.47}{10.5} = 1.3064$$

$$n'_{PT\min} = \frac{360}{400}\frac{400}{400}\frac{12.47}{10.5} = 1.0688$$

$$\Delta n'_{PT} = \frac{1.3064 - 1.0688}{20} = 0.01188$$

Each branch in equivalent Δ circuit can be represented by an equivalent π circuit shown in Figure 4.5. Referring to (4.11) the parameters of π circuit are

PS branch:

$$y_1 = \frac{1}{n'_{PS}} y'_{PS} = \frac{1}{0.9167}\frac{1}{0.0004 + j0.0145} = 2.0762 - j75.2619$$

$$y_2 = \frac{1}{n'_{PS}}\left(\frac{1}{n'_{PS}} - 1\right) y'_{PS} = 0.1868 - j6.7736$$

$$y_3 = \left(1 - \frac{1}{n'_{PS}}\right) y'_{PS} = \left(1 - \frac{1}{0.9167}\right)\frac{1}{0.0004 + j0.0145} = -0.1714 + j6.2143$$

ST branch:

$$y_1 = \frac{1}{n'_{ST}} y'_{ST} = \frac{1}{1.2956} \frac{1}{0.0002 + j0.0097} = 1.6421 - j79.6432$$

$$y_2 = \frac{1}{n'_{ST}} \left(\frac{1}{n'_{ST}} - 1 \right) y'_{ST} = -0.3747 + j18.1746$$

$$y_3 = \left(1 - \frac{1}{n'_{ST}} \right) y'_{ST} = 0.4855 - j23.5483$$

PT branch:

$$y_1 = \frac{1}{n'_{PT}} y'_{PT} = \frac{1}{1.1876} \frac{1}{-0.0008 - j0.0254} = -1.0443 + j33.1578$$

$$y_2 = \frac{1}{n'_{PT}} \left(\frac{1}{n'_{PT}} - 1 \right) y'_{PT} = 0.1649 - j5.2389$$

$$y_3 = \left(1 - \frac{1}{n'_{PT}} \right) y'_{PT} = -0.1959 + j6.222$$

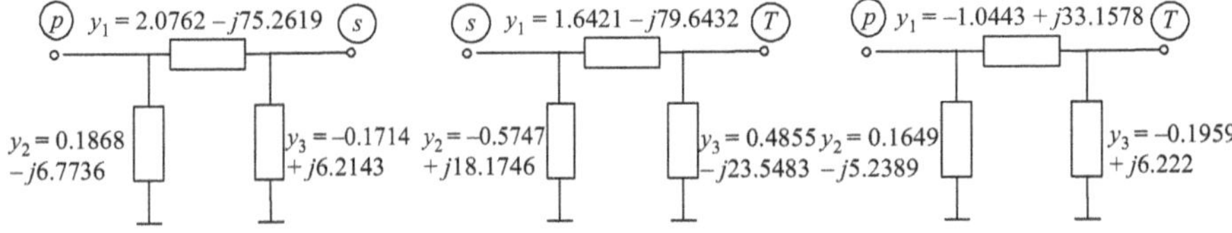

4.1.2 Modelling of phase-shifting transformers

The phase-shifting transformer can be represented by a series admittance with an ideal transformer connected between bus p and bus q and having a complex turns ratio, $\boldsymbol{n} = n \angle\alpha$ (Figure 4.7), where α is the phase shift from bus p to bus q. In case of load flow and transient stability studies, it is reasonable to consider equal phase angle step size at different tap positions.

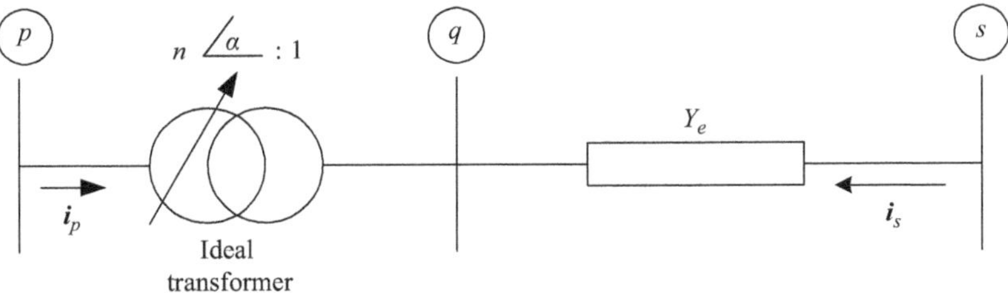

Figure 4.7 Representation of phase-shifting transformer

The turns ratio $\boldsymbol{n} = n \angle \alpha$ as a complex quantity comprises real and imaginary components and, therefore, it can be mathematically written as

$$n \angle \alpha = \frac{\boldsymbol{v}_p}{\boldsymbol{v}_q} = a_s + jb_s = n(\cos\alpha + j\sin\alpha) \tag{4.14}$$

Considering the transformer as ideal (no power loss) and considering α as a positive shift-angle, i.e. $\boldsymbol{v}_p$ leads $\boldsymbol{v}_q$, the rated power on the primary side is related to that on the secondary side by the relation:

$$\boldsymbol{v}_p \boldsymbol{i}_p^* = -\boldsymbol{v}_q \boldsymbol{i}_s^* \tag{4.15}$$

From (4.14) and (4.15) the current on the primary side, at bus p, is given by

$$\begin{aligned} \boldsymbol{i}_p &= -\frac{1}{a_s - jb_s}\boldsymbol{i}_s = \frac{\boldsymbol{Y}_e}{a_s - jb_s}(\boldsymbol{v}_q - \boldsymbol{v}_s) = \frac{\boldsymbol{Y}_e}{a_s - jb_s}\left(\frac{1}{a_s + jb_s}\boldsymbol{v}_p - \boldsymbol{v}_s\right) \\ &= \frac{\boldsymbol{Y}_e}{a_s^2 + b_s^2}\left[\boldsymbol{v}_p - (a_s + jb_s)\boldsymbol{v}_s\right] \end{aligned} \tag{4.16}$$

Similarly,

$$\boldsymbol{i}_s = \frac{\boldsymbol{Y}_e}{a_s + jb_s}\left[(a_s + jb_s)\boldsymbol{v}_s - \boldsymbol{v}_p\right] \tag{4.17}$$

The relations of the currents $\boldsymbol{i}_p$ and $\boldsymbol{i}_s$ in terms of the voltages $\boldsymbol{v}_p$ and $\boldsymbol{v}_s$ given in (4.16) and (4.17) can be rewritten in matrix form as

$$\begin{bmatrix} \boldsymbol{i}_p \\ \boldsymbol{i}_s \end{bmatrix} = \begin{bmatrix} \dfrac{\boldsymbol{Y}_e}{a_s^2 + b_s^2} & \dfrac{-\boldsymbol{Y}_e}{a_s - jb_s} \\ \dfrac{-\boldsymbol{Y}_e}{a_s + jb_s} & \boldsymbol{Y}_e \end{bmatrix} \begin{bmatrix} \boldsymbol{v}_p \\ \boldsymbol{v}_s \end{bmatrix} \tag{4.18}$$

It is found from (4.18) that the transfer admittance from bus p to bus s is not the same as that from bus s to bus p as the admittance matrix is not symmetrical. Consequently, the model cannot be expressed by π-equivalent circuit. It is to be noted that the model of the ideal transformer given by the π circuit in Figure 4.5 can be verified by (4.18) when substituting a by n and b by zero.

Example 4.3 Data for a two-winding phase-shifting transformer are given below. Neglecting the resistance per phase, find the elements of admittance matrix in (4.18) when $\alpha = 0°$ and 10° 'at 8th step'.

MVA rating = 42 MVA
Primary/secondary base voltage 110/110 kV
Leakage reactance per phase, $X_e = 0.1633$ pu
Phase-shift range and steps = 30°, 24 steps
System voltage base = 110/115 kV
System MVA base = 100 MVA

Solution:

X_e in pu on system voltage and MVA base: $X_e = 0.1633\left(\frac{110}{115}\right)^2 \frac{100}{42} = 0.3557$ pu

$$\text{ONR: } n = \frac{110}{110}\frac{115}{110} = 1.04545$$

The phase shift angle is in a range between $\alpha_{max} = 30°$ and $\alpha_{min} = -30°$

For $\alpha = 0°$

$$\boldsymbol{Y}_e = \frac{1}{j0.3557} = -j2.81136 \text{ pu}$$

The turns ratio $= a_s + jb_s = n(\cos\alpha + j\sin\alpha) = 1.04545 + j0$

$$\frac{\boldsymbol{Y}_e}{a_s^2 + b_s^2} = -j2.5722 \frac{-\boldsymbol{Y}_e}{a_s - jb_s} = j2.6891$$

$$\frac{-\boldsymbol{Y}_e}{a_s + jb_s} = j2.6891$$

Therefore, the admittance matrix is

$$\boldsymbol{Y}_s = j\begin{bmatrix} -2.5722 & 2.6891 \\ 2.6891 & -2.81136 \end{bmatrix}$$

For $\alpha = 10°$

As the impedance changes with the phase-shift angle, the manufacturer provides an impedance multiplier '*m*' at each desired angle. Thus, in general, $\boldsymbol{Y}_e = m/Z_e$

$$\boldsymbol{Y}_e = -jm2.81136, \text{ the turns ratio} = a_s + jb_s = n(\cos 10° + j\sin 10°)$$
$$= 1.0295 + j0.1815, \quad \text{and}$$

$$\boldsymbol{Y}_s = m\begin{bmatrix} -j2.5728 & -0.5103 + j2.6487 \\ 0.5103 + j2.6487 & -j2.81136 \end{bmatrix}$$

4.2 Transmission lines

Transmission lines are characterised by distributed parameters: (i) series impedance Z comprising conductor resistance R and inductance L, (ii) shunt conductance G due to leakage currents between phases and ground and (iii) shunt capacitance C due to the electric field between conductors.

The transmission line can be modelled by a π-equivalent circuit with lumped parameters or a number of cascaded π circuits. This depends on the nature of study and the length of the line. However, π-equivalent circuit is an adequate model for studying power system stability. The model is designed based on assumptions such as (i) the line is transposed to consider that the three phases of the line are symmetric, i.e. equal self-impedances of all phases and equal mutual impedances between any two phases as well, and (ii) the line parameters are constant. In addition, the relationship between voltage and current in terms of line parameters should be defined as below [2].

4.2.1 Voltage and current relationship of a line

A transmission line of length l between receiving and sending ends can be represented as shown in Figure 4.8, noting that the voltages and currents are phasors representing time-varying quantities.

Considering a differential section of length ds at a distance s from the receiving end, the differential voltage across the incremental length ds is given by

$$\mathrm{d}\boldsymbol{v} = \boldsymbol{i}(z\mathrm{d}s)$$

Hence,

$$\frac{\mathrm{d}\boldsymbol{v}}{\mathrm{d}s} = z\boldsymbol{i} \tag{4.19}$$

The differential current flowing into the shunt admittance is

$$\mathrm{d}\boldsymbol{i} = \boldsymbol{v}(y\mathrm{d}s)$$

Hence,

$$\frac{\mathrm{d}\boldsymbol{i}}{\mathrm{d}s} = y\boldsymbol{v} \tag{4.20}$$

Differentiate (4.19) and (4.20) with respect to s to get

$$\frac{\mathrm{d}^2\boldsymbol{v}}{\mathrm{d}s^2} = z\frac{\mathrm{d}\boldsymbol{i}}{\mathrm{d}s} \quad \text{and} \quad \frac{\mathrm{d}^2\boldsymbol{i}}{\mathrm{d}s^2} = y\frac{\mathrm{d}\boldsymbol{v}}{\mathrm{d}s} \tag{4.21}$$

To solve these two second-order differential equations, the initial conditions need to be considered.

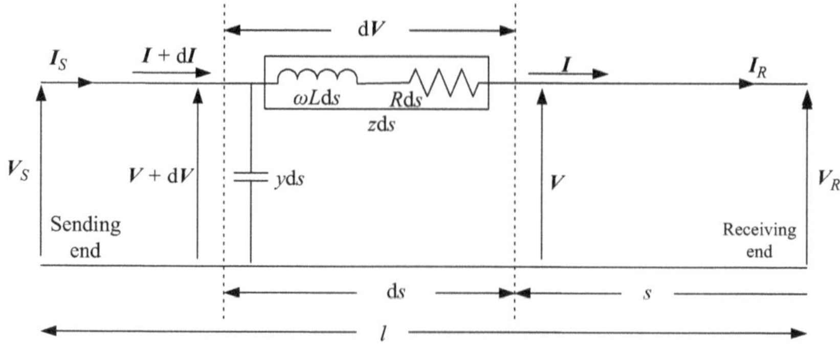

Figure 4.8 Distributed parameter line (z and y are the line impedance and admittance per unit length, respectively)

At $s = 0$: the voltage $\boldsymbol{v} = \boldsymbol{V}_R \angle\Phi_1$ and the current $\boldsymbol{i} = \boldsymbol{I}_R \angle\Phi_2$ where the voltage and current at the receiving end are assumed to be known. The general solution giving the voltage and current at a point distance s from the receiving end, as phasors ($\boldsymbol{V}_S$ and $\boldsymbol{I}_S$), is given by

$$\left.\begin{aligned} \boldsymbol{V}_S &= \frac{\boldsymbol{V}_R + Z_C \boldsymbol{I}_R}{2} e^{\gamma s} + \frac{\boldsymbol{V}_R - Z_C \boldsymbol{I}_R}{2} e^{-\gamma s} \\ \boldsymbol{I}_S &= \frac{(\boldsymbol{V}/Z_C) + \boldsymbol{I}_R}{2} e^{\gamma s} - \frac{(\boldsymbol{V}_R/Z_C) - \boldsymbol{I}_R}{2} e^{-\gamma s} \end{aligned}\right\} \tag{4.22}$$

where

$Z_C \triangleq$ the characteristic impedance $= \sqrt{\frac{z}{y}}$
$\gamma \triangleq$ the propagation constant $= \sqrt{zy} = \alpha + j\beta$
$\alpha \triangleq$ the attenuation constant
$\beta \triangleq$ the phase constant

The exponential terms may be written in the expanded form as

$$\mathrm{e}^{\gamma s} = \mathrm{e}^{(\alpha + j\beta)s} = \mathrm{e}^{\alpha s}(\cos\beta s + j\sin\beta s) \quad \text{and}$$

$$\mathrm{e}^{-\gamma s} = \mathrm{e}^{-(\alpha + j\beta)s} = \mathrm{e}^{-\alpha s}(\cos\beta s - j\sin\beta s)$$

In (4.22) both voltage and current consist of two terms: the first term is defined as the incident component and the second is called the reflected component.

Equation (4.22) can be re-arranged to be written in the form

$$\left.\begin{aligned} \boldsymbol{V}_S = \boldsymbol{V}_R \frac{\mathrm{e}^{\gamma s} + \mathrm{e}^{-\gamma s}}{2} + Z_C \boldsymbol{I}_R \frac{\mathrm{e}^{\gamma s} - \mathrm{e}^{-\gamma s}}{2} &= \boldsymbol{V}_R \cosh(\gamma s) + Z_C \boldsymbol{I}_R \sinh(\gamma s) \\ \text{Similarly, } \boldsymbol{I}_S &= \frac{1}{Z_C} \boldsymbol{V}_R \sinh(\gamma s) + \boldsymbol{I}_R \cosh(\gamma s) \end{aligned}\right\} \tag{4.23}$$

4.2.2 Modelling of transmission lines

A transmission line from the sending to the receiving ends can be represented by π-equivalent circuit as shown in Figure 4.9. It includes series equivalent impedance Z_e and two equivalent shunt admittances, $Y_e/2$ each. The voltage at the sending end, $\boldsymbol{V}_S$, in terms of the receiving end voltage, $\boldsymbol{V}_R$, is calculated by

$$\boldsymbol{V}_S = Z_e\left(\boldsymbol{I}_R + \frac{Y_e}{2}\boldsymbol{V}_R\right) + \boldsymbol{V}_R = \left(\frac{Z_e Y_e}{2} + 1\right)\boldsymbol{V}_R + Z_e \boldsymbol{I}_R \tag{4.24}$$

Voltage at the sending end can be obtained from (4.23) by substituting $s = l$:

$$\boldsymbol{V}_S = \boldsymbol{V}_R \cosh(\gamma l) + Z_C \boldsymbol{I}_R \sinh(\gamma l) \tag{4.25}$$

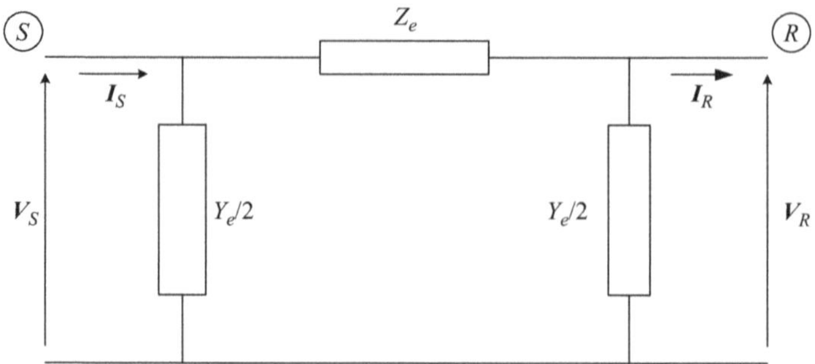

Figure 4.9 π-equivalent circuit for transmission line representation

Equating (4.24) and (4.25) gives

$$Z_e = Z_C \sinh(\gamma l) \quad \text{and} \quad \left(\frac{Z_e Y_e}{2} + 1\right) = \cosh(\gamma l)$$

Therefore,

$$\left.\begin{aligned} Z_e &= Z_C \sinh(\gamma l) \\ \frac{Y_e}{2} &= \frac{1}{Z_C} \frac{\cosh(\gamma l) - 1}{\sinh(\gamma l)} = \frac{1}{Z_C} \tanh\left(\frac{\gamma l}{2}\right) \end{aligned}\right\} \tag{4.26}$$

It is to be noted that if $\gamma l \ll 1$ then

$$Z_e = Z_C \sinh(\gamma l) \approx Z_C(l) \approx zl = Z \quad \text{and}$$

$$\frac{Y_e}{2} = \frac{1}{Z_C} \tanh\left(\frac{\gamma l}{2}\right) \approx \frac{1}{Z_C} \frac{\gamma l}{2} \approx \frac{yl}{2} = \frac{Y}{2}$$

In this case, the parameters of π-equivalent circuit are the total impedance and total admittance of the line. The equivalent circuit is called the 'nominal π-equivalent circuit', which is accepted for medium-length overhead lines (usually used for high and extra high voltage networks) of length in the range of 80–200 km.

In general, short overhead lines, $l < 80$ km, may be represented by their series impedance by ignoring the shunt admittance. Medium-length overhead lines, $80 < l < 200$ km, may be represented by a nominal π-equivalent circuit. Long overhead lines, $l > 200$ km, can be divided into a number of cascaded medium length sections, each section represented by a nominal π-equivalent circuit to partially take into consideration the effect of distributed nature of the line parameters.

Example 4.4 Parameters of a typical 230-kV overhead transmission line of 100 km length are $x = 0.488$ Ω/km, $r = 0.05$ Ω/km, $y = 3.371$ μmho/km. Calculate the characteristic impedance, Z_C, and propagation constant 'γ'. Find the π-equivalent circuit in pu of Z_C.

Solution:

$$Z_C = \sqrt{\frac{z}{y}} = \sqrt{\frac{r + jx}{y}} = \sqrt{\frac{0.05 + j0.488}{j3.371 \times 10^{-6}}} \approx 380\,\Omega$$

$$\gamma = \sqrt{zy} = \alpha + j\beta = j\sqrt{xy}\left(1 - j\frac{r}{2x}\right) = j1.2826 \times 10^{-3}\left(1 - j\frac{0.05}{0.976}\right)$$

$$= 0.0000657 + j00128$$

Hence,

$\alpha = 0.0000657$ nepers/km and $\beta = 00128$ rad/km
Neglecting the resistance, $X_L = 0.488 \times 100 = 48.8\,\Omega$
$Y_L = 3.371 \times 10^{-6} \times 100 = 0.000371$ mho
In pu of Z_C, the values of X_L and Y_L are
$X_L = 48.8/380 = 0.128$ pu
$Y_L = 0.000371 \times 380 = 0.141$ pu

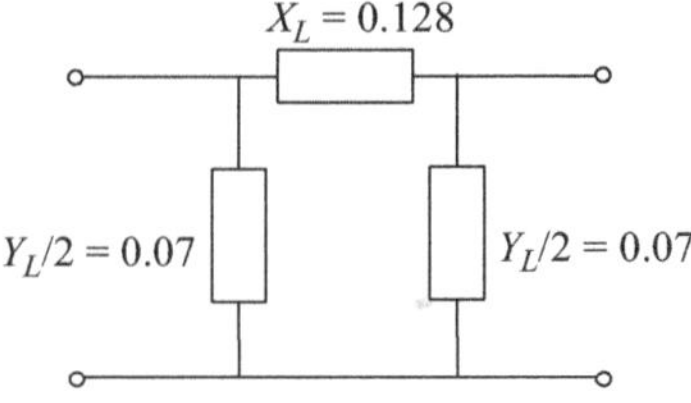

4.3 Loads

In general, to perform power system analysis, models must be developed for all pertinent system components. Inadequate modelling causing under/over-building of the system or degrading reliability may prove to be costly. For stability studies of the power system a balance between generated power and demand power by loads must be maintained to keep the system continuously in stable operation. Therefore, load characteristics are of crucial importance to be employed in system analysis as they have a significant effect on system performance and highly impact the stability results. To achieve that, load modelling must be determined in such a way that the model is relevant to the nature of study and helps obtain useful and, to a large extent, accurate results. Accurate modelling of loads is a difficult task as, for instance, the power system includes a huge number of diverse load components in different locations with different characteristics and their composition changes from time to time. In addition, lack of data regarding the loads all over the system and lack of a tool to develop models on a large-scale basis makes load modelling a formidable task.

Two main approaches to load model development have been considered by utilities: measurement-based and component-based [3]. The first approach involves monitors used at various load points to determine the load sensitivity (active and reactive power) to voltage and frequency variations to be used directly or to identify parameters for a more detailed load model. This approach has the advantage of producing load model parameters directly in the form needed for power flow and transient stability programmes through the direct monitoring of the true load. On the other hand, its cost is high because of acquiring and installing the measurement equipment and the need to monitor all system loads. As well, the measurements must be repeated as the load changes.

The second approach, component-based approach, implies building the load model from information on its constituent parts as shown in Figure 4.10. It requires three sets of data: (i) load class mix data that describe the percentage contribution of each of several load classes to the total active power load at the bus, (ii) load composition data that describe the percentage contribution of each of several load components to the active power consumption of a particular load class and (iii) load characteristics data that describe the electrical characteristics of each of the load components. This approach has the advantage of not requiring system measurements and consequently being more readily put into use, and the possibility of using standard model for each component as well. It is to be noted that the load class mix data need to be prepared for each bus or area, and updated for changes in the system load [4].

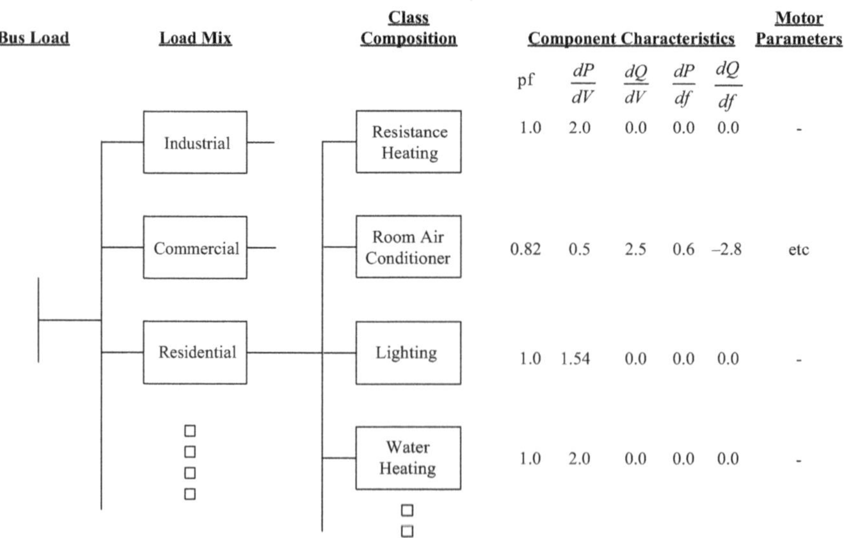

Figure 4.10 Terminology of component-based load modelling

Several efforts have been made to develop methods for constructing improved load models [5–9]. The basic load models can be divided into two classes: static and dynamic.

4.3.1 Static load models

Static load models express the active and reactive powers at any instant of time as functions of the bus voltage magnitude and frequency at the same instant. These models are used both for essentially static load components, e.g. resistive and lighting load, and as an approximation for dynamic load components, e.g. motor-driven loads. Polynomial and exponential representations are the two forms that can be used to perform static load models [10].

i. Polynomial representation

The power relationship to voltage magnitude is represented as a polynomial equation in the form below.

$$P = P_o\left[a_o + a_1\left(\frac{V}{V_o}\right) + a_2\left(\frac{V}{V_o}\right)^2\right] \tag{4.27}$$

$$Q = Q_o\left[b_o + b_1\left(\frac{V}{V_o}\right) + b_2\left(\frac{V}{V_o}\right)^2\right] \tag{4.28}$$

where V_o, P_o and Q_o are the initial values of voltage, power and reactive power, respectively (the initial system operating condition for study), when representing a bus load. If this model is used for representing a specific load device, V_o should be the rated voltage of the device and P_o and Q_o should be the power consumed at rated voltage. The model in this case is composed of sum of three terms; each term represents a model as (i) constant impedance, Z, model, where the load power varies directly with the square of the voltage magnitude. It may be also called a constant admittance model, (ii) constant current, I, model, where the load power varies directly with the voltage magnitude. It has been accepted that, in the absence of data, composite load can be approximated using a constant current load model and (iii) constant power, P, model, where the load power does not vary with changes in the voltage magnitude. It may be also called a constant MVA model. This type of load draws higher current under low-voltage conditions to maintain constant power. So, it has a problem of non-applicability for cases involving severe voltage drops. The coefficients a_o, a_1, a_2 and b_o, b_1, b_2 are the fractions of the constant power, constant current and constant impedance components in the active and reactive power loads, respectively. They have the relations as described below:

$$\left.\begin{aligned} a_o + a_1 + a_2 = 1 \\ b_o + b_1 + b_2 = 1 \end{aligned}\right\} \tag{4.29}$$

The composite load model represented by (4.27) and (4.28) is sometimes referred to as ZIP model and its parameters are the coefficients a_o, a_1, a_2 and b_o, b_1, b_2 and the power factor of the load.

ii. Exponential representation

The power relationship to voltage magnitude is represented as an exponential equation in the form below:

$$P = P_o\left(\frac{V}{V_o}\right)^{np} \quad \text{and} \quad Q = Q_o\left(\frac{V}{V_o}\right)^{nq} \tag{4.30}$$

The parameters of this model are the exponents *np* and *nq*. By setting these exponents to 0, 1 or 2, the load can be represented by constant power, constant current or constant impedance models, respectively. Other exponents can be used to represent the aggregate effect of different types of load components where exponents greater than 2 or less than 0 may be appropriate for some types of loads. Two or more terms with different exponents are sometimes included in the two relations of (4.30). For instance, when a bus in the power system is chosen as a load bus, to include both voltage dependence and the effect of frequency variations, the active power can be expressed as

$$P = P_o\left[C_1\left(\frac{V}{V_o}\right)^{np1}(1 + k_p\Delta f) + (1 - C_1)\left(\frac{V}{V_o}\right)^{np2}\right] \tag{4.31}$$

where

$C_1 \triangleq$ the frequency dependent fraction of active power load
$np_1 \triangleq$ the voltage exponent for frequency-dependent component of active power load
$np_2 \triangleq$ the voltage exponent for frequency-independent component of active power load
$\Delta f \triangleq$ the per unit frequency deviation from nominal
$k_p \triangleq$ frequency sensitivity coefficient for the active power load

To add the effect of load compensation, the reactive power is expressed as

$$Q = P_o\left[C_2\left(\frac{V}{V_o}\right)^{nq1}(1 + k_{q1}\Delta f) + \left(\frac{Q_o}{P_o} - C_2\right)\left(\frac{V}{V_o}\right)^{nq2}(1 + k_{q2}\Delta f)\right] \tag{4.32}$$

where

$C_2 \triangleq$ the reactive load coefficient-ratio of initial uncompensated reactive load to total initial active power load P_o
$n_{q1} \triangleq$ the voltage exponent for the uncompensated reactive load
$n_{q2} \triangleq$ the voltage exponent for the reactive compensation term
$k_{q1} \triangleq$ the frequency sensitivity coefficient for the uncompensated reactive power load
$k_{q2} \triangleq$ the frequency sensitivity coefficient for reactive compensation

The reactive power is normalised to P_o rather than Q_o to avoid difficulties when Q_o equals zero due to the cancellation of the load reactive consumption and reactive losses by shunt capacitance. The first term includes the reactive power consumption of all of the load components, and it is built up using the power factors of the individual load components. The second term, to a first approximation, represents the effect of reactive losses and compensation in the sub-transmission and distribution system between the bus and the various loads. The two terms include frequency sensitivity.

4.3.2 Dynamic load models

Static models explained above may be accepted for application to composite loads having fast response to voltage/frequency changes and reaching the steady state rapidly. Some cases necessitate considering the dynamics of load components such as discharge lamps, protective relays, thermostatic controlled loads, transformers with LTCs and motors. Motors represent the major portion of load components regardless of the class of the load (industrial, commercial or residential), and this section, thus, focuses on the dynamic modelling of motors, in particular, induction motors [11–15].

4.3.2.1 Induction motor model

The equivalent circuit of an induction machine in steady state can be in one of the two forms shown in Figure 4.11(a, b). The only difference between the two circuits is that the rotor power is represented by its two components, resistance loss and shaft power, in Figure 4.11(b). All quantities in the equivalent circuit are referred to the stator side. In motor operation the slip is positive and the directions of currents shown are positive. More details of equivalent circuit for double-cage induction machines with saturation effects and deep-bar induction machines as well can be found in [16, 17].

For stability studies, the DC component in the stator transient currents is neglected permitting representation of only fundamental frequency components. Neglecting stator transients and the rotor windings shortened, the per unit induction motor electrical equations of the simplest model can be written as below.

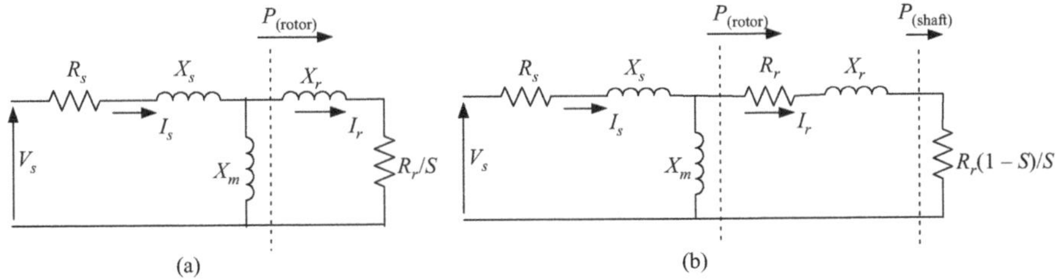

Figure 4.11 Induction machine equivalent circuit: (a) representation of total rotor power and (b) representation of rotor power components

Dynamics of the rotor inertia is described by

$$\frac{d\omega_r}{dt} = \frac{1}{2H}(T_e - T_m) \tag{4.33}$$

where ω_r is the per unit motor speed

T_m is the per unit mechanical torque and is a function of ω_r as given by the relation:

$$T_m = T_{mo}\left[A\omega_r^{\ 2} + B\omega_r + C\right] \tag{4.34}$$

T_e is the per unit electrical torque and is a function of the motor slip. It is computed from the steady-state equivalent circuit shown in Figure 4.11 as

$$T_e = \frac{I_r^2 R}{S} \tag{4.35}$$

where H is the motor inertia constant

Typical data of induction motors, coefficients A, B and C, and parameters of the equivalent circuit in different installations can be found in [18].

By including the rotor transients, the simplified equivalent circuit for stability studies is shown in Figure 4.12, where E' is a complex voltage source behind transient impedance, X_s', and is defined by

$$\frac{dE'}{dt} = -j2\pi f S E' - \frac{1}{T_o}\left[E' - j(X - X_s')I_t\right] \tag{4.36}$$

where f is the operating frequency

$$T_o = \frac{X_r + X_m}{2\pi f R_r}, \quad I_t = \frac{V - E_t'}{R_s + jX_s'} = i_q + ji_d$$
$$X = X_s + X_m \quad \text{and} \quad X_s' = X_s + \frac{X_m X_r}{X_m + X_r} \tag{4.37}$$

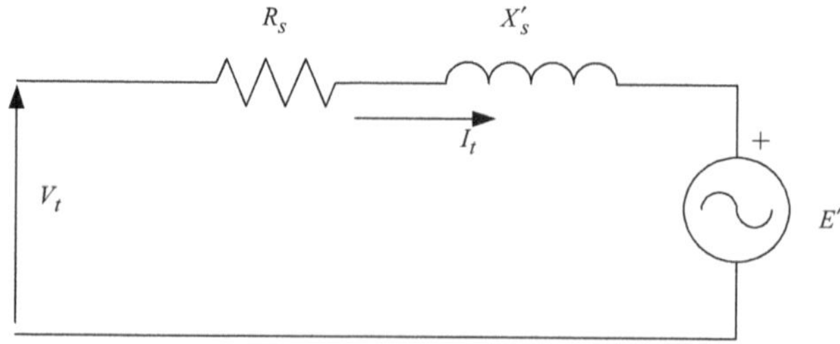

Figure 4.12 Transient-equivalent circuit of induction machine (V_t is the stator terminal voltage, E' is the voltage behind transient impedance)

Also, E' in (4.36) can be expressed by two real values E'_d and E'_q in the d–q frame of reference as

$$\left.\begin{aligned}\frac{\mathrm{d}E'_d}{\mathrm{d}t} &= -(\omega_s - \omega_r)E'_q + \frac{1}{T_o}(X - X'_s)i_q - \frac{1}{T_o}E'_d \\ \frac{\mathrm{d}E'_q}{\mathrm{d}t} &= (\omega_s - \omega_r)E'_d - \frac{1}{T_o}(X - X'_s)i_d - \frac{1}{T_o}E'_q\end{aligned}\right\} \tag{4.38}$$

where $\omega_s = 2\pi f$

The electrical torque can be calculated by using the relation: $T_e = E'_d i_d + E'_q i_q$

4.4 Remarks on load modelling for stability and power flow studies

- The common practice is to represent the composite load characteristics as seen from bulk power delivery points (Figure 4.13).
- To ensure accuracy, stability studies should employ good dynamic load models including the effect of motor rotor flux transients, discharge lighting discontinuities, effect of LTCs on load magnitude after a disturbance, saturation effects in transformers and motors and similar phenomena.
- Static load models may be adequate in such cases that give the same results as more detailed dynamic models. So, comparison of static load models and detailed dynamic models using typical data in both cases should be implemented to decide which model is considered in the study [19].
- Data gathering is essential to represent composite load characteristics. Two approaches can be applied to obtain the data: (i) by direct measurement of the voltage and frequency sensitivity of load P and Q at representative substations and feeders, (ii) by building up a composite load model from knowledge of the mix of load classes served by a substation, the composition of each class and typical characteristic of each load component. In general, both should be used as they are complementary and desirable to understand and predict load characteristics under varying conditions.

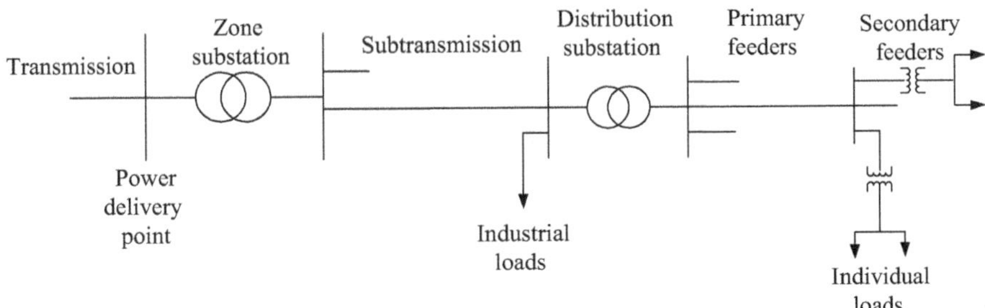

Figure 4.13 Illustration of a bulk power delivery point in a part of power system

References

1. Stagg G.W., El-Abiad A.H. *Computer Methods in Power System Analysis*. New York, USA: McGraw-Hill; 1968
2. Westinghouse Electric Corporation. *Electrical Transmission and Distribution Reference Book*. East Pittsburgh, PA, USA; 1964
3. Price W.W., Wirgua K.A., Murdoch A., Mitsche J.V., Vaahedi E., El-Kady M. 'Load modeling for power flow and transient stability computer studies'. *IEEE Transactions on Power Systems*. 1988;**3**(1):180–7
4. Shi J.H., Renmu H. (eds.). 'Measurement-based load modeling – model structure'. *IEEE Bologna Power Tech Conference Proceedings, 2003 IEEE Bologna*; Italy, vol. 2, June 2003
5. Pai M.A., Sauer P.W., Lesieutre B.C. 'Static and dynamic nonlinear loads and structural stability in power systems'. *Proceedings of the IEEE*. 1995; **83**(11):1562–72
6. Wang J.C., Ciang H.D., Chang C.L., Liu A.H. 'Development of a frequency-dependent composite load model using the measurement approach'. *IEEE Transactions on Power Systems*. 1994;**9**(3):1546–56
7. Song Y.H., Dang D.Y. (eds.). 'Load modeling in commercial power systems using neural networks'. *Industrial and Commercial Power Systems Technical Conference, 1994. Conference Record, Papers Presented at the 1994 Annual Meeting, 1994 IEEE*; Irvine, CA, USA, May 1994. pp. 1–6
8. Hsu C.T. 'Transient stability study of the large synchronous motors starting and operating for the isolated integrated steel-making facility'. *IEEE Transactions on Industry Applications*. 2003;**39**(5):1436–41
9. Shimada T., Agematsu S., Shoji T., Funabashi T., Otoguro H., Ametani A. (eds.). 'Combining power system load models at a busbar'. *IEEE Power Engineering Society Summer Meeting, 2000 IEEE*; Seattle, WA, USA, 2000. pp. 383–88
10. Coker M.L., Kgasoane H. (eds.). 'Load modeling'. *5th Africon Conference in Africa, 1999 IEEE Africon*; Cape Town, South Africa, Sep/Oct 1999, vol. 2. pp. 663–8
11. Kao W.S., Huang C.T., Chiou C.Y. 'Dynamic load modeling in Taipower system stability studies'. *IEEE Transactions on Power Systems*. 1995; **10**(2):907–14
12. Houlian C., Shande S., Shouzhen Z. (eds.). 'Radial basis function networks for power system dynamic load modeling'. *TENCON '93, Proceedings; Computer, Communication, Control and Power Engineering*. 1993 *IEEE Region 10 Conference on*; Beijing, China, Oct 1993. pp. 179–82
13. Karlsson D., Hill D.J. 'Modeling and identification of nonlinear dynamic loads in power systems'. *IEEE Transactions on Power Systems*. 1994; **9**(1):157–66
14. Zhu S.Z., Dong Z.Y., Wong K.P., Wang Z.H. (eds.). 'Power system dynamic load identification and stability'. *Power System Technology, 2000.*

Proceedings. PowerCon 2000. International Conference on; Perth, WA, USA, Dec 2000, vol. 1. pp. 13–18

15. Vaahedi E., Zein El-Din H.M.Z., Price W.W. 'Dynamic load modeling in large scale stability studies'. *IEEE Transactions on Power Systems*. 1988; **3**(3):1039–45
16. Hung R., Dommel H.W. 'Synchronous machine models for simulation of induction motor transients'. *IEEE Transactions on Power Systems*. 1996; **11**(2):833–8
17. Price W.W., Chiang H.D., Clark H.K., Concordia C., Lee D.C., Hsu J.C. *et al.* 'Load representation for dynamic performance analysis'. *IEEE Transactions on Power Systems*. 1993;**8**(2):472–82
18. IEEE Task Force on Load Representation for Dynamic Performance. 'Standard load models for power flow and dynamic performance simulation'. *IEEE Transactions on Power Systems*. 1995;**10**(3):1302–13
19. Kao W.S., Lin C.J., Huang C.T., Chen Y.T., Chiou C.Y. 'Comparison of simulated power system dynamics applying various load models with actual recorded data'. *IEEE Transactions on Power Systems*. 1994;**9**(1):248–54

Part II

Power flow

Chapter 5
Power flow analysis

Models of the individual components of the electric power system are described in Part I. The purpose of this chapter is to study mathematical relations between individual components to develop a model for the overall power system network, which is made of an interconnection of the various components. This model showing the currents, voltages, real power and reactive power flows at each bus in the network is known as Power Flow or Load Flow model. It is found that all relationships – between voltage and current at each bus, between the real and reactive power demand at a load bus or the generated real power and scheduled voltage magnitude at a generator bus – are non-linear. Therefore, power flow calculation implies the solution of a set of non-linear equations to give the electrical response of the transmission system to a particular set of loads and generator power outputs. In practice, the distribution system is not represented in power flow studies of bulk transmission systems and the loads are represented at substation levels. In addition, some assumptions regarding component modelling are made depending on the operating condition, whether it is in steady state or under a contingency, and should be consistent with the time period and purpose of study. A single-phase equivalent representation of the power network is used in power flow studies as the system is generally assumed to be balanced. Section 5.1 is focused on the general concepts of AC power flow calculation methods using bus admittance matrix because of its relevance to the needs of the stability studies. In particular, Newton–Raphson and fast-decoupled methods have been explained with illustrative examples because of their accuracy and fast convergence.

5.1 General concepts

For a network with n independent buses, applying Kirchhoff's law at each bus the following n equations can be written [1] as

$$\left.\begin{aligned} Y_{11}\boldsymbol{V}_1 + Y_{12}\boldsymbol{V}_2 + \cdots + Y_{1n}\boldsymbol{V}_n &= \boldsymbol{I}_1 \\ Y_{21}\boldsymbol{V}_1 + Y_{22}\boldsymbol{V}_2 + \cdots + Y_{2n}\boldsymbol{V}_n &= \boldsymbol{I}_2 \\ &\vdots \\ Y_{n1}\boldsymbol{V}_1 + Y_{n2}\boldsymbol{V}_2 + \cdots + Y_{nn}\boldsymbol{V}_n &= \boldsymbol{I}_n \end{aligned}\right\} \tag{5.1}$$

Equation (5.1) is expressed in matrix form as

$$\begin{bmatrix} Y_{11} & Y_{12} & \cdots & Y_{1n} \\ Y_{21} & Y_{22} & & Y_{2n} \\ \vdots & & \ddots & \vdots \\ Y_{n1} & Y_{n2} & \cdots & Y_{nn} \end{bmatrix} \begin{bmatrix} \boldsymbol{V}_1 \\ \boldsymbol{V}_2 \\ \vdots \\ \boldsymbol{V}_n \end{bmatrix} = \begin{bmatrix} \boldsymbol{I}_1 \\ \boldsymbol{I}_2 \\ \vdots \\ \boldsymbol{I}_n \end{bmatrix} \quad \text{or} \quad [Y][V] = [I] \tag{5.2}$$

where

$\boldsymbol{I} \triangleq$ bus current injection vector
$\boldsymbol{V} \triangleq$ bus voltage vector
$\boldsymbol{Y} \triangleq$ bus admittance matrix
$Y_{ii} \triangleq$ the diagonal element of the bus admittance matrix, called the 'self-admittance of bus i'. It equals the sum of all branch admittances connecting to bus i '$y_{io} + y_{i2} + \cdots + y_{in}$'
$y_{io} \triangleq$ the total capacitive susceptance at bus i
$Y_{ij} \triangleq$ the off-diagonal element of the bus admittance matrix, called 'mutual-admittance of bus i'. It equals the negative of branch admittance between buses i and j.

It is noted that the off-diagonal element Y_{ij} is zero if there is no line between buses i and j. Also, the bus admittance matrix is, in general, a sparse matrix.

The bus current in terms of bus voltage and power can be written as

$$\boldsymbol{I}_i = \frac{\boldsymbol{S}_i^*}{\boldsymbol{V}_i^*} = \frac{\left(P_{(\text{net})i} - jQ_{(\text{net})i}\right)}{\boldsymbol{V}_i^*} \tag{5.3}$$

where

$\boldsymbol{S} \triangleq$ the complex power injection vector, the superscript '*' denotes the conjugate vector
$P_{(\text{net})i} \triangleq$ the net real power injected to bus $i = P_{Gi} - P_{Li}$
$Q_{(\text{net})I} \triangleq$ the net reactive power injected to bus $i = Q_{Gi} - Q_{Li}$
$P_{Gi} \triangleq$ the real power output of the generator connected to bus i
$P_{Li} \triangleq$ the real power demand of the load connected to bus i
$Q_{Gi} \triangleq$ the reactive power output of the generator connected to bus i
$Q_{Li} \triangleq$ the reactive power demand of the load connected to bus i

From (5.1) and (5.3) the following relation can be obtained:

$$\frac{P_{(\text{net})i} - jQ_{(\text{net})i}}{\boldsymbol{V}_i^*} = Y_{i1}\boldsymbol{V}_1 + Y_{i2}\boldsymbol{V}_2 + \cdots + Y_{in}\boldsymbol{V}_n, \quad i = 1, 2, \ldots, n \tag{5.4}$$

or

$$P_{(\text{net})i} + jQ_{(\text{net})i} = \boldsymbol{V}_i \sum_{j=1}^{n} Y_{ij}^* \boldsymbol{V}_j^*, \quad i = 1, 2, \ldots, n \tag{5.5}$$

By equating real and imaginary parts in (5.5), two equations for each bus are obtained in terms of four variables P, Q, V and angle θ. Thus, two of these variables at each bus should be specified to solve the power flow equations and determine the other two variables. According to the known variables of the bus and the operating conditions of the power system as well, the buses are classified into three types:

Type 1 – PV bus: The real power P and voltage magnitude $|V|$ are known while the reactive power Q and voltage angle θ are unknown. The bus connected to the generator is usually a PV bus.

Type 2 – PQ bus: The real and reactive power, P and Q, respectively, are known and voltage magnitude and angle, V and θ, are unknown. The bus connected to a load is commonly a PQ bus as well as the bus at which a generator of constant or un-adjustable output power is connected.

Type 3 – Slack bus: It is also called Swing bus or Reference bus. During the power flow calculations the power loss of the network is unknown till the end of the power flow solution. So, a generator bus – called slack bus – is selected. The voltage magnitude and phase angle at this bus are specified so that the unknown power losses are also assigned to this bus in addition to the balance of generation. Traditionally, there is only one slack bus in the power flow calculations. As the voltage of the slack bus is given, only $n-1$ bus voltages need to be calculated and accordingly the number of power flow equations is $2(n-1)$.

The conventional methods used to solve the power flow equations are presented in Sections 5.2 through 5.4. These methods have some common features as they are iterative computational methods because of the non-linearity of equations and start with guessing an initial solution.

5.2 Newton–Raphson method

The general form of a set of non-linear equations with n variables is

$$\left.\begin{aligned} f_1(x_1, x_2, \ldots, x_n) &= 0 \\ f_2(x_1, x_2, \ldots, x_n) &= 0 \\ &\vdots \\ f_n(x_1, x_2, \ldots, x_n) &= 0 \end{aligned}\right\} \tag{5.6}$$

To solve this set of non-linear equations, an initial solution, x_i^o, $i = 1, 2, \ldots, n$, is selected. The difference between the initial value x_i^o and the final solution x will be Δx^o, i.e. $x = x^o + \Delta x^o$ is the solution of non-linear (5.6). Thus,

$$\left.\begin{aligned} f_1\left(x_1^0 + \Delta x_1^o, x_2^o + \Delta x_2^o, \ldots, x_n^o + \Delta x_n^o\right) &= 0 \\ f_2\left(x_1^0 + \Delta x_1^o, x_2^o + \Delta x_2^o, \ldots, x_n^o + \Delta x_n^o\right) &= 0 \\ &\vdots \\ f_n\left(x_1^0 + \Delta x_1^o, x_2^o + \Delta x_2^o, \ldots, x_n^o + \Delta x_n^o\right) &= 0 \end{aligned}\right\} \tag{5.7}$$

Applying Taylor series expansion to (5.7) and ignoring the second and higher derivatives gives

$$\left.\begin{aligned} f_1\left(x_1^o, x_2^o, \ldots, x_n^o\right) + \left.\frac{\partial f_1}{\partial x_1}\right|_{x_1^o} \Delta x_1^0 + \cdots + \left.\frac{\partial f_1}{\partial x_n}\right|_{x_n^o} \Delta x_n^0 = 0 \\ f_2\left(x_1^o, x_2^o, \ldots, x_n^o\right) + \left.\frac{\partial f_2}{\partial x_1}\right|_{x_1^o} \Delta x_1^0 + \cdots + \left.\frac{\partial f_2}{\partial x_n}\right|_{x_n^o} \Delta x_n^0 = 0 \\ \vdots \qquad\qquad \\ f_n\left(x_1^o, x_2^o, \ldots, x_n^o\right) + \left.\frac{\partial f_n}{\partial x_1}\right|_{x_1^o} \Delta x_1^0 + \cdots + \left.\frac{\partial f_n}{\partial x_n}\right|_{x_n^o} \Delta x_n^0 = 0 \end{aligned}\right\} \tag{5.8}$$

and in matrix form

$$\begin{bmatrix} f_1\left(x_1^o, x_2^o, \ldots, x_n^o\right) \\ f_2\left(x_1^o, x_2^o, \ldots, x_n^o\right) \\ \vdots \\ f_n\left(x_1^o, x_2^o, \ldots, x_n^o\right) \end{bmatrix} = -\begin{bmatrix} \left.\frac{\partial f_1}{\partial x_1}\right|_{x_1^o} & \left.\frac{\partial f_1}{\partial x_2}\right|_{x_2^o} & \cdots & \left.\frac{\partial f_1}{\partial x_n}\right|_{x_n^o} \\ \left.\frac{\partial f_2}{\partial x_1}\right|_{x_1^o} & \left.\frac{\partial f_2}{\partial x_2}\right|_{x_2^o} & & \left.\frac{\partial f_2}{\partial x_n}\right|_{x_n^o} \\ \vdots & \vdots & & \vdots \\ \left.\frac{\partial f_n}{\partial x_1}\right|_{x_1^o} & \left.\frac{\partial f_n}{\partial x_2}\right|_{x_2^o} & \cdots & \left.\frac{\partial f_n}{\partial x_n}\right|_{x_n^o} \end{bmatrix} \begin{bmatrix} \Delta x_1^o \\ \Delta x_2^o \\ \vdots \\ \Delta x_n^o \end{bmatrix} \tag{5.9}$$

Thus, $\Delta X^o = [\Delta x_1^o, \Delta x_1^o, \ldots, \Delta x_1^o]^t$ can be calculated from (5.9) and, therefore, the new solution is obtained. This solution is an approximate solution as the higher-order derivative terms of Taylor series are neglected. Therefore, the solution is an approximation as well, i.e. not the real solution. Consequently, further iterations are required. The iteration equations can be expressed as

$$\begin{bmatrix} f_1\left(x_1^k, x_2^k, \ldots, x_n^k\right) \\ f_2\left(x_1^k, x_2^k, \ldots, x_n^k\right) \\ \vdots \\ f_n\left(x_1^k, x_2^k, \ldots, x_n^k\right) \end{bmatrix} = -\begin{bmatrix} \left.\frac{\partial f_1}{\partial x_1}\right|_{x_1^k} & \left.\frac{\partial f_1}{\partial x_2}\right|_{x_2^k} & \cdots & \left.\frac{\partial f_1}{\partial x_n}\right|_{x_n^k} \\ \left.\frac{\partial f_2}{\partial x_1}\right|_{x_1^k} & \left.\frac{\partial f_2}{\partial x_2}\right|_{x_2^k} & & \left.\frac{\partial f_2}{\partial x_n}\right|_{x_n^k} \\ \vdots & \vdots & & \vdots \\ \left.\frac{\partial f_n}{\partial x_1}\right|_{x_1^k} & \left.\frac{\partial f_n}{\partial x_2}\right|_{x_2^k} & \cdots & \left.\frac{\partial f_n}{\partial x_n}\right|_{x_n^k} \end{bmatrix} \begin{bmatrix} \Delta x_1^k \\ \Delta x_2^k \\ \vdots \\ \Delta x_n^k \end{bmatrix} \tag{5.10}$$

and

$$x_i^{k+1} = x_i^k + \Delta x_i^k, \quad i = 1, 2, \ldots, n \tag{5.11}$$

The iteration can be stopped if $\max|\Delta xi| \leq \varepsilon$, $i = 1, 2, \ldots, n$, where ε is a small positive number that represents the permitted convergence precision.

In matrix form, (5.10) and (5.11) are written as

$$\left.\begin{aligned} F\left(X^k\right) &= -J^k \Delta X^k \\ X^{k+1} &= X^k + \Delta X^k \end{aligned}\right\} \tag{5.12}$$

where $J \triangleq n \times n$ Jacobian matrix

The mathematical principles explained above can be applied to solve the non-linear power flow equations expressed in polar coordinate system or rectangular coordinate system [2].

5.2.1 Power flow solution with polar coordinate system

The complex voltage and real and reactive powers (5.5) can be expressed in polar coordinates as

$$V_i = V_i\left(\cos\delta_i + j\sin\delta_j\right) \tag{5.13}$$

$$P_i = V_i \sum_{j=1}^{n} V_j (G_{ij}\cos\delta_{ij} + B_{ij}\sin\delta_{ij}) \tag{5.14}$$

$$Q_i = V_i \sum_{j=1}^{n} V_j (G_{ij}\sin\delta_{ij} - B_{ij}\cos\delta_{ij}) \tag{5.15}$$

where $\delta_{ij} = \delta_i - \delta_j \triangleq$ the angle difference between bus i and bus j.

For a network with n buses and according to the types of buses explained in Section 5.1, it is assumed that the network is composed of PQ buses ($1 \rightarrow m$), PV buses ($m+1 \rightarrow n-1$) and the nth bus is the slack bus. Therefore, the magnitude of voltages $V_{m+1} \rightarrow V_{n-1}$ are given as well as the voltage magnitude and angle at the slack bus, V_n and δ_n, are known. This results in the unknown variables being the voltage angle at $n-1$ buses and the voltage magnitude at m buses. For each bus in the network, the difference between the scheduled and produced real power, P_{sch} and P_i, respectively, is given by

$$\Delta P_i = P_{\text{sch}} - P_i = P_{\text{sch}} - V_i \sum_{j=1}^{n-1} V_j (G_{ij}\cos\delta_{ij} + B_{ij}\sin\delta_{ij}) \tag{5.16}$$

Similarly, for each PQ bus the difference of reactive power is

$$\Delta Q_i = Q_{\text{sch}} - Q_i = Q_{\text{sch}} - V_i \sum_{j=1}^{m} V_j (G_{ij}\sin\delta_{ij} - B_{ij}\cos\delta_{ij}) \tag{5.17}$$

Applying (5.12) the following equation can be obtained:

$$\begin{bmatrix} \Delta P \\ \Delta Q \end{bmatrix} = -J \begin{bmatrix} \Delta\delta \\ \dfrac{\Delta V}{V} \end{bmatrix} \quad \text{or} \quad \begin{bmatrix} \Delta P \\ \Delta Q \end{bmatrix} = -\begin{bmatrix} H & N \\ M & L \end{bmatrix} \begin{bmatrix} \Delta\delta \\ V_D^{-1}\Delta V \end{bmatrix} \tag{5.18}$$

where

$$\Delta P = \begin{bmatrix} \Delta P_1 \\ \Delta P_2 \\ \vdots \\ \Delta P_{n-1} \end{bmatrix}, \quad \Delta Q = \begin{bmatrix} \Delta Q_1 \\ \Delta Q_2 \\ \vdots \\ \Delta Q_m \end{bmatrix}, \quad \Delta\delta = \begin{bmatrix} \Delta\delta_1 \\ \Delta\delta_2 \\ \vdots \\ \Delta\delta_{n-1} \end{bmatrix}, \quad \Delta V = \begin{bmatrix} \Delta V_1 \\ \Delta V_2 \\ \vdots \\ \Delta V_m \end{bmatrix},$$

$$V_D = \begin{bmatrix} V_1 & & & \\ & V_2 & & \\ & & \ddots & \\ & & & V_m \end{bmatrix}$$

$H \triangleq (n-1) \times (n-1)$ matrix and $H_{ij} = \partial\Delta P_i/\partial\, \delta_j$
$N \triangleq (n-1) \times m$ matrix and $N_{ij} = V_j(\partial\Delta P_i/\partial\, V_j)$
$M \triangleq m \times (n-1)$ matrix and $M_{ij} = \partial\Delta Q_i/\partial\delta_j$
$L \triangleq m \times m$ matrix and $L_{ij} = V_j\,(\partial\, \Delta Q_i/\partial\, V_j)$

By definition, the off-diagonal elements '$i \neq j$' of the Jacobian matrix can be computed using the relations

$$\left.\begin{aligned} H_{ij} &= -V_iV_j\left(G_{ij}\sin\delta_{ij} - B_{ij}\cos\delta_{ij}\right) \\ N_{ij} &= -V_iV_j\left(G_{ij}\cos\delta_{ij} - B_{ij}\sin\delta_{ij}\right) \\ M_{ij} &= V_iV_j\left(G_{ij}\cos\delta_{ij} - B_{ij}\sin\delta_{ij}\right) \\ L_{ij} &= -V_iV_j\left(G_{ij}\sin\delta_{ij} - B_{ij}\cos\delta_{ij}\right) \end{aligned}\right\} \tag{5.19}$$

Similarly, the relations of diagonal elements '$i = j$' of the Jacobian matrix are

$$\left.\begin{aligned} H_{ii} &= V_i^2B_{ii} + Q_i \\ N_{ii} &= -V_i^2G_{ii} - P_i \\ M_{ii} &= V_i^2G_{ii} - P_i \\ L_{ii} &= V_i^2B_{ii} - Q_i \end{aligned}\right\} \tag{5.20}$$

The flow chart shown in Figure 5.1 depicts the steps of the Newton–Raphson power flow solution with polar coordinate system.

5.2.2 Power flow solution with rectangular coordinate system

From (5.5) the voltage and real and reactive power can be expressed in the rectangular coordinate system as

$$\left.\begin{aligned} V_i &= e_i + jf_i \\ P_i &= e_i\sum_{j=1}^{n}\left(G_{ij}e_j - B_{ij}f_j\right) + f_i\sum_{j=1}^{n}\left(G_{ij}f_j + B_{ij}e_j\right) \\ Q_i &= f_i\sum_{j=1}^{n}\left(G_{ij}e_j - B_{ij}f_j\right) - e_i\sum_{j=1}^{n}\left(G_{ij}f_j + B_{ij}e_j\right) \end{aligned}\right\} \tag{5.21}$$

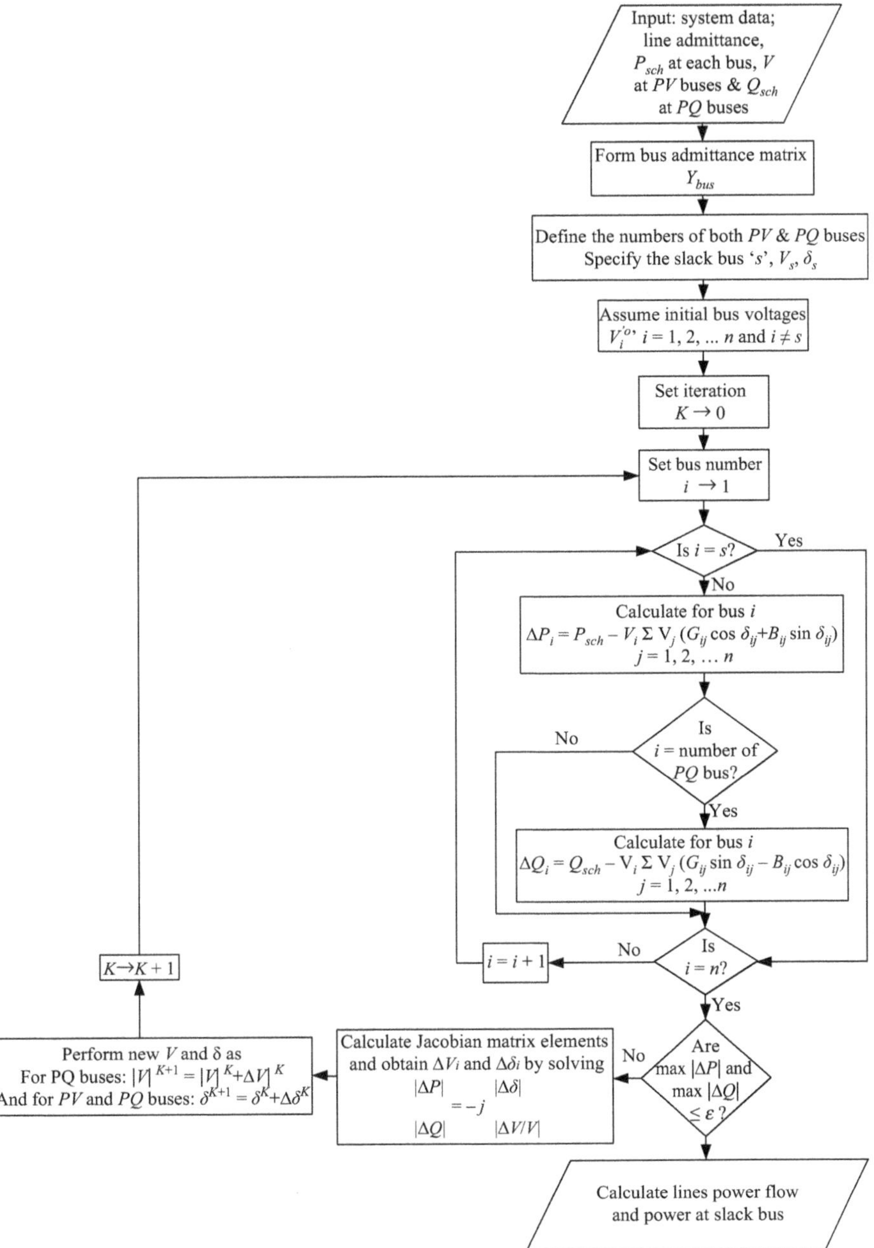

Figure 5.1 Flow chart for Newton–Raphson power flow solution with polar coordinates

For each PQ bus in the network, the differences between the scheduled and produced real and reactive power are given by

$$\left.\begin{aligned} \Delta P_i &= P_{\text{sch}} - P_I = P_{\text{sch}} - e_i\sum_{j=1}^{n}(G_{ij}e_j - B_{ij}f_j) - f_i\sum_{j=1}^{n}(G_{ij}f_j + B_{ij}e_j) \\ \Delta Q_i &= Q_{\text{sch}} - Q_i = Q_{\text{sch}} - f_i\sum_{j=1}^{n}(G_{ij}e_j - B_{ij}f_j) + e_i\sum_{j=1}^{n}(G_{ij}f_j + B_{ij}e_j) \end{aligned}\right\} \tag{5.22}$$

Similarly, for each PV bus the following relations can be written:

$$\left.\begin{aligned} \Delta P_i &= P_{\text{sch}} - P_I = P_{\text{sch}} - e_i\sum_{j=1}^{n}(G_{ij}e_j - B_{ij}f_j) - f_i\sum_{j=1}^{n}(G_{ij}f_j + B_{ij}e_j) \\ \Delta V_i^2 &= V_{\text{sch}}^2 - V_i^2 = V_{\text{sch}}^2 - (e_i^2 + f_i^2) \end{aligned}\right\} \tag{5.23}$$

Equations (5.22) and (5.23) include $2(n-1)$ equations: $(n-1)$ equations represent the real power at all buses excluding the slack bus and the other $(n-1)$ equations comprise m equations representing the reactive power for PQ buses and $(n-m-1)$ equations representing ΔV_i^2 for PV buses. Expanding the equations into Taylor series to be written in a linearised form, neglecting the derivative terms of second and higher order, and according to Newton–Raphson method the equation in the form $\Delta F = -J\Delta V$ can be written as

$$\begin{bmatrix} \Delta P_1 \\ \Delta P_2 \\ \vdots \\ \Delta P_{n-1} \\ \Delta Q_1 \\ \Delta Q_2 \\ \vdots \\ \Delta Q_m \\ \Delta V_{m+1}^2 \\ \vdots \\ \Delta V_{n-1}^2 \end{bmatrix} = -\begin{bmatrix} \frac{\partial \Delta P_1}{\partial e_1} & \cdots & \frac{\partial \Delta P_1}{\partial e_{n-1}} & \frac{\partial \Delta P_1}{\partial f_1} & \cdots & \frac{\partial \Delta P_{n-1}}{\partial f_{n-1}} \\ \vdots & \cdots & \vdots & \vdots & \cdots & \vdots \\ \frac{\partial \Delta P_{n-1}}{\partial e_1} & \cdots & \frac{\partial \Delta P_{n-1}}{\partial e_{n-1}} & \frac{\partial \Delta P_{n-1}}{\partial f_1} & \cdots & \frac{\partial \Delta P_{n-1}}{\partial f_{n-1}} \\ \frac{\partial \Delta Q_1}{\partial e_1} & \cdots & \frac{\partial \Delta Q_1}{\partial e_{n-1}} & \frac{\partial \Delta Q_1}{\partial f_1} & \cdots & \frac{\partial \Delta Q_1}{\partial f_{n-1}} \\ \vdots & \cdots & \vdots & \vdots & \cdots & \vdots \\ \frac{\partial \Delta Q_m}{\partial e_1} & \cdots & \frac{\partial \Delta Q_m}{\partial e_{n-1}} & \frac{\partial \Delta Q_m}{\partial f_1} & \cdots & \frac{\partial \Delta Q_m}{\partial f_{n-1}} \\ \frac{\partial \Delta V_{m+1}^2}{\partial e_1} & \cdots & \frac{\partial \Delta V_{m+1}^2}{\partial e_{n-1}} & \frac{\partial \Delta V_{m+1}^2}{\partial f_1} & \cdots & \frac{\partial \Delta V_{m+1}^2}{\partial f_{n-1}} \\ \vdots & \cdots & \vdots & \vdots & \cdots & \vdots \\ \frac{\partial \Delta V_{n-1}^2}{\partial e_1} & \cdots & \frac{\partial \Delta V_{n-1}^2}{\partial e_{n-1}} & \frac{\partial \Delta V_{n-1}^2}{\partial f_1} & \cdots & \frac{\partial \Delta V_{n-1}^2}{\partial f_{n-1}} \end{bmatrix} \begin{bmatrix} \Delta e_1 \\ \Delta e_2 \\ \vdots \\ \Delta e_{n-1} \\ \Delta f_1 \\ \vdots \\ \Delta f_{m+1} \\ \vdots \\ \Delta f_{n-1} \end{bmatrix} \tag{5.24}$$

Equation (5.24) may be written as

$$\begin{bmatrix} \Delta P \\ \Delta Q \\ \Delta V^2 \end{bmatrix} = \begin{bmatrix} J1 & J2 \\ J3 & J4 \\ J5 & J6 \end{bmatrix} \begin{bmatrix} \Delta e \\ \Delta f \end{bmatrix} \tag{5.25}$$

where

For all network buses: $(1 \rightarrow n-1)$ except the slack bus, nth bus:

$$\Delta P = [\Delta P_1, \Delta P_2, \ldots, \Delta P_{n-1}]^t, \Delta e = [\Delta e_1, \Delta e_2, \ldots, \Delta e_{n-1}]^t,$$
$$\Delta f = [\Delta f_1, \Delta f_2, \ldots, \Delta f_{n-1}]^t$$

$J1 \triangleq (n-1) \times (n-1)$ matrix with elements obtained by the expression:

$$\left.\begin{aligned} \frac{\partial \Delta P_i}{\partial e_j} &= -\sum_{j=1}^{n} (G_{ij} e_j - B_{ij} f_j) - G_{ii} e_i - B_{ii} f_i, && \text{for } i = j \\ &= -(G_{ij} e_i + B_{ij} f_i), && \text{for } i \neq j \end{aligned}\right\} \tag{5.26}$$

$J2 \triangleq (n-1) \times (n-1)$ matrix and its elements are given by

$$\left.\begin{aligned} \frac{\partial \Delta P_i}{\partial f_j} &= -\sum_{j=1}^{n} (G_{ij} e_j - B_{ij} f_j) - G_{ii} e_i - B_{ii} f_i, && \text{for } i = j \\ &= -(G_{ij} f_i - B_{ij} e_i), && \text{for } i \neq j \end{aligned}\right\} \tag{5.27}$$

For PQ buses: $(1 \rightarrow m)$

$$\Delta Q = [\Delta Q_1, \Delta Q_2, \ldots, \Delta Q_m]^t$$

$J3 \triangleq m \times (n-1)$ matrix with elements given by

$$\left.\begin{aligned} \frac{\partial \Delta Q_i}{\partial e_j} &= \sum_{j=1}^{n} (G_{ij} f_j + B_{ij} e_j) - G_{ii} f_i + B_{ii} e_i, && \text{for } i = j \\ &= (G_{ij} f_i - B_{ij} e_i), && \text{for } i \neq j \end{aligned}\right\} \tag{5.28}$$

$J4 \triangleq m \times (n-1)$ matrix with elements given by

$$\left.\begin{aligned} \frac{\partial \Delta Q_i}{\partial f_j} &= -\sum_{j=1}^{n} (G_{ij} e_j - B_{ij} f_j) + G_{ii} e_i + B_{ii} f_i, && \text{for } i = j \\ &= (G_{ij} e_i + B_{ij} f_i), && \text{for } i \neq j \end{aligned}\right\} \tag{5.29}$$

For PV buses: $(m+1 \rightarrow n-1)$

$$\Delta V^2 = \left[\Delta V_{m+1}^2, \Delta V_{m+2}^2, \ldots, \Delta V_{n-1}^2\right]^t$$

$J5 \triangleq (n-m-1) \times (n-1)$ matrix with elements given by

$$\left.\begin{aligned} \frac{\partial V_i^2}{\partial e_j} &= -2e_i, && \text{for } i = j \\ &= 0, && \text{for } i \neq j \end{aligned}\right\} \tag{5.30}$$

$J6 \triangleq (n-m-1)\times(n-1)$ matrix with elements given by

$$\left.\begin{aligned}\frac{\partial V_i^2}{\partial f_j} &= -2f_i, \quad \text{for } i=j\\ &= 0, \qquad \text{for } i\neq j\end{aligned}\right\} \tag{5.31}$$

The steps of solution are the same as that applied when using polar coordinates in Section 5.2.1.

Example 5.1 Using Newton–Raphson method, find the power flow solution of the three generators–nine bus system shown in Figure 5.2 [3]. The system data are given in Appendix II.

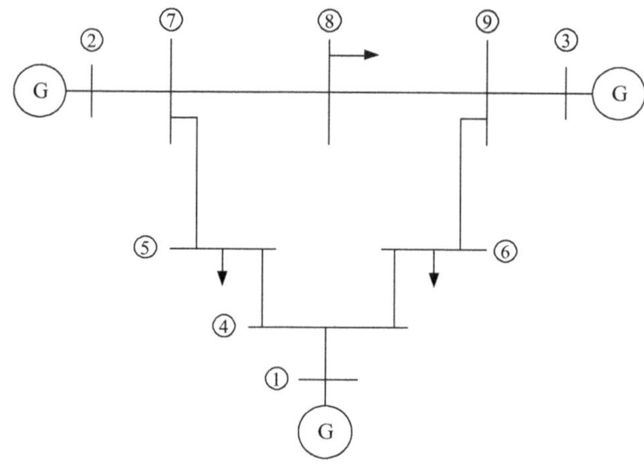

Figure 5.2 Three generators–nine bus system [3]

Solution:

The system buses are classified as bus #1 is a slack bus, buses #2, 3 are *PV* buses and buses #4 → 9 are *PQ* buses. Power and voltage set-points on the system base of 100 MVA are summarised in Table 5.1.

Table 5.1 Power and voltage set points

Bus no.	1	2	3	4	5	6	7	8	9
Gen/Load		G2	G3	–	L5	L6	–	L8	–
P (pu)	Slack	1.63	0.85	0	1.25	0.9	0	1.0	0
Q (pu)	bus	–	–	0	0	0	0	0	0
V (pu) 0°		1.025	1.025	1.02	–	–	–	–	–

By writing the program and downloading the data in PSAT Toolbox, v2.1.6 [4], to solve the iterative (5.18) in polar coordinates according to the flow chart (Figure 5.1) it is found that:

The program run terminated at the second iteration. The maximum convergence error for first iteration is 8.8336×10^{-4} and for the second iteration is 7.7519×10^{-7}.

The elements of bus admittance matrix, Y_{bus} =

0 − j17.361	0	0	0 + j17.361	0	0	0	0	0
0	0 − j14.388	0	0	0	0	0 + j14.388	0	0
0	0	0 − j17.065	0	0	0	0	0	0 + j17.065
0 + j17.361	0	0	2.9253 − j27.621	−0.98314 + j0.082739	−1.9422 + j10.511	0	0	0
0	0	0	−0.98314 + j0.082739	2.2015 − j5.5629	0	−1.2184 + j5.9622	0	0
0	0	0	−1.9422 + j10.511	0	3.2242 − j15.583	0	0	−1.282 + j5.5882
0	0 + j14.388	0	0	−1.2184 + j5.9622	0	2.8355 − j33.593	−1.6171 + j13.698	0
0	0	0	0	0	0	−1.6171 + j13.698	2.7722 − j23.124	−1.1551 + j9.7843
0	0	0 + j17.065	0	0	−1.282 + j5.5882		−1.1551 + j9.7843	2.4371 − j31.87

Results: The power and voltage at each bus are summarised in Table 5.2 and line flows are summarised in Table 5.3. The real power and reactive power losses in each line are illustrated in Table 5.4.

Table 5.2 Bus power and voltage

Bus no.	$\|V\|$ (pu)	Voltage angle (deg.)	P (pu)	Q (pu)
1	1.04	0	0.71641	0.71641
2	1.025	9.28	1.63	1.63
3	1.025	4.6648	0.85	0.85
4	1.0258	−2.2168	0	0
5	0.99563	−3.9888	−1.25	−1.25
6	1.0127	−3.6874	−0.9	−0.9
7	1.0258	3.7197	0	0
8	1.0159	0.72754	−1	−1
9	1.0324	1.9667	0	0

Table 5.3 Line flows

Bus no.	1	2	3	4	5	6	7	8	9
1	0	0	0	$0.71641 + j0.27046$	0	0	0	0	0
2	0	0	0	0	0	0	$1.63 + j0.06654$	0	0
3	0	0	0	0	0	0	0	0	$0.85 - j0.1086$
4	$-0.71641 - j0.23923$	0	0	0	$0.40937 + j0.22893$	$0.30704 + j0103$	0	0	0
5	0	0	0	$-0.4068 - j0.38687$	0	0	$-0.8432 - j0.11313$	0	0
6	0	0	0	$-0.30537 - j0.16543$	0	0	0	0	$-0.59463 - j0.13457$
7	0	$-1.63 + j0.09178$	0	0	$0.8662 - j0.08381$	0	0	$0.7638 - j0.00797$	0
8	0	0	0	0	0	0	$-0.75905 - j0.10704$	0	$-0.24095 - j0.24296$
9	0	0	$-0.85 + j0.14955$	0	0	$0.60817 - j0.18075$	0	$0.24183 + j0.0312$	0

Table 5.4 Line power losses in pu

Line$_{(ij)}$	**1–4**	**2–7**	**3–9**	**4–5**	**4–6**	**5–7**	**6–9**	**7–8**	**8–9**
$P_{\text{losses}(ij)}$	0	0	0	0.00258	0.00166	0.023	0.01354	0.00475	0.00088
$Q_{\text{losses}(ij)}$	0.031	0.158	0.041	−0.158	−0.155	−0.196	−0.315	−0.115	−0.212

$Line_{(ij)} \triangleq$ the line connecting bus i to bus j
$P_{\text{losses}(ij)} \triangleq$ real power losses in line ij
$Q_{\text{losses}(ij)} \triangleq$ reactive power losses in line ij
It is to be noted that $P_{\text{losses}(ij)} = P_{\text{losses}(ij)}$ and $Q_{\text{losses}(ij)} = Q_{\text{losses}(ij)}$

Total real power generation, $P_{G(\text{total})} = 3.1964$ pu
Total reactive power generation, $Q_{G(\text{total})} = 0.2284$ pu
Total load, real power, $P_{L(\text{total})} = 3.15$ pu
Total load, reactive power, $Q_{L(\text{total})} = 1.15$ pu
Total real power losses, $P_{\text{losses}} = 0.04641$ pu
Total reactive power losses, $Q_{\text{losses}} = -0.9216$ pu

5.3 Gauss–Seidel method

Again, a system of n non-linear equations in n unknown variables can be described generally by (5.32).

$$\left.\begin{aligned} f_1(x_1, x_2, \ldots, x_n) &= 0 \\ f_2(x_1, x_2, \ldots, x_n) &= 0 \\ &\vdots \\ f_n(x_1, x_2, \ldots, x_n) &= 0 \end{aligned}\right\} \quad (5.32)$$

Its solution can be formulated as

$$\left.\begin{aligned} x_1 &= g_1(x_1, x_2, \ldots, x_n) \\ x_2 &= g_2(x_1, x_2, \ldots, x_n) \\ &\vdots \\ x_n &= g_n(x_1, x_2, \ldots, x_n) \end{aligned}\right\} \quad (5.33)$$

At first iteration, the initial solution is assumed and substituted into the RHS of (5.33) to get a new solution that is considered as a primary solution for the next iteration. Thus, at the kth iteration, the new solution is expressed as

$$\left.\begin{aligned} x_1^{k+1} &= g_1\left(x_1^k, x_2^k, \ldots, x_n^k\right) \\ x_2^{k+1} &= g_2\left(x_1^k, x_2^k, \ldots, x_n^k\right) \\ &\vdots \\ x_n^{k+1} &= g_n\left(x_1^k, x_2^k, \ldots, x_n^k\right) \end{aligned}\right\} \quad (5.34)$$

The iterative solution is terminated when the convergence condition (5.35) is satisfied.

$$\max|x_i^{k+1} - x_i^k| \leq \varepsilon, \quad i = 1, 2, \ldots, n \tag{5.35}$$

The solution by following the above steps is called Gauss method. This method has been modified to be called Gauss–Seidel method. The modification is mainly concerned with speeding up the convergence and requires smaller number of iterations, i.e. less computation time is required. It is based on substituting immediately the new values of variables in the calculation of the next variable (in the same iteration) rather than waiting until the next iteration. Therefore, the modified formula of the iteration calculation is

$$\left.\begin{aligned} x_1^{k+1} &= g_1\left(x_1^k, x_2^k, \ldots, x_n^k\right) \\ x_2^{k+1} &= g_2\left(x_1^{k+1}, x_2^k, \ldots, x_n^k\right) \\ &\vdots \\ x_n^{k+1} &= g_n\left(x_1^{k+1}, x_2^{k+1}, \ldots, x_{n-1}^{k+1}, x_n^k\right) \end{aligned}\right\} \tag{5.36}$$

or

$$x_i^{k+1} = g_i\left(x_1^{k+1}, x_2^{k+1}, \ldots, x_{i-1}^{k+1}, x_i^k, \ldots, x_n^k\right) \tag{5.37}$$

Applying this method to a network with n-buses ordered as buses $1 \rightarrow m$ are PQ buses; buses $m \rightarrow n-1$ are PV buses and the slack bus is the nth bus. From (5.5) the voltage at each bus can be written as

$$V_i = \frac{1}{Y_{ii}}\left[\frac{P_i - jQ_i}{V_i^*} - \sum_{\substack{j=1 \\ j \neq i}}^{n} Y_{ij} V_j\right], \quad i = 1, 2, \ldots, n-1 \tag{5.38}$$

Thus, by Gauss–Seidel method, (5.38) at the kth iteration is

$$V_i^{k+1} = \frac{1}{Y_{ii}}\left[\frac{P_i - jQ_i}{V_i^{*k}} - \sum_{j=1}^{i-1} Y_{ij} V_j^{k+1} - \sum_{j=i+1}^{n} Y_{ij} V_j^k\right], \quad i = 1, 2, \ldots, n-1 \tag{5.39}$$

For the PQ bus: The real and reactive powers are known. Equation (5.39) can be used to perform iteration calculations as the initial bus voltages are assumed.

For the PV bus: The bus real power and voltage magnitude are known. Thus, the voltage magnitude must be fixed at the scheduled value and its angle is computed from the estimated voltage as below.

The voltage as a complex quantity is expressed as $V_i = e_i + jf_i$.

Thus, the relation $|V_{i(\text{sch})}|^2 = |e_i|^2 + |f_i|^2$ must be satisfied. This necessitates that the phase angle of the scheduled voltage is equal to that of the estimated voltage δ_i.

Also, the components of the estimated voltage are adjusted accordingly. At the *k*th iteration δ_i is given by

$$\delta_i^k = \tan^{-1}\left[\frac{f_i^k}{e_i^k}\right] \tag{5.40}$$

Then, the adjusted components of estimated voltage are

$$\left.\begin{aligned} e_{i(\text{adj})}^k &= |V_{i(\text{sch})}| \cos \delta_i^k \\ f_{i(\text{adj})}^k &= |V_{i(\text{sch})}| \sin \delta_i^k \end{aligned}\right\} \tag{5.41}$$

and the corresponding reactive power is calculated by

$$Q_i^k = \text{Im}\left[V_i^k I_i^{*k}\right] = \text{Im}\left[V_i^k \sum_{j=1}^{i-1} Y_{ij}^* V_j^{*k+1} + \sum_{j=i}^{n} Y_{ij}^* V_j^{*k}\right] \tag{5.42}$$

Using the adjusted voltage components (5.41) and the corresponding reactive power (5.42) the new estimated voltage V_i^{k+1} can be computed.

It is to be noted that the limits of the reactive power source must be considered in such a way that in case of violating the limits the value of the reactive power is fixed at the violated limit. The *PV* bus is treated as a *PQ* bus ignoring the desired voltage magnitude.

Finally, when the convergence is satisfied, and the iteration is terminated, the values of voltage (magnitude and angle) and real and reactive power at all buses are obtained. Then, the power flow into a line with a shunt admittance y_{io} and an admittance y_{ij} between bus *i* and bus *j* can be calculated by

$$S_{ij} = P_{ij} + jQ_{ij} = V_i I_{ij}^* = V_i^2 y_{io} + V_i\left(V_i^* - V_j^*\right) y_{ij}^* \tag{5.43}$$

and the power at the slack bus is given by

$$P_n + jQ_n = V_n \sum_{j=1}^{n} Y_{nj}^* V_j^* \tag{5.44}$$

The steps of Gauss–Seidel program are depicted by the flow chart shown in Figure 5.3.

5.4 Decoupling method

This method is based on some simplifications to make power flow iteration very easy, to expedite solution computation and satisfying accepted accuracy of results. For instance, in Newton–Raphson power flow calculation, to obtain a solution with high accuracy, there is no simplification as well as the elements of Jacobian matrix must be recalculated each iteration. So, more iteration, more storage and more computation time are required.

The simplification used in the decoupling method mainly depends on the availability of neglecting the impact of both voltage magnitude change on the real

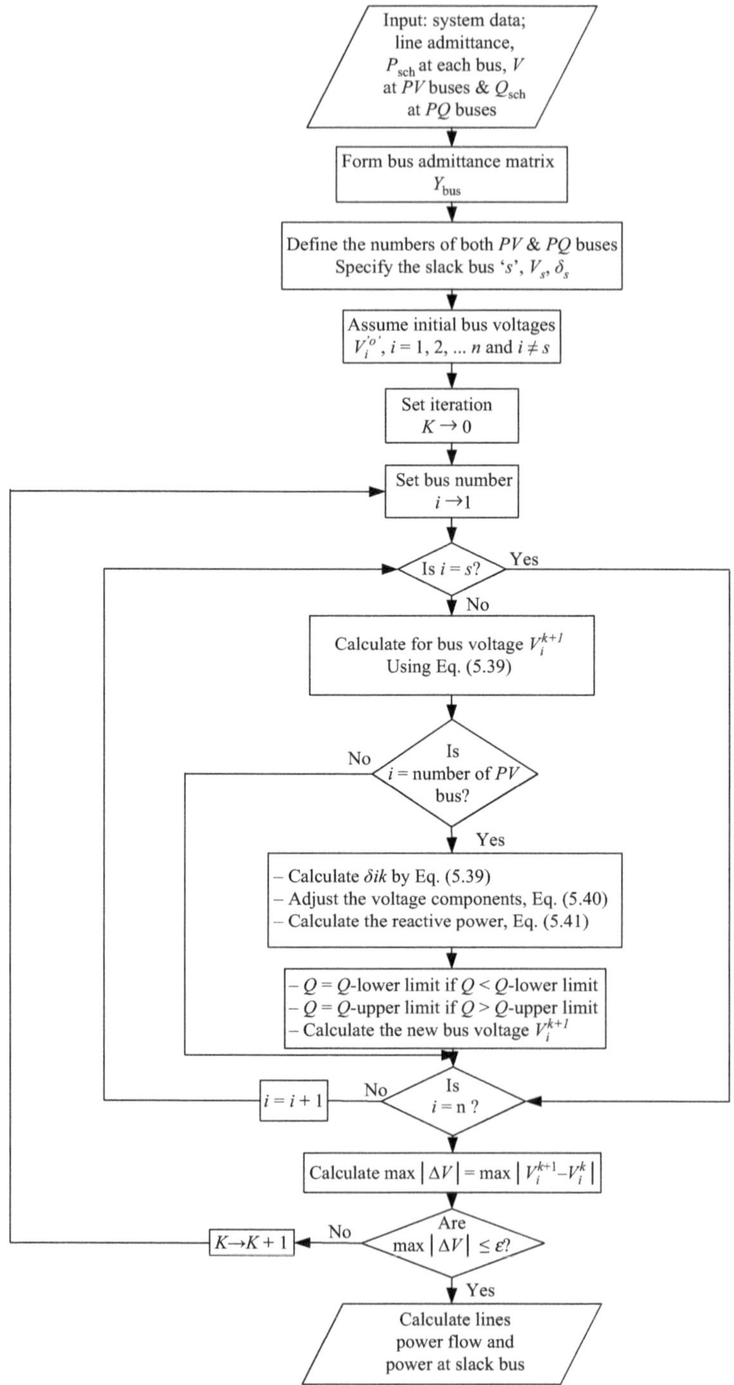

Figure 5.3 Flow chart for Gauss–Seidel power flow method using Y_{bus}

power and angle change on the reactive power. This assumption, to some extent, is reasonable as in practical power systems the branches have much higher reactance than the resistance, i.e. the coupling between the real power and voltage magnitude is weak and the same is for the coupling between the reactive power and voltage angle. That is weak as well [5].

Neglecting weak couplings implies that $\frac{\partial \Delta P_i}{\partial V_i} \approx 0$ and $\frac{\partial \Delta Q_i}{\partial \delta_i} \approx 0$, consequently, the sub-matrices N and M in (5.18) vanish. Thus,

$$\begin{bmatrix} \Delta P \\ \Delta Q \end{bmatrix} = -\begin{bmatrix} H & 0 \\ 0 & L \end{bmatrix} \begin{bmatrix} \Delta\delta \\ V_D^{-1}\Delta V \end{bmatrix}$$

Hence,

$$\Delta P = -H\Delta\delta \quad \text{and} \quad \Delta Q = -LV_D^{-1}\Delta V \tag{5.45}$$

5.4.1 Fast-decoupled method

Further simplification can be applied to (5.45) as the difference between the voltage angles at the two ends of a line ij is small. Then, $\cos\delta_{ij} = \cos(\delta_i - \delta_j) \cong 1$ and $G_{ij}\sin\delta_{ij} << B_{ij}$. Assuming that $Q_i << V_i^2 B_{\text{ii}}$, the elements of the two sub-matrices H and L are given by

$$\left.\begin{aligned} H_{ij} &= -\frac{\partial P_i}{\partial \delta_j} = V_i V_j B_{ij} \quad i,j = 1,2,\ldots,n-1 \\ L_{ij} &= -\frac{\partial Q_i}{\left(\dfrac{\partial V_j}{V_j}\right)} = V_i V_j B_{ij} \quad i,j = 1,2,\ldots,m \end{aligned}\right\} \tag{5.46}$$

Thus, the matrices $[H]$ and $[L]$ can be formulated as

$$[H] = \begin{bmatrix} V_1 & & & \\ & V_2 & & \\ & & \ddots & \\ & & & V_{n-1} \end{bmatrix} \begin{bmatrix} B_{11} & B_{12} & \cdots & B_{1,n-1} \\ B_{21} & B_{22} & \cdots & B_{2,n-1} \\ \vdots & \vdots & \ddots & \vdots \\ B_{n-1,1} & B_{n-1,2} & \cdots & B_{n-1,n-1} \end{bmatrix} \begin{bmatrix} V_1 & & & \\ & V_2 & & \\ & & \ddots & \\ & & & V_{n-1} \end{bmatrix}$$
$$= VB'V$$

and

$$[L] = \begin{bmatrix} V_1 & & & \\ & V_2 & & \\ & & \ddots & \\ & & & V_m \end{bmatrix} \begin{bmatrix} B_{11} & B_{12} & \cdots & B_{1,m} \\ B_{21} & B_{22} & \cdots & B_{2,m} \\ \vdots & \vdots & \ddots & \vdots \\ B_{m,1} & B_{m,2} & \cdots & B_{m,m} \end{bmatrix} \begin{bmatrix} V_1 & & & \\ & V_2 & & \\ & & \ddots & \\ & & & V_m \end{bmatrix} \tag{5.47}$$
$$= VB''V$$

where

$$[B'] = \begin{bmatrix} B_{11} & B_{12} & \cdots & B_{1,n-1} \\ B_{21} & B_{22} & \cdots & B_{2,n-1} \\ \vdots & \vdots & \ddots & \vdots \\ B_{n-1,1} & B_{n-1,2} & \cdots & B_{n-1,n-1} \end{bmatrix} \quad \text{and} \quad [B''] = \begin{bmatrix} B_{11} & B_{12} & \cdots & B_{1,m} \\ B_{21} & B_{22} & \cdots & B_{2,m} \\ \vdots & \vdots & \ddots & \vdots \\ B_{m,1} & B_{m,2} & \cdots & B_{m,m} \end{bmatrix}$$

Substitute (5.47) into (5.45) to obtain

$$\left.\begin{aligned} \frac{\Delta P}{V} &= -B' V \Delta\delta \\ \frac{\Delta Q}{V} &= -B'' \Delta V \end{aligned}\right\} \tag{5.48}$$

Equation (5.48) can be written in matrix form as

$$\left.\begin{aligned} \begin{bmatrix} \frac{\Delta P_1}{V_1} \\ \frac{\Delta P_2}{V_2} \\ \vdots \\ \frac{\Delta P_{n-1}}{V_{n-1}} \end{bmatrix} &= -\begin{bmatrix} B_{11} & B_{12} & \cdots & B_{1,n-1} \\ B_{21} & B_{22} & \cdots & B_{2,n-1} \\ \vdots & \vdots & \ddots & \vdots \\ B_{n-1,1} & B_{n-1,2} & \cdots & B_{n-1,n-1} \end{bmatrix} \begin{bmatrix} V_1\Delta\delta_1 \\ V_2\Delta\delta_2 \\ \vdots \\ V_{n-1}\Delta\delta_{n-1} \end{bmatrix} \\ \begin{bmatrix} \frac{\Delta Q_1}{V_1} \\ \frac{\Delta Q_2}{V_2} \\ \vdots \\ \frac{\Delta Q_m}{V_m} \end{bmatrix} &= -\begin{bmatrix} B_{11} & B_{12} & \cdots & B_{1,m} \\ B_{21} & B_{22} & \cdots & B_{2,m} \\ \vdots & \vdots & \ddots & \vdots \\ B_{m,1} & B_{m,2} & \cdots & B_{m,m} \end{bmatrix} \begin{bmatrix} \Delta V_1 \\ \Delta V_2 \\ \vdots \\ \Delta V_m \end{bmatrix} \end{aligned}\right\} \tag{5.49}$$

It is found that the elements of matrices $-B'$ and $-B''$ in (5.49) are the imaginary part of the corresponding elements of bus admittance matrix. Therefore, for a specific configuration of a power system, matrices B' and B'' are constant, symmetrical, real and sparse matrices. In addition, they need to be triangularised only once at the beginning of the study. Thus, it is called the 'fast decoupled power flow model'.

The fast decoupled power flow solution requires more iterations than the Newton–Raphson method, but requires considerably less time per iteration and a power flow is obtained very rapidly. This technique is very useful in contingency analysis where a power flow solution is required for online control [6].

An illustrative example to show the application of power flow methods to the power system and a comparison between them is given below.

Example 5.2 Repeat Example 5.1 using fast-decoupled method.

Solution:

The program terminated at the fourth iteration with maximum convergence error as below:

Iteration = 1: Maximum Convergence Error = 0.017302
Iteration = 2: Maximum Convergence Error = 0.00046404
Iteration = 3: Maximum Convergence Error = 1.8107×10^{-5}
Iteration = 4: Maximum Convergence Error = 5.9507×10^{-7}

Power and voltage set-points on the system base of 100MVA are the same as used in Example 5.1 and Table 5.1.

Results: The power and voltage at each bus are summarised in Table 5.5 and line flows are summarised in Table 5.6. The real power and reactive power losses in each line are illustrated in Table 5.7.

Table 5.5 Bus power and voltage

Bus no.	$\lvert V\rvert$ (pu)	Voltage angle (deg.)	P (pu)	Q (pu)
1	1.04	0	0.71641	0.27046
2	1.025	9.28	1.63	0.06654
3	1.025	4.6648	0.85	−1.1086
4	1.0258	−2.2168	0	0
5	0.99563	−3.9888	−1.25	−0.5
6	1.0127	−3.6874	−0.9	−0.3
7	1.0258	3.7197	0	0
8	1.0159	0.72754	−1	−0.35
9	1.0324	1.9667	0	0

$line_{(ij)} \triangleq$ the line connecting bus i to bus j
$P_{\text{losses}(ij)} \triangleq$ real power losses in line ij
$Q_{\text{losses}(ij)} \triangleq$ reactive power losses in line ij

It is to be noted that $P_{\text{losses}(ij)} = P_{\text{losses}(ji)}$ and $Q_{\text{losses}(ij)} = Q_{\text{losses}(ji)}$
Total real power generation, $P_{G(\text{total})} = 3.1964$ pu
Total reactive power generation, $Q_{G(\text{total})} = 0.2284$ pu
Total load, real power, $P_{L(\text{total})} = 3.15$pu
Total load, reactive power, $Q_{L(\text{total})} = 1.15$ pu
Total real power losses, $P_{\text{losses}} = 0.04641$ pu
Total reactive power losses, $Q_{\text{losses}} = -0.9216$ pu

It is found that the results obtained from Examples 5.1 and 5.2 are very close to each other. On the other hand, the fast-decoupled power flow for Example 5.2 has taken four iterations with maximum convergence error of 5.9507×10^{-7} compared to the Newton–Raphson method (Example 5.1) that took only two iterations with

Table 5.6 Line flows

Bus no.	1	2	3	4	5	6	7	8	9
1	0	0	0	0.71641 + *j*0.27046	0	0	0	0	0
2	0	0	0	0	0	0	1.63 + *j*0.06654	0	0
3	0	0	0	0	0	0	0	0	0.85 − *j*0.1086
4	−0.71641− *j*0.23923	0	0	0	0.40937 + *j*0.22893	0.30704 + *j*0103	0	0	0
5	0	0	0	−0.4068 − *j*0.38687	0	0	−0.8432 − *j*0.11313	0	0
6	0	0	0	−0.30537 − *j*0.16543	0	0	0	0	−0.59463 − *j*0.13457
7	0	−1.63 + *j*0.09178	0	0	0.8662 − *j*0.08381	0	0	0.7638 − *j*0.00797	0
8	0	0	0	0	0	0	−0.75905 − *j*0.10704	0	−0.24095 − *j*0.24296
9	0	0	−0.85 + *j*0.14955	0	0	0.60817 − *j*0.18075	0	0.24183 + *j*0.0312	0

Table 5.7 Line power losses in pu

$Line_{(ij)}$	1–4	2–7	3–9	4–5	4–6	5–7	6–9	7–8	8–9
$P_{losses(ij)}$	0	0	0	0.0026	0.0017	0.023	0.014	0.005	0.001
$Q_{losses(ij)}$	0.031	0.158	0.041	−0.158	−0.155	−0.197	−0.315	−0.115	−0.212

the maximum convergence error of 7.7519×10^{-7}. Regarding the computation time, it is found that with running the program on a PC, Processor INSPIRON I5R N5110 Core i7, the time for the two iterations of Newton–Raphson and the four iterations of fast-decoupled method is 24 ms and 39 ms, respectively. This illustrates that the fast-decoupled solution requires less time per iteration than that required by Newton–Raphson solution.

References

1. El-Hawary M.E., Christensen G.S. *Optimal Economic Operation of Electric Power Systems*. New York, NY, US: Academic Press; 1979
2. Murty P.S.R. *Operation and Control in Power Systems*. Hyderabad, India: BS Publications; 2008
3. Anderson P.M., Fouad A.A. *Power System Control and Stability*. 2nd edn. Hoboken, NJ, US: Wiley-IEEE Press; 2003
4. Milano F. *Power System Analysis Toolbox PSAT* [online]. 2014. Available from http://www.power.uwaterloo.ca/~fmilano/psat.htm [Accessed 12 Jul 2014]
5. Zhu J. *Optimization of Power System Operation*. Hoboken, NJ, US: Wiley-IEEE Press; 2009
6. Saadat H., *Power System Analysis*. 3rd edn. New York, NY, US: McGraw-Hill; 2010

Chapter 6
Optimal power flow

One of the important tools for power system planning and operation is the optimal power flow (OPF). It is a power flow problem in which some control variables are adjusted to minimise or maximise an objective function, while satisfying physical and operating limits as constraints on various controls, dependent variables and functions of variables. In power system analysis, the control variables that must be accommodated with OPF are active and reactive power generation from power plants, generator terminal voltage, reactive power compensation, LTC of transformers and phase shift angles. The objective function includes minimisation of losses or costs in the case of studying economic power dispatch. As the controls include reactive power devices, the problem is characterised by a non-separable objective function and consequently the problem solution gets more difficult [1]. The constraints are either equality constraints such as power balance or inequality constraints, e.g. the generated power must be within a maximum and minimum permissible output power of the generator. However, the problem expressed in mathematical notation can be compactly formulated as

$$\text{Minimise } f(\boldsymbol{x}, \boldsymbol{u}) \tag{6.1}$$

$$\text{Subject to } h(\boldsymbol{x}, \boldsymbol{u}) = 0 \tag{6.2}$$

$$\text{and } g(\boldsymbol{x}, \boldsymbol{u}) \leq 0 \tag{6.3}$$

where $\boldsymbol{x}$ and $\boldsymbol{u}$ are the dependent and control variable vectors, respectively.

6.1 Problem formulation

The OPF problem is mainly formulated as an optimisation problem. The formulation is based on three elements: objective function, control variables and constraints.

The conventional OPF model is described below where the control variables include real and reactive power generation and control voltage setting.

$$\text{minimise } P_{loss} = \sum_{i=1}^{n} |V_i| \sum_{j=1}^{n} |V_j| \left(G_{ij} \cos \delta_{ij} + B_{ij} \sin \delta_{ij} \right) \tag{6.4}$$

subject to

$$\text{equality constraints}\begin{cases} P_{Gi} - P_{Di} - V_i \sum_{j=1}^{n} |V_j|(G_{ij}\cos\delta_{ij} + B_{ij}\sin\delta_{ij}) = 0 \\ Q_{Gi} - Q_{Di} - V_i \sum_{j=1}^{n} |V_j|(G_{ij}\sin\delta_{ij} - B_{ij}\cos\delta_{ij}) = 0 \end{cases} \tag{6.5}$$

$i, j = 1, 2, \ldots, n$

$$\text{and (inequality constraints)}\begin{cases} P_{Gi}^{\min} \le P_{Gi} \le P_{Gi}^{\max} & (i \in S_G) \\ Q_{Ri}^{\min} \le Q_{Ri} \le Q_{Ri}^{\max} & (i \in S_R) \\ V_i^{\min} \le V_i \le V_i^{\max} & (i \in S_B) \\ S_{Li}^{\min} \le S_{Li} \le S_{Li}^{\max} & (i \in S_L) \end{cases} \tag{6.6}$$

where

$P_{Gi} \triangleq$ active power output of the ith generator in the controllable generator set S_G
$Q_{Ri} \triangleq$ reactive output of the ith reactive source in the reactive source set S_R
$V_i \triangleq$ voltage magnitude of the ith bus in the bus set S_B
$P_{Di} \triangleq$ load of the ith bus
G_{ij} and $B_{ij} \triangleq$ conductance and susceptance between the ith and the jth bus, respectively
$S_{Li} \triangleq$ apparent power across the ith branch in the branch set S_L
$n \triangleq$ number of buses
$G_{ij} + jB_{ij} \triangleq$ element of nodal admittance matrix
$\delta_{ij} \triangleq$ voltage angle difference at the two ends of the tie-line ij

Depending on the type of study, other constraints may be considered such as boiler constraints, running reserve constraints, phase shift transformer constraints, line outage or security constraints.

6.2 Problem solution

The solution is based on two steps: First, applying Newton's method the normal static load flow as a feasible solution is calculated. Second, using the gradient method and Lagrange multiplier the optimum solution can be obtained [2, 3].

- Equation (6.5) for convenience can be rewritten as

$$\left.\begin{aligned} P_{i(net)} - P_i(V, \delta) = 0 \\ Q_{i(net)} - Q_i(V, \delta) = 0 \end{aligned}\right\} \quad i = 1, 2, \ldots, n \tag{6.7}$$

where

$P_{i(net)} = P_{Gi} - P_{Di}, Q_{i(net)} = Q_{Gi} - Q_{Di}, P_{Gi}$ and Q_{Gi} have positive sign when entering and P_{Di} and Q_{Di} have positive sign when leaving the ith bus

$$\left.\begin{aligned} P_i(V,\delta) &= V_i\sum_{j=1}^{n}|V_j|\left(G_{ij}\cos\delta_{ij} + B_{ij}\sin\delta_{ij}\right) \\ Q_i(V,\delta) &= V_i\sum_{j=1}^{n}|V_j|\left(G_{ij}\sin\delta_{ij} - B_{ij}\cos\delta_{ij}\right) \end{aligned}\right\} \tag{6.8}$$

It is seen that each bus in the power system is characterised by four variables: the net real and reactive power entering the bus and the bus voltage (magnitude and phase angle), which are $P_{(net)}$, $Q_{(net)}$, V and δ, respectively. Thus, to solve (6.7) that is a set of $2n$ non-linear equations, two of the four variables at each bus must be specified. Determination of which two variables are specified depends on the type of the bus as summarised in Table 6.1. Usually, the phase angle δ_s of the voltage V_s at the slack bus is taken as zero and considered as reference. The control variables are regarded as specified variables in the power flow solution. $P_{(net)}$ and $Q_{(net)}$ can be calculated directly from (6.7), but if V and/or δ are unknown the solution is not direct and is explained below.

Assume **X** and **Y** are the vectors of unknown (V and δ) and known variables, respectively. Thus, $\mathbf{X}^t = [V_1, \ldots, V_m, \delta_1, \ldots, \delta_m, \delta_{m+1}, \ldots, \delta_{n-1}]$ and $\mathbf{Y}^t = [P_1, \ldots, P_{n-1}, Q_1, \ldots, Q_m, V_{m+1}, \ldots, V_n, \delta_s]$.

The vector **X** includes $n + m - 1$ elements (unknown). Then, a set of $n + m - 1$ relations must be selected from (6.7) that has $2n$ relations to form a vector of functions in terms of **X** and **Y** elements, $\mathbf{h}(x, y)$.

Using Newton's method described in Chapter 5 and assuming the initial solution $\mathbf{X}^{(o)}$ is improved successively by $\Delta\mathbf{X}$, the solution can be found by solving the set of equations:

$$\left[\frac{\partial h}{\partial x}\left(x^k, y\right)\right][\Delta\mathbf{X}] = -\left[h\left(x^k, y\right)\right] \tag{6.9}$$

where $(\partial h/\partial x)$ is the Jacobian matrix.

Table 6.1 Variables characterising network buses

Bus type	**Specified variables**	**Unknown variables**	**Bus no.**
PQ-bus	$P_{(net)}$, $Q_{(net)}$	V, δ	$1 \rightarrow m$
PV-bus	$P_{(net)}$, V	$Q_{(net)}$, δ	$m + 1 \rightarrow n - 1$
Slack-bus	V, δ	$P_{(net)}$, $Q_{(net)}$	N

The solution obtained (as a first step) represents the static feasible non-OPF solution. Therefore, the second step is to optimise that solution in two cases: without and with inequality constraints.

OPF without inequality constraints: The objective function to be minimised is the total system losses given by (6.4). It is a function of the independent and dependent variables and can generally be expressed as $f(x, y)$.

Using the classical Lagrange-multipliers method to minimise $f(x, y)$ subject to the equality constraints $h(x, y) = 0$ as in (6.7), the solution is found by introducing the Lagrangian function

$$\mathcal{L}(x,y) = f(x,y) + [\lambda]^t h(x,y) \tag{6.10}$$

where the elements of $[\lambda]$ are called Lagrangian multipliers, and satisfy the following three necessary conditions for a minimum:

$$\left.\begin{aligned} \left[\frac{\partial \mathcal{L}}{\partial x}\right] &= \left[\frac{\partial f}{\partial x}\right] + \left[\frac{\partial h}{\partial x}\right]^t [\lambda] = 0 \\ \left[\frac{\partial \mathcal{L}}{\partial y}\right] &= \left[\frac{\partial f}{\partial y}\right] + \left[\frac{\partial h}{\partial y}\right]^t [\lambda] = 0 \\ \left[\frac{\partial \mathcal{L}}{\partial \lambda}\right] &= h(x,y) = 0 \end{aligned}\right\} \tag{6.11}$$

Any feasible solution, such as the solution obtained by the Newton's method in the first step, satisfies the third condition while the first condition can be verified by calculating λ as

$$[\lambda] = -\left[\frac{\partial h}{\partial x}\right]^{t^{-1}} \left[\frac{\partial f}{\partial x}\right] \tag{6.12}$$

It is noted that the Jacobian $(\partial h/\partial x)$ has already been calculated in (6.9). To satisfy the second condition, $(\partial \mathcal{L}/\partial y)$ is determined by using the gradient method with incremental change of controls 'y' given by

$$y_i^{k+1} = y_i^k + c\frac{\partial \mathcal{L}}{\partial y_i} = y_i^k - c\nabla f \tag{6.13}$$

where c is a scalar factor and k is the number of iterations.

The computation is repeated iteratively until the minimum is reached.

OPF with inequality constraints: Actually, the control variables have permissible values such as given in (6.6), and then they cannot be assumed to have any value. These inequality constraints can easily be treated by setting the control variables at their limits if (6.13) gives violating values, i.e. beyond the permissible limits. Thus,

$$y_i^{k+1} = \begin{cases} y_i^{\max}, & \text{if } y_i^k + \Delta y_i > y_i^{\max} \\ y_i^{\min}, & \text{if } y_i^k + \Delta y_i < y_i^{\min} \\ y_i^k + \Delta y_i, & \text{otherwise} \end{cases} \tag{6.14}$$

It is to be noted that when a control variable has reached its limit its component in the gradient vector must be computed in the following cycles because it might eventually back off from the limit. At the minimum, Kuhn–Tucker theorem proves that the components $(\partial f/\partial y_i)$ will be as in (6.15) as necessary conditions.

$$\left.\begin{aligned} \frac{\partial f}{\partial y_i} &= 0, \quad \text{if } y_i^{\min} < y_i < y_i^{\max} \\ \frac{\partial f}{\partial y_i} &\leq 0, \quad \text{if } y_i = y_i^{\max} \\ \frac{\partial f}{\partial y_i} &\geq 0, \quad \text{if } y_i = y_i^{\min} \end{aligned}\right\} \qquad (6.15)$$

Example 6.1 Find the OPF in the nine-bus system given in Appendix II. Formulate the objective function as minimising the generation cost subject to the constraints.

Solution:

The objective function is considered as

$$F = \sum_{i=1}^{N_G} F_i(P_{Gi})$$

where F is the total generated power fuel cost, $F_i(P_{Gi})$ is the ith generating unit fuel cost that is a function of the active power generation output, P_{Gi}, in MW of the ith generator. N_G is the total number of generating units.

The fuel cost curve is considered as a quadratic cost curve (Fq) as

$$F_{qi}(P_{Gi}) = \text{a}_i + \text{b}_i P_{Gi} + \text{c}_i P_{Gi}^2$$

where the cost coefficients are summarised in Table 6.2.

The constraints are the equations of active and reactive power balance

$$\text{equality constraints}\begin{cases} P_{Gi} - P_{Di} - V_i \sum_{j=1}^{n} |V_j| \left(G_{ij}\cos\theta_{ij} + B_{ij}\sin\theta_{ij}\right) = 0 \\ Q_{Gi} - Q_{Di} - V_i \sum_{j=1}^{n} |V_j| \left(G_{ij}\sin\theta_{ij} - B_{ij}\cos\theta_{ij}\right) = 0 \end{cases}$$

Table 6.2 Cost coefficients

Generator 1			Generator 2			Generator 3		
a ($/hr)	b ($/MW/hr)	c ($/MW²/hr)	a ($/hr)	b ($/MW/hr)	c ($/MW²/hr)	a ($/hr)	b ($/MW/hr)	c ($/MW²/hr)
150	5	0.11	600	1.2	0.085	335	1	0.1225

and the limits of active and reactive power, voltage, transformer tap setting and line loading.

$$\text{inequality constraints}\begin{cases} P_{Gi}^{\min} \le P_{Gi} \le P_{Gi}^{\max} & (i = 1, 2, \ldots, N_G) \\ Q_{Gi}^{\min} \le Q_{Gi} \le Q_{Gi}^{\max} & (i = 1, 2, \ldots, N_G) \\ V_i^{\min} \le V_i \le V_i^{\max} & (i = 1, 2, \ldots, N) \\ T_{i,\min} \le T_i \le T_{i,\max} & (i = 1, 2, \ldots, N_T) \\ |S_{Li}| \le S_{Li}^{\max} & (i = 1, 2, \ldots, N) \end{cases}$$

where

P_{Gi} and Q_{Gi} are the total active and reactive power generation at bus i. P_{Di} and Q_{Di} are the total active and reactive power demands at bus i. V_i and V_j are the voltage magnitudes at buses i and j

G_{ij} and B_{ij} are the real and imaginary parts of the ijth element of admittance matrix (Y_{bus}); θ_{ij} is the difference of voltage angles between buses i and j

$P_{Gi,\max}$ and $P_{Gi,\min}$ are the upper and lower limits of active power output of the ith generator

$Q_{Gi,\max}$ and $Q_{Gi,\min}$ are the upper and lower limits of reactive power output of the ith generator

$V_{i,\max}$ and $V_{i,\min}$ are the upper and lower limits of voltage magnitude at bus i

T_i is the tap setting of the ith transformer; $T_{i,\max}$ and $T_{i,\min}$ are the lower and upper limits of the tap setting of the ith transformer

$|S_{Li}|$ is the line loading in MVA at line i

$S_{Li,\max}$ is the line loading limit in MVA at line i (the line flow when the system is fully loaded)

N, N_L and N_T are the total number of buses, lines and transformers, respectively.

The limits of P, Q, V and T at generation buses are summarised in Table 6.3.

Applying MATPOWER Version 4.1, 14-Dec-2011-AC Optimal Power Flow, MATLAB® Interior Point Solver – MIPS, Version 1.0 and using PC-Intel® Core™ i5-2430M, CPU@2.4 GHz, RAM: 8 GB, System type: 64-bit OS, it is

Table 6.3 Limits of parameters: P, Q, V and T

Generator no.	**Active power generation**		**Reactive power generation**		**Voltage at all buses (pu)**	
	Min (MW)	**Max (MW)**	**Min (MVAr)**	**Max (MVAr)**	$V_{\min} = 0.9$	$V_{\max} = 1.1$
1	10	250	−300	300	The tap setting of transformers ranges from 0 to 1	
2	10	300	−300	300		
3	10	270	−300	300		

Table 6.4 Voltage, power and cost at system buses

Bus no.	Voltage		Generation		Load		Lambda ($/MVA-hr)	
	Mag (pu)	Ang (deg)	*P* (MW)	*Q* (MVAr)	*P* (MW)	*Q* (MVAr)	*P*	*Q*
1	1.087	0	87.05	−22.20			24.150	
2	1.1	0.044	136.16	6.68			24.347	
3	1.1	−6.212	97.45	20.89			24.875	
4	1.1	−2.403					24.150	
5	1.098	−4.248			90	30	24.384	−0.002
6	1.011	−17.468			100	35	26.976	0.848
7	1.099	−3.993					24.350	0.045
8	1.076	−8.530			125	50	24.842	0.166
9	1.090	−8.941					24.880	0.120
Total			320.65	5.37	315	115		

Table 6.5 Power flow in the lines

Line	From bus no.	To bus no.	From bus injection		Losses	
			P (MW)	*Q* (MVAr)	*P* (MW)	*Q* (MVAr)
1	1	4	87.05	−22.20	0	3.93
2	4	5	41.49	−13.82	0.244	1.32
3	6	9	−100.00	−35.00	3.918	17.08
4	3	9	97.45	20.89	0	4.81
5	9	8	−6.47	3.58	0.03	0.25
6	8	7	−131.50	−22.15	1.283	10.87
7	7	2	−136.16	2.92	0	9.60
8	7	5	3.37	−18.31	0.003	0.02
9	5	4	−45.39	−7.47	0.172	1.46
Total					5.650	49.34

found that the program is converged in 0.07 s and the objective function value = 5353.58 $/hr. The voltage, power and cost at system buses as well as the power flow in system lines are obtained as summarised in Tables 6.4 and 6.5, respectively.

6.3 OPF with dynamic security constraint

There is increasing interest and need to take into account the dynamic security constraints in the OPF computations to protect the system against disturbances. As the issue of system stability, small signal stability and transient stability is of an increasing concern in relation to the security of system operation, the conventional OPF should be adapted to find an optimal solution that minimises a cost function while maintaining certain stability criteria. This may need to modify the objective function or add some constraints to the problem formulated in Section 6.1.

Small signal stability, as is known, is defined as the ability of a power system to maintain the generators in synchronism when subjected to small disturbances. To analyse power system small signal stability, the state equation describing the system should be linearised at the operating point [4].

Assuming zero input, the power system can be described by

$$\dot{\boldsymbol{X}} = f(\boldsymbol{X}) \tag{6.16}$$

where $\boldsymbol{X}$ is the state vector of the power system, f is a set of non-linear functions and the derivative is w.r.t. time.

By Taylor's series expansion and assuming a small deviation of the state vector, $\Delta\boldsymbol{X}$, the linearised form of (6.16) is

$$\frac{\mathrm{d}\Delta\boldsymbol{X}}{\mathrm{d}t} = [A]\Delta\boldsymbol{X} \tag{6.17}$$

where $[A]$ is the state matrix. The small signal stability is determined by the eigenvalues of matrix A, which can be written in a general form as

$$\lambda = \sigma \pm j\omega \tag{6.18}$$

where the real and imaginary components give the damping and frequency of the corresponding mode, respectively. Accordingly, determination of stability is determined as explained below:

- The system is stable if all eigenvalues have negative real parts.
- The system is unstable if at least one eigenvalue has positive real part.
- The system is oscillatory if at least one eigenvalue has zero real part.

The decay rate of oscillation can be defined by a common index (damping ratio ζ) deduced in terms of eigenvalues components by (6.19). The system is considered to have wider stability margin as ζ gets larger.

$$\zeta = \frac{-\sigma}{\sqrt{\sigma^2 + \omega^2}} \tag{6.19}$$

Therefore, to consider small signal stability in the OPF, mode analysis described above should be conducted for each candidate operating point so that the eigenvalue real parts or damping ratios can be achieved to compare these operating points. These indices can be involved in the objective function or constraints in the OPF process.

Transient stability means the ability of the power system to maintain its generators in synchronism when subjected to a large severe disturbance. To keep the system operating satisfactorily it is of importance to develop the OPF calculations by incorporating transient stability constraints in addition to static security constraints. Different approaches have been proposed for this purpose. However, these approaches can be categorised into three categories [5]. The first category includes the methods based on transient energy function or energy balance. They have limitations pertaining to power system modelling with the difficulty in forming the function expressing the stability margin in terms of generator active power. The second category comprises the methods that express the transient stability boundary by an approximated non-linear function in terms of the control variables. These methods are exposed to the difficulty of determining the required non-linear function. In the third category the methods simulate the power system in time-domain with detailed dynamic models. Then, stability constraints based on relative rotor angles at each time interval and a set of algebraic equations for all time intervals are included in the OPF. Of course, high accuracy and robustness may be achieved, but, on the other hand, the practical implementation is time consuming as well as the computation requires high memory storage. This is due to implying large numbers of additional variables associated with power system dynamical model and constraints in various time intervals in the OPF formulation.

Power system dynamics is described by a set of differential-algebraic equations as [6]:

$$\left.\begin{aligned} \dot{\boldsymbol{X}} &= f_1[\boldsymbol{X}(t), \mathfrak{B}(t), \boldsymbol{Y}] \\ 0 &= f_2[\boldsymbol{X}(t), \mathfrak{B}(t), \boldsymbol{Y}] \end{aligned}\right\} \tag{6.20}$$

where $\boldsymbol{X}(t)$ is the state variables vector including generator rotor angles and speeds, and $\mathfrak{B}(t)$ is the vector of algebraic variables including network-related variables such as bus voltages and angles. $\boldsymbol{Y}$ represents a set of control variables such as generator active power output, size of capacitor banks installed at a bus and so on that is generally time independent for transient stability analysis and can be viewed as parameters of (6.20).

The initial values of ith generator rotor angles δ_i^o and emf E_i' are obtained from the system pre-fault steady-state conditions as below:

$$\left.\begin{aligned} \frac{E_i' V_t \sin\left(\delta_i^o - \delta_t\right)}{X'_{di}} - P_{Gi} = 0 \\ \frac{E_i' V_t \cos\left(\delta_i^o - \delta_t\right)}{X'_{di}} - Q_{Gi} = 0 \end{aligned}\right\} \tag{6.21}$$

where the generator is represented by a constant voltage source E' behind direct-axis transient reactance X'_d, and δ_t and V_t represent the voltage angle and magnitude at generator terminal bus, respectively. In addition, as the pre-fault is a steady-state condition the initial angular speed

$$(\omega_i^o) = 1 \text{ pu} \tag{6.22}$$

The swing equations for the generator as given in Chapter 2 are represented by

$$\left.\begin{aligned} \dot{\delta}_\iota &= \omega_i - 1 \\ \dot{\omega}_i &= \frac{1}{2H_i\omega_{\text{B}}}(P_{mi} - P_{ei}) \end{aligned}\right\} \tag{6.23}$$

and must be discretised using the trapezoidal rule to be converted into algebraic equations. Hence, generator rotor angles and speeds for a generic time interval $(k+1)$ are defined by the following equations:

$$\left.\begin{aligned} \delta_i^{k+1} - \delta_i^k - \frac{\Delta t}{2}\left(\omega_i^{k+1} + \omega_i^k\right) = 0 \\ \omega_i^{k+1} - \omega_i^k - \frac{\Delta t}{2}\frac{1}{2H_i\omega_B}\left(P_{ai}^{k+1} - P_{ai}^k\right) = 0 \end{aligned}\right\} \tag{6.24}$$

where $P_{ai}^k = P_{mi} - P_{ei}^k =$ accelerating power at time interval k.

The dynamical constraints for the transient stability can be considered as the angle deviation between any two generators is less than the appointed value $\delta_{\max}$ in the whole system trajectory.

$$|\delta_i^k(t) - \delta_j(t)| < \delta_{\max} \quad t \in [0, T] \tag{6.25}$$

where T denotes the concerned time range.

Additional constraints may be added, such as

$$\left.\begin{aligned} -\pi \le \delta_i \le \pi \\ \delta_{ref} = 0 \end{aligned}\right\} \tag{6.26}$$

Therefore, the formulation of stability-constrained OPF problem can be summarised as below [7].

Minimise (6.4)
subject to

- power flow (6.5)
- the limits in (6.6)
- initial values of generator rotor angles and emf's, (6.21) and (6.22)
- discrete swing (6.24)
- transient stability limit (6.25)
- additional constraints (6.26)

Several studies are reported in the literature to solve the stability-constrained OPF problem aiming to minimise the costs of generating power output as an objective function [8] or considering various constraints [9, 10]. Different methods, e.g. differential evolution, inexact Newton method and primal-dual interior, have been proposed to formulate various OPF problems [11–19] as well as attempts to reduce the computation time of solution [20] using appropriate algorithms [21, 22]. Numerical discretisation of dynamic security constraints should be implemented appropriately [23]. On the other hand, it has been found that the consideration of stability constraints (steady-state and/or transient stability constraints) has remarkable impact on security pricing and solution results [24–26].

Example 6.2 Solve the problem in Example 6.1 taking into account the transient stability constraints.

Solution:

The difference between the bus voltage angle θ_f at the '*from* end' of a branch and the angle θ_t at the '*to* end' can be bounded above and below to act as a proxy for a transient stability limit. MATPOWER creates the corresponding constraints on the voltage angle variables. So, using MATPOWER Version 4.1, AC Optimal Power Flow – MATLAB Interior Point Solver – MIPS, Version 1.0, it is seen that the program is converged in 0.12 s and the objective function value is 6095$/hr. The voltage, power and cost at system buses as well as the power flow in system lines are obtained as summarised in Tables 6.6 and 6.7, respectively.

Table 6.6 Voltage, power and cost at system buses

Bus no.	**Voltage**		**Generation**		**Load**		**Lambda ($/MVA-hr)**	
	Mag (pu)	**Ang (deg)**	***P* (MW)**	***Q* (MVAr)**	***P* (MW)**	***Q* (MVAr)**	***P***	***Q***
1	1.089	0.000	90.00	−18.61			17.035	
2	1.096	−1.673	87.76	9.42			24.035	
3	1.073	−1.120	142.62	14.83			15.000	
4	1.100	−2.480					23.773	−0.292
5	1.096	−4.380			90	30	23.999	−0.247
6	0.985	−14.227			100	35	16.312	0.477
7	1.092	−4.300					24.035	
8	1.063	−7.255			125	50	24.365	0.205
9	1.068	−5.303					15.000	
Total			320.38	5.64	315	115		

Table 6.7 Power flow in the lines

Line	**From bus**	**To bus**	**From bus injection**		**Losses**	
			***P* (MW)**	***Q* (MVAr)**	***P* (MW)**	***Q* (MVAr)**
1	1	4	90.00	−18.61	0.000	4.10
2	4	5	42.95	−12.23	0.260	1.41
3	6	9	−100.00	−35.83	4.147	18.08
4	3	9	142.62	14.83	0.000	10.46
5	9	8	38.47	−10.94	0.155	1.31
6	8	7	−86.68	−38.53	0.633	5.36
7	7	2	−87.76	−5.37	0.000	4.05
8	7	5	0.45	−21.22	0.002	0.01
9	5	4	−46.86	−9.19	0.183	1.55
Total					5.380	46.34

References

1. Sun D.I., Ashley B., Brewer B., Hughes A., Tinney W.F. 'Optimal power flow by Newton approach'. *IEEE Transactions on Power Apparatus and Systems.* 1984;**PAS-103**(10):2864–80
2. Maria G.A., Findlay J.A. 'A Newton optimal power flow program for Ontario hydro EMS'. *IEEE Transactions on Power Systems.* 1987;**2**(3):576–82
3. Dommel H.W., Tinney W.F. 'Optimal power flow solutions'. *IEEE Transactions on Power Apparatus and Systems.* 1968;**PAS-87**(10):1866–76
4. Su C., Chen Z. (eds.). 'An optimal power flow (OPF) method with improved power system stability'. *Universities Power Engineering Conference (UPEC) 2010 45th International*; Cardiff, Wales, Aug/Sep 2010. pp. 1–6
5. Nguyen T.T., Nguyen V.L., Karimishad A. 'Transient stability-constrained optimal power flow for online dispatch and nodal price evaluation in power systems with flexible AC transmission system devices'. *IET Generation Transmission & Distribution.* 2011;**5**(3):332–46
6. Xin H., Gan D., Huang Z., Zhuang K., Cao L. 'Applications of stability-constrained optimal power flow in the East China system'. *IEEE Transactions on Power Systems.* 2010;**25**(3):1423–33
7. Miñano R.Z., Cutsem T.V., Milano F., Conejo A.J. 'Securing transient stability using time-domain simulations within an optimal power flow'. *IEEE Transactions on Power Systems.* 2010;**25**(1):243–53
8. Kdsi S.K.M., Canizares C.A. (eds.). 'Stability-constrained optimal power flow and its application to pricing power system stabilizers'. *Power Symposium, 2005 Proceedings of the 37th Annual North American*, Oct 2005. pp. 120–6
9. Wen S., Fang D.Z., Shiqiang Y. (eds.). 'Sensitivity-based approach for optimal power flow with transient stability constraints'. *International Conference on Energy and Environment Technology, 2009 ICEET'09*, Oct 2009. pp. 267–70
10. Layden D., Jeyasurya B. (eds.). 'Integrating security constraints in optimal power flow studies'. *PES General Meeting, IEEE, Proceedings*; Denver, CO, US, Jun 2004, vol. 1. pp. 125–9
11. Cai H.R., Chung C.Y., Wong K.P. 'Application of differential evolution algorithm for transient stability constrained optimal power flow'. *IEEE Transactions on Power Systems.* 2008;**23**(2):719–28
12. Xu Y., Dong Z.Y., Meng K., Zhao J.H., Wong K.P. 'A hybrid method for transient stability-constrained optimal power flow computation'. *IEEE Transactions on Power Systems.* 2012;**27**(4):1769–77
13. Huang Y., Liu M. (eds.). 'Transient stability constrained optimal power flow based on trajectory sensitivity, one-machine infinite bus equivalence and differential evolution'. *International Conference on Power System Technology (Power Con), 2010*; Hangzhou, China, Oct 2010. pp. 1–6
14. Li R., Chen L., Yokoyama R. (eds.). 'Stability constrained optimal power flow by inexact Newton method'. *Power Tech Proceedings, 2001 IEEE Porto*; Porto, Portugal, Sep 2001. pp. 1–6

15. Xia Y., Chan K.W., Liu M. 'Direct nonlinear primal–dual interior-point method for transient stability constrained optimal power flow'. *IEE Proceedings – Generation, Transmission and Distribution*. 2005;**152**(1):11–6
16. Bhattacharya A., Chattopadhyay P.K. 'Application of biogeography-based optimization to solve different optimal power flow problems'. *IET Proceedings Generation, Transmission and Distribution*. 2011;**5**(1):70–80
17. Alam A., Makram E.B. (eds.). 'Transient stability constrained optimal power flow'. *Power Engineering Society General Meeting, 2006 IEEE PES*; Montreal, Canada, Jul 2006. pp. 1–6
18. Xia Y., Chan K.W., Liu M. (eds.). 'Improved BFGS method for optimal power flow calculation with transient stability constraints'. *Power Engineering Society General Meeting, 2005 IEEE*; San Francisco, CA, US, Jun 2005. pp. 434–9
19. Chen L., Ono A., Tada Y., Okamoto H., Tanabeb R. (eds.). 'Optimal power flow constrained by transient stability'. *International Conference on Power System Technology, Proceedings of PowerCon 2000*; Perth, Australia, Dec 2000, vol. 1. pp. 1–6
20. Sun Y., Xinlin Y., Wang H.F. 'Approach for optimal power flow with transient stability constraints'. *IEE Proceedings – Generation Transmission and Distribution*. 2004;**151**(1):8–18
21. Chen L., Yasuyuki T., Okamoto H., Tanabe R., Ono A. 'Optimal operation solutions of power systems with transient stability constraints'. *IEEE Transactions on Circuits and Systems—I: Fundamental Theory and Applications*. 2001;**48**(3):327–39
22. Martínez A.P., Esquivel C.R.F., Vega D.R. 'A New practical approach to transient stability-constrained optimal power flow'. *IEEE Transactions on Power Systems*. 2011;**26**(3):1686–96
23. Jiang Q., Huang Z. 'An enhanced numerical discretization method for transient stability constrained optimal power flow'. *IEEE Transactions on Power Systems*. 2010;**25**(4):1790–7
24. Uaahedi E., Zein El-Din H.M. 'Considerations in applying optimal power flow to power system operation'. *IEEE Transactions on Power Systems*. 1989;**4**(2):694–703
25. Liu H., Miao Y. (eds.). 'A Novel OPF-Based security pricing method with considering effects of transient stability and static voltage stability'. *IEEE T&D Conference & Exposition: Asia and Pacific*; Seoul, Oct 2009. pp. 1–5
26. Condren J., Gedra T.W. 'Expected-security-cost optimal power flow with small-signal stability constraints'. *IEEE Transactions on Power Systems*. 2006;**21**(4):1736–43

Part III

Stability analysis

Chapter 7
Small signal stability

7.1 Basic concepts

Power systems are subjected to either small or large disturbances. Small disturbances, e.g. load changes, occur continually. Thus, the system must have the ability to withstand the effect of such disturbances that implies change of conditions and restoration of the normal operating conditions. On the other hand, large disturbances (disturbances of a severe nature) such as loss of a large generator or short circuit on a transmission line may lead to structural changes as the faulted elements are isolated by actuating the protective relays. Some generators and loads may be disconnected to preserve the continuity of operation of bulk of the system. Interconnected systems, for certain severe disturbances, may also be intentionally split into a number of islands to preserve as much of the generation and load as possible, permitting the actions of automatic controls to eventually restore the system to a normal state. Otherwise, the system may become unstable, particularly, if the normal operation of the system could not be restored. In this case, a progressive increase in angular deviation between generator rotors or a progressive decrease in bus voltages (speed-up and speed-down situation) may occur. The instability condition of a power system could lead to cascading outages and a shutdown of a major portion of the system.

Analysing the power system with a goal of determining its stability is based on models of system components encompassing adequate assumptions to formulate an appropriate mathematical model in the time scale that properly describes the phenomenon under study. Then, by selecting an analytical method the stability of the power system can be determined when a specific disturbance occurs. The results can be examined to test the adequacy of model assumptions.

Referring to the equation of motion of the rotor, swing equation, discussed in Sections 2.5.1 and 2.5.2, Chapter 2, and considering the damping term it can be written as

$$\frac{2\mathrm{H}}{\omega_{\mathrm{B}}}\ddot{\delta} + D\dot{\delta} = P_{\mathrm{m}} - P_{\mathrm{e}} \tag{7.1}$$

where t is the time in seconds, H is the inertia constant (s), ω is the angular speed in elec. rad/s, D is the damping coefficient, P_{m} is the mechanical input power and P_{e} is the electrical output power. The difference, $P_{\mathrm{m}} - P_{\mathrm{e}}$, is called the accelerating power, P_{a}. The damping coefficient D is assumed to be small and positive.

Thus, the damping term $D\dot{\delta}$ can be ignored when solving the equation for a short period following a disturbance. Consequently, D can be neglected in transient stability but should be considered in small disturbance analysis.

Solution of (7.1) as a second-order non-linear ordinary differential equation gives the change of angle δ versus time to decide whether the generator is stable. As there is no analytic solution in general, numerical methods have to be used. This entails rewriting (7.1) in the form $\dot{\boldsymbol{x}} = f(\boldsymbol{x}, \boldsymbol{u})$ as below:

$$\left.\begin{aligned} \dot{\delta} &= \omega \\ \dot{\omega} &= \frac{\omega_{\mathrm{B}}}{2\mathrm{H}}[P_{\mathrm{m}} - P_{\mathrm{e}} - D\omega] \end{aligned}\right\} \tag{7.2}$$

The variables in (7.2) depend on machine parameters, network topology, nature of disturbance and characteristics of prime movers and automatic voltage regulators. Therefore, the response of the power system to a disturbance may involve much of the equipment. For example, a fault on a critical element followed by its isolation will cause variations in power flow, network bus voltages and machine rotor speeds. The voltage variations will actuate both generator and transmission network voltage regulators while the generator speed variations will actuate prime mover governors. As well the voltage and frequency variations will affect the system loads to varying degrees depending on their individual characteristics. Further, devices used to protect individual equipment may respond to variations in system variables and cause tripping of the equipment, thereby weakening the system and possibly leading to system instability.

Accordingly, (7.2) will be incorporated with mathematical equations representing the dynamic characteristics of both voltage regulators and excitation system to determine the electrical power delivered by generators as well as the characteristics of prime mover governors to calculate the input mechanical power. Additional equations have been included to consider some constraints such as excitation limits and voltage limits. This leads to modelling each generator in the power system by a system of algebraic-differential equations. The order of the model gets higher as more detailed models of involved devices are considered. Thus, a typical power system is a high-order non-linear multi-variable process whose dynamic response is influenced by a wide array of devices with different characteristics and response rates. It operates in a constantly changing environment (loads, generator output, operating parameters change continually). The system stability when subjected to a disturbance depends on initial operating conditions and the nature of disturbance as well.

It is seen that the stability problem is a high dimensional and complicated problem. So, some simplifying assumptions may help analysing specific types of problems with a condition of using appropriate degree of detail of system representation and adequate analytical techniques.

For instance, the simplified second-order model of a generator used in the classical methods of analysis is expressed by an equivalent voltage source, E_g, behind impedance, X_g, with considering some assumptions: (i) manual excitation control is used with absence of voltage regulators, i.e. in steady state the magnitude of the voltage source is determined by the field current, which is constant; (ii) damper circuits are neglected; (iii) transient stability is judged by the first swing; (iv) saliency has little effect and can be neglected; (v) flux decay in the field circuit is neglected for short period of time less than the field time constant; (vi) losses are neglected as the impedance is purely reactive; and (vii) response of prime mover governor is neglected, i.e. the input mechanical power to the generator is constant, Thus, (7.2) becomes

$$\left.\begin{aligned} \dot{\delta} &= \omega \\ \dot{\omega} &= \frac{\omega_{\mathrm{B}}}{2\mathrm{H}}[P_{\mathrm{m}} - P_{\mathrm{max}} \sin \delta] \end{aligned}\right\} \tag{7.3}$$

where $P_{\mathrm{max}} = \frac{E_g E_t}{X_g}$ as given in (1.1) and E_t is the terminal machine voltage.

If the generator is connected to an infinite bus, the external reactance is added to X_g and E_t is replaced by the infinite bus voltage to get the angle deviations with respect to the infinite bus voltage. On the other hand, if the generator is connected to a multi-machine system, the electrical output power delivered by the generator is calculated by applying load flow techniques that may contain some constraints [1].

7.1.1 Equilibrium points

Stability is a condition of equilibrium between opposing forces. For an electric power system, it is a property of the system motion around the initial operating condition when the system is subjected to a disturbance, small or large [2]. In the case of a small disturbance, at the equilibrium points, different opposing forces that exist in the system are equal instantaneously or over a cycle as in the case of slow cyclic variations. For large disturbances, it is impractical and uneconomical to satisfy equilibrium points at which the various sets of opposing forces are balanced for every possible disturbance. Therefore, design contingencies are selected on the basis that they have a reasonable probability of occurrence. Hence, large disturbance stability refers to a specified disturbance scenario. On the other hand, the various sets of opposing forces may experience sustained imbalance leading to different forms of instability, depending on system operating condition, type of disturbance and network topology.

Various equilibrium points may get involved in the stability analysis of power systems. In the case of large disturbances, the perturbations of interest are specified and all post-disturbance equilibrium points relevant for a given pre-disturbance equilibrium are assumed to be determined. Numerical methods are usually used to solve a large non-linear set of differential equations in the form:

$$\dot{x} = f(x, u, t) \tag{7.4}$$

where x is the state vector and a function of time t of dimension n, $\dot{x}$ is its derivative, f is differentiable function of dimension n *with* a domain including the origin (called vector field) and u can be viewed, generally, as control input vector of dimension r.

As the state vector x is a function of time, the explicit writing of time argument t can be omitted and commonly (7.4) can be rewritten as

$$\dot{x} = f(x, u) \tag{7.5}$$

For small disturbances, the perturbation is characterised by size (small signal analysis) and linearised system equations are valid for such analysis. The same equilibrium point typically characterises the system pre and post disturbance. The linearised form of (7.5) is

$$\dot{x} = \mathrm{A}x + \mathrm{B}u \tag{7.6}$$

and the stability analysis is based on studying its characteristic equation as explained in the forthcoming sections.

The system described by (7.4) is said to be autonomous if u is a constant vector. Otherwise, if the elements of u are explicit functions of time t, the system is said to be non-autonomous.

For autonomous systems, the solution of (7.5) for a specified initial condition $x(t_o) = x_o$ can be expressed as $\Phi_t(x_o)$ to illustrate explicitly the dependence on initial condition. $\Phi_t\,(x_o)$ is called the trajectory through x_o while $\Phi_t(x)$ where $x \in R^n$ is called the flow. The vector u is not explicitly written as it is constant and can be considered a parameter. Meanwhile, for non-autonomous systems the trajectory is also a function of time t and can be expressed as $\Phi_t(x_o,\ t_o)$ to illustrate that the solution passes through x_o at t_o.

As the power system can be modelled as an autonomous system, the solution of (7.5) for autonomous systems is of interest taking some consideration into account such as (i) the presence of solution for all t, (ii) with respect to the initial condition the derivative of a trajectory exists and is non-singular, i.e. $\Phi_t(x_o)$ is continuous with respect to initial state x_o and (iii) $\Phi_t(x) = \Phi_t(y)$ at any time t if and only if $\Phi_{t1+t2} = \Phi_{t1}\Phi_{t2}$. Consequently, a trajectory of an autonomous system is uniquely specified by its initial condition and that distinct trajectories do not intersect.

Therefore, an equilibrium point x_{eq} of an autonomous system is a constant solution $\Phi_t(x_{eq})$ and satisfies the relation

$$0 = f(x_{eq}, u) \tag{7.7}$$

It is to be noted that the real solutions of (7.7) give several equilibrium points.

7.1.2 Stability of equilibrium point

The equilibrium point that satisfies (7.7) is said to be an asymptotically stable equilibrium point (SEP) if all trajectories lying initially in a sufficiently small spherical neighbourhood of radius ρ can be forced for all time $t \geq t_o$ to lie entirely in a given cylinder of radius $\mathcal{E}$. Figure 7.1 depicts the asymptotic stability for the case of a two-dimensional system in the space of real variables [3]. Thus, for each $\mathcal{E} > 0$, $\rho = \rho(\mathcal{E},\ t_o)$, such that

$$||x(t_o)|| < \rho \Rightarrow ||x(t)|| < \mathcal{E}\, \forall\, t \geq t_o > 0 \tag{7.8}$$

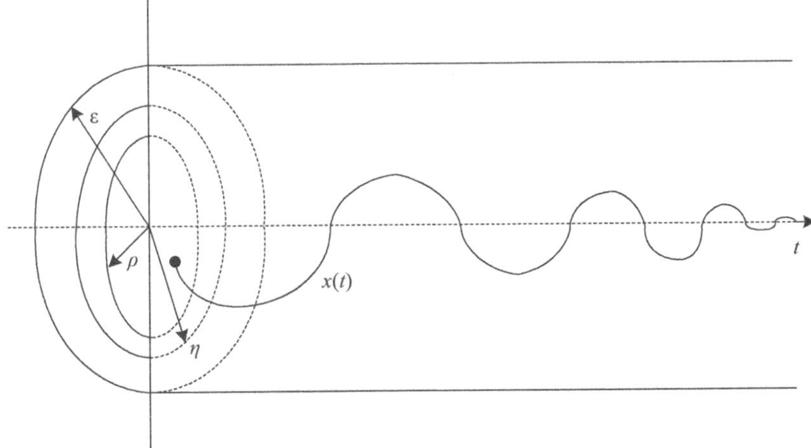

Figure 7.1 Asymptotic stable trajectories

and $\eta(t_o) > 0$, such that

$$||x(t_o)|| < \eta(t_o) \Rightarrow x(t) \rightarrow 0 \quad \text{as } t \rightarrow \infty \tag{7.9}$$

The stability of an equilibrium point should be decided [4]. For small disturbances, it can be determined by the solution of the linearised differential equations describing the system behaviour at x_{eq}. Assuming small changes in system quantities, linearisation can be found by making a Taylor series expansion about x_{eq} and neglecting higher-order terms as below.

Assuming $x = x_{eq} + \Delta x$ and substituting in (7.5) gives

$$\dot{x} = \dot{x}_{eq} + \Delta\dot{x} = f(x_{eq}, u) + \left[\frac{\partial f(x,u)}{\partial x}\right]_{x=x_{eq}} \Delta x \tag{7.10}$$

From (7.7) and (7.10) the following relation can be obtained:

$$\Delta\dot{x} = [A(x_{eq}, u)]\Delta x \tag{7.11}$$

where A is a matrix of dimension $n \times n$. Its elements are functions of x_{eq} and u. The ijth element is given by

$$A_{ij}(x_{eq}, u) = \frac{\partial f_i}{\partial x_j}(x_{eq}, u) \tag{7.12}$$

The matrix A is a constant matrix for a given x_{eq} and u. The solution of (7.11) can be found as

$$\Delta x(t) = e^{A(t-t_o)}\Delta x(t_o) = c_1 e^{\lambda_1 t} v_1 + c_2 e^{\lambda_2 t} v_2 + \cdots + c_n e^{\lambda_n t} v_n \tag{7.13}$$

where $c_1, c_2, \ldots, c_n$ are constants depending on the initial conditions. λ_i and v_i are the ith eigenvalue and corresponding eigenvector of matrix $\boldsymbol{A}$, respectively.

By examining (7.13) it is found that (i) if $\Re e(\lambda_i) < 0$ for all λ_i, $i = 1, 2, \ldots, n$, then for any sufficiently small disturbance from the equilibrium point x_{eq}, the trajectories tend to x_{eq} as $t \to \infty$. Thus, the equilibrium point x_{eq} is said to be asymptotically stable; (ii) if $\Re e(\lambda_i) > 0$ for all λ_i, then any disturbance leads to a trajectory leaving the neighbourhood of x_{eq} and the equilibrium point is unstable; (iii) the equilibrium point is a saddle point if there are two eigenvalues λ_i and λ_j such that $\Re e(\lambda_i) < 0$ and $\Re e(\lambda_j) > 0$.

It is concluded that an equilibrium point is asymptotically stable if all nearby trajectories approach x_{eq} as $t \to \infty$. It will be an unstable equilibrium point (UEP) if no trajectories in proximity remain nearby. Of course, an UEP is asymptotically stable in reverse time, i.e. as $t \to -\infty$. An equilibrium point is unstable and called a 'saddle point' if there is a nearby trajectory that approaches x_{eq} as $t \to \infty$ and another trajectory approaches x_{eq} as $t \to -\infty$.

7.1.3 Phasor diagrams of synchronous machines

The stability study concerns with the examination of machine behaviour when the system is subjected to a disturbance and its state is investigated (stable or not) after clearing the disturbance. This can be achieved by solving system equations [5]. As explained in Chapters 2 and 3 and in Section 7.1, the system including synchronous generators is represented by a set of differential equations to describe its behaviour as a function of time. Solution of these equations entails determination of the initial conditions that can be considered the steady-state condition prior to the disturbance. Consequently, phasor diagrams representing phasor equations help determine the initial conditions as in steady state the differential equations are not needed where the variables are constants or varying sinusoidaly with time.

At steady state, the currents in (3.16) are constants.

Thus,

$$\mathrm{p}i_d = \mathrm{p}i_q = \mathrm{p}i_f = \mathrm{p}i_{kd} = \mathrm{p}i_{kq} = 0 \tag{7.14}$$

and

$$R_{kd}i_{kd} = R_{kq}i_{kq} = 0 \text{ (as } i_{kd} = i_{kq} = 0 \text{ at steady state)} \tag{7.15}$$

Therefore, the voltages v_d and v_q from (3.16) can be written as

$$v_d = -R_a i_d - \omega L_q i_q \quad \text{and} \quad v_q = -R_a i_q + \omega L_d i_d + kM_f \omega i_f \tag{7.16}$$

Using (2.16) and substituting $v_o = 0$ at balanced conditions, the phase voltage v_a is

$$v_a = \sqrt{\frac{2}{3}}(v_d \cos\theta + v_q \sin\theta) \tag{7.17}$$

where θ is defined as $\omega_o t + \delta + \pi/2$

From (7.16) and (7.17) the equation below can be obtained:

$$v_a = \sqrt{\frac{2}{3}}\Big[-(R_a i_d + \omega L_q i_q)\cos\Big(\omega_o t + \delta + \frac{\pi}{2}\Big) + (-R_a i_q + \omega L_d i_d + kM_f \omega i_f)\sin\Big(\omega_o t + \delta + \frac{\pi}{2}\Big)\Big]$$

Hence,

$$v_a = \sqrt{\frac{2}{3}}\Big[-(R_a i_d + \omega L_q i_q)\cos\Big(\omega_o t + \delta + \frac{\pi}{2}\Big) + (-R_a i_q + \omega L_d i_d + kM_f \omega i_f)\cos(\omega_o t + \delta)\Big] \tag{7.18}$$

At steady state, it is to be noted that (i) the angular speed ω is constant and equals ω_o; (ii) the direct and quadrature axis reactances 'X_d and X_q' equal ωL_d and ωL_q, respectively; (iii) $\omega k M_f i_f = \sqrt{3}E$, where E is the rms of stator EMF as a line-to-neutral value corresponding to the field current i_f as given in (I.11), Appendix I. Accordingly, the rms voltage phasor $\boldsymbol{V}_a$ can be found as

$$\boldsymbol{V}_a = -R_a\left(\frac{i_d}{\sqrt{3}}\angle\Big(\delta + \frac{\pi}{2}\Big) + \frac{i_q}{\sqrt{3}}\angle\delta\right) - X_q\frac{i_q}{\sqrt{3}}\angle\Big(\delta + \frac{\pi}{2}\Big) + X_d\frac{i_d}{\sqrt{3}}\angle\delta + E\angle\delta \tag{7.19}$$

Using the relation $j = 1\angle\pi/2$, (7.19) becomes

$$\boldsymbol{V}_a = -R_a\left(\frac{i_q}{\sqrt{3}}\angle\delta + j\frac{i_d}{\sqrt{3}}\angle\delta\right) - jX_q\frac{i_q}{\sqrt{3}}\angle\delta + X_d\frac{i_d}{\sqrt{3}}\angle\delta + E\angle\delta \tag{7.20}$$

where $\boldsymbol{V}_a$ and E are stator rms phase voltages in pu, using the rated phase quantities as base values (Appendix I); i_d and i_q are the currents obtained from the Park's transformation. The rms equivalent d and q axes currents are defined as

$$I_d \triangleq \frac{i_d}{\sqrt{3}} \quad \text{and} \quad I_q \triangleq \frac{i_q}{\sqrt{3}} \tag{7.21}$$

and the stator current i_a as a phasor will have the two rectangular components I_d and I_q. Thus, the relation below can be written if the q-axis is a phasor reference:

$$\boldsymbol{I}_a = (I_q + jI_d)e^{j\delta} \tag{7.22}$$

Incorporating (7.21) and (7.22) into (7.20) gives

$$\boldsymbol{E}_I = \boldsymbol{V}_a + R_a\boldsymbol{I}_a + jX_q\boldsymbol{I}_q + jX_d\boldsymbol{I}_d \tag{7.23}$$

where $\boldsymbol{E}_I = E_I\angle\delta$, $\boldsymbol{I}_q = I_q\angle\delta$, $\boldsymbol{I}_d = jI_d\angle\delta$

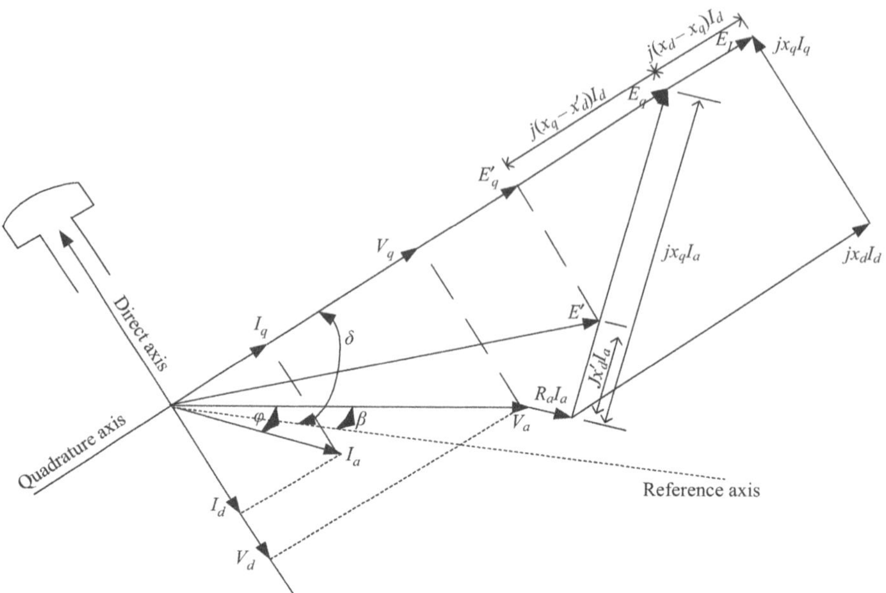

Figure 7.2 Phasor diagram of synchronous machine to determine steady state parameters

From (7.16) the rms stator voltages can be computed as

$$V_d \triangleq \frac{v_d}{\sqrt{3}} = -R_a I_d - X_q I_q \quad \text{and} \quad V_q \triangleq \frac{v_q}{\sqrt{3}} = -R_a I_q + X_d I_d + E \qquad (7.24)$$

Equation (7.23) can be represented by the phase diagram shown in Figure 7.2. It is noted from Figure 7.2 that the components of voltages and currents in the direction of d-axis are negative quantities as the current I_a lags the voltage V_a and the direct axis leads the quadrature axis by 90 electrical degrees. The position of the quadrature axis can be determined by calculating a fictitious voltage located on this axis. The voltage E' is defined as the internal machine voltage behind transient direct axis reactance, X'_d. Its projection in q-axis direction is known as E'_q. The difference between E_q and X'_q is given by $j(X_q - X'_d)I_d$. Also, the difference between E_I 'equals E_{fd} in steady state' and E_q is $j(X_d - X_q)I_d$.

E_I is a voltage proportional to field current, E_{fd} is a term representing the field voltage acting along the quadrature axis, E_q is the voltage behind quadrature axis synchronous reactance and X'_q is the voltage proportional to the field flux linkages resulting from the combined effect of the field and armature currents.

7.2 Small signal stability

Regarding the rotor angle stability, small signal stability is the ability of power system to maintain synchronism under small disturbances. Some analyses have been done to determine this type of system stability [6–8]. The disturbances are considered to be sufficiently small so that linearisation of system equations is allowed for analysis.

To mathematically explain the small signal stability problem, the dynamic behaviour of a machine connected to an infinite bus is considered. The equation of motion, swing equation, of the synchronous machine (7.25) is a non-linear function of the power angle δ.

$$\left.\begin{aligned}\frac{2\mathrm{H}}{\omega_{\mathrm{B}}}\ddot{\delta} &= P_{\mathrm{m}} - P_{\mathrm{e}}\\ &= P_{\mathrm{m}} - P_{\max}\sin\delta\end{aligned}\right\} \tag{7.25}$$

For small disturbances, linearisation of (7.14) can be applied with acceptable accuracy.

Assuming $\Delta\delta$ is the small deviations in power angle from its initial operating point δ_o. Then, (7.25) becomes

$$\frac{2\mathrm{H}}{\omega_{\mathrm{B}}}\frac{\mathrm{d}^2(\delta_o+\Delta\delta)}{\mathrm{d}t^2} = P_{\mathrm{m}} - P_{\max}\sin(\delta_o+\Delta\delta)$$

i.e.

$$\frac{2\mathrm{H}}{\omega_{\mathrm{B}}}\frac{\mathrm{d}^2\delta_o}{\mathrm{d}t^2} + \frac{2\mathrm{H}}{\omega_{\mathrm{B}}}\frac{\mathrm{d}^2\Delta\delta}{\mathrm{d}t^2} = P_{\mathrm{m}} - P_{\max}(\sin\delta_o\cos\Delta\delta + \cos\delta_o\sin\Delta\delta)$$

Substituting $\sin\Delta\delta \cong \Delta\delta$ and $\cos\Delta\delta \cong 1$ as $\Delta\delta$ is small, gives

$$\frac{2\mathrm{H}}{\omega_{\mathrm{B}}}\frac{\mathrm{d}^2\delta_o}{\mathrm{d}t^2} + \frac{2\mathrm{H}}{\omega_{\mathrm{B}}}\frac{\mathrm{d}^2\Delta\delta}{\mathrm{d}t^2} = P_{\mathrm{m}} - P_{\max}\sin\delta_o - P_{\max}\cos\delta_o\Delta\delta \tag{7.26}$$

At the initial state

$$\frac{2\mathrm{H}}{\omega_{\mathrm{B}}}\frac{\mathrm{d}^2\delta_o}{\mathrm{d}t^2} = P_{\mathrm{m}} - P_{\max}\sin\delta_o$$

Thus, (7.26) in the linearised form and written as

$$\frac{2\mathrm{H}}{\omega_{\mathrm{B}}}\frac{\mathrm{d}^2\Delta\delta}{\mathrm{d}t^2} + P_{\max}\cos\delta_o\Delta\delta = 0 \tag{7.27}$$

is the power–angle curve. At δ_o, i.e. $\frac{\mathrm{d}P_e}{\mathrm{d}\delta}|_{\delta_o}$, its value is known as the 'synchronising power coefficient' P_s. This coefficient has a prominent role in system stability determination.

Equation (7.27) can be rewritten as

$$\frac{2H}{\omega_B}\frac{d^2\Delta\delta}{dt^2} + P_s\Delta\delta = 0 \tag{7.28}$$

and has the solution depending on the roots of its characteristic equation $\frac{2H}{\omega_B}s^2 + P_s = 0$. The roots are given by

$$s = \pm j\sqrt{\frac{P_s\omega_B}{2H}} \tag{7.29}$$

In case of a negative value of P_s, the response is exponentially increasing and stability is lost as one root lies in the right-half of the s-plane. When P_s is positive the two roots lie on the imaginary axis and the motion is un-damped oscillatory. The system is marginally stable with a natural frequency of oscillation, ω_n, given by

$$\omega_n = \sqrt{\frac{P_s\omega_B}{2H}} \tag{7.30}$$

It can be seen that the synchronising power coefficient, $P_s = dP_e/dt$, is positive when δ lies between 0° and 90° with a maximum value at $\delta_o = 0°$.

The damping torque is a component of electrical torque. It is proportional to the speed change and will be set up on the rotor tending to minimise the difference between the rotor angular velocity and the angular velocity of the resultant rotating air gap field. The damping power 'P_d' is approximately proportional to the speed deviation and can be expressed as

$$P_d = D\frac{d\delta}{dt} \tag{7.31}$$

where D is the damping coefficient in pu.

The damping coefficient D has a small value and can be neglected in transient analysis as the solution of swing equation is required for a short period, e.g. 1–2 s following a disturbance. However, in small signal stability analysis it should be considered as the damping power oscillations may damp out eventually when P_s is positive and the operation at the equilibrium angle will be restored. Therefore, a damping power term is added to (7.28) to become

$$\frac{2H}{\omega_B}\frac{d^2\Delta\delta}{dt^2} + D\frac{d\Delta\delta}{dt} + P_s\Delta\delta = 0 \tag{7.32}$$

and the roots of its characteristic equation determine the system response.

Further mathematical explanation is given below by writing (7.32) in state space form to make it possible to extend the analysis to multi-machine systems.

Equation (7.32) is rewritten as

$$\frac{d^2\Delta\delta}{dt^2} + \frac{\omega_B}{2H} D \frac{d\Delta\delta}{dt} + \frac{\omega_B}{2H} P_s \Delta\delta = 0 \tag{7.33}$$

Assuming $x_1 = \Delta\delta$ and $x_2 = \Delta\omega = \Delta\dot{\delta}$, then

$$\left.\begin{aligned} \dot{x}_1 &= x_2 \\ \dot{x}_2 &= -\frac{\omega_B}{2H} P_s x_1 - \frac{\omega_B}{2H} D x_2 \end{aligned}\right\} \tag{7.34}$$

and in matrix form, (7.34) is written as

$$\begin{bmatrix} \dot{x}_1 \\ \dot{x}_2 \end{bmatrix} = \begin{bmatrix} 0 & 1 \\ -\dfrac{P_s}{\hat{H}} & -\dfrac{D}{\hat{H}} \end{bmatrix} \begin{bmatrix} x_1 \\ x_2 \end{bmatrix} \tag{7.35}$$

where $\hat{H} = \frac{2H}{\omega_B}$

or

$$\dot{\mathbf{X}}(t) = \mathbf{A}\mathbf{X}(t) \tag{7.36}$$

where

$$\mathbf{A} = \begin{bmatrix} 0 & 1 \\ -\dfrac{P_s}{\hat{H}} & -\dfrac{D}{\hat{H}} \end{bmatrix}$$

Equation (7.36) is a homogeneous state equation, unforced state variable equation, as it is assumed that the disturbances causing the changes disappear. When the state variables are the desired response the output vector $\boldsymbol{y}(t)$ is defined as $\boldsymbol{y}(t) = \mathbf{C}\mathbf{x}(t)$, where $\mathbf{C}$ is a unit matrix of dimension 2×2.

Applying Laplace transform gives

$$s\mathbf{X}(s) - \mathbf{x}(o) = \mathbf{A}\mathbf{X}(s)$$

or

$$\mathbf{X}(s) = (s\mathbf{I} - \mathbf{A})^{-1}\mathbf{x}(o) \tag{7.37}$$

where

$$(s\mathbf{I} - \mathbf{A}) = \begin{bmatrix} s & -1 \\ -\dfrac{P_s}{\hat{H}} & s + \dfrac{D}{\hat{H}} \end{bmatrix}$$

Hence,

$$\mathbf{X}(s) = \frac{\begin{bmatrix} s + \dfrac{D}{\hat{\mathrm{H}}} & 1 \\ -\dfrac{P_s}{\hat{\mathrm{H}}} & s \end{bmatrix} \mathbf{x}(o)}{s^2 + \dfrac{D}{\hat{\mathrm{H}}} s + \dfrac{P_s}{\hat{\mathrm{H}}}}$$

If the rotor is suddenly disturbed by a small angle $\Delta\delta_o$, the state variables $x_1(o) = \Delta\delta_o$ and $x_2(o) = \Delta\omega_o = 0$. Then

$$\Delta\delta(s) = \frac{\left(s + \dfrac{D}{M\hat{\mathrm{H}}}\right)\Delta\delta_o}{s^2 + \dfrac{D}{\hat{\mathrm{H}}} s + \dfrac{P_s}{\hat{\mathrm{H}}}} \quad \text{and} \quad \Delta\omega(\mathrm{s}) = \frac{\dfrac{P_s}{\hat{\mathrm{H}}}\Delta\delta_o}{s^2 + \dfrac{D}{\hat{\mathrm{H}}} s + \dfrac{P_s}{\hat{\mathrm{H}}}} \tag{7.38}$$

The eigenvalues are the roots of the characteristic equation $s^2 + \frac{D}{\hat{\mathrm{H}}} s + \frac{P_s}{\hat{\mathrm{H}}} = 0$ and given by

$$\lambda = -\frac{D}{2\hat{\mathrm{H}}} \pm \sqrt{\frac{D^2}{4\hat{\mathrm{H}}^2} - \frac{P_s}{\hat{\mathrm{H}}}} \tag{7.39}$$

It can be seen that both eigenvalues have negative real parts when P_s is positive. If P_s is negative one of the eigenvalues is positive real. For small D and $P_s > 0$ the eigenvalues are complex and given by

$$\lambda = -\sigma \pm j\omega_d \tag{7.40}$$

where

$$\sigma = \left(D/2\hat{\mathrm{H}}\right) \quad \text{and} \quad \omega_d = \sqrt{\tfrac{P_s}{\hat{\mathrm{H}}} - \tfrac{D^2}{4\hat{\mathrm{H}}^2}} \equiv \text{damping frequency of oscillation}$$

Then, (7.38) can be rewritten as

$$\Delta\delta(s) = \frac{(s + 2\sigma)\Delta\delta_o}{s^2 + 2\sigma s + \omega_n^2} \quad \text{and} \quad \Delta\omega(s) = \frac{\omega_n^2 \Delta\delta_o}{s^2 + 2\sigma s + \omega_n^2} \tag{7.41}$$

Taking inverse Laplace transform results in zero-input response:

$$\Delta\delta = \frac{\omega_n}{\omega_d}\Delta\delta_o \mathrm{e}^{-\sigma t}\sin(\omega_d t + \theta) \quad \text{and} \quad \Delta\omega = \frac{\omega_n^2}{\omega_d}\Delta\delta_o \mathrm{e}^{-\sigma t}\sin\omega_d t \tag{7.42}$$

where $\theta = \cos^{-1}(\sigma/\omega_n)$

$\Delta\delta$ and $\Delta\omega$ are added to δ_o and ω_o, respectively, to give the rotor angle and angular frequency.

The response time constant, τ, and the damping ratio, D_R, are defined as

$$\tau = 1/\sigma \quad \text{and} \quad D_R = \frac{\sigma}{\omega_n} = \frac{\sigma}{\sqrt{\sigma^2 + \omega_d^2}} = \sqrt{\frac{D^2}{4\hat{\mathrm{H}}P_s}} \tag{7.43}$$

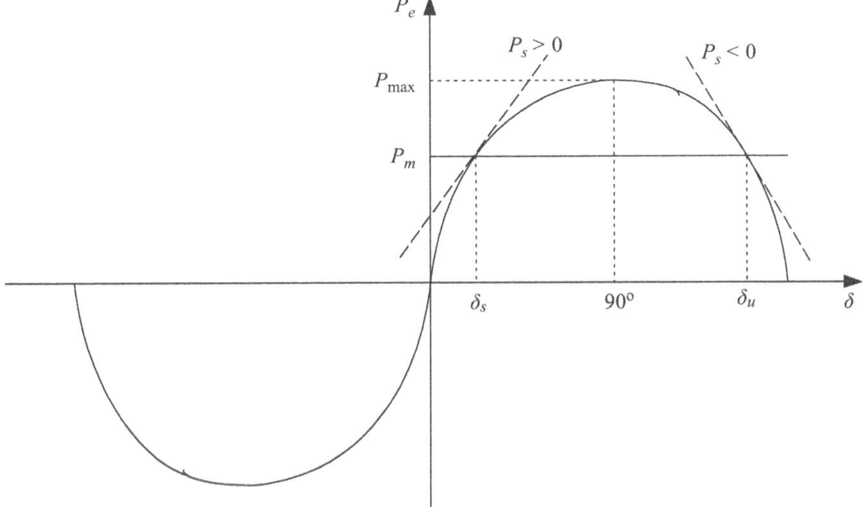

Figure 7.3 Power–angle curve with stable and unstable equilibrium points

Therefore, for stability of an equilibrium point, the necessary condition $P_s > 0$ must be satisfied. This is illustrated by the power–angle curve, shown in Figure 7.3. When $P_m < P_{\max}$ there are two values of δ corresponding to a specified value of P_m and $-180° < \delta < 180°$. Thus, there are two equilibrium points as

$$\left.\begin{aligned} x_{\rm s} &= (\delta_{\rm s}, 0) \triangleq SEP \quad \text{at } P_{\rm s} > 0 \\ x_{\rm u} &= (\delta_{\rm u}, 0) \triangleq UEP \quad \text{at } P_{\rm s} < 0 \end{aligned}\right\} \tag{7.44}$$

where the subscripts s and u denote stable and unstable states, respectively.

The maximum electrical power output, $P_{\max}$, at $\delta = 90°$ is called the 'steady-state stability limit'. This stability limit is the boundary at which the system becomes unstable if any small disturbance occurs.

Example 7.1 Referring to (7.40) find the eigenvalues loci in the s-plane as P_m is varied.

Solution:

At a specific value of P_m there are two equilibrium points: one of them is SEP and the other is UEP. These two equilibrium points come closer as P_m is increased.

As shown in Figure 7.4(a), at SEP and $0 < P_m < P_{\max}$: It is found that P_s is positive and the eigenvalues are complex quantities with negative real parts (points #1 and #2). As P_m increases, P_s decreases and both eigenvalues reach the *Re-axis* at $-\sigma$ when $\frac{P_{\rm s}}{\hat{\rm H}} = \frac{D^2}{4\hat{\rm H}^2}$ (point #3). Continuing the increase in P_m, P_s continues to

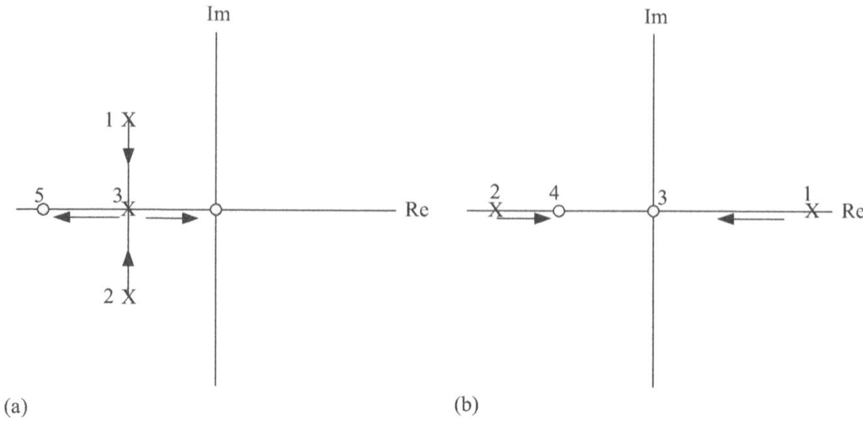

Figure 7.4 Eigenvalues loci: (a) at SEP, (b) at UEP

decrease until being zero at $P_m = P_{\max}$ and then one of the eigenvalues reaches the origin while the other approaches $-D/\hat{H}'$, points #4 and #5, respectively.

At UEP, Figure 7.4(b), P_s is negative and both eigenvalues are real and move towards the origin (points #1 and #2). When $P_m = P_{\max}$ one of the values is zero, i.e. the eigenvalue reaches the origin, point #3, and the other equals $(-D/\hat{H})$, point #4.

Example 7.2 A synchronous generator is connected to an infinite bus through a transformer and a transmission line with data in pu as shown in Figure 7.5 and delivers a power of 0.8 pu. Find the equilibrium points when (i) neglecting the system resistance and (ii) considering the system resistance.

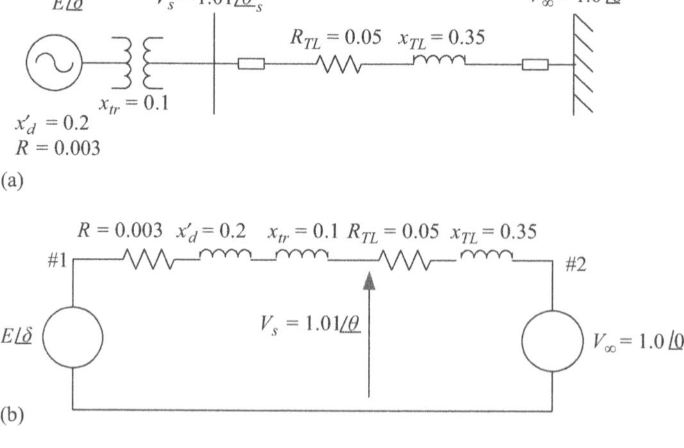

Figure 7.5 Machine connected to an infinite bus: (a) single-line diagram, (b) the equivalent circuit

Solution:

The power delivered by the machine, P_e, can be calculated by (3.45), that is,

$$P_e = E^2 G_{11} + EV_\infty Y_{12} \sin(\delta - \gamma)$$

where

$$G_{11} \triangleq Y_{11} \cos\theta_{11}, \quad \gamma = \theta_{12} - \pi/2, \quad Y_{11}\angle\theta_{11} = y_{12} + y_{1o}$$
$$Y_{12}\angle\theta_{12} = -y_{12} \quad \text{and} \quad \gamma = \theta_{12} - \pi/2$$

This relation indicates that the power–angle curve of the synchronous generator is a sine curve shifted from the origin vertically by an amount $E^2 G_{11}$ and horizontally by the angle 'γ'.

(i) The system resistance is neglected: In this case, the transmission network is reactive.

$y_{1o} = 0$ as there is no shunt admittance at the sending bus, $y_{12} = 1/j0.65 = -j1.538$, $Y_{11} = -j1.538$, $Y_{12} = j1.538$, $\theta_{11} = -\pi/2$, $\theta_{12} = \pi/2$

It can be seen that both $E^2 G_{11}$ and γ are equal to zero. Thus, the power at the sending bus, P_1, is $P_1 = 0.8 = V_s V_\infty Y_{TL} \sin\theta_s = (1.01/0.35) \sin\theta_s$

Then $\theta_s = 16.09°$

The current flow, I, can be calculated as

$$I = (V_s - V_a)/Z_{TL} = (1.01\angle 16.09° - 1.0\angle 0°)/j0.35 = 0.804\angle 1.88°$$

The internal machine voltage is

$$E\angle\delta = 1.01\angle 16.09° + (0.804\angle 1.88°)(0.3\angle 90°) = 1.09\angle 28.4°$$

Therefore, the power delivered by the machine to the infinite bus, P_e, is given by

$$P_e = [(1.09 \times 1.0)/0.65] \sin\delta = 1.677 \sin\delta$$

Hence, at $P_e = 0.8$, the power angle at the two equilibrium points is 26.986° and 153.014°. The first represents SEP while the second is UEP, i.e. $\delta_s = 26.986°$ and $\delta_u = 153.014°$

(ii) Considering the system resistance: From the sending bus to the infinite bus it is found that

$$Y_{s\infty} = -1/(0.05 + j0.35) = -0.4 + j2.8 = 2.828\angle 98.13°$$
$$Y_{ss} = 0.4 - j2.8 = 2.828\angle -81.87°$$
$$\theta_{s\infty} = 98.13°, \quad \theta_{ss} = -81.87°, \quad \gamma = 98.13 - 90 = 8.13°$$
$$G_{ss} = 2.828 \times \cos(-81.87) = 0.4$$

The power delivered at the sending bus, $P_1 = 0.8 = 0.4(1.01)^2 + 1.01 \times 2.828 \sin(\theta_s - 8.13)$

Hence, $\theta_s = 16.02°$

The current, I, is given by

$$I = (1.01\angle 16.02° - 1.0\angle 0°)/(0.05 + j0.35) = 0.768 + j0.192 = 0.79\angle 14.04°$$

The internal machine voltage $E\angle\delta = 1.01\angle 16.02° + 0.79\angle 14.04°\ (0.003 + j0.3) = 0.915 + j0.509 = 1.04\angle 29.08°$

For the system shown by its equivalent circuit (Figure 7.4(b)):

$$Y_{12} = -1/(0.035 + j0.65) = -0.0813 + j0.951 = 0.9545\angle 94.9°$$

$$Y_{11}\angle\theta_{11} = 0.0813 - j0.951 = 0.9545\angle - 85.1°$$

$$\theta_{12} = 94.9°, \quad \theta_{11} = -85.1°, \quad \gamma = 94.9 - \pi/2 = 4.9° \quad \text{and}$$
$$G_{11} = 0.9545 \times 0.085 = 0.081$$

The power delivered from the machine to the infinite bus, P_e, is given by

$$P_e = 0.09 + \left(1.04/\sqrt{0.053^2 + 0.65^2}\right)\sin(\delta - \gamma)$$
$$= 0.09 + 1.6\ \sin(\delta - 4.9°)$$

At $P_e = 0.8 = 0.09 + 1.6\ \sin(\delta - 4.9°)$

Therefore, the rotor angle at SEP, $\delta_s = 29°$, and at UEP, $\delta_u = 151°$.

7.2.1 Forced state variable equation

On the other hand, if the system is subjected to an increase of power input by a small amount ΔP the system response can be determined by a linearised forced swing equation that becomes

$$\frac{d^2\Delta\delta}{dt^2} + \frac{\omega_B}{2H}D\frac{d\Delta\delta}{dt} + \frac{\omega_B}{2H}P_s\Delta\delta = \frac{\omega_B}{2H}\Delta P \tag{7.45}$$

Incorporating (7.30), (7.40) and (7.45) gets

$$\frac{d^2\Delta\delta}{dt^2} + 2\sigma\frac{d\Delta\delta}{dt} + \omega_n^2\Delta\delta = \Delta u \tag{7.46}$$

where

$$\Delta u = \frac{\omega_B}{2H}\Delta P$$

Assuming $x_1 = \Delta\delta$ *and* $x_2 = \Delta\omega = \Delta\dot{\delta}$, (7.46) is given in the state space form as

$$\left.\begin{aligned} \dot{x}_1 &= x_2 \\ \dot{x}_2 &= -\omega_n^2 x_1 - 2\sigma x_2 + \Delta u \end{aligned}\right\} \tag{7.47}$$

and in matrix form

$$\begin{bmatrix} \dot{x}_1 \\ \dot{x}_2 \end{bmatrix} = \begin{bmatrix} 0 & 1 \\ -\omega_n^2 & -2\sigma \end{bmatrix} \begin{bmatrix} x_1 \\ x_2 \end{bmatrix} + \begin{bmatrix} 0 \\ 1 \end{bmatrix} \Delta u \tag{7.48}$$

or

$$\dot{\mathbf{X}}(t) = \mathbf{A}\mathbf{X}(t) + \mathbf{B}\Delta u(t) \tag{7.49}$$

Equation (7.49) represents a forced state variable equation with x_1 and x_2 the desired response, and the output vector $\boldsymbol{y}(t)$ is given by $\boldsymbol{y}(t) = \mathbf{C}\mathbf{x}(t)$, where $\mathbf{C}$ is a unit matrix of dimension 2×2.

The Laplace transform of (7.49) is

$$s\mathbf{X}(s) = \mathbf{A}\mathbf{X}(s) + \mathbf{B}\Delta U(s)$$

or

$$\mathbf{X}(s) = (s\mathbf{I} - \mathbf{A})^{-1}\mathbf{B}\Delta U(s) \tag{7.50}$$

where $\Delta U(s) = \Delta u/s$

By substituting for $(s\mathbf{I} - \mathbf{A})^{-1}$ and $\mathbf{B}$ in (7.49) obtain

$$\mathbf{X}(s) = \frac{\begin{bmatrix} s + 2\sigma & 1 \\ -\omega_n^2 & s \end{bmatrix} \begin{bmatrix} 0 \\ 1 \end{bmatrix} \dfrac{\Delta u}{s}}{s^2 + 2\sigma s + \omega_n^2}$$

Hence,

$$\Delta\delta(s) = \frac{\Delta u}{s\left(s^2 + 2\sigma s + \omega_n^2\right)} \quad \text{and} \quad \Delta\omega(s) = \frac{\Delta u}{s^2 + 2\sigma s + \omega_n^2}$$

In the step response inverse Laplace transform gives

$$\Delta\delta = \frac{\Delta u}{\omega_n^2}\left[1 - \frac{\omega_n}{\omega_d} \mathrm{e}^{-\sigma t} \sin(\omega_d t + \theta)\right] \quad \text{and} \quad \Delta\omega = \frac{\Delta u}{\omega_d} \mathrm{e}^{-\sigma t} \sin \omega_d t \tag{7.51}$$

where $\theta = \cos^{-1}(\sigma/\omega_n)$

In (7.46) Δu is defined as $\Delta u = \frac{\omega_B}{2H}\Delta P = \frac{\Delta P}{\hat{H}}$. Substituting its value into (7.51) to obtain $\Delta\delta$ and $\Delta\omega$, then adding to δ_o and ω_o, respectively, gives the rotor angle in electrical radians and rotor angular frequency in radians per second as

$$\left.\begin{aligned} \delta &= \delta_o + \frac{\Delta P}{\hat{H}\omega_n^2}\left[1 - \frac{\omega_n}{\omega_d} \mathrm{e}^{-\sigma t} \sin(\omega_d t + \theta)\right] \\ \omega &= \omega_0 + \frac{\Delta P}{\hat{H}\omega_d} \mathrm{e}^{-\sigma t} \sin \omega_d t \end{aligned}\right\} \tag{7.52}$$

The criterion of stability, $P_s < 0$, explained above is a simple algebraic relation and does not need the computation of eigenvalues. It is derived from dynamic analysis based on some assumptions. Therefore, as the complexity of system dynamics increases, a detailed linearised model of the synchronous generator is required and the assumptions should be avoided. Two models, current and flux linkage state space models, have been described in Chapter 2. Their linearisation is explained as below.

7.3 Linearised current state space model of a synchronous generator

To linearise the current model given in (2.104), the state space vector $\mathbf{x}$ is assumed to have an initial state $\mathbf{x}_o = \mathbf{x}(t_o)$ that is known and constant for a specific dynamic study [9]. Hence,

$$\mathbf{x}_o^t = \left[i_{do}, i_{fo}, i_{kdo}, i_{qo}, i_{kqo}, \omega_o, \delta_o\right] \tag{7.53}$$

When a small disturbance occurs, the states will change slightly from their initial values and become

$$\mathbf{x} = \mathbf{x}_o + \Delta\mathbf{x} \tag{7.54}$$

The general form of the state space model is

$$\dot{\mathbf{x}} = \mathbf{f}(\mathbf{x}, t) \tag{7.55}$$

and by substituting '(7.54)' it gives

$$\dot{\mathbf{x}}_o + \Delta\dot{\mathbf{x}} = \mathbf{f}(\mathbf{x}_o + \Delta\mathbf{x}, t) \tag{7.56}$$

By Taylor series expansion and neglecting the second- and higher-order terms, it is found that

$$\mathbf{f}(\mathbf{x}_o + \Delta\mathbf{x}, t) = \mathbf{f}(\mathbf{x}_o, t) + \left.\frac{\partial \mathbf{f}}{\partial x_1}\right|_{x_o} \Delta x_1 + \left.\frac{\partial \mathbf{f}}{\partial x_2}\right|_{x_o} \Delta x_2 + \cdots + \left.\frac{\partial \mathbf{f}}{\partial x_n}\right|_{x_o} \Delta x_n \tag{7.57}$$

Incorporating '(7.55)' at $\mathbf{x}_o$, '(7.56)' and '(7.57)' obtain

$$\Delta\dot{\mathbf{x}} = \mathbf{A}(\mathbf{x}_o)\Delta\mathbf{x} \tag{7.58}$$

where

$$\mathbf{A}(\mathbf{x}_o) = \left[\frac{\partial \mathbf{f}}{\partial x_1}\, \frac{\partial \mathbf{f}}{\partial x_2} \cdots \frac{\partial \mathbf{f}}{\partial x_n}\right]_{\mathbf{x}_o} \tag{7.59}$$

Equation (2.53) can be used in the expanded form to deduce the change of each state as below:

$$\begin{aligned} v_{do} + \Delta v_d = {} & -R_a(i_{do} + \Delta i_d) - (\omega_o + \Delta\omega)L_q\left(i_{qo} + \Delta i_q\right) \\ & - (\omega_o + \Delta\omega)kM_{kq}\left(i_{kqo} + \Delta i_{kq}\right) - L_d(\mathrm{p}i_{do} + \mathrm{p}\Delta i_d) \\ & - kM_f\left(\mathrm{p}i_{fo} + \mathrm{p}\Delta i_f\right) - kM_{kd}(\mathrm{p}i_{kdo} + \mathrm{p}\Delta i_{kd}) \end{aligned}$$

Dropping the second-order terms, e.g. $\Delta x_i \Delta x_j$, as they are very small, it is seen that

$$\begin{aligned} v_{do} + \Delta v_d = {} & \left(-R_a i_{do} - \omega_o L_q i_{qo} - \omega_o k M_{kq} i_{kqo} - L_d \mathrm{p} i_{do} - k M_f \mathrm{p} i_{fo} - k M_{kd} \mathrm{p} i_{kdo}\right) \\ & - R_a \Delta i_d - \omega_o L_q \Delta i_q - i_{qo} L_q \Delta\omega - \omega_o k M_{kq} \Delta i_{kq} - i_{kqo} k M_{kq} \Delta\omega \\ & - L_d \mathrm{p} \Delta i_d - k M_f \mathrm{p} \Delta i_f - k M_{kd} \mathrm{p} \Delta i_{kd} \end{aligned}$$

It is noted that v_{do} is equal to the quantity between parentheses on the RHS, thus,

$$\begin{aligned} \Delta v_d = {} & -R_a \Delta i_d - \omega_o L_q \Delta i_q - \omega_o k M_{kq} \Delta i_{kq} - \left(L_q i_{qo} + k M_{kq} i_{kqo}\right)\Delta\omega \\ & - L_d \mathrm{p} \Delta i_d - k M_f \mathrm{p} \Delta i_f - k M_{kd} \mathrm{p} \Delta i_{kd} \end{aligned} \tag{7.60}$$

As $\Psi_{qo} = L_q i_{qo} + k M_{kq} i_{kqo}$, '(7.60)' can be rewritten as

$$\begin{aligned} \Delta v_d = {} & -R_a \Delta i_d - \omega_o L_q \Delta i_q - \omega_o k M_{kq} \Delta i_{kq} - \Psi_{qo} \Delta\omega - L_d \mathrm{p} \Delta i_d \\ & - k M_f \mathrm{p} \Delta i_f - k M_{kd} \mathrm{p} \Delta i_{kd} \end{aligned} \tag{7.61}$$

Similarly, the q-axis voltage change Δv_q can be found as

$$\begin{aligned} \Delta v_q = {} & \omega_o L_d \Delta i_d + \omega_o k M_f \Delta i_f + \omega_o k M_{kd} \Delta i_{kd} + \left(i_{do} L_d + i_{fo} k M_f + i_{kdo} k M_{kd}\right)\Delta\omega \\ & - R_a \Delta i_q - L_q \mathrm{p} \Delta i_q - k M_{kq} \mathrm{p} \Delta i_{kq} \end{aligned}$$

Thus,

$$\begin{aligned} \Delta v_q = {} & \omega_o L_d \Delta i_d + \omega_o k M_f \Delta i_f + \omega_o k M_{kd} \Delta i_{kd} \\ & + \Psi_{do} \Delta\omega - R_a \Delta i_q - L_q \mathrm{p} \Delta i_q - k M_{kq} \mathrm{p} \Delta i_{kq} \end{aligned} \tag{7.62}$$

and the change of the field voltage, Δv_f, can be deduced as

$$-\Delta v_f = -R_f \Delta i_f - k M_f \mathrm{p} \Delta i_d - L_f \mathrm{p} \Delta i_f - L_{fkd} \mathrm{p} \Delta i_{kd} \tag{7.63}$$

For the d- and q-axis damper windings '*KD* and *KQ*', respectively, the linearised equations are

$$\left.\begin{aligned} 0 &= -R_{kd} \Delta i_{kd} - k M_{kd} \mathrm{p} \Delta i_d - L_{fkd} \mathrm{p} \Delta i_f - L_{kd} \mathrm{p} \Delta i_{kd} \\ 0 &= -R_{kq} \Delta i_{kq} - k M_{kq} \mathrm{p} \Delta i_q - L_{kq} \mathrm{p} \Delta i_{kq} \end{aligned}\right\} \tag{7.64}$$

To compute the linearised torque equation, referring to the relations in Section 2.8, the following equations can be written again for convenience.

$$\left.\begin{aligned} \dot{\omega} &= \frac{1}{\hat{\mathrm{H}}}[T_\mathrm{m} - T_\mathrm{e} - T_\mathrm{d}] \\ T_\mathrm{e} &= i_q \Psi_d - i_d \Psi_q \\ \Psi_d &= L_d i_d + k M_f i_f + k M_{kd} i_{kd} \\ \Psi_q &= L_q i_q + k M_{kq} i_{kq} \\ T_\mathrm{d} &= D\omega \end{aligned}\right\} \tag{7.65}$$

Hence,

$$\dot{\omega} = \frac{1}{\hat{H}}(T_m - D\omega) - \frac{1}{\hat{H}}\left[i_q\left(L_d i_d + kM_f i_f + kM_{kd} i_{kd}\right) - i_d\left(L_q i_q + kM_{kq} i_{kq}\right)\right] \quad (7.66)$$

and the linearised form can be computed as

$$\begin{aligned}\Delta\dot{\omega} = \frac{1}{\hat{H}}\Big[& L_d i_{qo}\Delta i_d - L_d i_{do}\Delta i_q - kM_f i_{qo}\Delta i_f - kM_f i_{fo}\Delta i_q - kM_{kd} i_{qo}\Delta i_{kd} - kM_{kd} i_{kdo}\Delta i_q \\ & + L_q i_{do}\Delta i_q + L_q i_{qo}\Delta i_d + kM_{kq} i_{do}\Delta i_{kq} + kM_{kq} i_{kqo}\Delta i_d - D\Delta\omega + \Delta T_{\text{m}}\Big]\end{aligned} \quad (7.67)$$

In terms of flux linkages in *d*- and *q*-axis, '(7.67)' can be written as

$$\begin{aligned}\Delta\dot{\omega} = \frac{1}{\hat{H}}[& \Delta T_{\text{m}} - \left(L_d i_{qo} - \Psi_{qo}\right)\Delta i_d - \left(\Psi_{do} - L_q i_{do}\right)\Delta i_q \\ & - kM_f i_{qo}\Delta i_f - kM_{kd} i_{qo}\Delta i_{kd} + kM_{kq} i_{do}\Delta i_{kq} - D\Delta\omega]\end{aligned} \quad (7.68)$$

The linearised form of torque angle (2.103) may be written as

$$\Delta\dot{\delta} = \Delta\omega \quad (7.69)$$

Equations '(7.61)' through '(7.69)' form the linearised system equations for a synchronous machine *excluding the load equation*. In matrix form and dropping 'Δ', as the variables are considered as small changes, the equations can be written as below:

$$\begin{bmatrix} v_d \\ -v_f \\ 0 \\ \hline v_q \\ 0 \\ \hline T_m \\ 0 \end{bmatrix} = -\left[\begin{array}{ccc|cc|cc} R_a & 0 & 0 & \omega_o L_q & \omega_o kM_{kq} & \Psi_{qo} & 0 \\ 0 & R_f & 0 & 0 & 0 & 0 & 0 \\ 0 & 0 & R_{kd} & 0 & 0 & 0 & 0 \\ \hline -\omega_o L_d & -\omega_o kM_f & -\omega_o kM_{kd} & R_a & 0 & -\Psi_{do} & 0 \\ 0 & 0 & 0 & 0 & R_{kq} & 0 & 0 \\ \hline \Psi_{qo} - L_d i_{qo} & -kM_f i_{qo} & -kM_{kd} i_{qo} & -\Psi_{do} + L_q i_{do} & kM_{kq} i_{do} & -D & 0 \\ 0 & 0 & 0 & 0 & 0 & -1 & 0 \end{array}\right] \begin{bmatrix} i_d \\ i_f \\ i_{kd} \\ \hline i_q \\ i_{kq} \\ \hline \omega \\ \delta \end{bmatrix}$$

$$- \left[\begin{array}{ccc|cc|cc} L_d & kM_f & kM_{kd} & 0 & 0 & 0 & 0 \\ kM_f & L_f & L_{fkd} & 0 & 0 & 0 & 0 \\ kM_{kd} & L_{fkd} & L_{kd} & 0 & 0 & 0 & 0 \\ \hline 0 & 0 & 0 & L_q & kM_{kq} & 0 & 0 \\ 0 & 0 & 0 & kM_{kq} & L_{kq} & 0 & 0 \\ \hline 0 & 0 & 0 & 0 & 0 & -\hat{H} & 0 \\ 0 & 0 & 0 & 0 & 0 & 0 & 1 \end{array}\right] \begin{bmatrix} \text{p}i_d \\ \text{p}i_f \\ \text{p}i_{kd} \\ \hline \text{p}i_q \\ \text{p}i_{kq} \\ \hline \text{p}\omega \\ \text{p}\delta \end{bmatrix} \quad (7.70)$$

or

$$\mathbf{v} = -\mathbf{L}_1\mathbf{x} - \mathbf{L}_2\dot{\mathbf{x}} \tag{7.71}$$

Therefore, the state equation can be derived from '(7.71)' to be in the linear form:

$$\dot{\mathbf{x}} = \mathbf{Ax} + \mathbf{Bu} \tag{7.72}$$

where

$$[\mathbf{A}] = -\mathbf{L}_2^{-1}\mathbf{L}_1, \quad [\mathbf{B}] = -\mathbf{L}_2^{-1} \quad \text{and} \quad \mathbf{u} = \mathbf{v}$$

To *include the load equation*, the relation of both v_d and v_q must be linearised to obtain Δv_d and Δv_q and then substituting their values into '(7.70)' obtains the current model of one machine problem as below.

Referring to (3.9) the voltages v_d and v_q can be computed by

$$\begin{aligned} v_d &= -\sqrt{3}V_\infty \sin(\delta - \alpha) + R_e i_d + L_e \mathrm{p} i_d + \omega L_e i_q \\ v_q &= \sqrt{3}V_\infty \cos(\delta - \alpha) + R_e i_q + L_e \mathrm{p} i_q - \omega L_e i_d \end{aligned} \tag{7.73}$$

The trigonometric non-linearities are treated as

$$\sin(\delta_o + \Delta\delta) = \sin\delta_o \cos\Delta\delta + \cos\delta_o \sin\Delta\delta \cong \sin\delta_o + (\cos\delta_o)\Delta\delta$$

where

$$\cos\Delta\delta \cong 1 \quad \text{and} \quad \sin\Delta\delta \cong \Delta\delta$$

Then, the incremental change in $\sin\delta \triangleq \sin(\delta_o + \Delta\delta) - \sin\delta_o \cong (\cos\delta_o)\Delta\delta$

$$\tag{7.74}$$

Similarly, $\cos(\delta_o + \Delta\delta) = \cos\delta_o \cos\Delta\delta - \sin\delta_o \sin\Delta\delta \cong \cos\delta_o - (\sin\delta_o)\Delta\delta$

Thus, the incremental change in $\cos\delta \triangleq \cos(\delta_o + \Delta\delta) - \cos\delta_o = -(\sin\delta_o)\Delta\delta$

$$\tag{7.75}$$

Linearisation of '(7.73)' leads to

$$\left.\begin{aligned} \Delta v_d &= -\sqrt{3}V_\infty \cos(\delta_o - \alpha)\Delta\delta + R_e\Delta i_d + \omega_o L_e \Delta i_q + i_{qo} L_e \Delta\omega + L_e \mathrm{p}\Delta i_d \\ \Delta v_q &= -\sqrt{3}V_\infty \sin(\delta_o - \alpha)\Delta\delta + R_e\Delta i_q - \omega_o L_e \Delta i_d - i_{do} L_e \Delta\omega + L_e \mathrm{p}\Delta i_q \end{aligned}\right\} \tag{7.76}$$

Substituting '(7.76)' into '(7.61)' and '(7.62)'

$$\begin{aligned} &-\sqrt{3}V_\infty \cos(\delta_o - \alpha)\Delta\delta + R_e\Delta i_d + \omega_o L_e\Delta i_q + i_{qo}L_e\Delta\omega + L_e\mathrm{p}\Delta i_d \\ &\quad = -R_a\Delta i_d - \omega_o L_q \Delta i_q - \omega_o k M_{kq}\Delta i_{kq} - \Psi_{qo}\Delta\omega - L_d \mathrm{p}\Delta i_d - kM_f \mathrm{p}\Delta i_f \\ &\qquad - kM_{kd}\mathrm{p}\Delta i_{kd} - \sqrt{3}V_\infty \sin(\delta_o - \alpha)\Delta\delta + R_e\Delta i_q - \omega_o L_e \Delta i_d - i_{do}L_e\Delta\omega + L_e\mathrm{p}\Delta i_q \\ &\quad = \omega_o L_d \Delta i_d + \omega_o k M_f \Delta i_f + \omega_o k M_{kd}\Delta i_{kd} + \Psi_{do}\Delta\omega \\ &\qquad - R_a \Delta i_q - L_q \mathrm{p}\Delta i_q - kM_{kq}\mathrm{p}\Delta i_{kq} \end{aligned} \tag{7.77}$$

Assuming $\overline{R} = R_a + R_e, \overline{L}_q = L_q + L_e, \overline{L}_d = L_d + L_e, \overline{\Psi}_d = \Psi_d + L_e i_d, \overline{\Psi}_q = \Psi_q + L_e i_q$ and rearranging '(7.77)' by dropping Δ, as explained above, gives

$$\begin{aligned} 0 &= -\overline{R}i_d - \omega_o\overline{L}_q i_q - \omega_o kM_{kq} i_{kq} - \overline{\Psi}_{qo}\omega \\ &\quad + \sqrt{3}V_\infty \cos(\delta_o - \alpha)\delta - \overline{L}_d \mathrm{p}i_d - kM_f \mathrm{p}i_f - kM_{kd}\mathrm{p}i_{kd} \\ 0 &= -\overline{R}i_q + \omega_o\overline{L}_d i_d + \omega_o kM_f i_f + \omega_o kM_{kd} i_{kd} \\ &\quad + \overline{\Psi}_{do}\omega + \sqrt{3}V_\infty \sin(\delta_o - \alpha)\delta - \overline{L}_q \mathrm{p}i_q - kM_{kq}\mathrm{p}i_{kq} \end{aligned} \tag{7.78}$$

Incorporating '(7.63)', '(7.64)', '(7.68)', '(7.69)' and '(7.78)', the linearised set of system equations with constant coefficients is

$$\begin{bmatrix} 0 \\ -v_f \\ 0 \\ 0 \\ 0 \\ T_m \\ 0 \end{bmatrix} = -\begin{bmatrix} \overline{R} & 0 & 0 & \omega_o\overline{L}_q & \omega_o kM_{kq} & \overline{\Psi}_{qo} & \mathcal{K}\cos(\delta_o - \alpha) \\ 0 & R_f & 0 & 0 & 0 & 0 & 0 \\ 0 & 0 & R_{kd} & 0 & 0 & 0 & 0 \\ -\omega_o\overline{L}_d & -\omega_o kM_f & -\omega_o kM_{kd} & \overline{R} & 0 & -\overline{\Psi}_{do} & \mathcal{K}\sin(\delta_o - \alpha) \\ 0 & 0 & 0 & 0 & R_{kq} & 0 & 0 \\ \Psi_{qo} - L_d i_{qo} & -kM_f i_{qo} & -kM_{kd} i_{qo} & -\Psi_{do} + L_q i_{do} & kM_{kq} i_{do} & -D & 0 \\ 0 & 0 & 0 & 0 & 0 & -1 & 0 \end{bmatrix}$$

$$\times \begin{bmatrix} i_d \\ i_f \\ i_{kd} \\ i_q \\ i_{kq} \\ \omega \\ \delta \end{bmatrix} - \begin{bmatrix} \overline{L}_d & kM_f & kM_{kd} & 0 & 0 & 0 & 0 \\ kM_f & L_f & L_{fkd} & 0 & 0 & 0 & 0 \\ kM_{kd} & L_{fkd} & L_{kd} & 0 & 0 & 0 & 0 \\ 0 & 0 & 0 & \overline{L}_q & kM_{kq} & 0 & 0 \\ 0 & 0 & 0 & kM_{kq} & L_{kq} & 0 & 0 \\ 0 & 0 & 0 & 0 & 0 & -\hat{\mathrm{H}} & 0 \\ 0 & 0 & 0 & 0 & 0 & 0 & 1 \end{bmatrix} \begin{bmatrix} \mathrm{p}i_d \\ \mathrm{p}i_f \\ \mathrm{p}i_{kd} \\ \mathrm{p}i_q \\ \mathrm{p}i_{kq} \\ \mathrm{p}\omega \\ \mathrm{p}\delta \end{bmatrix} \tag{7.79}$$

where

$$\mathcal{K} = -\sqrt{3}V_\infty$$

and in matrix form it can be written as

$$\mathbf{v} = -\mathbf{L}_3\mathbf{x} - \mathbf{L}_4\dot{\mathbf{x}}$$

Therefore, the state equations can be expressed by the general linear form as

$$\dot{\mathbf{x}} = \mathbf{A}\mathbf{x} + \mathbf{B}\mathbf{u} \tag{7.80}$$

where

$$\mathbf{A} = -\mathbf{L}_4^{-1}\mathbf{L}_3 \quad \text{and} \quad \mathbf{B} = -\mathbf{L}_4^{-1}\mathbf{v}$$

Example 7.3 For machine data given below find:

- The matrices $\mathbf{L}_1$ and $\mathbf{L}_2$ of the linearised current model for unloaded machine.
- The linearised current model for loaded machine that is connected to an infinite bus through a transmission line with resistance, $R_e = 0.05$, and reactance, $X_e = 0.35$. Determine the machine stability when the system is subjected to a small disturbance (H = 3.5 s and $D = 0$). The power delivered at machine terminal is 0.8 pu at power factor = 0.85 (lagging)

$L_d = 1.81$, $L_q = 1.76$, $L_f = 1.75$, $L_{md} = 1.66$, $L_{mq} = 1.61$, $l_d = l_q = l_a = 0.15$, $L_{kd} = 1.72$, $L_{kq} = 1.63$, $kM_f = kM_{kd} = L_d - l_d = 1.66$, $L_{fkd} = kM_f = 1.66$, $kM_{kq} = 1.59$, $R_a = 0.003$, $R_f = 0.009$, $R_{kd} = 0.0284$, $R_{kq} = 0.006$, $l_f = L_f - kM_f = 0.09$, $l_{kd} = L_{kd} - kM_{kd} = 0.06$, $l_{kq} = L_{kq} - kM_{kq} = 0.03$

Using (2.77) and (2.80), L_{Md} and L_{Mq} can be obtained as $L_{Md} = 0.05$, $L_{Mq} = 0.042$.

Solution:

(i) From '(7.70)' and using the power per phase as a base quantity, it can be found that

$$\mathbf{L}_1 = \left[\begin{array}{ccc|cc|cc} 0.003 & 0 & 0 & 1.76 & 1.59 & \Psi_{qo} & 0 \\ 0 & 0.009 & 0 & 0 & 0 & 0 & 0 \\ 0 & 0 & 0.028 & 0 & 0 & 0 & 0 \\ \hline -1.81 & -1.66 & -1.66 & 0.003 & 0 & -\Psi_{do} & 0 \\ 0 & 0 & 0 & 0 & 0.006 & 0 & 0 \\ \hline \frac{1}{3}(\Psi_{qo} - L_d i_{qo}) & \frac{1}{3}(-kM_f i_{qo}) & \frac{1}{3}(-kM_{kd} i_{qo}) & \frac{1}{3}(-\Psi_{do} + L_q i_{do}) & \frac{1}{3}(kM_{kq} i_{do}) & -D & 0 \\ 0 & 0 & 0 & 0 & 0 & -1 & 0 \end{array}\right]$$

$$\mathbf{L}_2 = \left[\begin{array}{ccc|cc|cc} 1.81 & 1.66 & 1.66 & 0 & 0 & 0 & 0 \\ 1.66 & 1.75 & 1.66 & 0 & 0 & 0 & 0 \\ 1.66 & 1.66 & 1.72 & 0 & 0 & 0 & 0 \\ \hline 0 & 0 & 0 & 1.76 & 1.59 & 0 & 0 \\ 0 & 0 & 0 & 1.59 & 1.63 & 0 & 0 \\ \hline 0 & 0 & 0 & 0 & 0 & -2637.6 & 0 \\ 0 & 0 & 0 & 0 & 0 & 0 & 1 \end{array}\right]$$

It is to be noted that $\hat{\mathrm{H}} = 2\mathrm{H}\omega_o = 2637.6$ as the form of (2.36) is used and $\omega_o = \omega_B$. The numeric quantities in matrix $\mathbf{L}_1$ are constants while the other non-numeric values depend on the load data as explained in case (ii).

(ii) As a first step the steady-state operating conditions are calculated. The phasor diagram of the machine is depicted in Figure 7.6: if $\mathbf{V}_\infty = 1\angle 0°$ is taken as a reference, then $\alpha = 0°$. The current components in the direction of reference axis

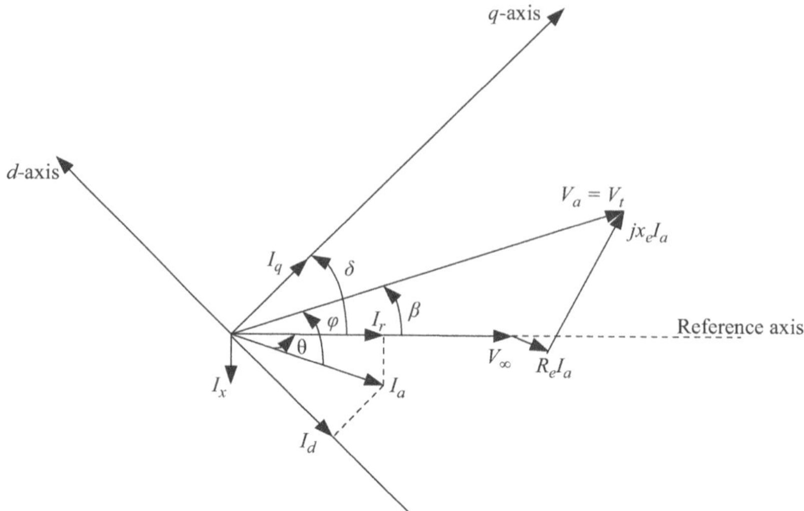

Figure 7.6 Phasor diagram of a synchronous machine connected to an infinite bus

'in phase with V∞' and its perpendicular are I_r and I_x, respectively. As an approximation, the power loss in the transmission line is estimated based on assumed current of 1 pu. Then the power loss equals $(1.0)^2R_e = 0.05$ pu and $P_\infty = 0.75$ pu. Therefore, $I_r = 0.75$ pu.

$$\Phi = \cos^{-1} 0.85 = 31.788° = \beta + \theta$$

$$\tan \Phi = \frac{\tan \beta + \tan \theta}{1 - \tan \beta \tan \theta}, \quad \tan \beta = \frac{X_eI_r + R_eI_x}{V_\infty - X_eI_x + R_eI_r} \quad \text{and} \quad \tan \theta = 1.333I_x$$

$$0.62 = \frac{-1.333I_x(1.002 - 0.35I_x) + (0.263 + 0.05I_x)}{(1.002 - 0.35I_x) + 1.333I_x(0.263 + 0.05I_x)}$$

From which $I_x = -0.256$ pu

$\tan \theta = 1.333I_x$ and then $\theta = 18.84°$ 'lagging V_∞'

and $\tan \beta = (0.25)/(1.092) = 0.2289$, then, $\beta = 12.89°$

$$V_t = (V_\infty - X_eI_x + R_eI_r) + j(X_eI_r + R_eI_x) = 1.127 + j0.25 = 1.154\angle 12.5°$$

$$\delta = \tan^{-1} \frac{(X_q + X_e)I_r + (R_a + R_e)I_x}{V_\infty - (X_q + X_e)I_x + (R_a + R_e)I_r} = \tan^{-1} 0.81 = 39°$$

$$I_a = 0.75 - j0.256 = 0.793 \angle -18.84°$$

It can be seen that $I_{do} = -I_a \sin(\delta_o - \beta + \Phi) = -0.793 \sin 57.9 = -0.672$ and $I_{qo} = I_a \sin(\delta_o - \beta + \Phi) = 0.421$

Using (7.21) to get $i_{do} = \sqrt{3}I_{do} = -1.164$ pu and $i_{qo} = \sqrt{3}I_{qo} = 0.729$ pu

$$V_d = -V_a \sin(\delta - \beta) = -1.154 \sin(39 - 12.5) = -0.515 \text{ pu}$$

$$V_d = V_a \cos(\delta - \beta) = 1.154 \cos(39 - 12.5) = 1.033 \text{ pu}$$

Applying (7.24) to obtain

$$v_{do} = \sqrt{3}V_{do} = 0.892 \quad \text{and} \quad v_{qo} = \sqrt{3}V_{qo} = 1.789\ \text{pu}$$

$$E = V_q + R_a I_q - X_d I_d = 1.789 + 0.001 + 1.21 = 2.999(=E_{fd})$$

$$i_{fo} = (\sqrt{3}E)/kM_f = \sqrt{3} \times 2.999/1.66 = 3.129\ \text{pu}$$

$$\Psi_{do} = L_d i_{do} + kM_f i_{fo} = 3.087, \quad \Psi_{qo} = L_q i_{qo} = 1.283$$

$$\Psi_{fo} = kM_f i_{do} + L_f i_{fo} = 3.544, \quad \Psi_{kdo} = kM_{kd} i_{do} + L_{fkd} i_{fo} = 3.262$$

$$\Psi_{kqo} = kM_{kq} i_{qo} = 1.159$$

$$\overline{L}_q = L_q + L_e = 1.76 + 0.35 = 2.11, \quad \overline{L}_d = L_d + L_e = 1.81 + 0.35 = 2.16$$

$$\overline{R} = R_a + R_e = 0.053$$

$$\overline{\Psi}_{do} = \Psi_{do} + L_e i_{do} = 1.411 + 0.35 \times -0.22 = 2.68$$

$$\overline{\Psi}_{qo} = \Psi_{qo} + L_e i_{qo} = 1.355 + 0.35 \times 0.77 = 1.538$$

$$\sqrt{3}V_\infty \cos\delta_o = 1.346\sqrt{3}V_\infty \sin\delta_o = 1.09$$

Thus,

$$\mathbf{L}_3 = \left[\begin{array}{ccc:cc:cc} 0.053 & 0 & 0 & 2.11 & 1.59 & 1.538 & -1.346 \\ 0 & 0.009 & 0 & 0 & 0 & 0 & 0 \\ 0 & 0 & 0.028 & 0 & 0 & 0 & 0 \\ \hdashline -2.16 & -1.66 & -1.66 & 0.053 & 0 & -2.68 & -1.09 \\ 0 & 0 & 0 & 0 & 0.725 & 0 & 0 \\ \hdashline -0.012 & -0.403 & -0.403 & -1.711 & -0.617 & 0 & 0 \\ 0 & 0 & 0 & 0 & 0 & -1 & 0 \end{array}\right]$$

$$\mathbf{L}_4 = \left[\begin{array}{ccc:cc:cc} 2.16 & 1.66 & 1.66 & 0 & 0 & 0 & 0 \\ 1.66 & 1.75 & 1.66 & 0 & 0 & 0 & 0 \\ 1.66 & 1.66 & 1.72 & 0 & 0 & 0 & 0 \\ \hdashline 0 & 0 & 0 & 2.11 & 1.59 & 0 & 0 \\ 0 & 0 & 0 & 1.59 & 0.725 & 0 & 0 \\ \hdashline 0 & 0 & 0 & 0 & 0 & -2637.6 & 0 \\ 0 & 0 & 0 & 0 & 0 & 0 & 1 \end{array}\right]$$

$\mathbf{A} = -\mathbf{L}_4^{-1}\mathbf{L}_3$

$$
= \begin{bmatrix}
-0.1005 & 0.0052 & 0.0362 & -4.002 & -3.0157 & -2.9171 & 2.5529 \\
0.0304 & -0.0713 & 0.2009 & 1.2112 & 0.9127 & 0.8828 & -0.7726 \\
0.0685 & 0.0646 & -0.248 & 2.7251 & 2.0535 & 1.9864 & -1.7384 \\
-1.5686 & -1.2055 & -1.2055 & 0.0385 & -1.1547 & -1.9462 & -0.7916 \\
3.4401 & 2.6438 & 2.6438 & -0.0844 & 1.5323 & 4.2682 & 1.7360 \\
-0.0000 & -0.0002 & -0.0002 & -0.0006 & -0.0002 & 0 & 0 \\
0 & 0 & 0 & 0 & 0 & 1.0000 & 0
\end{bmatrix}
$$

The computation of eigenvalues of matrix **A** using MATLAB® 2012a gives the values

$$-0.0535 \pm j0.9841, \quad -0.0039 \pm j0.0265, \quad -0.3043, \quad -0.0084, \quad 1.5785$$

It is to be noted that the system is unstable as one of the eigenvalues has a positive real component.

7.4 Linearised flux linkage state space model of a synchronous generator

The same procedure used to linearise the current model can be applied to obtain the linear flux model of a synchronous machine. The linearised form of (2.89) and (2.90) can be written as (7.81) and (7.82), respectively.

$$
\left.
\begin{aligned}
\Delta\dot{\Psi}_d &= -R_a\left(1-\frac{L_{Md}}{\ell_a}\right)\frac{\Delta\Psi_d}{\ell_a} + R_a\frac{L_{Md}}{\ell_a}\frac{\Delta\Psi_f}{\ell_f} + R_a\frac{L_{Md}}{\ell_a}\frac{\Delta\Psi_{kd}}{\ell_{kd}} \\
&\quad - \omega_o\Delta\Psi_q - \Psi_{qo}\Delta\omega - \Delta v_d \\
\Delta\dot{\Psi}_f &= R_f\frac{L_{Md}}{\ell_f}\frac{\Delta\Psi_d}{\ell_a} - R_f\left(1-\frac{L_{Md}}{\ell_f}\right)\frac{\Delta\Psi_f}{\ell_f} + R_f\frac{L_{Md}}{\ell_f}\frac{\Delta\Psi_{kd}}{\ell_{kd}} + \Delta v_f \\
\Delta\dot{\Psi}_{kd} &= R_{kd}\frac{L_{Md}}{\ell_{kd}}\frac{\Delta\Psi_d}{\ell_a} + R_{kd}\frac{L_{Md}}{\ell_{kd}}\frac{\Delta\Psi_f}{\ell_f} - R_{kd}\left(1-\frac{L_{Md}}{\ell_{kd}}\right)\frac{\Delta\Psi_{kd}}{\ell_{kd}} \\
\Delta\dot{\Psi}_q &= -R_a\left(1-\frac{L_{Mq}}{\ell_a}\right)\frac{\Delta\Psi_q}{\ell_a} + R_a\frac{L_{Mq}}{\ell_a}\frac{\Delta\Psi_{kq}}{\ell_{kq}} + \omega_o\Delta\Psi_d + \Psi_{do}\Delta\omega - \Delta v_q \\
\Delta\dot{\Psi}_{kq} &= R_{kq}\frac{L_{Mq}}{\ell_{kq}}\frac{\Delta\Psi_q}{\ell_a} - R_{kq}\left(1-\frac{L_{Mq}}{\ell_q}\right)\frac{\Delta\Psi_{kq}}{\ell_{kq}}
\end{aligned}
\right\}
\tag{7.81}
$$

$$\Delta T_e = \left(\frac{L_{Md} - L_{Mq}}{\ell_a^2}\Psi_{qo} - \frac{L_{Mq}}{\ell_a \ell_{kq}}\Psi_{kqo}\right)\Delta\Psi_d + \frac{L_{Md}}{\ell_a \ell_f}\Psi_{qo}\Delta\Psi_f + \frac{L_{Md}}{\ell_a \ell_{kd}}\Psi_{qo}\Delta\Psi_{kd}$$
$$+ \left(\frac{L_{Md} - L_{Mq}}{\ell_a^2}\Psi_{qo} + \frac{L_{Md}}{\ell_a \ell_f}\Psi_{fo} + \frac{L_{Md}}{\ell_a \ell_{kd}}\Psi_{kdo}\right)\Delta\Psi_q - \frac{L_{Mq}}{\ell_a \ell_{kq}}\Psi_{do}\Delta\Psi_{kq} \tag{7.82}$$

The angular speed equation becomes

$$\Delta\dot{\omega} = \frac{1}{\hat{\mathrm{H}}}\left[\left(\frac{L_{Mq}}{\ell_a \ell_{kq}}\Psi_{qo} - \frac{L_{Md} - L_{Mq}}{\ell_a^2}\Psi_{qo}\right)\Delta\Psi_d - \left(\frac{L_{Md}}{\ell_a \ell_f}\Psi_{qo}\right)\Delta\Psi_f\right.$$
$$- \left(\frac{L_{Md}}{\ell_a \ell_{kd}}\Psi_{qo}\right)\Delta\Psi_{kd} - \left(\frac{L_{Md} - L_{Mq}}{\ell_a^2}\Psi_{do} + \frac{L_{Md}}{\ell_a \ell_f}\Psi_{fo} + \frac{L_{Md}}{\ell_a \ell_{kd}}\Psi_{kdo}\right)\Delta\Psi_q$$
$$\left. + \left(\frac{L_{Mq}}{\ell_a \ell_{kq}}\Psi_{do}\right)\Delta\Psi_{kq} - D\Delta\omega + \Delta T_{\mathrm{m}}\right] \tag{7.83}$$

and the torque angle equation is

$$\Delta\dot{\delta} = \Delta\omega \tag{7.84}$$

If the machine is connected to a simple power system, Chapter 3, Section 3.1, the voltages v_d and v_q can be computed by (3.11) and (3.12), respectively. Then, substituting these voltages into (2.89) gives the load equations as

$$\left[1 + \frac{L_e}{\ell_a}\left(1 - \frac{L_{Md}}{\ell_a}\right)\right]\dot{\Psi}_d - \frac{L_e L_{Md}}{\ell_a \ell_f}\dot{\Psi}_f - \frac{L_e L_{Md}}{\ell_a \ell_{kd}}\dot{\Psi}_{kd}$$
$$= -\frac{\overline{R}}{\ell_a}\left(1 - \frac{L_{Md}}{\ell_a}\right)\Psi_d + \frac{\overline{R}L_{Md}}{\ell_a \ell_f}\Psi_f + \frac{\overline{R}L_{Md}}{\ell_a \ell_{kd}}\Psi_{kd}$$
$$- \omega\left[1 + \frac{L_e}{\ell_a}\left(1 - \frac{L_{Mq}}{\ell_a}\right)\right]\Psi_q + \frac{\omega L_e L_{Mq}}{\ell_a \ell_{kq}}\Psi_{kq} + \sqrt{3}V_\infty \sin(\delta - \alpha) \tag{7.85}$$

$$\left[1 + \frac{L_e}{\ell_a}\left(1 - \frac{L_{Mq}}{\ell_a}\right)\right]\dot{\Psi}_q - \frac{L_e L_{Mq}}{\ell_a \ell_{kq}}\dot{\Psi}_{kq}$$
$$= -\frac{\overline{R}}{\ell_a}\left(1 - \frac{L_{Mq}}{\ell_a}\right)\Psi_q + \frac{\overline{R}L_{Mq}}{\ell_a \ell_{kq}}\Psi_{kq} + \omega\left[1 + \frac{L_e}{\ell_a}\left(1 - \frac{L_{Md}}{\ell_a}\right)\right]\Psi_d$$
$$- \frac{\omega L_e L_{Md}}{\ell_a \ell_f}\Psi_f - \frac{\omega L_e L_{Md}}{\ell_a \ell_{kd}}\Psi_{kd} - \sqrt{3}V_\infty \cos(\delta - \alpha) \tag{7.86}$$

Equations (7.85) and (7.86) are then linearised to obtain

$$\left[1+\frac{L_e}{\ell_a}\left(1-\frac{L_{Md}}{\ell_a}\right)\right]\Delta\dot{\Psi}_d-\frac{L_eL_{Md}}{\ell_a\ell_f}\Delta\dot{\Psi}_f-\frac{L_eL_{Md}}{\ell_a\ell_{kd}}\Delta\dot{\Psi}_{kd}$$
$$=-\frac{\overline{R}}{\ell_a}\left(1-\frac{L_{Md}}{\ell_a}\right)\Delta\Psi_d+\frac{\overline{R}L_{Md}}{\ell_a\ell_f}\Delta\Psi_f+\frac{\overline{R}L_{Md}}{\ell_a\ell_{kd}}\Delta\Psi_{kd}$$
$$+\omega_o\frac{L_eL_{Mq}}{\ell_a\ell_{kq}}\Delta\Psi_{kq}-\omega_o\left[1+\frac{L_e}{\ell_a}\left(1-\frac{L_{Mq}}{\ell_a}\right)\right]\Delta\Psi_q$$
$$-\overline{\Psi}_{qo}\Delta\omega+\sqrt{3}V_\infty\cos(\delta-\alpha)\Delta\delta \tag{7.87}$$

and

$$\left[1+\frac{L_e}{\ell_a}\left(1-\frac{L_{Mq}}{\ell_a}\right)\right]\Delta\dot{\Psi}_q-\frac{L_eL_{Mq}}{\ell_a\ell_{kq}}\Delta\dot{\Psi}_{kq}$$
$$=\omega_o\left[1+\frac{L_e}{\ell_a}\left(1-\frac{L_{Md}}{\ell_a}\right)\Delta\Psi_d\right]-\omega_o\frac{L_eL_{Md}}{\ell_a\ell_f}\Delta\Psi_f$$
$$-\omega_o\frac{L_eL_{Md}}{\ell_a\ell_{kd}}\Delta\Psi_{kd}-\frac{\overline{R}}{\ell_a}\left(1-\frac{L_{Mq}}{\ell_a}\right)\Delta\Psi_q+\frac{\overline{R}L_{Mq}}{\ell_a\ell_{kq}}\Delta\Psi_{kq}$$
$$+\overline{\Psi}_{do}\Delta\omega+\sqrt{3}V_\infty\sin(\delta-\alpha)\Delta\delta \tag{7.88}$$

where

$$\overline{R}=R_a+R_e$$
$$\overline{\Psi}_{qo}=\left[1+\frac{L_e}{\ell_a}\left(1-\frac{L_{Mq}}{\ell_a}\right)\right]\Psi_{qo}-\frac{L_eL_{Mq}}{\ell_a\ell_{kq}}\Psi_{kqo}$$
$$\overline{\Psi}_{do}=\left[1+\frac{L_e}{\ell_a}\left(1-\frac{L_{Md}}{\ell_a}\right)\right]\Psi_{do}-\frac{L_eL_{Md}}{\ell_a\ell_f}\Psi_{fo}-\frac{L_eL_{Md}}{\ell_a\ell_{kd}}\Psi_{kdo}$$

Therefore, the linearised flux linkage model can be derived from (7.81) through (7.84) and (7.87) through (7.88) in the matrix form $\mathbf{C}_1\dot{\mathbf{x}}=\mathbf{C}_2\mathbf{x}+\mathbf{u}$ as in (7.89).

$$C_1 \begin{bmatrix} \dot{\Psi}_d \\ \dot{\Psi}_f \\ \dot{\Psi}_{kd} \\ \dot{\Psi}_q \\ \dot{\Psi}_{kq} \\ \dot{\omega} \\ \dot{\delta} \end{bmatrix} = \left[\begin{array}{ccc:cc:cc} -\frac{\bar{R}_a}{\ell_a}\left(1-\frac{L_{Md}}{\ell_a}\right) & \frac{\bar{R}_a}{\ell_a}\frac{L_{Md}}{\ell_f} & \frac{\bar{R}_a}{\ell_a}\frac{L_{Md}}{\ell_{kd}} & -\omega_o\left[1+\frac{L_e}{\ell_a}\left(1-\frac{L_{Mq}}{\ell_a}\right)\right] & 0 & -\bar{\Psi}_{qo} & \sqrt{3}V_\infty\cos(\delta-\alpha) \\ \frac{R_f}{\ell_f}\frac{L_{Md}}{\ell_a} & -\frac{R_f}{\ell_f}\left(1-\frac{L_{Md}}{\ell_f}\right) & \frac{R_f}{\ell_f}\frac{L_{Md}}{\ell_{kd}} & 0 & 0 & 0 & 0 \\ \frac{R_{kd}}{\ell_{kd}}\frac{L_{Md}}{\ell_a} & \frac{R_{kd}}{\ell_{kd}}\frac{L_{Md}}{\ell_f} & -\frac{R_{kd}}{\ell_{kd}}\left(1-\frac{L_{Md}}{\ell_{kd}}\right) & 0 & 0 & 0 & 0 \\ \hdashline -\omega_o\left[1+\frac{L_e}{\ell_a}\left(1-\frac{L_{Md}}{\ell_a}\right)\right] & -\frac{\omega_o L_e L_{Md}}{\ell_a\ell_f} & -\frac{\omega_o L_e L_{Md}}{\ell_a\ell_{kd}} & -\frac{\bar{R}_a}{\ell_a}\left(1-\frac{L_{Mq}}{\ell_a}\right) & \frac{\bar{R}_a}{\ell_q}\frac{L_{Mq}}{\ell_{kq}} & -\bar{\Psi}_{do} & \sqrt{3}V_\infty\sin(\delta-\alpha) \\ 0 & 0 & 0 & \frac{R_{kq}}{\ell_{kq}}\frac{L_{Mq}}{\ell_a} & -\frac{R_{kq}}{\ell_{kq}}\left(1-\frac{L_{Mq}}{\ell_{kq}}\right) & 0 & 0 \\ \hdashline \frac{-1}{3\hat{\mathrm{H}}\ell_a}\left(\Psi_{Aqo}-\frac{L_{Md}\Psi_{qo}}{\ell_a}\right) & -\frac{L_{Md}}{3\hat{\mathrm{H}}\ell_a\ell_f}\Psi_{qo} & -\frac{L_{Md}}{3\hat{\mathrm{H}}\ell_a\ell_{kd}}\Psi_{qo} & \frac{-1}{3\hat{\mathrm{H}}\ell_a}\left(\Psi_{Ado}-\frac{L_{Mq}\Psi_{do}}{\ell_a}\right) & \frac{L_{Mq}}{3\hat{\mathrm{H}}\ell_a\ell_{kq}}\Psi_{qo} & -\text{Đ} & 0 \\ 0 & 0 & 0 & 0 & 0 & 1 & 0 \end{array}\right]$$

$$\times \begin{bmatrix} \Psi_d \\ \Psi_f \\ \Psi_{kd} \\ \Psi_q \\ \Psi_{kq} \\ \omega \\ \delta \end{bmatrix} + \begin{bmatrix} \sqrt{3}V_\infty \sin(\delta-\alpha) \\ v_f \\ 0 \\ -\sqrt{3}V_\infty \cos(\delta-\alpha) \\ 0 \\ \frac{T_m}{\hat{\mathrm{H}}} \\ -1 \end{bmatrix} \tag{7.89}$$

where

$$
\mathbf{C}_1 = \left[\begin{array}{ccc:cc:cc}
1+\frac{L_e}{\ell_a}\left(1-\frac{L_{Md}}{\ell_a}\right) & -\frac{L_e}{\ell_a}\frac{L_{Md}}{\ell_f} & -\frac{L_e}{\ell_a}\frac{L_{Md}}{\ell_{kd}} & 0 & 0 & 0 & 0 \\
0 & 1 & 0 & 0 & 0 & 0 & 0 \\
0 & 0 & 1 & 0 & 0 & 0 & 0 \\
\hdashline
0 & 0 & 0 & 1+\frac{L_e}{\ell_a}\left(1-\frac{L_{Mq}}{\ell_a}\right) & -\frac{L_e}{\ell_a}\frac{L_{Md}}{\ell_{kq}} & 0 & 0 \\
0 & 0 & 0 & 0 & 1 & 0 & 0 \\
\hdashline
0 & 0 & 0 & 0 & 0 & 1 & 0 \\
0 & 0 & 0 & 0 & 0 & 0 & 1
\end{array}\right]
$$

To conform to the form $\dot{\mathbf{x}} = \mathbf{Ax} + \mathbf{Bu}$, it can be found that

$$\mathbf{A} = \mathbf{C}_1^{-1}\mathbf{C}_2 \quad \text{and} \quad \mathbf{B} = \mathbf{C}_1^{-1}$$

Example 7.4 Derive the matrices $\mathbf{C}_1$ and $\mathbf{C}_2$ of the linearised flux linkage state space model for the system in Example 7.3 and determine its stability

Solution:

The computation of the elements of matrix $\mathbf{C}_1$ gives

$$
\mathbf{C}_1 = \left[\begin{array}{ccc:cc:cc}
2.556 & -1.296 & -1.944 & 0 & 0 & 0 & 0 \\
0 & 1 & 0 & 0 & 0 & 0 & 0 \\
0 & 0 & 1 & 0 & 0 & 0 & 0 \\
\hdashline
0 & 0 & 0 & 2.68 & -3.267 & 0 & 0 \\
0 & 0 & 0 & 0 & 1 & 0 & 0 \\
\hdashline
0 & 0 & 0 & 0 & 0 & 1 & 0 \\
0 & 0 & 0 & 0 & 0 & 0 & 1
\end{array}\right]
$$

From (2.78) and (2.79), Ψ_{Ado} and Ψ_{Aqo} are

$$\Psi_{Ado} = \frac{L_{Md}}{\ell_a}\Psi_{do} + \frac{L_{Md}}{\ell_f}\Psi_{fo} + \frac{L_{Md}}{\ell_{kd}}\Psi_{kdo} = 5.716$$

$$\Psi_{Aqo} = \frac{L_{Mq}}{\ell_a}\Psi_{qo} + \frac{L_{Mq}}{\ell_{kq}}\Psi_{kqo} = 1.982$$

The matrix $\mathbf{C}_2$ can be derived and written as below:

$$\mathbf{C}_2 = \left[\begin{array}{ccc|cc|cc} -0.236 & 0.196 & 0.294 & -2.68 & 3.267 & -1.538 & 1.346 \\ 0.006 & -0.004 & 0.008 & 0 & 0 & 0 & 0 \\ 0.016 & 0.263 & -0.118 & 0 & 0 & 0 & 0 \\ \hline 2.556 & -1.296 & -1.944 & -0.254 & 0.495 & 2.68 & 1.09 \\ 0 & 0 & 0 & 0.056 & -0.036 & 0 & 0 \\ \hline 0.0013 & -0.0006 & -0.0009 & -0.004 & 0.0036 & 0 & 0 \\ 0 & 0 & 0 & 0 & 0 & 1 & 0 \end{array}\right]$$

From the relation $\mathbf{A} = \mathbf{C}_1^{-1}\mathbf{C}_2$, matrix $\mathbf{A}$ can be computed as

$$\mathbf{A} = \begin{pmatrix} -0.0771 & 0.2747 & 0.0293 & -1.0485 & 1.2782 & -0.6017 & 0.5266 \\ 0.0060 & -0.0040 & 0.0080 & 0 & 0 & 0 & 0 \\ 0.0160 & 0.2630 & -0.1180 & 0 & 0 & 0 & 0 \\ 0.9537 & -0.4836 & -0.7254 & -0.0265 & 0.1408 & 1.0000 & 0.4067 \\ 0 & 0 & 0 & 0.0560 & -0.0360 & 0 & 0 \\ 0.0013 & -0.0006 & -0.0009 & -0.0040 & 0.0036 & 0 & 0 \\ 0 & 0 & 0 & 0 & 0 & 1.0000 & 0 \end{pmatrix}$$

and the corresponding eigenvalues are

$$-0.0938 \pm j0.9987, \quad 0.0070 \pm j0.0481, \quad -0.1319, \quad 0.0258, \quad 0.0182$$

Therefore, the system is unstable since some of the eigenvalues have positive real components.

7.5 Small signal stability of multi-machine systems

The multi-machine system can be viewed as a number of synchronous generators in different locations connected to a transmission network to feed various loads. Therefore, to study the small signal stability of such a system the desired mathematical relations describing the interconnected machines must be obtained. Also, the impact of other system components is involved. In this section, without loss of generality, the loads are represented by constant impedances. The transmission network can be represented by its impedance or admittance matrix. Each machine is modelled by using detailed or classic model as a unit connected to the rest of system components. Then, all generators in the system as interconnected machines can be completely described by mathematical relations in terms of system parameters. Finally, these relations are linearised to be in the form

$\dot{\mathbf{x}} = \mathbf{A}\mathbf{x} + \mathbf{B}\mathbf{u}$, and consequently the system stability can be determined by examining the eigenvalues of matrix $\mathbf{A}$ [10].

It is essential to note that, to obtain the relationships of system components, their various quantities to the same reference as a common frame of reference is desirable.

Network and load representation

Figure 7.7(a) depicts a schematic diagram of a transmission network with n generators feeding m loads. The loads are represented by constant impedances that can be calculated from pre-fault conditions in the system. Therefore, the network has only n active sources and can be reduced to n-node network as shown in Figure 7.7(b). The phasor currents and phasor terminal voltages are denoted by $\boldsymbol{I}_1, \boldsymbol{I}_2, \ldots, \boldsymbol{I}_n$ and $\boldsymbol{V}_1, \boldsymbol{V}_2, \ldots, \boldsymbol{V}_n$, respectively. These phasors are expressed in terms of frames of reference that are different for each generator node.

Thus, the currents $\boldsymbol{I}_i$ and voltages $\boldsymbol{V}_i$, $i = 1, 2, \ldots, n$ can be converted to phasors to a common frame of reference, $\overline{\boldsymbol{I}}_i$ and $\overline{\boldsymbol{V}}_i$ at steady state as below:

$$\overline{\boldsymbol{I}} = \boldsymbol{Y}\overline{\boldsymbol{V}} \tag{7.90}$$

where

$$\overline{\boldsymbol{I}} \triangleq \begin{bmatrix} \overline{\boldsymbol{I}}_1 \\ \overline{\boldsymbol{I}}_2 \\ \vdots \\ \overline{\boldsymbol{I}}_n \end{bmatrix}, \quad \overline{\boldsymbol{V}} = \begin{bmatrix} \overline{\boldsymbol{V}}_1 \\ \overline{\boldsymbol{V}}_2 \\ \vdots \\ \overline{\boldsymbol{V}}_n \end{bmatrix}$$

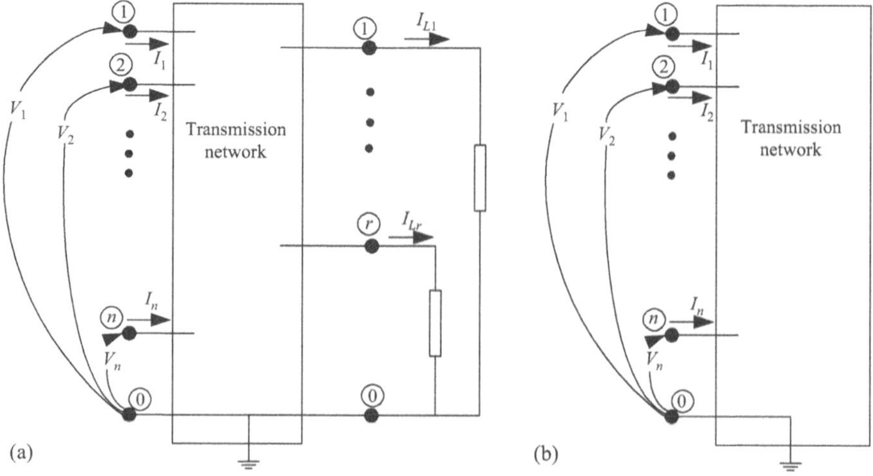

Figure 7.7 Representation of multi-machine system: (a) transmission system with n generators and m equivalent load impedances, (b) the reduced network

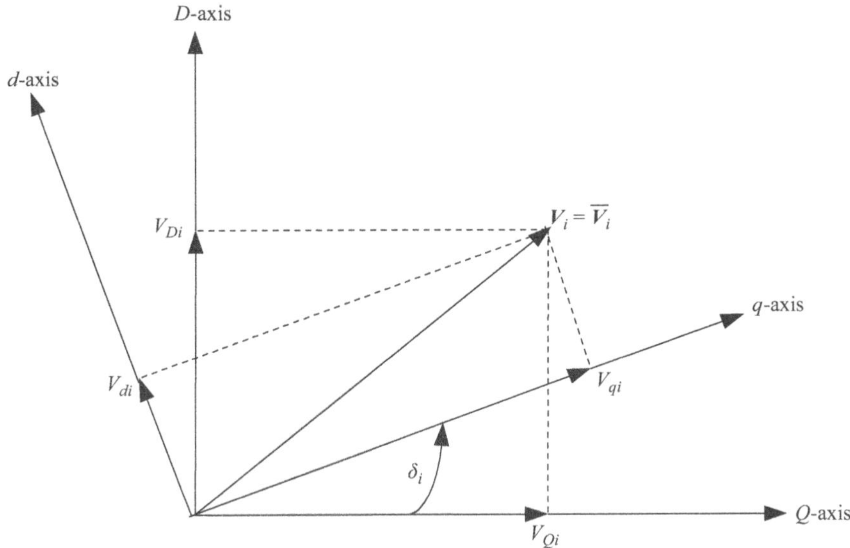

Figure 7.8 Phasor quantities to two frames of reference (d–q and D–Q)

and **Y** is the short circuit admittance matrix of the reduced network that comprises a number of branches '$k = 1, 2, \ldots, b$' between any two nodes in the network. More details of its calculation are found in References 11 and 12.

Both $\boldsymbol{V}_i$ and $\boldsymbol{I}_i$ can be converted to $\overline{\boldsymbol{V}}_i$ and $\overline{\boldsymbol{I}}_i$, respectively, as below.

Assume d_i–q_i is the frame of reference of machine node i and D–Q is the common frame of reference rotating at synchronous speed. The phasor $\boldsymbol{V}_i = V_{qi} + jV_{di}$ where the reference is q-axis of rotor i located at angle δ_i (Figure 7.8). Thus, this phasor to the common frame of reference can be expressed as $\overline{\boldsymbol{V}}_i = V_{Qi} + jV_{Di}$. By inspection of Figure 7.8 it can be found that

$$\begin{aligned}\overline{\boldsymbol{V}}_i &= V_{Qi} + jV_{Di} = \left(V_{qi}\cos\delta_i - V_{di}\sin\delta_i\right) + j\left(V_{qi}\sin\delta_i + V_{di}\cos\delta_i\right)\\ &= \boldsymbol{V}_i \mathrm{e}^{j\delta i}\end{aligned} \tag{7.91}$$

Similarly,

$$\overline{\boldsymbol{I}}_i = \boldsymbol{I}_i \mathrm{e}^{j\delta i} \tag{7.92}$$

Machine representation

As explained in Chapter 2, the synchronous machine can be represented by either detailed model or classical model. The detailed model is presented by current state space model (2.104) or flux state space model (2.93). It is to be noted that the general form of the models can be written as

$$\dot{\mathbf{x}} = f(\mathbf{x}, \mathbf{u}, T_{\mathrm{m}}, t) \tag{7.93}$$

where $\mathbf{x}$ is a vector of state variables (currents or flux linkages), ω and δ; $\mathbf{u}$ is a vector of voltages (v_d, v_q and v_f); and T_m is the mechanical torque. The value of v_f is determined by presentation of excitation system mathematical model, i.e. additional state variables are added to **x** [13]. In this analysis, this presentation is not included and v_f is assumed to be known. Consequently, (7.93) can be expanded as a set of seven first-order differential equations for each machine in nine unknown variable; five currents or flux linkages, ω, and δ in addition to two voltages v_d and v_q. If the system comprises n machines, then a set of $7n$ differential equations with $9n$ unknowns is obtained. Thus, a set of $2n$ additional equations is required to completely describe the system. This additional set can be obtained by deriving the algebraic relations between machine terminal voltages, currents and angles for n machines interconnected to the network and loads.

Each machine in the reduced network (Figure 7.7(b)) is represented by an internal node at a voltage V connected to the network through the machine equivalent impedance. Hence, the vector of terminal voltages of machines to d–q frame of reference of each machine is given by

$$V = \begin{bmatrix} V_{q1} + jV_{d1} \\ V_{q2} + jV_{d2} \\ \vdots \\ V_{qn} + jV_{dn} \end{bmatrix} \tag{7.94}$$

and can be transformed to a common frame of reference D–Q moving at synchronous speed as

$$\overline{V} = \begin{bmatrix} V_{Q1} + jV_{D1} \\ V_{Q2} + jV_{D2} \\ \vdots \\ V_{Qn} + jV_{Dn} \end{bmatrix} \tag{7.95}$$

satisfying the relation

$$\overline{\mathbf{V}} = \mathbf{TV} \tag{7.96}$$

where

$$\mathbf{T} = \begin{bmatrix} e^{j\delta_1} & 0 & \cdots & 0 \\ 0 & e^{j\delta_2} & \cdots & 0 \\ & \vdots & \ddots & \vdots \\ 0 & 0 & \cdots & e^{j\delta_n} \end{bmatrix} \tag{7.97}$$

Similarly for the node currents, the relation below is given as

$$\overline{\mathbf{I}} = \mathbf{TI} \tag{7.98}$$

Applying (7.90) by using (7.96) and (7.98) the relation between machine currents **I** and voltages **V** can be found as below:

$$\mathbf{TI} = \mathbf{YTV} \tag{7.99}$$

Pre-multiplying (7.99) by $\mathbf{T}^{-1}$ to obtain

$$\mathbf{I} = (\mathbf{T}^{-1}\mathbf{YT})\mathbf{V} = \mathbf{MV} \tag{7.100}$$

where

$$\mathbf{M} = \mathbf{T}^{-1}\mathbf{YT} \tag{7.101}$$

Hence,

$$\mathbf{V} = \mathbf{M}^{-1}\mathbf{I} \text{ (assuming } \mathrm{M}^{-1} \text{ exists)} \tag{7.102}$$

It can be seen from (7.97) that

$$\mathbf{T}^{-1} = \begin{bmatrix} e^{-j\delta_1} & 0 & \cdots & 0 \\ 0 & e^{-j\delta_2} & \cdots & 0 \\ \vdots & \vdots & \ddots & \vdots \\ 0 & 0 & \cdots & e^{-j\delta_n} \end{bmatrix} \tag{7.103}$$

The form of network matrix **Y** can be written as

$$\mathbf{Y} = \begin{bmatrix} Y_{11}e^{j\theta_{11}} & Y_{12}e^{j\theta_{12}} & \cdots & Y_{1n}e^{j\theta_{1n}} \\ Y_{21}e^{j\theta_{21}} & Y_{22}e^{j\theta_{22}} & \cdots & Y_{2n}e^{j\theta_{2n}} \\ \vdots & \vdots & \ddots & \vdots \\ Y_{n1}e^{j\theta_{n1}} & Y_{n2}e^{j\theta_{n2}} & \cdots & Y_{nn}e^{j\theta_{nn}} \end{bmatrix} \tag{7.104}$$

Using (7.97), (7.101) and (7.103), the matrix **M** is given by

$$\mathbf{M} \triangleq \begin{bmatrix} Y_{11}e^{j\theta_{11}} & Y_{12}e^{j(\theta_{12}-\delta_{12})} & \cdots & Y_{1n}e^{j(\theta_{1n}-\delta_{1n})} \\ Y_{21}e^{j(\theta_{21}-\delta_{21})} & Y_{22}e^{j\theta_{22}} & \cdots & Y_{2n}e^{j(\theta_{2n}-\delta_{2n})} \\ \vdots & \vdots & \ddots & \vdots \\ Y_{n1}e^{j(\theta_{n1}-\delta_{n1})} & Y_{n2}e^{j(\theta_{n2}-\delta_{n2})} & \cdots & Y_{nn}e^{j\theta_{nn}} \end{bmatrix} \tag{7.105}$$

It is to be noted that the off-diagonal elements m_{ij} can be calculated by

$$m_{ij} = Y_{ij}e^{j(\theta_{ij}-\delta_{ij})} = (G_{ij}\cos\delta_{ij} + B_{ij}\sin\delta_{ij}) + j(B_{ij}\cos\delta_{ij} - G_{ij}\sin\delta_{ij}) \tag{7.106}$$

where

$$G_{ij} = Y_{ij}\cos\theta_{ij} \quad \text{and} \quad B_{ij} = Y_{ij}\sin\theta_{ij}$$

Thus, (7.100) can be rewritten in expanded form as

$$\begin{bmatrix} I_{q1}+jI_{d1} \\ I_{q2}+jI_{d2} \\ \vdots \\ I_{qn}+jI_{dn} \end{bmatrix} = \begin{bmatrix} Y_{11}e^{j\theta_{11}} & Y_{12}e^{j(\theta_{12}-\delta_{12})} & \cdots & Y_{1n}e^{j(\theta_{1n}-\delta_{1n})} \\ Y_{21}e^{j(\theta_{21}-\delta_{21})} & Y_{22}e^{j\theta_{22}} & \cdots & Y_{2n}e^{j(\theta_{2n}-\delta_{2n})} \\ \vdots & \vdots & \ddots & \vdots \\ Y_{n1}e^{j(\theta_{n1}-\delta_{n1})} & Y_{n2}e^{j(\theta_{n2}-\delta_{n2})} & \cdots & Y_{nn}e^{j\theta_{nn}} \end{bmatrix} \begin{bmatrix} V_{q1}+jV_{d1} \\ V_{q2}+jV_{d2} \\ \vdots \\ V_{qn}+jV_{dn} \end{bmatrix} \tag{7.107}$$

Equation (7.107) gives a set of $2n$ real algebraic relations that are needed to be incorporated with (2.104) to complete the description of the system of n interconnected machines by $9n$ relations with $9n$ unknowns.

Linearisation of these $9n$ relations is required to study the small signal stability. The set of $7n$ differential equations given by (2.104) have been linearised as explained in Section 7.3 and the rest of $2n$ algebraic relations given by (7.100) or (7.107) can be linearised as below.

Linearisation of (7.100) gives

$$\Delta\mathbf{I} = \mathbf{M}_o\Delta\mathbf{V} + \Delta\mathbf{M}\mathbf{V}_o \tag{7.108}$$

where $\mathbf{M}_o$ is calculated at the initial angles δ_{io}, $i = 1, 2, \ldots, n$, and $\mathbf{V}_o$ is the initial value of the vector $\mathbf{V}$. Assuming $\delta_i = \delta_{io} + \Delta\delta_i$. The matrix $\mathbf{M}$ becomes

$$\mathbf{M} = \begin{bmatrix} Y_{11}e^{j\theta_{11}} & Y_{12}e^{j(\theta_{12}-\delta_{12o}-\Delta\delta_{12})} & \cdots & Y_{1n}e^{j(\theta_{1n}-\delta_{1no}-\Delta\delta_{1n})} \\ Y_{21}e^{j(\theta_{21}-\delta_{21o}-\Delta\delta_{21})} & Y_{22}e^{j\theta_{22}} & \cdots & Y_{2n}e^{j(\theta_{2n}-\delta_{2no}-\Delta\delta_{2n})} \\ \vdots & \vdots & \ddots & \vdots \\ Y_{n1}e^{j(\theta_{n1}-\delta_{n1o}-\Delta\delta_{n1})} & Y_{n2}e^{j(\theta_{n2}-\delta_{n2o}-\Delta\delta_{no})} & \cdots & Y_{nn}e^{j\theta_{nn}} \end{bmatrix} \tag{7.109}$$

Thus, $Y_{ij}e^{j\left(\theta_{ij}-\delta_{ijo}-\Delta\delta_{ij}\right)} = m_{ij} \triangleq$ the general term of matrix $\mathbf{M}$ can be written as $m_{ij} = Y_{ij}e^{j\left(\theta_{ij}-\delta_{ijo}\right)}e^{-j\Delta\delta_{ij}}$ and considering $\cos\Delta\delta_{ij} \cong 1$, $\sin\Delta\delta_{ij} \cong \Delta\delta_{ij}$ it becomes

$$m_{ij} \cong Y_{ij}e^{j\left(\theta_{ij}-\delta_{ijo}\right)}\left(1 - j\Delta\delta_{ij}\right) \tag{7.110}$$

Consequently, the general term in the matrix $\Delta\mathbf{M}$ is given by

$$\begin{aligned} \Delta m_{ij} &\cong -jY_{ij}e^{j\left(\theta_{ij}-\delta_{ijo}\right)}\Delta\delta_{ij} \quad \text{for } i \neq j \\ &\cong 0 \quad \text{for } i = j \end{aligned} \tag{7.111}$$

i.e. the matrix $\Delta\mathbf{M}$ has off-diagonal elements only, with all diagonal elements equal to zero.

The second term of RHS in (7.108) is

$$\Delta\mathbf{MV}_o = -j\begin{bmatrix} 0 & \cdots & Y_{ij}e^{j(\theta_{ij}-\delta_{ijo})}\Delta\delta_{ij} \\ Y_{ij}e^{j(\theta_{ij}-\delta_{ijo})}\Delta\delta_{ij} & \cdots & Y_{ij}e^{j(\theta_{ij}-\delta_{ijo})}\Delta\delta_{ij} \\ \vdots & \ddots & \vdots \\ Y_{ij}e^{j(\theta_{ij}-\delta_{ijo})}\Delta\delta_{ij} & \cdots & 0 \end{bmatrix}\begin{bmatrix} V_{1o} \\ V_{2o} \\ \cdots \\ V_{no} \end{bmatrix}$$

$$= -j\begin{bmatrix} \sum_{k=1}^{n} Y_{1k}e^{j(\theta_{1k}-\delta_{1ko})}V_{ko}\Delta\delta_{1k} \\ \sum_{k=1}^{n} Y_{2k}e^{j(\theta_{2k}-\delta_{2ko})}V_{ko}\Delta\delta_{2k} \\ \vdots \\ \sum_{k=1}^{n} Y_{nk}e^{j(\theta_{nk}-\delta_{nko})}V_{ko}\Delta\Delta_{nk} \end{bmatrix} \tag{7.112}$$

Then substituting into (7.108) gives the linearised equation as

$$\begin{bmatrix} \Delta I_1 \\ \Delta I_2 \\ \vdots \\ \Delta I_n \end{bmatrix} = \begin{bmatrix} Y_{11}e^{j\theta_{11}} & \cdots & Y_{1n}e^{j(\theta_{1n}-\delta_{1no})} \\ Y_{21}e^{j(\theta_{21}-\delta_{21o})} & \cdots & Y_{2n}e^{j(\theta_{2n}-\delta_{2no})} \\ \vdots & \ddots & \vdots \\ Y_{n1}e^{j(\theta_{n1}-\delta_{n1o})} & \cdots & Y_{nn}e^{j\theta_{nn}} \end{bmatrix}\begin{bmatrix} \Delta V_1 \\ \Delta V_2 \\ \vdots \\ \Delta V_n \end{bmatrix}$$

$$-j\begin{bmatrix} \sum_{k=1}^{n} Y_{1k}e^{j(\theta_{1k}-\delta_{1ko})}V_{ko}\Delta\delta_{1k} \\ \sum_{k=1}^{n} Y_{2k}e^{j(\theta_{2k}-\delta_{2ko})}V_{ko}\Delta\delta_{2k} \\ \vdots \\ \sum_{k=1}^{n} Y_{nk}e^{j(\theta_{nk}-\delta_{nko})}V_{ko}\Delta\Delta_{nk} \end{bmatrix} \tag{7.113}$$

In (7.113) Δ can be dropped for convenience. Then substituting into (7.70) obtains the linearised set of equations in the form $\dot{\mathbf{x}} = \mathbf{A}\mathbf{x} + \mathbf{B}\mathbf{u}$. By examining the eigenvalues of $\mathbf{A}$, the system stability can be determined.

References

1. Condren J., Gedra T.W. 'Expected-security-cost optimal power flow with small-signal stability constraints'. *IEEE Transactions on Power Systems*. 2006;**21**(4):1736–43
2. Tayora C.J., Smith O.J.M. 'Equilibrium analysis of power systems'. *IEEE Transactions on Power Apparatus and Systems*. 1972;**PAS-91**(3):1131–7

3. Kundur P., Paserba J., Ajjarapu V., Anderson G. 'Definition and classification of power system stability IEEE/CIGRE joint task force on stability terms and definitions'. *IEEE Transactions on Power Systems*. 2004;**19**(3):1387–401
4. Chen L., Min Y., Xu F., Wang K.P. 'A continuation-based method to compute the relevant unstable equilibrium points for power system transient stability analysis'. *IEEE Transactions on Power Systems*. 2009;**24**(1): 165–72
5. Rueda J.L., Colome D.G., Erlich I. 'Assessment and enhancement of small signal stability considering uncertainties'. *IEEE Transactions on Power Systems*. 2009;**24**(1):198–207
6. Byerly R.T., Sheman D.E., McLain D.K. 'Normal modes and mode shapes applied to dynamic stability analysis'. *Transactions on Power Apparatus and Systems*. 1975;**94**(2):224–9
7. Gross G., Imparato C.F., Look P.M. 'A tool for the comprehensive analysis of power system dynamic stability'. *IEEE Transactions on Power Apparatus and Systems*. 1982;**101**(1):226–34
8. Ewart D.N., Demello F.P. 'A digital computer program for the automatic determination of dynamic stability limits'. *IEEE Transactions on Power Apparatus and Systems*. 1967;**PAS-86**(7):867–75
9. Anderson P.M., Fouad A.A. *Power System Control and Stability*. 2nd edn. Piscataway, NJ, US: IEEE Press; 2003
10. Ma J., Dong Z.Y., Zhang P. 'Comparison of BR and QR eigenvalue algorithms for power system small signal stability analysis'. *IEEE Transactions on Power Systems*. 2006;**21**(4):1848–55
11. Stagg and El-Abiad A. *Computer Methods in Power System Analysis*. New York, NY, US: McGraw-Hill; 1968
12. Anderson P.M. *Analysis of Faulted Power Systems*. Ames, IA, US: Iowa State University Press; 1973
13. Arcidiacono V., Ferrari E., Saccomanno F. 'Studies on damping of electromechanical oscillations in multimachine systems with longitudinal structure'. *IEEE Transactions on Power Apparatus and Systems*. 1976; **95**(2):450–60

Chapter 8

Transient stability

The objective of transient stability study is to determine whether the system generators remain in synchronism when subjected to large disturbances. The transient stability is evaluated by studying the system dynamic response during the transient period that usually lasts up to a few seconds taking into account the rapid change of electrical variables, including relative swinging between generators. A longer transient period may be covered in the study when the behaviour of some controls is of interest.

Because of the nature of transient disturbances, the non-linear system equations cannot be linearised and must be solved in stability evaluation. Significant simplifications are required to obtain analytical solutions. Therefore, numerical integration techniques are applied.

To form the system equations, adequate models of system components are needed to implement stability study with the desired accuracy. Models of system components, such as synchronous generators with associated controls, excitation system and prime mover, transformers, transmission lines and loads [1, 2], have been discussed in Part I. It is to be noted that the most important component is the synchronous generator with its associated controls. On the other hand, in stability analysis, load frequency controllers and prime mover models are often neglected without loss of accuracy. Per unit equations of current or flux linkage models given in Chapter 3 completely describe the dynamic performance of a synchronous machine. However, these equations cannot be used directly for system transient stability studies. Some simplifications and approximations are required to represent the synchronous machine in stability studies. Therefore, the performance equations of a synchronous machine are developed based on the following assumptions:

- Effect of mutual inductance between stator and rotor is considered assuming sinusoidal distribution of the stator windings around the air gap.
- Effect of stator slots on rotor inductances with rotor position is neglected.
- Magnetic hysteresis is negligible.
- Magnetic saturation effects are negligible.
- Stator transients are ignored.

As reported in [3], the number of rotor windings and corresponding state variables can vary from one to six depending on the degree of detail used. The suggested models are defined based on the degree of complexity and are denoted by

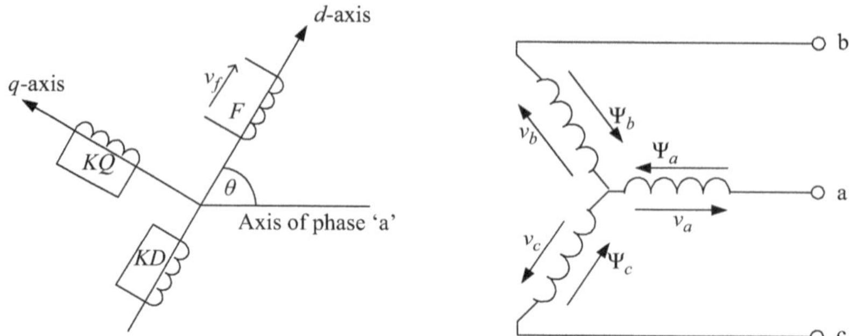

Figure 8.1 Synchronous machine Model 2.1. Stator phase windings: a, b, c: and rotor windings F, KD on d-axis, KQ on q-axis

Model *x.y*, where *x* and *y* are the number of rotor windings on *d* and *q* axes, respectively. Thus,

- Model 0.0: damper circuits and field flux decay are neglected, i.e. all state variables for rotor coils are ignored.
- Model 1.0: a field circuit only is considered on the *d*-axis.
- Model 1.1: a field circuit and one damper on *q*-axis
- Model 2.1: a field circuit and one damper on the *d*-axis plus one damper on the *q*-axis are considered.
- Model 2.2: a field circuit and one damper on the *d*-axis plus two dampers on the *q*-axis are considered.
- Model 3.2: a field circuit and two dampers on the *d*-axis plus two dampers on the *q*-axis are considered.
- Model 3.3: a field circuit and two dampers on the *d*-axis plus three dampers on the *q*-axis are considered.

As an application, Model 2.1 (Figure 8.1) is used in the following to represent the synchronous machine.

8.1 Synchronous machine model

The stator equations expressed in per unit are given by (I.4). As all quantities are in per unit, the subscript *u* can be dropped and the equation is rewritten as

$$\left.\begin{aligned} v_d &= -\frac{1}{\omega_B}\frac{d\Psi_d}{dt} - \frac{\omega}{\omega_B}\Psi_q - R_a i_d \\ v_q &= -\frac{1}{\omega_B}\frac{d\Psi_q}{dt} + \frac{\omega}{\omega_B}\Psi_d - R_a i_q \end{aligned}\right\} \tag{8.1}$$

assuming that the zero sequence current in the stator does not exist, i.e. $v_o = 0$. In transient stability studies, it is common to neglect the transformer voltage terms $\frac{\mathrm{d}\Psi_d}{\mathrm{d}t}$ and $\frac{\mathrm{d}\Psi_q}{\mathrm{d}t}$, and the effect of speed variations as well. Accordingly (8.1) becomes

$$\left.\begin{aligned} v_d &= -\Psi_q - R_a i_d \\ v_q &= \Psi_d - R_a i_q \end{aligned}\right\} \tag{8.2}$$

Similarly, the voltage equations for *F*, *KD* and *KQ* rotor windings, (I.21)–(I.23), become:

$$v_f - R_f i_f = \frac{1}{\omega_{\mathrm{B}}}\frac{\mathrm{d}\Psi_f}{\mathrm{d}t} \tag{8.3}$$

$$-R_{kd} i_{kd} = \frac{1}{\omega_{\mathrm{B}}}\frac{\mathrm{d}\Psi_{kd}}{\mathrm{d}t} \tag{8.4}$$

$$-R_{kq} i_{kq} = \frac{1}{\omega_{\mathrm{B}}}\frac{\mathrm{d}\Psi_{kq}}{\mathrm{d}t} \tag{8.5}$$

It is noted that neglecting stator transients, the stator equations become algebraic. Consequently, it is not possible to choose stator currents i_d and i_q as state variables, as these currents can be discontinuous functions due to any sudden changes in the network. On the other hand, the flux linkages of rotor windings, field and dampers cannot change suddenly. This implies that with sudden changes of i_d, the field and damper currents also change suddenly in order to maintain the field and damper flux linkages continuous, i.e. immediately after a disturbance flux linkages remain constant at the value just prior to the disturbance. Accordingly, rotor winding currents cannot be treated as state variables. Hence, rotor flux linkages can be chosen as state variables.

From (I.5) the stator flux linkages on *d*–*q* axes are

$$\Psi_d = L_d i_d + kM_f i_f + kM_{kd} i_{kd} = X_d i_d + X_{ad}(i_f + i_{kd}) \tag{8.6}$$

$$\Psi_q = L_q i_q + kM_{kq} i_{kq} = X_q i_q + X_{aq} i_{kq} \tag{8.7}$$

and the rotor circuits flux linkages are

$$\Psi_f = L_f i_f + kM_f i_d + L_{fkd} i_{kd} = X_f i_f + X_{ad} i_d + X_{fkd} i_{kd} \tag{8.8}$$

$$\Psi_{kd} = L_{kd} i_{kd} + kM_{kd} i_d + L_{fkd} i_f = X_{kd} i_{kd} + X_{ad} i_d + X_{fkd} i_f \tag{8.9}$$

$$\Psi_{kq} = L_{kq} i_{kq} + kM_{kq} i_{kq} = X_{kq} i_{kq} + X_{aq} i_q \tag{8.10}$$

where $kM_f = kM_{kd} = X_{ad}$ and $kM_{kq} = X_{aq}$

It is to be noted that by choosing angular frequency, ω_{B}, as the base, the per unit values L_d, L_q, L_f, L_{kd}, L_{kq} equal X_d, X_q, X_f, X_{kd}, X_{kq}, respectively.

Solving (8.8)–(8.10) gives

$$i_f = \frac{\Psi_f}{X_f} - \frac{X_{ad}}{X_f} i_d - \frac{X_{fkd}}{X_f} i_{kd} \tag{8.11}$$

$$i_{kd} = \frac{\Psi_{kd}}{X_{kd}} - \frac{X_{ad}}{X_{kd}} i_d - \frac{X_{fkd}}{X_{kd}} i_f \tag{8.12}$$

$$i_{kq} = \frac{\Psi_{kq}}{X_{kq}} - \frac{X_{aq}}{X_{kq}} i_q \tag{8.13}$$

Then by substitution in (8.6) and (8.7) obtains

$$\Psi_d = X'_d i_d + E'_q \tag{8.14}$$

$$\Psi_q = X'_q i_q - E'_d \tag{8.15}$$

where

$$X'_d = X_d - \frac{X_{ad}^2}{x_f} \tag{8.16}$$

$$X'_q = X_q - \frac{X_{aq}^2}{X_{kq}} \tag{8.17}$$

$$E'_q = \frac{X_{ad}}{X_f}\lceil \Psi_f + (L_f - L_{fkd}) i_{kd} \rceil = \frac{x_{ad}}{X_f}\overline{\Psi}_f$$

$$= \frac{X_{ad}}{X_f}\Psi_f \text{ when the second term is neglected} \tag{8.18}$$

$$E'_d = -\frac{X_{aq}\Psi_{kq}}{X_{kq}} \tag{8.19}$$

Substituting (8.11) and (8.18) into (8.3) gives

$$\frac{1}{\omega_{\mathrm{B}}} \frac{X_f}{X_{ad}} \frac{\mathrm{d}E'_q}{\mathrm{d}t} = -\frac{R_f E'_q}{X_{ad}} + \frac{R_f X_{ad}}{X_f} i_d + v_f \tag{8.20}$$

Hence,

$$\frac{\mathrm{d}E'_q}{\mathrm{d}t} = \frac{\omega_{\mathrm{B}} R_f}{X_f}\left(-E'_q + \frac{X_{ad}^2}{X_f} i_d + \frac{X_{ad}}{R_f} v_f\right) \tag{8.21}$$

or

$$\frac{\mathrm{d}E'_q}{\mathrm{d}t} = \frac{1}{T'_{do}}\left[-E'_q + (X_d - X'_d) i_d + E_{fd}\right] \tag{8.22}$$

where

$$E_{fd} = \frac{X_{ad}}{R_f} v_f \tag{8.23}$$

$$T'_{do} = \frac{X}{\omega_B R_f} \tag{8.24}$$

Substituting (8.13) and (8.19) into (8.5) the relation below can be found:

$$\frac{dE'_d}{dt} = \frac{1}{T'_{qo}} \left[-E'_d - \left(X_q - X'_q \right) i_q \right] \tag{8.25}$$

where

$$T'_{qo} = \frac{X_{kq}}{\omega_B R_{kq}} \tag{8.26}$$

When considering the stator voltage and torque equations, it is convenient to define the equivalent voltage sources E'_d and E'_q as state variables rather than rotor flux linkages.

Substituting (8.14) and (8.15) into (8.2) gives

$$\left. \begin{aligned} v_q &= E'_q + X'_d i_d - R_a i_q \\ v_d &= E'_d - X'_q i_q - R_a i_d \end{aligned} \right\} \tag{8.27}$$

Assuming $X'_d = X'_q = X'$ in case of neglecting transient saliency, (8.27) can be written as

$$v_q + jv_d = \left(E'_q + jE'_d \right) - \left(R_a + jX' \right) \left(i_q + ji_d \right) \tag{8.28}$$

In vector notation, (8.28) can be expressed as

$$\boldsymbol{V}_t = \boldsymbol{E}' - \left(R_a + jX' \right) \boldsymbol{I}_t \tag{8.29}$$

where

$\boldsymbol{V}_t \triangleq$ machine terminal voltage $= v_q + jv_d$
$\boldsymbol{E}' \triangleq$ voltage behind transient reactance $= E'_q + jE'_d$
$\boldsymbol{I}_t \triangleq$ machine terminal current $= i_q + ji_d$
$R_a \triangleq$ armature resistance
$X' \triangleq$ transient reactance

The equivalent circuit of the stator corresponding to (8.29) is shown in Figure 8.2. It shows a voltage source $\boldsymbol{E}'$ behind equivalent impedance $(R_a + jx')$.

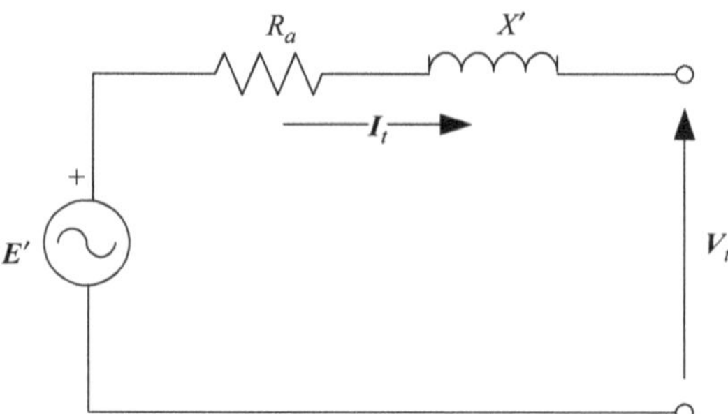

Figure 8.2 Stator equivalent circuit

The rotor mechanical (2.59), i.e. swing equation, can be expressed as two first-order differential equations as

$$\left.\begin{aligned}\dot{\delta} &= \omega - \omega_o \\ \dot{\omega} &= \frac{\omega_{\mathrm{B}}}{2\mathrm{H}}[T_{\mathrm{m}} - T_{\mathrm{e}} - D\omega]\end{aligned}\right\} \tag{8.30}$$

The electrical torque T_e is given by $T_e = \Psi_d i_q - \Psi_q i_d$. Substituting from (8.14) and (8.15) gives

$$T_e = E'_d i_d + E'_q i_q + \left(X'_d - X'_q\right) i_d i_q \tag{8.31}$$

The third term in (8.31) is zero if the transient saliency is ignored as $X'_d = X'_q$. The variables i_d and i_q can be obtained from the stator algebraic (8.27) and the network equations or power flow solution.

Therefore, for synchronous machine Model 2.1 in addition to rotor (8.30) and (8.31) the following stator equations are used to represent the machine.

$$\left.\begin{aligned}v_q &= E'_q + X'_d i_d - R_a i_q \\ v_d &= E'_d - X'_q i_q - R_a i_d \\ \frac{\mathrm{d}E'_q}{\mathrm{d}t} &= \frac{1}{T'_{do}}[-E'_q + (X_d - X'_d)i_d + E_{fd}] \\ \frac{\mathrm{d}E'_d}{\mathrm{d}t} &= \frac{1}{T'_{qo}}[-E'_d - (X_q - X'_q)i_q]\end{aligned}\right\} \tag{8.32}$$

It is noted that the d–q axis transient effects require differential equations 'sE'_q and sE'_d'. A block diagram representation is shown in Figure 8.3 [4].

The field voltage E_{fd} and the mechanical power P_m ($\approx T_m$ in pu) can be held constant in the transient calculations for a period of analysis less than one second as the effects of the exciter and governor control systems on power system response are neglected. When a more detailed evaluation of system response is required or

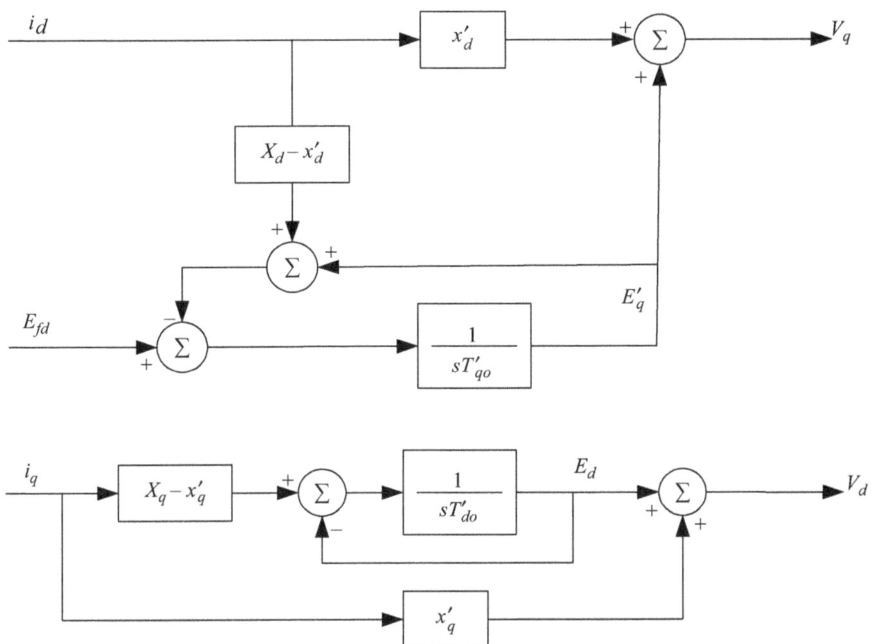

Figure 8.3 Block diagram representation for Model 2.1

the period of analysis extends beyond 1 s, it is important to consider the effects of the exciter and governor systems.

The exciter control system provides the proper field voltage to maintain a desired system voltage. An important characteristic of an exciter control system is its ability to respond rapidly to voltage deviations during both normal and emergency system operation. Different types of excitation systems and their block diagrams that relate the input and output variables through transfer functions have been explained in Chapter 3. Thus, the differential equations relating the input and output variables of excitation system components must be solved simultaneously with the stator and rotor equations.

Similarly, the effects of the speed governor control that provide the mechanical power P_m during transient periods can be taken into account by using the representation of the selected governor control system as described in Chapter 3. This representation includes a transfer function describing the system components. The differential equations relating the input and output variables of these transfer functions are solved simultaneously with the stator and rotor equations.

As explained above, in transient stability analysis, the synchronous machine with its associated controllers can be represented by a set of equations that comprises a combination of algebraic and non-linear differential equations. To solve these equations simultaneously the differential equations are converted to algebraic equations and one of the numerical integration methods is applied to obtain the solution step by step.

8.2 Numerical integration techniques

Many integration techniques have been applied to the power system transient stability analysis such as trapezoidal method, Euler's method, modified Euler–Cauchy method and Runge–Kutta methods [5–8]. More details about these methods are given in Appendix III and a summary is given below.

Assuming, for instance, two simultaneous ordinary differential equations (ODEs) of the form

$$\dot{x} = f_1(t, x, y; h)$$
$$\dot{y} = f_2(t, x, y; h)$$

where x and y are the state variables, e.g. δ and ω, that are calculated numerically versus time t. Considering the step size h equals the time increment Δt, the rules of some numerical solution methods are summarised in Table 8.1.

Table 8.1 Summary of some numerical integration methods formulae

Method	Rules
Euler's method	$x_{,i+1} = x_i + hf_1(x_i, y_i)$ $y_{,i+1} = y_i + hf_2(x_i, y_i)$
Modified Euler–Cauchy method	$x_{i+1} = x_i + hf_1\left(t_i + \frac{h}{2}, x_i + \frac{h}{2}f_1(x_i, y_i), y_i + \frac{h}{2}f_2(x_i, y_i)\right)$ $y_{i+1} = y_i + hf_2\left(t_i + \frac{h}{2}, x_i + \frac{h}{2}f_1(x_i, y_i), y_i + \frac{h}{2}f_2(x_i, y_i)\right)$
Trapezoidal method	$x_{i+1} = x_i + \frac{h}{2}[f_1(t_i, y_i)$ $+ f_1(t_{i+1}, x_i + hf_1(x_i, y_i), y_i + hf_2(x_i, y_i))]$ $y_{i+1} = y_i + \frac{h}{2}[f(t_i, y_i)$ $+ f(t_{i+1}, x_i + hf_1(x_i, y_i), y_i + hf_2(x_i, y_i))]$
Second-order Runge–Kutta method	$x_{i+1} = x_i + \frac{h}{2}[K_{11} + K_{21}]$ $y_{i+1} = y_i + \frac{h}{2}[K_{12} + K_{22}]$
Third-order Runge–Kutta method	$x_{i+1} = x_i + \frac{h}{6}[K_{11} + 4K_{21} + K_{31}]$ $y_{i+1} = y_i + \frac{h}{6}[K_{12} + 4K_{22} + K_{32}]$
Fourth-order Runge–Kutta method	$x_{i+1} = x + \frac{h}{6}[K_{11} + 2K_{21} + 2K_{31} + K_{41}]$ $y_{i+1} = y_i + \frac{h}{6}[K_{12} + 2K_{22} + 2K_{32} + K_{42}]$

The coefficients K_{ij} of Runge–Kutta methods are given in Appendix III.

8.3 Transient stability assessment of a simple power system

A single machine connected to an infinite bus through a transmission line is defined as a simple power system. For example, a remote power station connected to a load through a long transmission line can be represented by a simple system comprising one machine that is equivalent to all generators in the power station. This is acceptable for disturbances external to the power station and it needs to model the system elements – generator, transmission line and infinite bus – for studying the transient stability.

Two main points should be taken into account when writing the system equations. First, the system equations must be referred to a common frame of reference. Second, the non-state variables must be eliminated from the system equations and expressed as parameters and/or in terms of state variables.

The machine can be represented by (8.29)–(8.31). The transmission line is represented by its π-equivalent circuit as explained in Chapter 4. The transmission line is considered as a two-port external network: one port is connected to the generator terminals and the second port is connected to the infinite bus. The infinite bus, representing a large stiff system, may be modelled by a voltage source of constant magnitude and phase angle $E_b \angle\theta$. The angle θ is usually assumed as zero where the bus is taken as a common reference (Figure 8.4).

Assuming that the simple system consists of only series impedance $Z_e = R_e + jX_e$ and taking the system axis shown in Figure 8.5 as a common frame of reference, it can be seen that

$$\left(v_q + jv_d\right)e^{j\delta} = (R_e + jX_e)\left(i_q + ji_d\right)e^{j\delta} + \boldsymbol{E}_b \tag{8.33}$$

Multiplying both sides by $e^{-j\delta}$ and equating real and imaginary parts gives

$$\left.\begin{aligned} v_q &= R_e i_q - X_e i_d + E_b \cos\delta \\ v_d &= X_e i_q + R_e i_d - E_b \sin\delta \end{aligned}\right\} \tag{8.34}$$

As a further simplification, R_e is assumed to be zero, thus,

$$\left.\begin{aligned} v_q &= -X_e i_d + E_b \cos\delta \\ v_d &= X_e i_q - E_b \sin\delta \end{aligned}\right\} \tag{8.35}$$

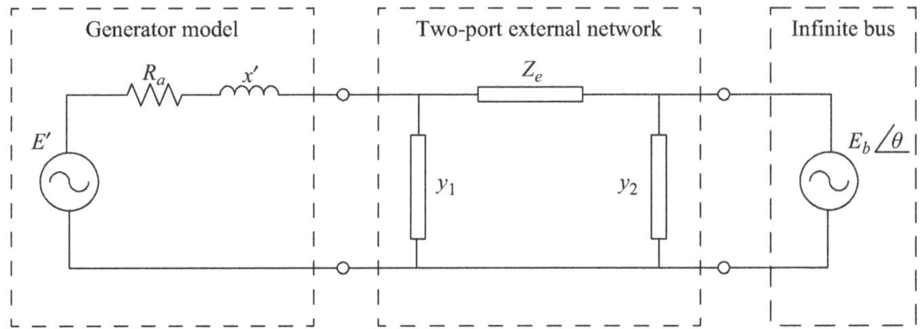

Figure 8.4 Equivalent circuit of a simple power system

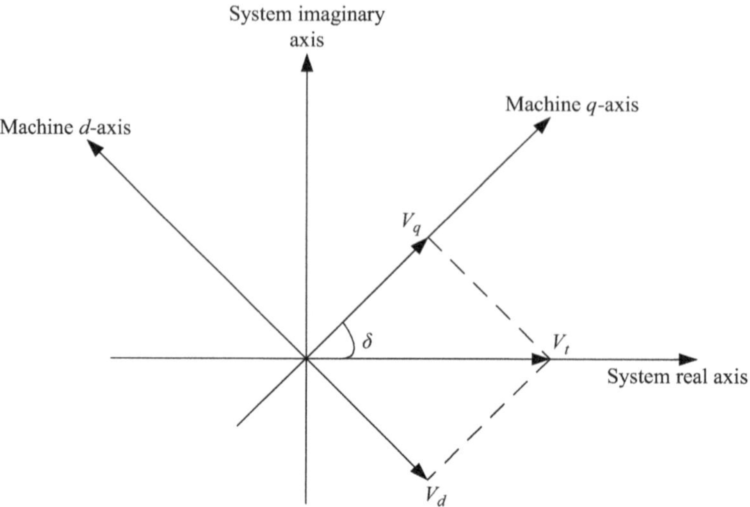

Figure 8.5 Synchronous machine and power system frames of reference

where i_d and i_q are non-state variables and must be eliminated. This can be obtained by expressing v_d and v_q 'machine terminal voltages' as written in (8.32) and assuming $R_a = 0$ as

$$\left.\begin{aligned} v_q &= E'_q + X'_d i_d \\ v_d &= E'_d - X'_q i_q \end{aligned}\right\} \tag{8.36}$$

From (8.35) and (8.36), i_d and i_q can be obtained by

$$\left.\begin{aligned} i_d &= \frac{E_b \cos\delta - E'_q}{(X_e + X'_d)} \\ i_q &= \frac{E_b \sin\delta + E'_d}{\left(X_e + X'_q\right)} \end{aligned}\right\} \tag{8.37}$$

Then, substitute these values of i_d and i_q, in (8.31) and (8.32) for $\frac{dE'_q}{dt}$ and $\frac{dE'_d}{dt}$ to get the system equations in the form $\dot{X} = f(x, u)$ as

$$\left.\begin{aligned} \frac{dE'_q}{dt} &= \frac{1}{T'_{do}}\left[-E'_q + X_{r1}E_1 + E_{fd}\right] \\ \frac{dE'_d}{dt} &= \frac{1}{T'_{qo}}\left[-E'_d - X_{r2}E_{r2}\right] \\ \dot{\delta} &= \omega - \omega_o \\ \dot{\omega} &= \frac{\omega_B}{2H}\left[T_m - T_e - D\omega\right] \end{aligned}\right\} \tag{8.38}$$

in addition to the algebraic equation

$$T_e = \frac{1}{X_1 X_2}\left[E'_d E_1 X_2 + E'_q E_2 X_1 + E_1 E_2\left(X'_d - X'_q\right)\right] \tag{8.39}$$

where

$$X_{r1} = \frac{(X_d - X'_d)}{(X_e + X'_d)}, \quad X_{r2} = \frac{\left(X_q - X'_q\right)}{\left(X_e + X'_q\right)}$$

$$X_1 = (X_e + X'_d),\ X_2 = \left(X_e + X'_q\right),\ E_1 = E_b\cos\delta - E'_q,\ E_2 = E_b\sin\delta + E'_d$$

It is concluded that the simple power system can be represented by a set of differential-algebraic (8.38) and (8.39). It is noted that E_b is treated as a parameter. E_{fd} and T_m are inputs from the excitation and governor control systems, respectively. They are treated as parameters if the dynamics of the controllers are ignored. Otherwise, the dynamics of the controllers represented by differential equations are to be appended to (8.38) to determine their outputs E_{fd} and T_m.

The set of equations representing the power system – machine stator equations, rotor mechanical equations and network equations – can be solved simultaneously by applying one of the numerical integration methods to obtain the change of state variables versus time, and then the stability can be determined. The initial conditions required to solve the ODEs are calculated from system behaviour at steady state prior to the disturbance as explained in Section 7.1.3.

Derivation of stator equations depends on the degree of complexity required to model the machine. For instance,

- Representing the machine as a constant voltage magnitude behind d-axis transient reactance X'_d requires no differential equations. Only the following algebraic equation is used:

$$E' = V_t + R_a I_t + jX'_d I_t \tag{8.40}$$

- If d-axis transient effects are considered one differential equation is required and the set of stator equations is

$$\left.\begin{aligned} E'_q &= v_q - X'_d i_d + R_a i_q \\ E'_d &= v_d + X'_q i_q + R_a i_d \\ \frac{dE'_q}{dt} &= \frac{1}{T'_{do}}\left[-E'_q + (X_d - X'_d) i_d + E_{fd}\right] \end{aligned}\right\} \tag{8.41}$$

- Representation of d- and q-axis sub-transient effects requires three differential equations as below:

$$\left.\begin{aligned}
E_q'' &= v_q - X_d'' i_d + R_a i_q \\
E_d'' &= v_d + X_q'' i_q + R_a i_d \\
\frac{dE_q'}{dt} &= \frac{1}{T_{do}'}\left[-E_q' + \left(X_d - X_d'\right)i_d + E_{fd}\right] \\
\frac{dE_q''}{dt} &= \frac{1}{T_{do}''}\left[E_q' - E_q'' + \left(X_d' - X_d''\right)i_d\right] \\
\frac{dE_q''}{dt} &= \frac{1}{T_{qo}''}\left[E_d' - E_q'' - \left(X_q' - X_d''\right)i_q\right]
\end{aligned}\right\} \tag{8.42}$$

- Representation of d- and q-axis sub-transient effects by four differential equations is as below:

$$\left.\begin{aligned}
E_q'' &= v_q - X_d'' i_d + R_a i_q \\
E_d'' &= v_d + X_q'' i_q + R_a i_d \\
\frac{dE_q'}{dt} &= \frac{1}{T_{do}'}\left[-E_q' + (X_d - X_d')i_d + E_{fd}\right] \\
\frac{dE_d'}{dt} &= \frac{1}{T_{qo}'}\left[-E_d' - \left(X_q - X_q'\right)i_q\right] \\
\frac{dE_q''}{dt} &= \frac{1}{T_{do}''}\left[E_q' - E_q'' + \left(X_d' - X_d''\right)i_d\right] \\
\frac{dE_d''}{dt} &= \frac{1}{T_{qo}''}\left[E_d' - E_d'' - \left(X_q' - X_q''\right)i_q\right]
\end{aligned}\right\} \tag{8.43}$$

The steps to assess the transient stability of a simple power system are summarised below:

- Derive machine equations as a model adequate to the degree of complexity required for system study.
- Derive network equations.
- Form system differential-algebraic equations: stator equations, rotor swing equation and network equations.
- Calculate initial conditions for the system at steady state prior to the disturbance.
- Calculate power delivered from the machine to the infinite bus during and after the fault period. This entails determination of an equivalent circuit between the sending and receiving ends during each period or using power flow analysis.
- Select a numerical integration method to solve the system algebraic-differential equations and plot the torque angle and machine speed versus time.

Example 8.1 A synchronous generator is connected to an infinite bus through a transformer and two parallel identical transmission lines with data in pu as shown in Figure 8.6 and delivers a power of 0.8 pu. A three-phase to earth fault occurs at a point, *F*, near the beginning of a TL and is cleared at 0.08 s by isolating the faulty TL. Neglecting all resistances and velocity damping coefficient, compute the variation of δ and ω of the machine versus time by solving the problem using a numerical integration method with a step size of 0.02 s when:

(i) The machine is represented by a voltage source E' 'constant magnitude and varying δ' behind the transient reactance X'_d.

$$E' = V_t + R_a I_t + jX'_d I_t$$

(ii) The machine is represented by a voltage source E_q behind X_q with system equations as

$$E'_q = v_q - X'_d i_d + R_a i_q$$

$$E'_d = v_d + X'_q i_q + R_a i_d$$

$$\frac{dE'_q}{dt} = \frac{1}{T'_{do}}\left(E_{fd} - E_I\right)$$

$$= \frac{1}{T'_{do}}\left[-E'_q + \left(X_d - X'_d\right) i_d + E_{fd}\right]$$

$$E_q = V_t + R_a I_t + jX_q I_t$$

$$E_I = V_t + R_a I_t + jX_d I_d + jx_q I_q$$

$$E'_q = E_q - j\left(X_q - X'_d\right) I_d$$

Other machine data are given below:

$$T'_{do} = 0.4, \quad X_d = 1.9, \quad X_q = 1.75, \quad X'_q = 0.24, \quad \mathrm{H} = 3.5$$

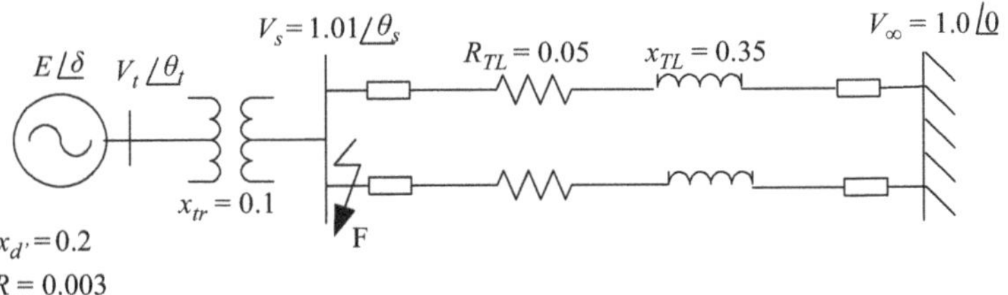

Figure 8.6 Simple system for Example 8.1

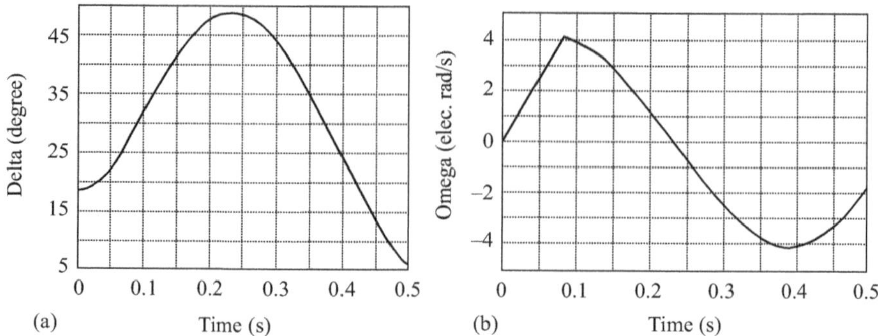

Figure 8.7 (a) Variation of δ versus time and (b) variation of ω versus time with fault clearing time 0.08 s

Solution:

(i) The machine is represented by a voltage source E' 'constant magnitude and varying δ' behind the transient reactance X_d'.

The initial values of system parameters are

$$\begin{aligned}
Z_L \text{ pre-fault} &= (R_L + jX_L)/2 = 0.0250 + j0.1750\\
Z_L \text{ during fault} &= \infty\\
Z_L \text{ post-fault} &= (R_L + jX_L) = 0.0500 + j0.35\\
\theta_o &= 18^\circ\\
\omega_o &= 314.1593 \text{ elec. rad/s}\\
\theta_s &= \sin^{-1}[(P_m \,\text{abs}(Z_L \text{ pre}))/(V_s V_\infty)] = 8.05^\circ\\
V_s &= V_s(\cos(\theta_s) + j\sin(\theta_s)) = 1.0000 + j0.1414\\
I_{gen} &= (V_s - V_\infty)/Z_L \text{ pre-fault} = 0.7920 + j0.1129\\
S &= (P_m/pf)(\exp(-i\cos^{-1}(pf))) = 0.8000 - j0.4958\\
|V_t| &= |S|/|I_{gen}| = 1.1765\\
\theta_t &= \sin^{-1}[(P_m(\text{abs}(Z_L \text{ pre}) + 0.1)/(V_t V_\infty))] = 10.85^\circ\\
V_t &= V_t e^{\theta t j} = 1.1554 + j0.2214\\
E' &= V_t + R_a I_{gen} + jX_d' I_{gen} = 1.1329 + j0.3798
\end{aligned}$$

Clearing time = 0.08: Using the initial values of system parameters to apply PSAT/MATLAB® toolbox to solve the swing equation, the variations of power angle and machine speed versus time are shown in Figure 8.7(a and b), respectively.

If the clearing time is decreased to be 0.02 s or increased to be 0.3 s, the variations of δ and ω versus time are depicted in Figures 8.8 and 8.9, respectively.

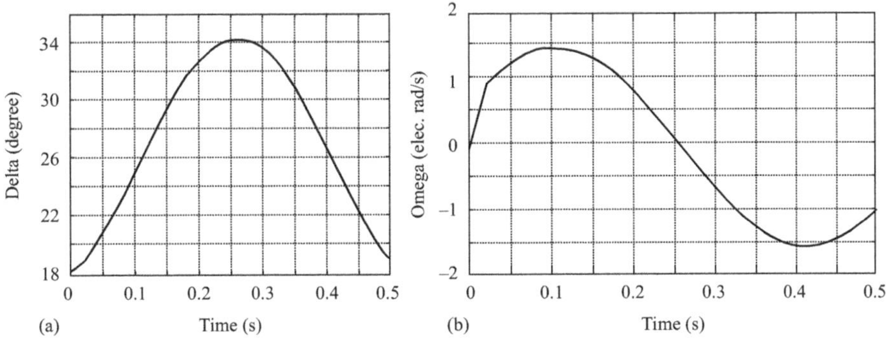

Figure 8.8 (a) Variation of δ versus time and (b) variation of ω versus time with fault clearing time 0.02 s

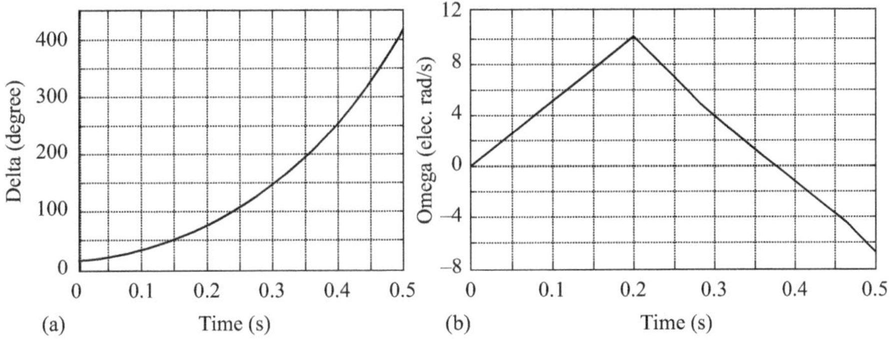

Figure 8.9 (a) Variation of δ versus time and (b) variation of ω versus time with fault clearing time 0.3 s

(ii) The machine is represented by a voltage source E_q behind X_q with system parameters as

$$T'_{do} = 0.4, \quad X_d = 1.9, \quad X_q = 1.75, \quad X'_q = 0.24, \quad \mathrm{H} = 3.5$$

Z_L pre-fault $= 0.0250 + j0.1750$
Z_L during fault $= \infty$ Z_L post-fault $= 0.0500 + j0.3500$

$$\theta_o = 73.1134°, \quad \omega_o = 314.1593 \text{ elec. rad/s}$$
$$E_{fdo} = 0.4258 + j2.4262$$
$$E_{do} = 0.8963 + j1.0621$$
$$E_{qo} = 0.6891 + j2.2699$$

By solving the swing equation the variations of δ and ω versus time at different values of clearing time are shown in Figures 8.10–8.12.

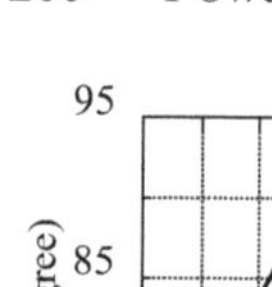
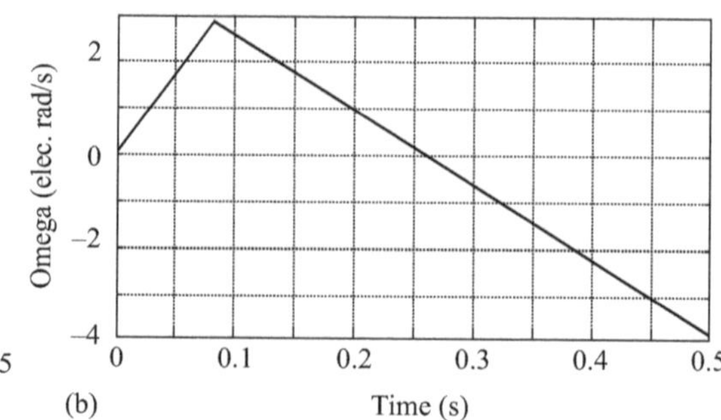

Figure 8.10 (a) Variation of δ versus time and (b) variation of ω versus time with fault clearing time 0.08 s

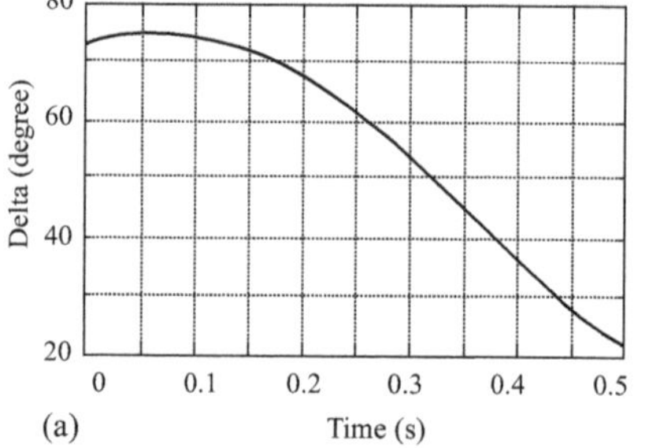

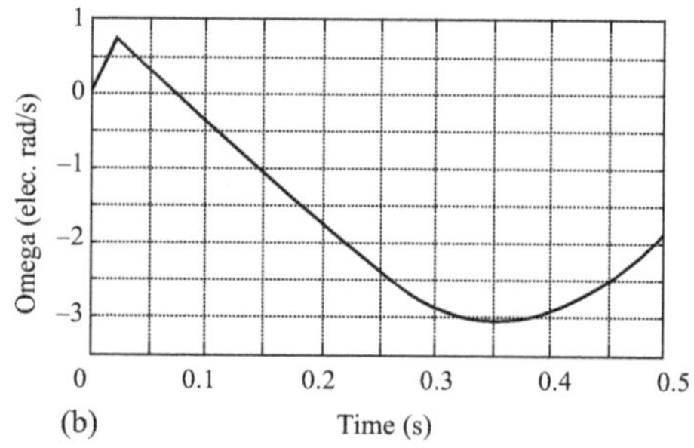

Figure 8.11 (a) Variation of δ versus time and (b) variation of ω versus time with fault clearing time 0.02 s

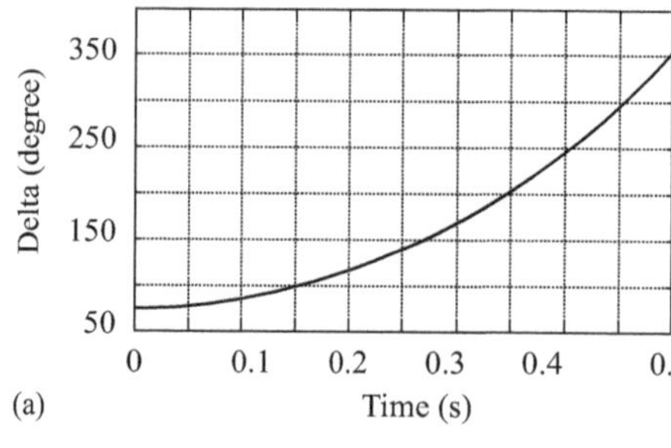

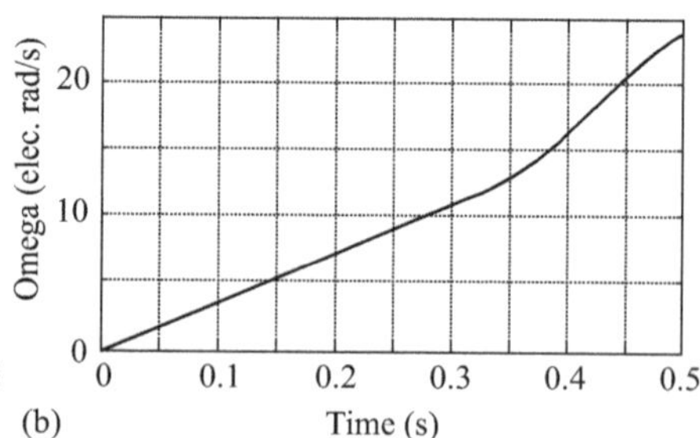

Figure 8.12 (a) Variation of δ versus time and (b) variation of ω versus time with fault clearing time 0.3 s

It is seen that the system is stable when the fault clearing time is 0.02 s or 0.08 s while it is unstable for fault clearing time of 0.3 s. Thus, as the fault clearing time decreases the stability is better and the system is more secure.

8.4 Transient stability analysis of a multi-machine power system

A multi-machine power system encompasses interconnected generators to feed loads with electrical power through a network (Figure 8.13). The first step to study transient stability is to model the system by incorporating the equations representing each component in the system. Stability studies involve short periods of analysis of the order of a second or less. Thus, the synchronous machine can be represented by a voltage source behind transient reactance. As a simplified representation, the voltage source is assumed to be constant in magnitude and varies in angular position as the saturation and saliency effects are neglected as well as constant flux linkages and small speed change are assumed. There is no need to involve stator differential equations and the voltage source is denoted by E' that is given by (8.40), repeated below:

$$E' = V_t + R_a I_t + jX'_d I_t \tag{8.44}$$

where

E' = voltage behind transient reactance
V_t = machine terminal voltage
I_t = machine terminal current
R_a = armature resistance
X'_d = transient reactance

Accordingly, the synchronous machine representation used for network solution is shown in Figure 8.14(a) and its phasor diagram is depicted in Figure 8.14(b).

When the effects of saliency and changes in field flux linkages need to be taken into account in a study, the synchronous machine can be represented by a voltage E_q behind quadrature-axis synchronous reactance X_q and is determined from

$$E_q = V_t + R_a I_t + jX_q I_t \tag{8.45}$$

The machine representation used for network solution and its phasor diagram are shown in Figure 8.15(a and b), respectively. The field current acting along d-axis produces a sinusoidal flux. The induced voltage E_I lags this flux by 90° and acts along the q-axis. The voltage E_I is determined by the summation of terminal voltage V_t, the voltage drop across the armature resistance and the voltage drops across X_d and X_q, representing the demagnetising effects Figure 8.15(b).

Thus,

$$E_I = V_t + R_a I_t + jX_d I_d + jX_q I_q \tag{8.46}$$

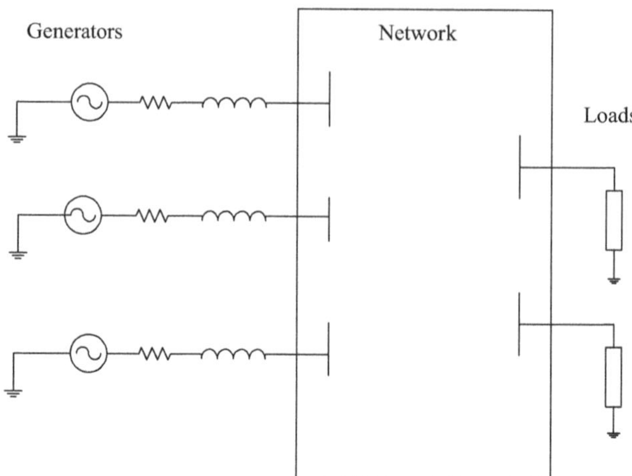

Figure 8.13 Schematic diagram of a multi-machine power system

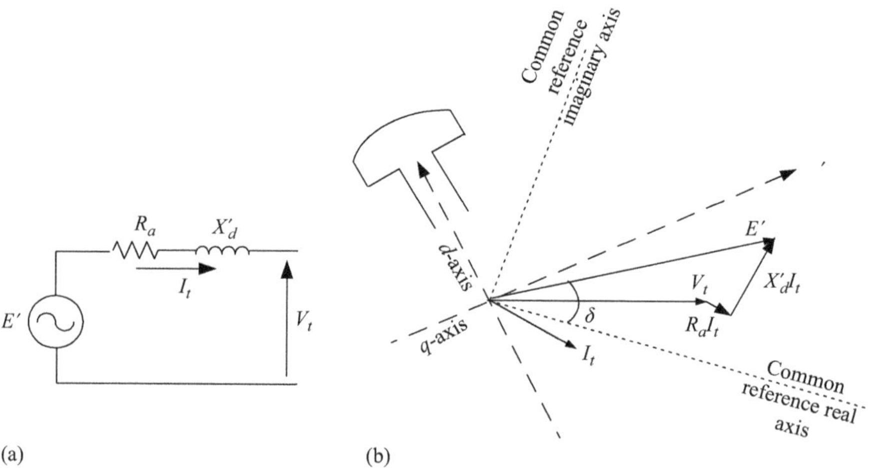

Figure 8.14 Machine simplified representation and phasor diagram: (a) machine representation, (b) the phasor diagram

As explained in Section 7.1.3 and illustrated in Figure 7.2, the relation between E_q and E'_q can be written as

$$E'_q = E_q - j(X_q - X'_d)I_d \tag{8.47}$$

And, as given by (8.21) the rate of change of E'_q is

$$\frac{\mathrm{d}E'_q}{\mathrm{d}t} = \frac{1}{T'_{do}}(E_{fd} - E_I) = \frac{1}{T'_{do}}\left[-E'_q + (X_d - X'_d)i_d + E_{fd}\right] \tag{8.48}$$

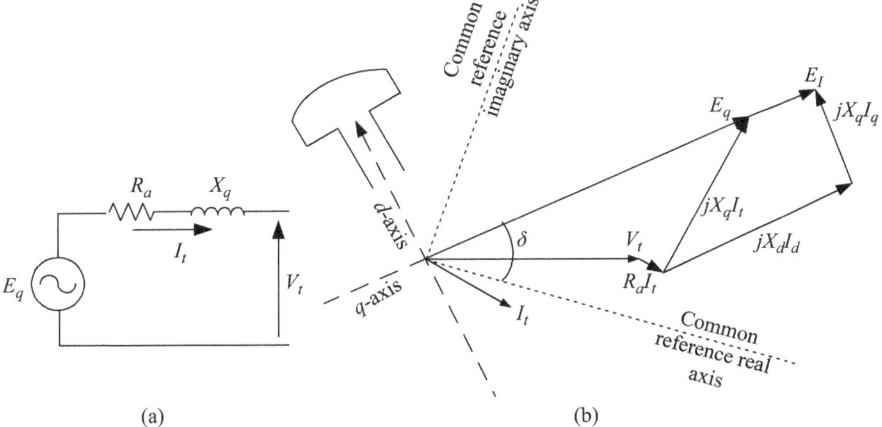

Figure 8.15 Machine representations by E_q: (a) machine representation, (b) the phasor diagram

Therefore, (8.45)–(8.48) can be incorporated to give the stator equations. Then, adding the rotor swing equation to obtain the synchronous machine model describes the dynamic behaviour when considering the effects of saliency and change of field flux linkages.

Network equations

To develop equations representing the network, models of power system loads must be determined. Motors as loads are represented by equivalent circuits; otherwise, the loads during the transient period can commonly be represented by static impedance or admittance to ground, constant current at fixed power factor, constant real and reactive power or a combination of these representations as explained in Chapter 4, Section 4.3.1.

The constant power load is either equal to the scheduled real and reactive bus load or a percentage of specified values in case of a combined representation. For a constant current load representation, the current is calculated as the initial value from the scheduled bus loads and voltages obtained from the load flow solution for the power system prior to a disturbance. Thus, at bus i the load current I_{io} is obtained from

$$I_{io} = \frac{P_{\mathrm{L}i} - jQ_{\mathrm{L}i}}{E_i^*} \tag{8.49}$$

where $P_{\mathrm{L}i}$ and $Q_{\mathrm{L}i}$ are the scheduled real and reactive bus loads, respectively. E_i is the calculated bus voltage. The current I_{io} flows from bus i to ground. Its magnitude and power factor remain constant.

The static admittance y_{io} representing the load at bus i can be obtained from

$$y_{io} = \frac{I_{io}}{E_p} = g_{io} - jb_{io} \tag{8.50}$$

where the ground voltage is zero and

$$g_{io} = \frac{P_{Li}}{e_i^2 + f_i^2} \quad \text{and} \quad b_{io} = \frac{Q_{Li}}{e_i^2 + f_i^2}$$

e_i and f_i are the real and imaginary components of the bus voltage E_i.

The network equations used for load flow calculations, explained in Chapter 5, can be applied to describe the performance of the network during a transient period. The calculation of bus admittance matrix, $\mathbf{Y}_{bus}$, at different states – prior to a disturbance, instant of fault occurrence and post-fault clearance – is required.

Elements of $\mathbf{Y}_{bus}$ are identified as diagonal elements $\triangleq \mathrm{Y}_{ii}$ = sum of all the admittances connected to bus i and off-diagonal elements $\triangleq \mathrm{Y}_{ij}$ = the negative of the admittance between bus i and bus j. The matrix $\mathbf{Y}_{bus}$ can be reduced by considering only the internal generator buses and eliminating all other buses in the network. The reduced matrix can be obtained by recalling that all buses have zero injection currents except for the internal generator buses. Accordingly, the relations below can be written as

$$\mathbf{I} = \mathbf{YV} \tag{8.51}$$

where

$$\mathbf{I} = \begin{bmatrix} \mathbf{I}_g \\ 0 \end{bmatrix}$$

Thus, both the matrix $\mathbf{Y}$ and vector $\mathbf{V}$ in (8.51) are partitioned to get

$$\begin{bmatrix} \mathbf{I}_{gg} \\ 0 \end{bmatrix} = \begin{bmatrix} \mathbf{Y}_{gg} & \mathbf{Y}_{gb} \\ \mathbf{Y}_{bg} & \mathbf{Y}_{bb} \end{bmatrix} \begin{bmatrix} \mathbf{V}_g \\ \mathbf{V}_b \end{bmatrix}. \tag{8.52}$$

where the subscript g denotes the internal generator buses and subscript b denotes the other network buses. The vectors $\mathbf{V}_g$ and $\mathbf{V}_b$ have the dimensions $n_g \times 1$ and $n_b \times 1$, respectively.

Equation (8.52) can be rewritten in expanded form as

$$\mathbf{I}_g = \mathbf{Y}_{gg}\mathbf{V}_g + \mathbf{Y}_{gb}\mathbf{V}_b \quad \text{and} \quad 0 = \mathbf{Y}_{bg}\mathbf{V}_g + \mathbf{Y}_{bb}\mathbf{V}_b$$

Thus, $\mathbf{V}_b$ can be eliminated to get

$$\mathbf{I}_g = \left(\mathbf{Y}_{gg} - \mathbf{Y}_{gb}\mathbf{Y}_{bb}^{-1}\mathbf{Y}_{bg}\right)\mathbf{V}_g = \mathbf{Y}_{\text{reduced}}\mathbf{V}_g \tag{8.53}$$

where $\mathbf{Y}_{\text{reduced}}$ is the required reduced matrix. It has the dimensions $n_g \times n_g$ 'n_g is the number of generators'. It is noted that this procedure of reduction is only viable when the loads are treated as constant shunt admittance.

Changes in system parameters that simulate the disturbance must be specified. The initial values of system parameters such as bus voltages and power can be obtained by applying load flow techniques to the system at steady state. The internal machine bus voltages are calculated according to the model used. Then, the system differential equations are solved numerically along with the algebraic equations and load flow calculations at each step. Details of the solution procedure are described by an example below.

Example 8.2 For a nine-bus test system shown in Figure 8.16, evaluate the system transient stability when subjected to a three-phase short circuit at bus no. 7 'at the beginning of line 7-5' for durations of 0.08 s and 0.20 s. The fault is cleared by isolating line 7-5. The system data are given in Appendix II.

Solution:

It is assumed that the system is operating at steady state for 1 s, then the fault occurs for a duration of 0.08 s, Case (I), and 0.20 s, Case (II). Elements of matrix $\mathbf{Y}_{bus}$ (in pu) as calculated at steady state are summarised in Table 8.2. Bus voltages, generation and loads and line flow in pu at steady state are calculated by load flow analysis and are given in Tables 8.2–8.4.

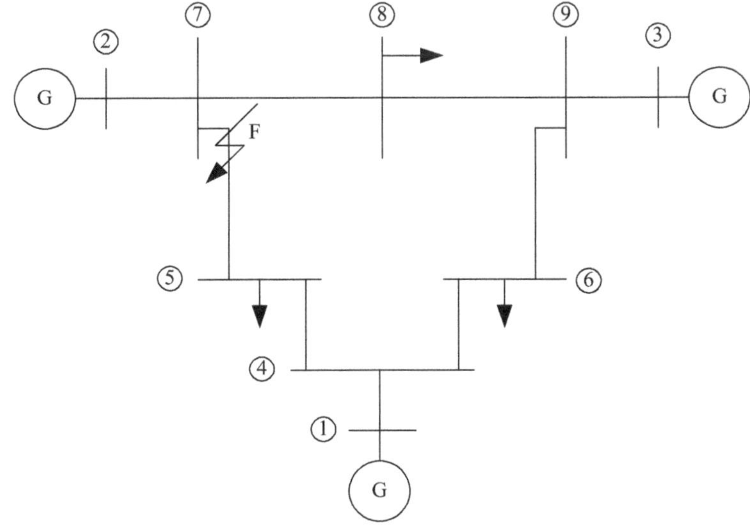

Figure 8.16 Nine-bus test system

Table 8.2 Bus admittance matrix in steady state

Bus	1	2	3	4	5	6	7	8	9
1	$0 - j4.8459$	0	0	$0 + j8.4459$	0	0	0	0	0
2	0	$0 - j5.4855$	0	0	0	0	$0 + j5.4855$	0	0
3	0	0	$0 - j4.1684$	0	0	0	0	0	$0 + j4.1684$
4	$0 + j8.4459$	0	0	$3.3074 - j30.3937$	$-1.3652 + j11.6041$	$-1.9422 + j10.5107$	0	0	0
5	0	0	0	$-1.3652 + j11.6041$	$2.5528 - j17.3382$	0	$-1.1876 + j5.9751$	0	0
6	0	0	0	$-1.3652 + j11.6041$	0	$3.2242 - j15.8409$	0	0	$-1.2820 + j5.5882$
7	0	$0 + j5.4855$	0	0	$-1.1876 + j5.9751$	0	$2.8047 - j24.9311$	$-1.6171 + j13.6980$	0
8	0	0	0	0	0	0	$-1.6171 + j13.6980$	$2.7722 - j23.3032$	$-1.1551 + j9.7843$
9	0	0	$0 + j4.1684$	0	0	$-1.2820 + j5.5882$	0	$-1.1551 + j9.7843$	$2.4371 - j19.2574$

Table 8.3 Line flow from load flow calculation prior to fault

Bus no.	Bus voltage (E_p)		Generation		Load	
	V	Phase angle	P	Q	P	Q
1	1.04	0	0.71641	0.27046	0	0
2	1.025	0.16197	1.63	0.06654	0	0
3	1.025	0.08142	0.85	−0.1086	0	0
4	1.026	−0.0387	0	0	0	0
5	0.996	−0.0696	0	0	1.25	0.5
6	1.013	−0.0644	0	0	0.9	0.3
7	1.026	0.06492	0	0	0	0
8	1.016	0.0127	0	0	1	0.35
9	1.032	0.03433	0	0	0	0

Table 8.4 Line flow from load flow calculation prior to fault

From bus	To bus	P flow	Q flow	P loss	Q loss
4	1	−0.71641	−0.23923	0	0.03123
7	2	−1.63	0.09178	0	0.15832
9	3	−0.85	0.14955	0	0.04096
7	8	0.7638	−0.00797	0.00475	−0.11502
9	8	0.24183	0.0312	0.00088	−0.21176
7	5	0.8662	−0.08381	0.023	−0.19694
9	6	0.60817	−0.18075	0.01354	−0.31531
5	4	−0.4068	−0.38687	0.00258	−0.15794
6	4	−0.30537	−0.16543	0.00166	−0.15513

At the instant of fault occurrence:
Each machine is represented by a voltage source 'constant magnitude' E' behind direct-axis transient reactance X'_d. Each load is represented by constant shunt admittance y_L (8.50). Based on steady-state load flow results, the calculated values of E' and y_L for each generator and load, respectively, are given below:

$$E_1' = 1.0565\angle 19.55^\circ, \quad E_2' = 1.0264\angle 12.93^\circ, \quad E_3' = 1.0319\angle 2.37^\circ$$
$$y_{L5} = 1.261 - j0.5044, \quad y_{L6} = 0.877 - j0.2926, \quad y_{L8} = 0.969 - j0.3391$$

The system admittance matrix for transient stability study can be formulated as summarised in Table 8.5. The admittance of each load is considered as a shunt element to the corresponding bus, i.e. it will be added to all admittances connected to that bus to calculate its self-admittance element.

The values of system state variables and parameters at steady state are taken as initial conditions for numerical integration. The second-order Runge–Kutta

Table 8.5 Bus admittance matrix for transient study

Bus	1	2	3	4	5	6	7	8	9
1	$0-j8.4459$	0	0	$0+j8.4459$	0	0	0	0	0
2	0	$0-j5.4855$	0	0	0	0	$0+j5.4855$	0	0
3	0	0	$0-j4.1684$	0	0	0	0	0	$0+j4.1684$
4	$0+j8.4459$	0	0	$3.3074-j30.3937$	$-1.3652+j11.6041$	$-1.9422+j10.5107$	0	0	0
5	0	0	0	$-1.3652+j11.6041$	$3.813-j17.826$	0	$-1.1876+j5.9751$	0	0
6	0	0	0	$-1.9422+j10.5107$	0	$4.019-j16.1355$	0	0	$-1.2820+j5.5882$
7	0	$0+j5.4855$	0	0	$-1.1876+j5.9751$	0	$2.8047-j24.9311$	$-1.6171+j13.6980$	0
8	0	0	0	0	0	0	$-1.6171+j13.6980$	$3.7412-j23.642$	$-1.1551+j9.7843$
9	0	0	$0+j4.1684$	0	0	$-1.2820+j5.5882$	0	$-1.1551+j9.7843$	$2.4371-j19.2574$

Table 8.6 Bus voltages, generation and loads at the instant of fault occurrence

Bus code (p)	Bus Voltage (E_p)		Generation		Load	
	V	Phase	P	Q	P	Q
1	0.83441	−0.00472	0.7169	0.1843	0	0
2	0.36933	0.33849	1.63	0.1214	0	0
3	0.65588	0.80722	0.85	−0.0548	0	0
4	0.62722	0.11775	0	0	0	0
5	0.64996	−0.07416	0	0	1.25	0.5
6	0.58798	−0.09371	0	0	0.9	0.3
7	0	0	0	0	0	0
8	0.21255	0.01174	0	0	1	0.35
9	0.50141	0.04652	0	0	0	0

Table 8.7 Line flow at the instant of fault occurrence

From bus	To bus	P flow	Q flow	P loss	Q loss
4	1	−0.65326	−2.0587	0	0.63609
7	2	−0.00807	−0.06205	0	2.0564
9	3	−0.38196	−1.0629	0	0.29736
7	8	0.00287	−0.03025	0.0659	0.55486
9	8	0.20363	−1.3872	0.09653	0.78664
7	5	0	0	0	0
9	6	0.17834	−0.32423	0.01703	−0.03266
5	4	−0.42051	−1.0905	0.07583	0.59193
6	4	−0.14917	−0.39506	0.00774	−0.01877

Table 8.8 State variables (δ and ω) and algebraic variables (P and Q) at the instant of fault occurrence

Generator 1				Generator 2				Generator 3			
δ	ω	P	Q	δ	ω	P	Q	δ	ω	P	Q
0.34097	1	0.0080	2.1185	0.2255	1	0.3819	1.360	0.0414	1	0.653	2.694

method and PSAT/MATLAB® toolbox are used for the transient analysis as below. The time interval is taken as 0.02 s. At the instant of fault occurrence bus voltages, generation, loads, line flow and state variables are summarised in Tables 8.6–8.8.

It is to be noted that in all tables, δ is given in elec. rad.; ω in elec. rad./s; and V, P and Q in pu.

Table 8.9 Bus voltages, generation and loads at the middle of first interval of fault occurrence

Bus code (p)	Bus voltage		Generation		Load	
	V	Phase	P	Q	P	Q
1	0.8344	−0.0051	0.7169	0.18434	0	0
2	0.3693	0.3399	1.63	0.12138	0	0
3	0.6272	0.1182	0.85	−0.0548	0	0
4	0.6499	−0.0745	0	0	0	0
5	0.4192	−0.1663	0	0	1.25	0.5
6	0.5879	−0.0939	0	0	0.9	0.3
7	0.0109	0.2137	0	0	0	0
8	0.2125	0.0121	0	0	1.00	0.3
9	0.5014	0.0469	0	0	0	0

Table 8.10 Line flow middle of first interval of fault occurrence

From bus	To bus	P flow	Q flow	P loss	Q loss
4	1	−0.65242	−2.0589	0	0.63606
7	2	−0.00811	−0.06204	0	2.0564
9	3	−0.38273	−1.0628	0	0.29745
7	8	0.00288	−0.03024	0.0659	0.55481
9	8	0.20359	1.3871	0.09652	0.78657
7	5	0	0	0	0
9	6	0.17914	−0.32425	0.01708	−0.03244
5	4	−0.42046	−1.0905	0.07583	0.5919
6	4	−0.14838	−0.39529	0.00774	−0.01878

Table 8.11 State variables (δ and ω) and algebraic variables (P and Q) middle of first interval of fault occurrence

Generator 1				Generator 2				Generator 3			
δ	ω	P	Q	δ	ω	P	Q	δ	ω	P	Q
0.34245	1.001	0.0081	2.1185	0.226	1	0.382	1.360	0.040	1	0.652	2.6949

Case (I) Fault duration = 0.08 s

At each interval the load flow calculation is applied two times for the second-order Runge–Kutta 'in general k times for kth-order Runge–Kutta'. The first time is using the load flow for calculating the coefficients K_{11} and K_{21} at the beginning of the interval and the second time for calculating K_{12} and K_{22} at the middle of interval.

For instance, at the first interval just after fault occurrence the results of load flow in Tables 8.6–8.8 are used to calculate K_{11} and K_{21}, and then applying load flow analysis at the middle of this interval to calculate K_{12} and K_{22}. The results are tabulated in Tables 8.9–8.11.

Table 8.12 Bus voltages, generation and loads as initial values for second interval of fault occurrence

Bus code (*p*)	Bus voltage		Generation		Load	
	V	Phase	*P*	*Q*	*P*	*Q*
1	0.83439	−0.00645	0.7169	0.18434	0	0
2	0.36932	0.34436	1.63	0.12138	0	0
3	0.62711	0.1197	0.85	−0.05481	0	0
4	0.64988	−0.07553	0	0	0	0
5	0.41914	−0.16736	0	0	1.25	0
6	0.58784	−0.0943	0	0	0.9	0
7	0.0109	0.21607	0	0	0	0
8	0.21251	0.01314	0	0	1	0
9	0.50131	0.04787	0	0	0	0

Table 8.13 Initial values of line flow for second interval of fault occurrence

From bus	To bus	*P* flow	*Q* flow	*P* loss	*Q* loss
4	1	−0.64989	−2.0593	0	0.63596
7	2	−0.00824	−0.062	0	2.0565
9	3	−0.38503	−1.0624	0	0.29773
7	8	0.00292	−0.03022	0.06588	0.55467
9	8	0.2035	1.3867	0.09649	0.78637
7	5	0	0	0	0
9	6	0.18153	−0.32431	0.01722	−0.03177
5	4	−0.4203	−1.0903	0.07582	0.59181
6	4	−0.14603	−0.39598	0.00774	−0.0188

Table 8.14 State variables (δ and ω) and algebraic variables (P and Q) for second interval of fault occurrence

Generator 1				Generator 2				Generator 3			
δ	ω	*P*	*Q*	δ	ω	*P*	*Q*	δ	ω	*P*	*Q*
0.34689	1.0025	0.0082	2.1185	0.228	1.0016	0.3850	1.3601	0.0394	1	0.6498	2.6953

The angles and speeds of the generators at the end of the interval are calculated in terms of the average values of *K* coefficients. Consequently, the new bus voltages and line flow are calculated to express the initial values for the next interval Tables 8.12–8.14.

Table 8.15 Bus voltages, generation and loads at the last interval after fault clearing

Bus code (p)	Bus voltage (E_p)		Generation		Load	
	V	Phase	P	Q	V	Q
1	0.9504	−0.22136	0.7169	0.18434	0	0
2	0.95206	0.91582	1.63	0.12138	0	0
3	0.91456	0.53713	0.85	−0.05481	0	0
4	0.87361	−0.19968	0	0	0	0
5	0.82835	−0.29988	0	0	1.25	0
6	0.81676	−0.0214	0	0	0.9	0
7	0.92093	0.77827	0	0	0	0
8	0.885	0.61164	0	0	1	0
9	0.88667	0.46614	0	0	0	0

Table 8.16 Line flow at the last interval after fault clearing

From bus	To bus	P flow	Q flow	P loss	Q loss
4	1	0.31256	−1.1613	0	0.10916
7	2	−1.9235	−0.32619	0	0.2805
9	3	−0.98157	−0.38716	0	0.08299
7	8	1.9235	0.32619	0.0386	0.20544
9	8	−1.1019	0.14489	0.01916	−0.00172
7	5	0	0	0	0
9	6	2.0835	0.24227	0.22261	0.7102
5	4	−0.88627	−0.35451	0.01271	−0.01952
6	4	1.2617	−0.66763	0.05021	0.15871

Table 8.17 State variables (δ and ω) and algebraic variables (P and Q) at the last interval after fault clearing

Generator 1				Generator 2				Generator 3			
δ	ω	P	Q	δ	ω	P	Q	δ	ω	P	Q
1.15	1.026	1.923	0.607	0.728	1.025	0.982	0.470	−0.240	1.023	−0.312	1.270

By repeating the same procedure for the next intervals until reaching the last interval of a total period of 5 s, the results obtained at the end of the period are summarised in Tables 8.15–8.17.

It is important to note that at the instant of fault clearance, the bus admittance matrix is changed as the line 7-5 is isolated. This admittance matrix is calculated as summarised in Table 8.18. The reduced matrices $\mathbf{Y}_{\text{reduced}}$ pre, during and post the fault are summarised in Table 8.19.

Table 8.18 Bus admittance matrix after clearing the fault

Bus	1	2	3	4	5	6	7	8	9
1	$0 - j8.4459$	0	0	$0 + j8.4459$	0	0	0	0	0
2	0	$0 - j5.4855$	0	0	0	0	$0 + j5.4855$	0	0
3	0	0	$0 - j4.1684$	0	0	0	0	0	$0 + j4.1684$
4	$0 + j8.4459$	0	0	$3.3074 - j30.3937$	$-1.3652 + j11.6041$	$-1.9422 + j10.5107$	0	0	0
5	0	0	0	$-1.3652 + j11.6041$	$2.5528 - 17.3382i$	0	0	0	0
6	0	0	0	$-1.9422 + j10.5107$	0	$3.2242 - j15.8409$	0	0	$-1.2820 + j5.5882$
7	0	$0 + j5.4855$	0	0	$-1.1876 + j5.9751$	0	$2.8047 - j24.9311$	$-1.6171 + j13.6980$	0
8	0	0	0	0	0	0	$-1.6171 + j13.6980$	$2.7722 - j23.3032$	$-1.1551 + j9.7843$
9	0	0	$0 + j4.1684$	0	0	$-1.2820 + j5.5882$	0	$-1.1551 + j9.7843$	$2.4371 - j19.2574$

Table 8.19 Reduced matrices $\mathbf{Y}_{reduced}$

Network type	Generator bus	1	2	3
Pre-fault	1	$0.846 - j2.988$	$0.287 + j1.513$	$0.210 + j1.226$
	2	$0.287 + j1.513$	$0.420 - j2.724$	$0.213 + j1.088$
	3	$0.210 + j1.226$	$0.213 + j1.088$	$0.277 - j2.368$
During fault	1	$0.657 - j3.816$	$0.000 + j0.000$	$0.070 + j0.631$
	2	$0.000 + j0.000$	$0.000 - j5.486$	$0.000 + j0.000$
	3	$0.070 + j0.631$	$0.000 + j0.000$	$0.174 - j2.796$
Post-fault	1	$1.181 - j2.229$	$0.138 + j0.726$	$0.191 + j1.079$
	2	$0.138 + j0.726$	$0.389 - j1.953$	$0.199 + j1.229$
	3	$0.191 + j1.079$	$0.199 + j1.229$	$0.273 - j2.342$

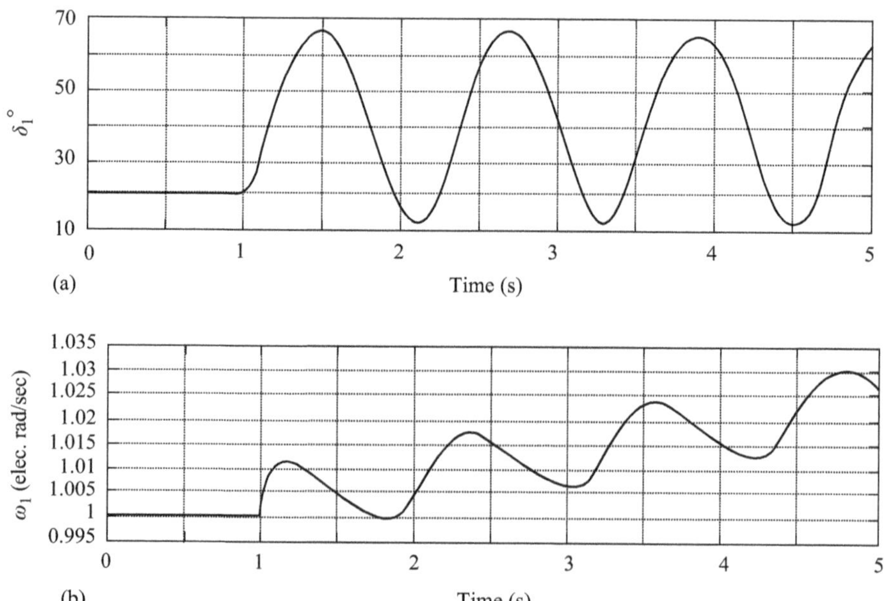

Figure 8.17 (a) Power angle versus time and (b) angular speed versus time for machine 1 for fault clearing time 0.08 s

The changes of δ and ω for each generator are shown in Figures 8.17 through 8.19. It is clear that the power angles of all machines in the system increase and then decrease. Therefore, the system is stable.

Case (II) Fault duration = 0.20 s

In this case, the fault clearing time becomes 0.20 s rather than 0.08 s. The same procedures as above are carried out. The initial conditions are the same in both cases. The results obtained are plotted as shown in Figures 8.20–8.22. It is found

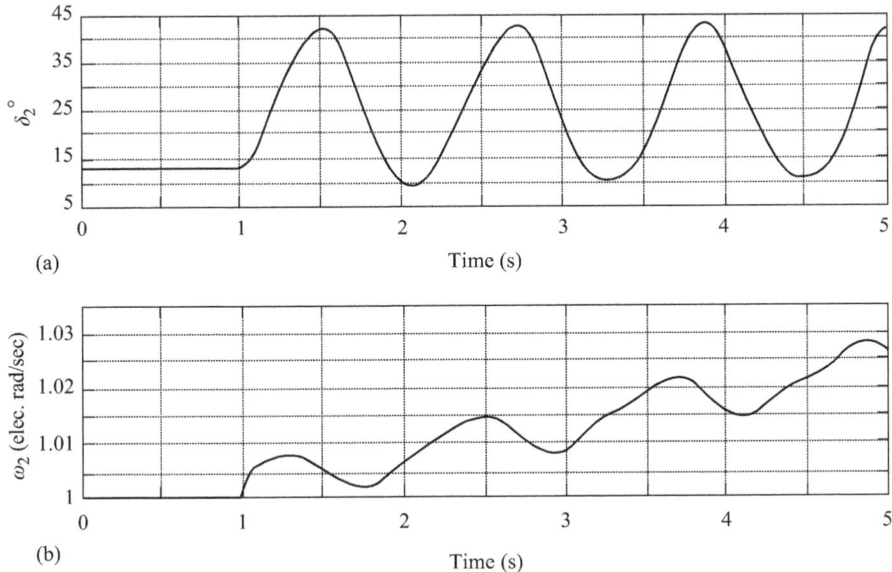

Figure 8.18 *(a) Power angle versus time and (b) angular speed versus time for machine 2 for fault clearing time 0.08 s*

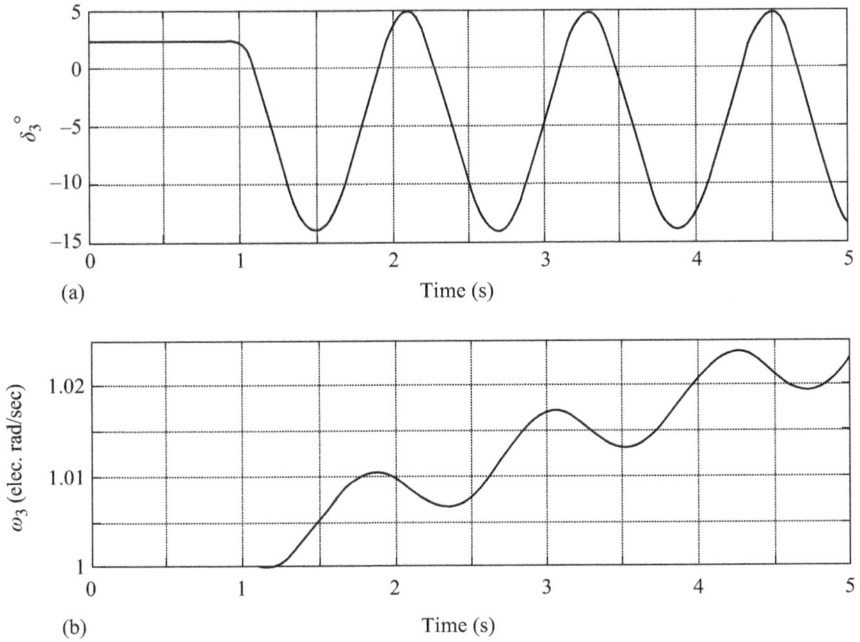

Figure 8.19 *(a) Power angle versus time and (b) angular speed versus time for machine 3 for fault clearing time 0.08 s*

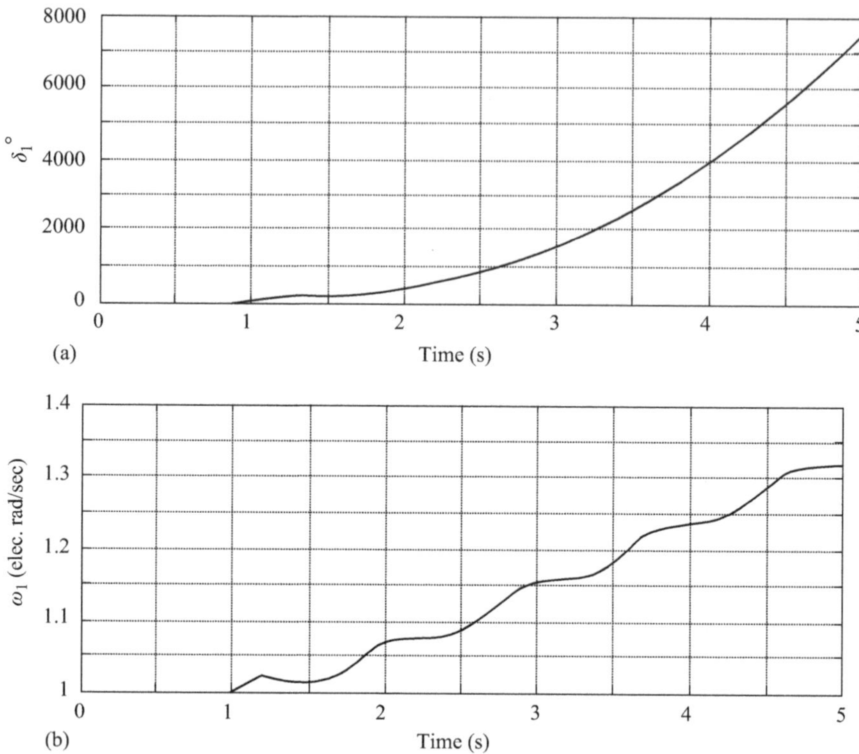

Figure 8.20 (a) Power angle versus time and (b) angular speed versus time for fault clearing time = 0.20 s

that the power angles of the machines in the system are continuously increasing or decreasing. Thus, the system is unstable because of delayed fault clearing.

From the figures of Case (I) and Case (II), it is found that the system is stable when the fault duration is 0.08 s. On the contrary, the system is unstable if the fault duration is 0.20 s. Therefore, the duration of fault clearing must be investigated carefully. It depends on several parameters such as system topology, selected machine model, type of fault, location of fault and characteristics of protective gears.

It can be concluded that the main steps to determine the transient stability of a multi-machine power system can be summarised as below:

- Define the data of each element in the power system.
- Define the period of study.
- Specify the fault: type, location and, clearing time and how the fault is cleared.
- Select an adequate model for each machine 'it may differ from one machine to another'.

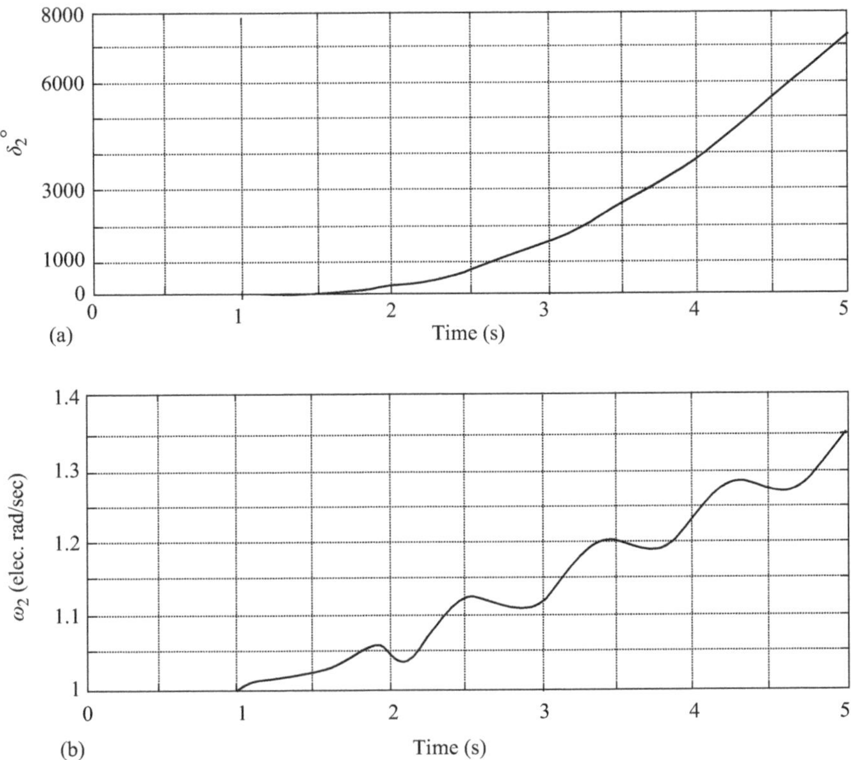

Figure 8.21 (a) Power angle versus time and (b) angular speed versus time for machine 2 at fault clearing time = 0.20 s

- Model the other components of the system 'transmission lines, transformers, etc.'.
- Define the frame of reference to which all parameters of component models are referred.
- Construct the bus-admittance matrix, $\mathbf{Y}_{bus}$, and apply the load flow analysis to obtain the system parameters at steady state and calculate the initial values of parameters necessary to solve the swing equation.
- Calculate the elements of $\mathbf{Y}_{bus}$ for transient analysis as well as for post-fault operation.
- Calculate the reduced matrices $\mathbf{Y}_{\text{reduced}}$ pre, during and post fault.
- Select the numerical integration method to solve the system differential equations.
- Define the time interval and start solving the system equations.
- During the numerical solution load flow analysis is applied whenever machine angles are changed. This is to calculate the power delivered from each machine, which is involved in the swing equation.

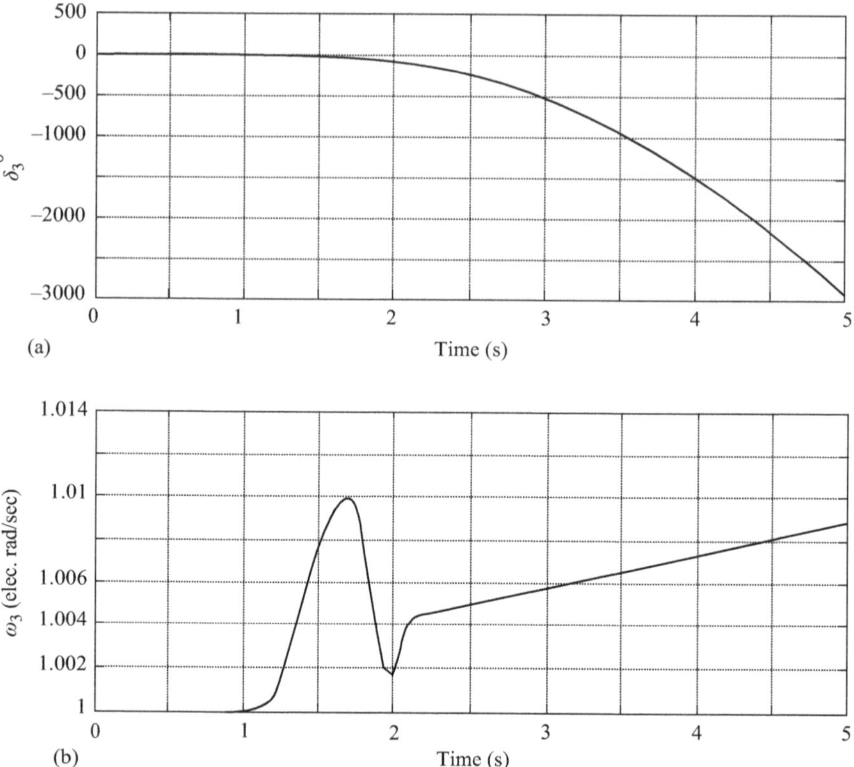

Figure 8.22 (a) Power angle versus time and (b) angular speed versus time for machine 3 at fault clearing time = 0.20 s

- When the clearing time is reached the admittance matrix is switched to $\mathbf{Y}_{bus}$ post fault.
- The numerical solution is continued until reaching the end of period of study.
- The system is stable if the change of power angle with time for each machine increases and then decreases. Otherwise the system is unstable.

References

1. Roderick J, Podmere F.R., Waldron M. 'Synthesis of dynamic load models for stability studies'. *IEEE Transactions on Power Apparatus and Systems*. Jan 1982;**101**(1):127–35
2. Price W.W., Wirgau K.A., Murdoch A., Mitsche J.V., Vaahedi E, El-Kady M.A. 'Load modelling for power flow and transient stability computer studies'. *IEEE Transactions on Power Systems*. Feb 1988;**3**(1):180–8

3. IEEE Task Force. 'Current usage and suggested practices in power system stability simulations for synchronous machines'. *IEEE Transactions on Energy Conversion.* 1986;**1**(1):77–93
4. Momoh J.A., El-Hawary M.E. *Electric Systems. Dynamics, and Stability with Artificial Intelligence Applications*. Germany: Marcel Dekker; 2000
5. Nandakumar K. *Numerical Solutions of Engineering Problems*. Edmonton, Alberta, Canada: University of Alberta; 1998
6. Harder D.W., Khoury R. *Numerical Analysis for Engineering*. Saskatoon, Saskatchewan, Canada: University of Waterloo; 2010
7. Awrwjcewiz J. *Numerical Analysis: Theory and Application*. Rijeka, Croatia: InTech; 2011
8. Moursund D.G., Duris C.S. *Elementary Theory & Application of Numerical Analysis*. New York, US: McGraw-Hill; 1967

Chapter 9

Transient energy function methods

The power system is described mathematically, as explained in Chapter 8, by a set of differential-algebraic equations. The equations are large in numbers and non-linear in nature. A typical number of equations is in the order of hundreds, if not thousands, for a network of moderate size. The solution of system equations is required for time simulation to obtain generator power angles and other system parameters of interest at different time instants. This helps in examining the system behaviour and deciding whether its dynamic response will result in an acceptable performance. The solution is repeated for different scenarios such as changing fault type, fault location, network topology and considering various control devices. Consequently, the solution is time consuming. Therefore, time simulation techniques are not appropriate for online stability monitoring. For instance, transient stability studies for a typical system with detailed modelling for a 500-bus, 100-machine system may take up to an hour [1].

Accordingly, power engineers and system operators are motivated to look for an alternative method by which system stability can be determined directly and has the desired specification for online applications. The key idea that the alternative method is based on is to specify a certain function by which the system transient energy at the end of the disturbance period can be calculated. The calculated value is compared with a critical energy value to assess the transient stability, as the difference between the two values gives an indication of stability.

9.1 Definitions of stability concepts

As explained in Chapter 7, Sections 7.1.1 and 7.1.2, the power system can be modelled as an autonomous system and described by the ordinary differential equation:

$$\dot{x} = f(x) \tag{9.1}$$

that is assumed to have an equilibrium point at the origin, i.e. $\dot{x} = f(0) = 0$

The equilibrium point is either stable (SEP) or unstable (UEP). In the sense of Lyapunov, the equilibrium $x = 0$ is stable if for any given $\varepsilon > 0$, there exists a $\rho \leq \varepsilon$ such that $\|x_o\| < \rho$ satisfies $\|x(t)\| < \varepsilon$ for all t where $x_o = x(t_o) \triangleq$ initial state. This concept is pictorially described in Figure 9.1. It is shown that the initial state x_o has a magnitude less than ρ and the trajectory of x remains within a cylinder of

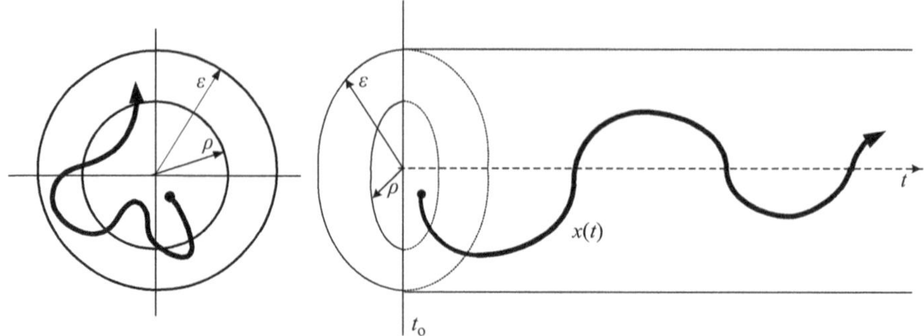

Figure 9.1 Stability concepts in the sense of Lyapunov

radius ε. It is noted that the stability concept in the sense of Lyapunov is a local concept as it does not specify how small ρ had to be chosen in the definition [2].

The origin $x = 0$ at time t_o is called unstable if it is not stable at t_o. Thus, the UEP $x = 0$ implies that for some $\varepsilon > 0$ there is no $\rho > 0$ satisfying the condition $\|x(t)\| < \varepsilon$ for all $t \geq t_o$ when $\|x_o\| < \rho$. The trajectory of x eventually leaves the cylinder of radius ε. Hence, intuitively speaking, an equilibrium point is stable if nearby trajectories stay nearby [3].

As discussed in Section 7.1.2, the equilibrium point $x = 0$ is asymptotically stable at t_o if it is stable at $t \rightarrow t_o$ and also if given $\|x_o\| < \rho$ satisfies $x \rightarrow 0$ as $t \rightarrow \infty$ (Figure 7.1). In this case, the requirement of 'nearby trajectories stay nearby' is not sufficient; it should be 'nearby trajectories stay nearby and all converge to the equilibrium point'.

Again, this concept is local as the region containing all initial conditions that converge to the equilibrium is some portion of the state space. Asymptotic stability is global if the equilibrium $x = 0$ is stable and satisfies the condition $\|x_o\| \rightarrow 0$ as $t \rightarrow \infty$ for any x_o in the whole space.

9.1.1 Positive definite function

A scalar and continuous function $v(x)$ is said to be positive definite in a region R if $v(x) > 0$ for $x \neq 0$ and $v(0) = 0$. All x satisfying $v(x) = C$ form a corresponding space surface called contour. With varying C different un-intersected contours are obtained.

9.1.2 Negative definite function

v is negative definite if $-v$ is positive definite.

9.1.3 Lemma

It states that there exists a sphere defined by $\|x\| = N$ in which $v(x)$ increases monotonically along radial vectors emanating from the origin. That is, $v(\beta_{\boldsymbol{u}})$ increases monotonically with β in $0 \leq \beta \leq N$ for any unit vector $\boldsymbol{u}$ started from the origin.

This can be illustrated by using the assumption of positive definiteness where $v(x) > 0$ for $x \neq 0$ and $v(0) = 0$. Assume that $v(\beta_{\boldsymbol{u}})$ increases monotonically with β in an interval $0 \leq \beta \leq \beta_u$ and begins to decrease after $\beta = \beta_u$. For a given u, there exists an associated β_u, which may be unbounded $\beta_u (= \infty)$. If $\boldsymbol{w}$ is among all the $\boldsymbol{u}$'s that has the smallest β_w, then $\|\beta_w w\| = \beta_w \|\boldsymbol{w}\| = \beta_w \leq \beta_u$. As $v(\beta_u)$ increases monotonically with β in the interval $0 \leq \beta \leq \beta_w \leq \beta_u$, the positive number $N = \beta_w$ can be identified.

9.1.4 Stability regions

For a SEP, x_s, a number $\rho > 0$ exists such that every point in the set $\|x_o - x_s\| < \rho$ implies that the trajectory starting from the initial point x_o converges to the SEP, x_s, i.e. $\Phi_i(x_0) \rightarrow \hat{x}$ as $t \rightarrow \infty$. If ρ is arbitrarily large, then $\hat{x}$ is called a global SEP. There are many systems containing SEPs but not globally SEPs. For these systems the stability region is also called the region of attraction. The stability region of a SEP, x_s, is the set of all points x such that $\lim_{t \rightarrow \infty} \emptyset_t(x) \rightarrow x_s$.

9.1.5 Lyapunov function theorem

The main concept of Lyapunov function theorem is to derive stability properties of the equilibrium point without numerically solving the system differential equations, i.e. without time-domain simulation. Assuming $\dot{v}(x)$ is the total derivative of $v(x)$ on the trajectory specified by (9.1),

$$\begin{aligned} \dot{v}(x) &= \frac{dv}{dt} = \sum_{i=1}^{n} \frac{\partial v(x)}{\partial x} f_i(x) \\ &= \nabla v(x)^T f(x) \end{aligned} \tag{9.2}$$

where $\nabla v(x)$ is the vector formed by the partial derivatives of $v(x)$ and can be performed without knowledge of the system trajectory; n is the dimension of the space.

Assuming regions R, R_1, R_2 such that $R_2 \leq R_1 \leq R$ and all regions contain the origin as an interior point, the theorem states that:

If $v(x)$ be a positive definite function with continuous partial derivatives in a region R, then

- the origin of the system described by (9.1) is stable if $\dot{v}(x) \leq 0$ in a sub-region $R_1 \leq R$.
- The system is asymptotically stable in the region if it is stable and $\dot{v}(x) = 0$ takes place only in a sub-region $R_2 \leq R_1$.
- The origin is globally asymptotically stable if the system is asymptotically stable: R_2 is the whole space and $v(x) \rightarrow \infty$ as $\|x\| \rightarrow 0$.

Example 9.1 The dynamics of a pendulum is chosen to illustrate how to use Lyapunov theorem for studying its stability. The motion of a pendulum can be described by

$$J\ddot{\theta} + D\dot{\theta} + \alpha \sin\theta = 0$$

where J and D are inertia and damping constant, respectively. α is a constant in terms of pendulum mass and θ is the angle. It is noted that this equation is analogous to the equation that describes the rotor motion. Multiplying by $\dot{\theta}$, then integrating with respect to t obtains

$$½J\dot{\theta}^2 + D\int\dot{\theta}^2\,dt + [a(-\cos\theta) + a] = C$$

The sum of kinetic and potential energy of the pendulum is the total energy and is given by

$$\Gamma_T = ½J\dot{\theta}^2 + a(1 - \cos\theta) = C - D\int\dot{\theta}^2\,dt > 0$$

Differentiation with respect to t yields

$$\dot{\Gamma}_T = \dot{\theta}(J\ddot{\theta} + a\sin\theta) = -D\dot{\theta}^2 \leq 0$$

The Lyapunov function, $v(x)$, is proposed to be the total energy per unit inertia of the pendulum system.

$$\text{Hence, } v(x) = \frac{E}{J} = ½\dot{\theta}_2 + a(1 - \cos\theta) = ½x_2^2 + a(1 - \cos x_1)$$

where $x_1 = \theta$ and $\dot{x}_1 = x_2$

The region R is defined by $R = \{-2\pi < x < 2\pi;\text{ free } x_2\}$.

From (9.2) the derivative of $v(x)$ is

$$\dot{v}(x) = \nabla v(x)^T f(x) = [a\sin x_1, x_2]\begin{bmatrix} x_2 \\ -Dx_2 - a\sin x_1 \end{bmatrix} = -Dx_2^2$$

where $f(x)$ in (9.1) can be obtained by writing the equation of motion in the form

$$\begin{bmatrix} \dot{x}_1 \\ \dot{x}_2 \end{bmatrix} = \begin{bmatrix} x_2 \\ -Dx_2 - a\sin x_1 \end{bmatrix} = f(x)$$

It is important to note that $\dot{\Gamma}_T \equiv \dot{v}(x)$. This means that the Lyapunov function is actually the sum of kinetic and potential energy per unit inertia.

By examining the functions $v(x)$ and $\dot{v}(x)$ and applying Lyapunov theorem, it is seen that:

- The origin of the system is stable as $\dot{v}(x) \leq 0$ in R_1, which equals the whole space $R = \{-2\pi < x < 2\pi;\text{ free } x_2\}$.
- The system is asymptotically stable in the region $R_2 \subset R_1$, where $R_2 = \{-2\pi < x < 2\pi;\text{ free } x_2\}$. This can be proved as below. Verification of this condition means that $\dot{v}(x) = -Dx_2^2 = 0$. It implies that $x_2 = 0$, then $\dot{x}_2 = 0 = -Dx_2 - a\sin x_1 = 0$. Consequently, $\sin x_1 = 0$ or $x_1 = n\pi$. Thus, $\dot{v}(x) = 0$ only at the origin by choosing $R_2 = \{-\pi < x < \pi;\text{ free } x_2\}$.
- Part iii of Lyapunov theorem is not satisfied as R_2 is not the whole space and if $\|x\| \rightarrow 0$, the function $v(x) = a(1 - \cos x_1) \leq 2a$, i.e. $v(x)$ does not approach ∞.

9.2 Stability of single-machine infinite-bus system

Transient energy-based stability assessment of a single machine connected to an infinite bus system leads to a method called the 'equal area criterion, EAC' as explained below.

The first step is to select the machine model in order to deduce the relation of electric power calculation. The machine model 0.0 'where only two stator algebraic equations are only used since there is no need to use differential equations and all state variables for rotor coils are ignored' is chosen to illustrate the idea behind EAC.

The machine is represented by a voltage source, $E' = E'_q + jE'_d$, with constant magnitude behind d-axis transient reactance, x'_d as shown in Figure 8.2. From (8.27) the two components of E' are expressed as

$$\left.\begin{aligned} E'_q &= V_q + R_a I_q - I_d X'_d \\ E'_d &= V_d + R_a I_d + I_q X'_q \end{aligned}\right\} \tag{9.3}$$

More simplifications are used such as (i) the armature resistance R_a is neglected and (ii) a direct axis rotor winding is only considered. Thus, $E'_d = 0, E' = E'_q$ and (9.3) becomes

$$\left.\begin{aligned} E' &= V_q - I_d X'_d \\ 0 &= V_d + I_q X'_q \end{aligned}\right\} \tag{9.4}$$

Hence, the current components in d–q axis can be found as

$$\left.\begin{aligned} I_d &= \frac{(V_q - E')}{X'_d} \\ I_q &= -\frac{V_d}{X'_q} \end{aligned}\right\} \tag{9.5}$$

The machine electric output power 'P_e' can be calculated by

$$P_e = V_d I_d + V_q I_q \tag{9.6}$$

Substituting (9.5) in (9.6) gives

$$\begin{aligned} P_e &= V_d \frac{V_q - E'}{X'_d} - V_q \frac{V_d}{X'_q} \\ &= V_d V_q \frac{X'_d - X'_q}{X'_d X'_q} - \frac{E' V_d}{X'_d} \end{aligned} \tag{9.7}$$

As shown in the phasor diagram (Figure 8.5), the terminal voltage is assumed to be taken as a reference as well as E' leads V_t by an angle δ. Thus, the following relations can be written as

$$\left.\begin{aligned} V_q &= V_t \cos\delta \\ V_d &= -V_t \sin\delta \end{aligned}\right\} \tag{9.8}$$

From (9.7) and (9.8) P_e can be expressed as

$$P_e = \frac{E'V_t}{X'_d}\sin\delta + V_t^2\frac{X'_d - X'_q}{2X'_dX'_q}\sin 2\delta \tag{9.9}$$

The second term in the RHS of (9.9) represents the saliency effect for salient pole machines where $X'_d \neq X'_q$. On the other hand, X'_d and X'_q are equal for round rotor machines, and then

$$P_e = \frac{E'V_t}{X'_d}\sin\delta \tag{9.10}$$

According to (9.9) and (9.10) the power–angle curve for salient pole and round rotor machines can be drawn as shown in Figure 9.2(a and b), respectively. It is to be noted that the P–δ curve for salient pole machines is not pure sinusoidal curve because of the presence of saliency effect.

In transient stability studies, the input mechanical power can be assumed as a constant. This assumption is accepted as the electric changes involved are much faster than the resulting mechanical changes produced by the generator/turbine speed control, e.g. the time constant of excitation control loop is much less than that of the governor control loop.

Considering a round rotor machine connected to an infinite bus through a transmission network with external reactance 'X_e' the reactance in (9.10) is replaced by the equivalent reactance, $X_{eq} = X'_d + X_{eq}$ and the electrical power is calculated by

$$P_e = \frac{E'V_\infty}{X_{eq}}\sin\delta \tag{9.11}$$

where $V_\infty \triangleq$ the infinite bus voltage at zero angle 'taken as a reference'
$\delta \triangleq$ the angle between E' and V_∞

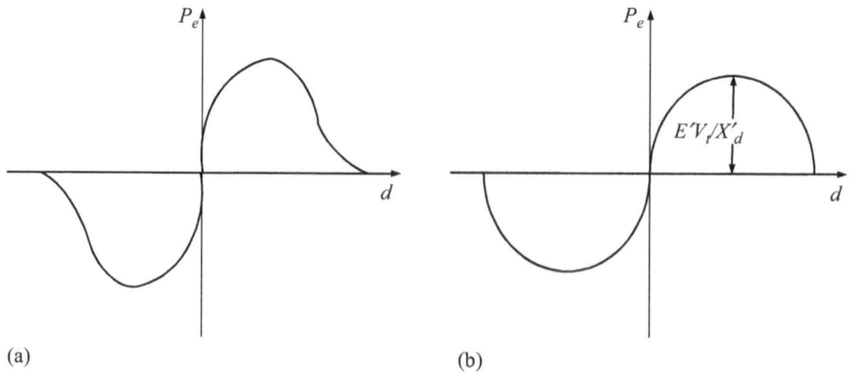

Figure 9.2 Power–angle curves for synchronous machines: (a) p–d curve for salient pole machines and (b) p–d curve for round rotor machines

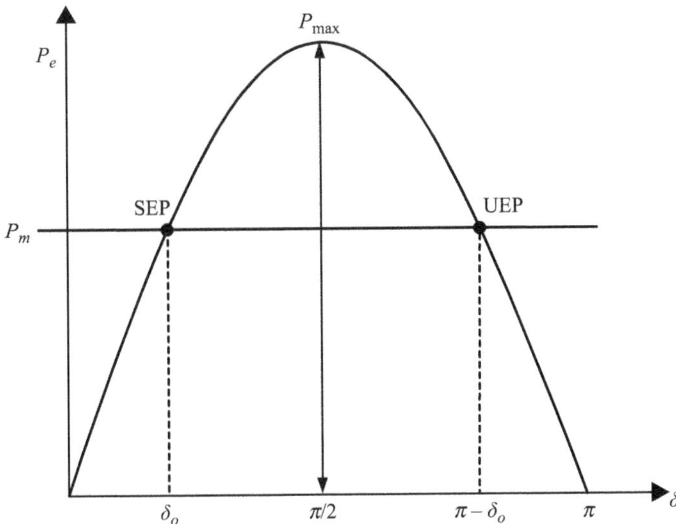

Figure 9.3 Power–angle curve illustrating stable and unstable equilibrium points

Equation (9.11) with constant mechanical power input P_m can be depicted as shown in Figure 9.3. The intersection between P_m and the power–angle curve gives two equilibrium points, δ_o and $\pi - \delta_o$. The first is SEP while the second is UEP as defined by (7.44) and illustrated by Figure 7.3. Physically, the stability and instability of equilibrium points can be interpreted by assuming that a small disturbance causes a change of δ by a small amount $\Delta\delta$. This change results in $P_e > P_m$ and $d\omega/dt$ becomes negative according to the swing equation. Consequently, δ is decreased until the system reaches its initial stable equilibrium point at δ_o. On the other hand, if this change occurs when the system is operating at $\pi - \delta_o$, the increase of δ continues as $P_m > P_e$ and $d\omega/dt$ is positive, i.e. δ moves further from $\pi - \delta_o$. That is why $\pi - \delta_o$ is called UEP.

As shown in Figure 9.4 the system is operating at an equilibrium state (δ_o, P_{eo}) where $P_{eo} = P_m$ and is subjected to a sudden change of P_e from P_{eo} to P_{e1} at which the power angle is δ_1. As P_m is greater than P_{e1} the rotor kinetic energy is increased. The accelerating power $P_a = P_m - P_e$ and $d\omega/dt$ are positive and δ increases until reaching the point (δ_o, P_{eo}) where both accelerating power and $d\omega/dt$ are zero. Because of the rotor inertia δ continues to increase beyond δ_o where rotor retardation starts until reaching the point (δ_s, P_{es}) from which point the retardation will bring δ down. At point 'δ_s, P_{es}' the areas A_1 and A_2 (Figure 9.4) are equal. The process continues on in the form of oscillations around the equilibrium point (δ_o, P_{eo}). If damping is present, the oscillations decrease, the system is stable and continues to operate at the equilibrium point.

The same oscillations may occur if the input mechanical power is changed suddenly. Assume P_m is increased at a fast rate from P_{mo} at initial equilibrium state (δ_o, P_{eo}) to P_{m1} (Figure 9.5). Accordingly, the angle δ will increase to δ_1 as the

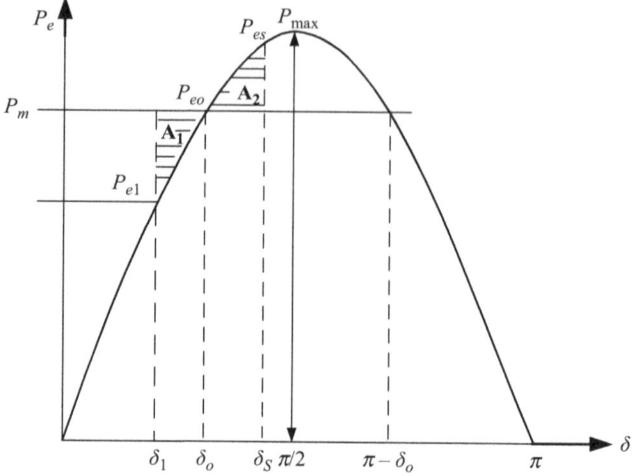

Figure 9.4 Power–angle curve in response to changing the electrical power

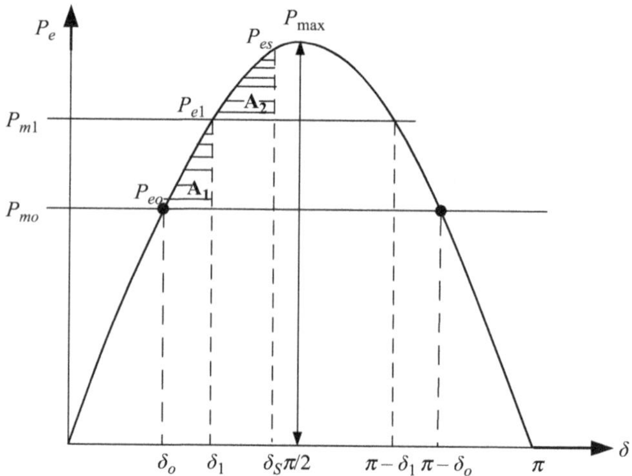

Figure 9.5 Power–angle curve with sudden change in mechanical input power

accelerating power is positive and the rotor kinetic energy is increased as well. At (δ_1, P_{e1}) the accelerating power P_a is zero but the speed deviation from the synchronous speed is not zero because of rotor inertia. Even though rotor retardation sets in at (δ_1, P_{e1}), the angle δ_1 continues to increase until reaching δ_s at which point the speed deviation is zero and the areas A_1 and A_2 (Figure 9.5) are equal. Rotor retardation brings δ down and the process continues on in the form of oscillations. If damping is present, the oscillations decrease and stable operation results at new equilibrium point (δ_1, P_{e1}).

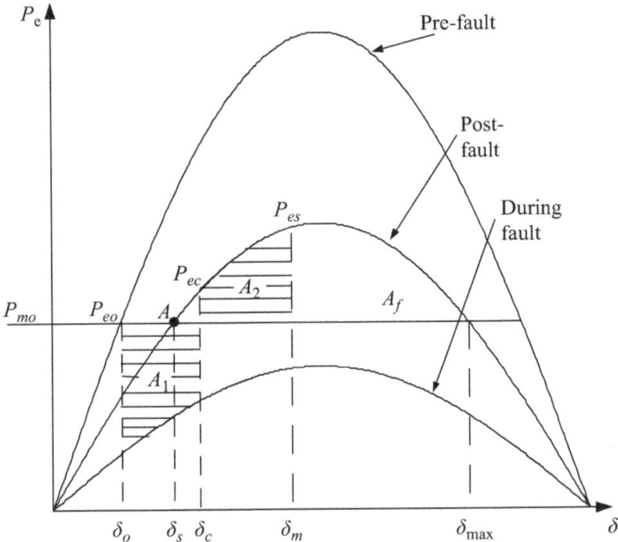

Figure 9.6 Power–angle curves for pre-fault, during fault and post-fault conditions

If the pre-fault, during fault and post-fault system configurations are not identical, then each case is represented by a power–angle curve corresponding to the parameters of the system operating at each specific state (Figure 9.6).

Assuming that the system is operating at initial steady state (δ_o, P_{eo}) a fault occurs. During the fault, the power–angle curve can be determined by calculating the transfer impedance between the internal machine voltage source and the infinite bus.

Accordingly, the electrical power, P_{eo}, is decreased and this results in positive accelerating power and positive $\mathrm{d}\omega/\mathrm{d}t$. Thus, the angle δ increases from δ_o to δ_c at which the fault is cleared and the corresponding post-fault transmitted electrical power is $P_{ec} > P_{mo}$. Both P_a and $\mathrm{d}\omega/\mathrm{d}t$ are negative causing rotor retardation but the speed error is not zero. Thus, δ continues to increase until reaching the point (δ_m, P_{es}) at which point the areas A_1 and A_2 (Figure 9.6) are equal: the rotor will momentarily stop and retardation will bring δ down. The process continues in the form of oscillations around the new equilibrium point (A) as shown in Figure 9.6. The system will be stable if the oscillations are damped.

Rotor oscillation when the system is subjected to sudden change of electrical power, sudden change of mechanical power and occurrence of a fault resulting in change of network configuration is shown in Figures 9.4–9.6, respectively. It can be concluded that in all cases the oscillation comprises two areas: A_1 below P_m line and A_2 above this line. A_1 represents the energy absorbed by the rotor in the form of kinetic energy causing rotor speed up and an increase in angle as $P_m > P_e$. A_2 represents the energy delivered from the rotor causing rotor slow down and a decrease in angle as $P_e > P_m$. Therefore, the stability condition in terms of A_1 and A_2 can be derived by going back to the swing equation as below.

The swing equation for the machine connected to the infinite bus is

$$M\ddot{\delta} = P_a \tag{9.12}$$

Substituting $\dot{\delta} = \omega$ gives the alternative form

$$\omega d\omega = \frac{P_a}{M}\mathrm{d}\delta$$

By definition, δ_o is the rotor angle when the machine is operated synchronously before the disturbance occurs, at which time $\mathrm{d}\delta/\mathrm{d}t = 0$. Thus, integrating both sides

$$\omega^2 = \frac{2}{M}\int_{\delta_o}^{\delta_m} P_a \mathrm{d}\delta$$

Hence, the relative speed ($\omega = \mathrm{d}\delta/\mathrm{d}t$) of the machine with respect to a frame of reference moving at a constant speed is given by

$$\frac{\mathrm{d}\delta}{\mathrm{d}t} = \sqrt{\frac{2}{M}\int_{\delta_o}^{\delta_m} P_a \mathrm{d}\delta} \tag{9.13}$$

The angle δ will cease to change and the machine will again be operating at synchronous speed after a disturbance when $\mathrm{d}\delta/\mathrm{d}t = 0$. From (9.13) the condition for stability can be expressed as

$$\int_{\delta_o}^{\delta_m} P_a \mathrm{d}\delta = 0 \tag{9.14}$$

When (9.14) is satisfied, the maximum value of δ is reached and $\mathrm{d}\delta/\mathrm{d}t = 0$. The area A_1 below P_m line in Figure 9.6 (the same procedure can be followed for Figures 9.4 and 9.5) is

$$A_1 = \int_{\delta_o}^{\delta_c} P_a \mathrm{d}\delta = \int_{\delta_o}^{\delta_c} (P_m - P_e)\mathrm{d}\delta \tag{9.15}$$

Similarly, area A_2 is

$$A_2 = \int_{\delta_c}^{\delta_m} P_a \mathrm{d}\delta = \int_{\delta_c}^{\delta_m} (P_e - P_m)\mathrm{d}\delta \tag{9.16}$$

Thus,

$$A_1 - A_2 = \int_{\delta_o}^{\delta_c} (P_m - P_e)\mathrm{d}\delta - \int_{\delta_c}^{\delta_m} (P_e - P_m)\mathrm{d}\delta = \int_{\delta_o}^{\delta_m} P_a \mathrm{d}\delta \tag{9.17}$$

It is found from (9.14) and (9.17) that $A_1 - A_2 = 0$, i.e. $A_1 = A_2$. The maximum angle of oscillations δ_m is located graphically so as to make A_2 equal to A_1.

Therefore, it is not necessary to assess stability by inspecting the swing curves. Stability can be determined by integrating the difference between the power angle curve and the constant mechanical power. This integral is interpreted as the area between P_e curve and P_m line. The area must equal zero as a condition for stability. So, the area must consist of two equal portions: one portion is positive, A_1, and the second is negative, A_2. This is the reason to call this method 'Equal Area Criterion, EAC'.

Based on EAC, the area A_1 represents the energy converted to rotor kinetic energy at clearance. This entails the existence of area A_2 of opposite sign with magnitude 'at least' equal to A_1 as a condition for stability. Determination of a function called 'transient energy function, TEF' to decide the capability of satisfying this condition can be used to directly assess system stability. The TEF can be derived as below.

Equation (9.10) can be rewritten as

$$P_e = A(x)\sin\delta \tag{9.18}$$

where $A(x) = \frac{E' V_\infty}{X_{eq}}$

Again for convenience, the alternative form of (9.12) is

$$M\omega \mathrm{d}\omega = (P_m - P_e)\mathrm{d}\delta \tag{9.19}$$

Assuming the states of the fault 'as depicted graphically in Figure 9.6' are

$S_o = (\delta_o, 0) \triangleq$ initial state, $S_c = (\delta_c, \omega_c) \triangleq$ fault clearing state, $S_m = (\delta_m, 0) \triangleq$ maximum state, $S = (\delta, \omega) \triangleq$ any state on the power–angle curve generated by x_{eq}, integration of (9.19) from S to S_c gives

$$\frac{1}{2}M\omega^2 - \frac{1}{2}M\omega_c^2 = P_m(\delta - \delta_c) - A(x)(\cos\delta_c - \cos\delta) \tag{9.20}$$

Hence, the rotor kinetic energy at clearance, $X_{eq} = X_c$, for any (δ, ω), is

$$\frac{1}{2}M\omega^2 = \Gamma - \Sigma \tag{9.21}$$

where

$$\Gamma = \frac{1}{2}M\omega_c^2 + P_m(\delta - \delta_c) \text{ and } \Sigma = A(x_c)(\cos\delta_c - \cos\delta)$$

and δ is substituted in radians.

It is to be noted that

(i) During the fault: $x_{eq} = x_f$ and for $S = S_o$, (9.20) becomes

$$\frac{1}{2}M\omega_c^2 = P_m(\delta_o - \delta_c) - A(x_f)(\cos\delta_c - \cos\delta_o) = A_1 \tag{9.22}$$

(ii) Post-fault clearance, $x_{eq}=x_c$ and for $S=S_m$, the relation below can be obtained from (9.20).

$$\frac{1}{2}M\omega_c^2 = A(x_c)(\cos\delta_c - \cos\delta_m) - P_m(\delta_m - \delta_c) = A_2 \tag{9.23}$$

As explained above, according to EAC (9.22) and (9.23) show that the relation $A_1=A_2$ can be solved to obtain δ_m to judge stability.

The RHS of (9.21), $(\Gamma - \Sigma)$ is the TEF that expresses the difference $A_1 - A_{2max}$ ($A_{2max}=A_2+A_f$), i.e. the total area above P_m line from δ_c to δ_{max} as shown in Figure 9.6. It can be directly used to assess system stability as follows:

(i) Evaluate TEF at the angle δ_{max}, 'the intersection of P_m line with post-fault power–angle curve at UEP' (Figure 9.6).
(ii) The system in transient state is stable if $(\Gamma - \Sigma) < 0$; large magnitude yields large margin from stability boundary, better stability and more secure system.
(iii) The system in transient state is unstable if $(\Gamma - \Sigma) \geq 0$.

Example 9.2 A single machine is connected to an infinite bus system through a double-circuit transmission line as shown in Figure 9.7. The system is delivering an apparent power of 1.1 pu at 0.8 power factor lagging. All reactances are given in pu on the machine rating as base. Find the source voltage and δ_o. If a three-phase short circuit occurs at the beginning of one circuit of the transmission line determine whether the system will be stable when the fault is cleared at $\delta_c=45°$ by isolating the faulty circuit.

Solution:

The equivalent reactance $x_{eq}=0.2+0.1+0.35/2=0.475$ pu
The power delivered $P_{eo}=1.1\times 0.8=0.88$ pu $=P_m$
The current flow in the circuit $I = 1.1\angle -36.87°$
The source voltage $E=V+jXI=1\angle 0° + (0.475\angle 90°)(1.1\angle -36.87°)=1.314+j0.418=1.38\angle 17.65°$
The maximum transmitted power $=EV/x_{eq}=1.38\times 1/0.475=2.9$ pu
The angle $\delta_o=17.65°$
The power–angle relation is $P_e=2.9\sin\delta$

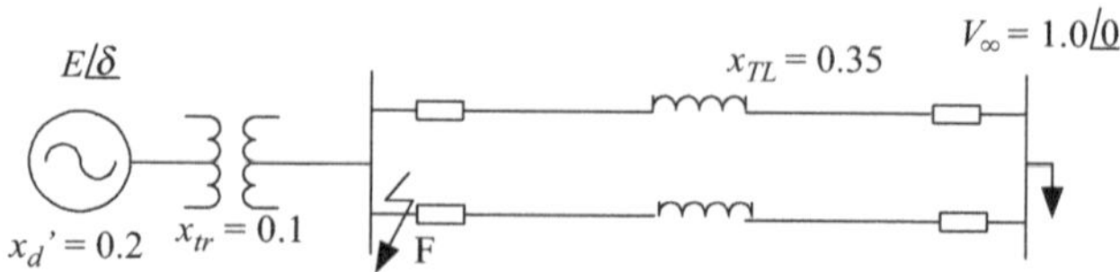

Figure 9.7 System for Example 9.2

When the fault occurs near the generator bus, the delivered power is zero. Thus, the horizontal axis represents the power–angle curve during the fault.

Post-fault clearance: $x_{eq} = 0.2 + 0.1 + 0.35 = 0.65$ pu
The maximum delivered power, $P_{\max(pf)} = 1.38/0.65 = 2.12$ pu
The power–angle relation is $P_e = 2.12 \sin \delta$
To determine the stability:

$$A_1 = P_m(\delta_c - \delta_o)\pi/180 = 0.88(45 - 17.65)\pi/180 = 0.42\ pu$$

$$A_2 = \int_{\delta_c}^{\delta_m} P_{\max(pf)} \sin \delta d\delta - P_m(\delta_m - \delta_c)$$

$$= 2.12(\cos \delta_c - \cos \delta_m) - 0.88(\delta_m - 0.785)$$

$A_1 = A_2$ if the system is stable, thus

$$0.42 = 2.12(\cos 45^\circ - \cos \delta_m) - 0.88(\delta_m - 0.785)$$

which in turn yields $2.41 \cos \delta_m + \delta_s = 1.88$

Solution of this non-linear equation by trial and error yields $\delta_m \approx 78^\circ$. Therefore, the system is stable as δ_m has a value less than $\delta_{\max}$. A schematic diagram for the solution is depicted in Figure 9.8.

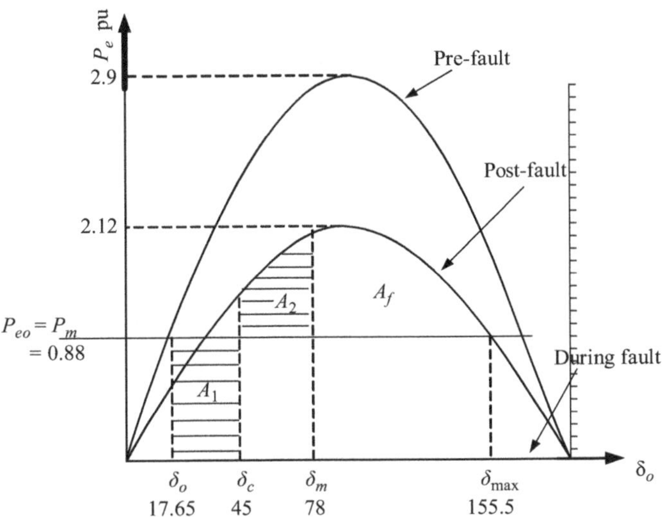

Figure 9.8 Power–angle relations: pre, during and post fault for Example 9.2

Example 9.3 Find the solution of Example 9.2 by using the TEF. Investigate the effect of fault clearing angle on system stability.

Solution:

The TEF is given by (9.21) as TEF $= \Gamma - \Sigma$
At $\delta c = 45°: \Gamma = A_1 + P_m(\delta_{max} - \delta_c)$ and $\Sigma = A(x_c)(\cos\delta_c - \cos\delta_{max})$
Thus,

$$\begin{aligned}\Gamma - \Sigma &= A_1 - [A(x_c)(\cos\delta_c - \cos\delta_{max}) - P_m(\delta_{max} - \delta_c)]\\ &= A_1 - A_{2max}\end{aligned}$$

and $A(x_c) = 2.12$ pu and $A_1 = 0.42$ as calculated in Example 9.2
$\delta_{max} = \sin^{-1}(P_m/P_{max(pf)}) = \sin^{-1}(0.415) = 180 - 24.5 = 155.5°$

$$\begin{aligned}\Gamma - \Sigma &= 0.42 - [2.12(\cos 45° - \cos 155.5°) - 0.88(2.713 - 0.785)]\\ &= -1.311 \text{ pu}\end{aligned}$$

It is noted that the TEF is negative, i.e. the system is stable.

Following the same procedure at different values of clearing angle δ_c the results can be obtained as

δ_c°	45	80	110	120
TEF	−1.311	−0.719	−0.089	0.097
System state	Stable	Stable	Stable	Unstable

It is seen that as the fault clearance is more delayed the magnitude of negative TEF decreases. This means that the system approaches the stability boundary with smaller margins. At $\delta_c = 120°$ the system is unstable where the TEF is positive. The critical clearing angle at which the TEF is zero lies between 110° and 120°.

9.3 Stability of multi-machine power system

In a multi-machine power system, the generators and loads are connected through a transmission network. Each generator, G_i, is connected to a certain number of network buses, j, through its terminal bus i as illustrated in Figure 9.9.

The following assumptions are made to model the system in transient conditions:

- The turbine dynamics are neglected, and the input mechanical power is constant.
- The loads are represented by constant impedances.

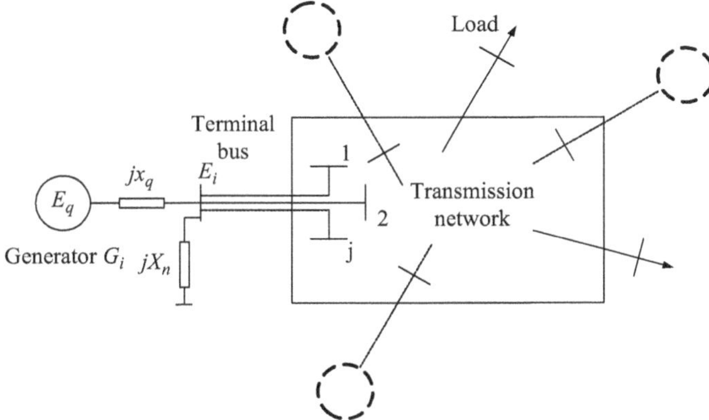

Figure 9.9 Representation of a generator G_i connected to an integrated power system

- The model is adequate only to examine the first swing stability, and so the damping torque is neglected.
- Each synchronous machine is represented by a voltage behind quadrature reactance to take into account the changes in field flux linkages when the bus voltage distribution is calculated by load flow techniques.
- The bus voltage distribution during and after the fault are time invariant.
- Resistances of the transmission elements are neglected in comparison to their reactances.

9.3.1 Energy balance approach

As explained in the former sections, the synchronism of a disturbed machine in a power system is maintained when a balance between the energies exists. The equal area criterion is based on this approach; it equates the change of kinetic energy produced during the fault to the post-fault change of potential energy as a condition for stability. From the instant of fault occurrence to that of fault clearing, a change of kinetic energy is created representing the total transient kinetic energy. If the machine is to survive the first swing, the transient kinetic energy must be totally converted into potential energy.

The energy balance approach can be extended to multi-machine power systems [4]. It necessitates an analytical justification. The basic idea of this justification is to study the individual machine stability. For each machine, the electrical power output 'potential energy' is derived as a function of

- the rotor angle δ referred to the common synchronously rotating reference axis of the network
- the bus voltage distribution

The electrical output power P_{ei} of generator G_i is

$$P_{ei} = Re\left[\boldsymbol{I}_i \boldsymbol{E}_i^*\right] \tag{9.24}$$

where

$\boldsymbol{I}_i \triangleq$ generator current flowing into terminal bus

$$= (E_q - \boldsymbol{E}_i)/jX_q$$

$\boldsymbol{E}_i \triangleq$ terminal bus voltage

$$= (-1/\boldsymbol{Y}_{ii}) \sum_{\substack{j=1 \\ \neq i}}^{N} \boldsymbol{Y}_{ij} E_i$$

and

N, number of network buses
n_b, plus the internal machine buses, n_g
$\boldsymbol{Y}_{ij}$, off-diagonal element of the admittance matrix
$\boldsymbol{E}_q$, machine internal voltage source behind quadrature reactance
Therefore,

$$\boldsymbol{E}_i = \frac{1}{(1/jX_n) + \sum\limits_{\substack{j=1 \\ \neq i}}^{N} (1/jX_{ij})} \sum_{\substack{j=1 \\ \neq i}}^{N} (\boldsymbol{E}_j/jX_{ij}) \tag{9.25}$$

Assuming the real and imaginary components of the voltage $\boldsymbol{E}$ to be e and f, respectively, substitute (9.24) into (9.23) to get

$$P_{ei} = \mathrm{Re}\left[\frac{e_q + jf_q}{jX_q}\left(\frac{1}{\frac{1}{jX_n} + \sum\limits_{\substack{j=1 \\ \neq i}}^{N} (1/jX_{ij})} \sum_{\substack{j=1 \\ \neq i}}^{N} \frac{e_j + jf_j}{jX_{ij}}\right)^* - \frac{1}{jX_q}(e_i^2 + f_i^2)\right]$$

i.e.

$$P_{ei} = \frac{(1/X_q)}{\frac{1}{X_n} + \sum\limits_{\substack{j=1 \\ \neq i}}^{N} (1/X_{ij})}\left[f_{qi} \sum_{\substack{j=1 \\ \neq i}}^{N} (e_j/X_{ij}) - e_{qi} \sum_{\substack{j=1 \\ \neq i}}^{N} (f_j/X_{ij})\right] \tag{9.26}$$

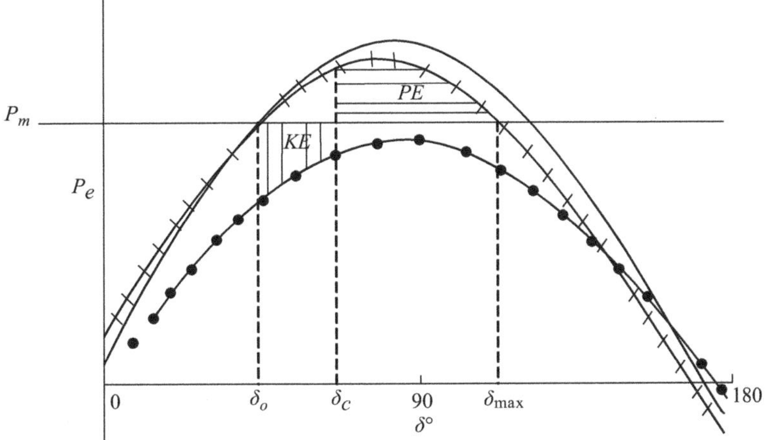

Figure 9.10 Power–angle curves at the three-fault periods
before the fault $P_e = \Lambda_{1B}\sin\delta - \Lambda_{2B}\cos\delta$ *(——)*
during the fault $P_e = \Lambda_{1D}\sin\delta - \Lambda_{2D}\cos\delta$ *(-•-•-)*
after the fault $P_e = \Lambda_{1A}\sin\delta - \Lambda_{2A}\cos\delta$ *(-+-+-)*

Let

$$\left.\begin{aligned}
\gamma_i &= (1/X_q)\left(1/X_n + \sum_{\substack{j=1\\ \neq i}}^{N}(1/X_{ij})\right)\\
\sigma_{1i} &= \sum_{\substack{j=1\\ \neq i}}^{N}(e_j/X_{ij})\\
\sigma_{2i} &= \sum_{\substack{j=1\\ \neq i}}^{N}(f_j/X_{ij})\\
\Lambda_{1i} &= \gamma_i E_q \sigma_{1i}\\
\Lambda_{2i} &= \gamma_i E_q \sigma_{2i}
\end{aligned}\right\} \tag{9.27}$$

Then (9.26) becomes

$$P_{ei} = \Lambda_{1i}\sin\delta_i - \Lambda_{2i}\cos\delta_i \tag{9.28}$$

where $\delta_i = \tan^{-1}(f_{qi}/e_{qi})_i \triangleq$ machine power angle referred to the common reference axis of the system.

It is seen from (9.28) that the generator output power is calculated in terms of the bus voltage distribution, the internal voltage source of the machine and the rotor angle. To determine the delivered power of each machine in a disturbed system, the coefficients Λ_1 and Λ_2 must be determined at the three states: pre-, during and post-fault (Figure 9.10). According to (9.27), these factors are dependent on the bus voltage distribution and the system configuration.

In practice, the bus voltages can be measured continually but, theoretically, the load flow techniques taking into account the assumptions mentioned above are used to compute these voltages, whatever the type and location of the fault. Equations (9.14) and (9.17) are used to examine the machine stability. The machine is stable if the kinetic energy generated during the fault is less than, or equal (totally converted) to, the potential energy during the post-fault period. The equality of both energies takes place in the critical case.

Example 9.4 Determine the system stability for 15-bus, 4-generator test system shown in Figure 9.11 with the data given in Appendix IV. A three-phase short circuit occurs at bus no. 15. At fault clearance, the generator G_2 is assumed to be disconnected and bus no. 15 is completely isolated from the rest of the system. The fault is cleared at $\delta_{c1} = 55°$, $\delta_{c3} = 57°$ and $\delta_{c4} = 60°$ for G_1, G_3 and G_4, respectively. The input mechanical power for each generator equals its rated active power.

Solution:

The bus voltage distribution, pre-, during and post-fault, is calculated by the load flow technique as explained in Chapter 5. The results are summarised in Table 9.1. Accordingly, the power delivered from each generator at states pre-, during and post-fault are calculated using (9.28) and are summarised in Table 9.2.

To judge the stability for each generator, the transient kinetic energy, *KE*, represented by the area under P_m line and the potential energy, *PE*, represented by the area above P_m line must be calculated. The machine is stable when the difference $(KE - PE)$ is a negative value, i.e. *KE* is fully converted into potential energy.

The intersection of P_m line with power–angle curve before the fault and after the fault gives the values of δ_o and δ_{max}, respectively. The parameters required to calculate *KE* and *PE* are summarised in Table 9.3.

The *KE* is given by

$$\begin{aligned} KE &= P_m(\delta_c - \delta_o) - \int_{\delta_o}^{\delta_c} (\Lambda_{1D} \sin\delta - \Lambda_{2D} \cos\delta) \\ &= P_m(\delta_c - \delta_o) - \Lambda_{1D}(\cos\delta_o - \cos\delta_c) - \Lambda_{2D}(\sin\delta_o - \sin\delta_c) \end{aligned} \quad (9.29)$$

and the potential energy is

$$\begin{aligned} PE &= \int_{\delta_c}^{\delta_{max}} (\Lambda_{1A} \sin\delta - \Lambda_{2A} \cos\delta)d\delta - P_m(\delta_{max} - \delta_c) \\ &= \Lambda_{1A}(\cos\delta_c - \cos\delta_{max}) + \Lambda_{2A}(\sin\delta_{max} - \sin\delta_c) - P_m(\delta_{max} - \delta_c) \end{aligned} \quad (9.30)$$

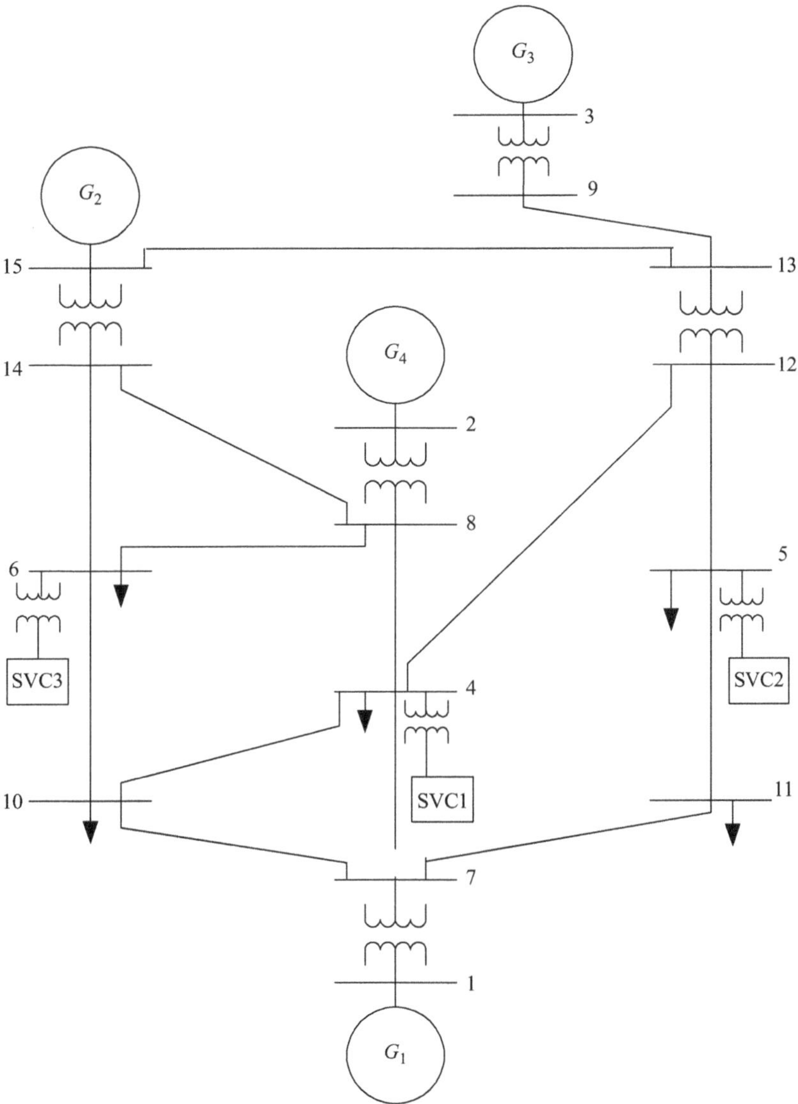

Figure 9.11 *Test system for Example 9.4*

By the data summarised in Tables 9.2 and 9.3 and using (9.29) and (9.30) the results of *KE*, *PE* and the difference between them are summarised in Table 9.4. It is seen that the difference '*KE* – *PE*' is negative, which means that *KE* is fully converted into *PE* and all generators are stable.

Equating *KE* and *PE* the critical clearing angle 'δ_{cr}' can be obtained from (9.29) and (9.30) as summarised in Table 9.4. The corresponding power–angle curves for generators G_1, G_3 and G_4 are shown in Figure 9.12.

Table 9.1 Bus voltage distribution of the test system pre-, during and post-fault occurrence

Fault period	Bus code	Voltage		Bus code	Voltage	
		mag. (pu)	angle (°)		mag. (pu)	angle (°)
Before the fault	1	1.0065	−2.4	9	1.0059	4.6
	2*	1.0000	0.0	10	1.0009	−7.1
	3	1.0082	8.3	11	0.9971	−7.6
	4	1.0068	−8.8	12	1.0062	−1.4
	5	1.0030	−11.7	13	1.0037	3.4
	6	1.0066	−7.6	14	1.0101	1.0
	7	1.0020	−6.5	15	1.0072	4.6
	8	1.0066	−2.8			
During the fault	1	0.6913	1.05	9	0.3414	7.5
	2	0.6521	5.08	10	0.5534	−4.7
	3	0.5025	13.5	11	0.5649	−5.0
	4	0.4981	−7.05	12	0.3650	−0.7
	5	0.4849	−10.59	13	0.2928	5.4
	6	0.4775	−6.16	14	0.1897	0.8
	7	0.5792	−4.00	15	0.0000	0.0
	8	0.5108	0.05			
After the fault	1	0.9844	−10.5	9	0.9797	−5.6
	2*	1.0000	0.0	10	0.9792	−15.4
	3	0.9818	−1.7	11	0.9740	−15.9
	4	0.9842	−16.7	12	0.9814	−11.0
	5	0.9767	−21.0	13	0.9776	−6.8
	6	0.9876	−15.8	14	0.9961	−11.4
	7	0.9797	−14.8	15	Isolated	
	8	0.9943	−7.9			

*Bus no. 2 is the slack bus.

Table 9.2 Electrical output power in terms of rotor angle for each generator in operation, pre-, during and post-fault period

Generator	P_e (pre)	P_e (during)	P_e (post)
G_1	$1.0615 \sin\delta + 0.12 \cos\delta$	$0.756 \sin\delta + 0.05 \cos\delta$	$1.009 \sin\delta + 0.26 \cos\delta$
G_3	$0.906 \sin\delta - 0.07 \cos\delta$	$0.432 \sin\delta - 0.06 \cos\delta$	$0.880 \sin\delta + 0.08 \cos\delta$
G_4	$0.956 \sin\delta + 0.05 \cos\delta$	$0.840 \sin\delta - 0.02 \cos\delta$	$0.936 \sin\delta + 0.13 \cos\delta$

Table 9.3 Parameters for calculating KE and PE

Generator	δ_o	δ_c	δ_{max}	P_m (pu)
G_1	46°	55°	111.5°	0.85
G_3	53°	57°	125.0°	0.68
G_4	45.7°	60°	122.5°	0.80

Table 9.4 Transient kinetic energy, potential energy, difference and δ_{cr}

Generator	***KE***	***PE***	***KE − PE***	δ_{cr}
G_1	0.0369	0.0819	−0.0450	60.5
G_3	0.0247	0.1790	−0.1543	62.1
G_4	0.0360	0.1016	−0.0656	64.4

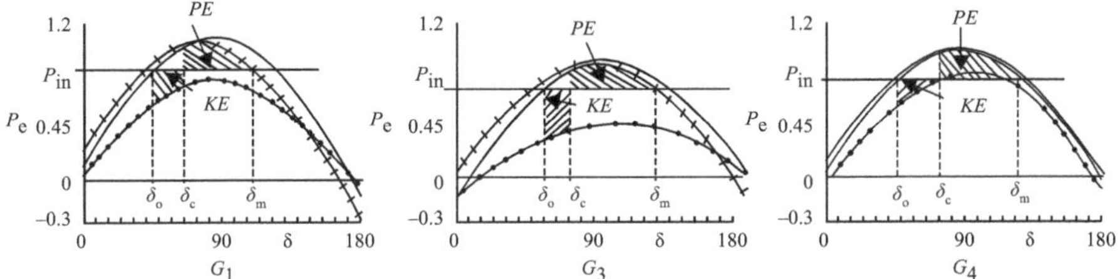

Figure 9.12 Power–angle curves at pre-, during and post-fault occurrence

9.3.2 TEF method

9.3.2.1 Formulation of centre of inertia

The equation of motion of the *i*th generator in a power system is

$$M_i\ddot{\delta}_i + D_i\dot{\delta}_i = P_{mi} - P_{ei} \tag{9.31}$$

where

$M_i \triangleq$ inertia constant of *i*th generator $= 2H_i/\omega_r$
$\dot{\delta}_i = \omega_i - \omega_r, i = 1, 2, \ldots, n_g$
$\omega_i \triangleq$ generator speed
$\omega_r \triangleq$ reference speed

$$P_{ei} = E_i^2 G_{ii} + \sum_{\substack{j=1 \\ \neq i}}^{n_g} E_i E_j Y_{ij} \cos\left(\theta_{ij} - \delta_i + \delta_j\right) \tag{9.32}$$

$Y_{ij}\theta_{ij} = G_{ij} + jB_{ij} =$ transfer admittance between nodes *i* and *j*

It is to be noted that the network includes only generator nodes; all other nodes are eliminated and loads are represented by constant impedance, i.e. included in transfer conductances.

The generator angle and speed in (9.31) are given with respect to a synchronous frame of reference. They can be referred to as centre of inertia (COI) coordinates that are defined by satisfying

$$\delta_o = \frac{1}{M_T}\sum_{i=1}^{n_g} M_i\delta_i \quad \text{and} \quad \dot{\delta}_o = \frac{1}{M_T}\sum_{i=1}^{n_g} M_i\dot{\delta}_i \tag{9.33}$$

where

$$M_T \triangleq \sum_{i=1}^{n_g} M_i$$

The equation of motion of COI is given by

$$M_T(\mathrm{p}+D)\dot{\delta}_o = \sum_{i=1}^{n_g} P_{mi} - \sum_{i=1}^{n_g} P_{ei} \triangleq P_{\mathrm{COI}} \tag{9.34}$$

where

$$\sum_{i=1}^{n_g} P_{ei} = \sum_{j=1}^{n_b} P_{lj} + P_{\mathrm{loss}}$$

For lossless network and constant active power loads, P_{COI} is constant as the mechanical input power P_{mi} is considered as constant. Thus, (9.34) becomes

$$M_T(\mathrm{p}+D)\dot{\delta}_o = P_{\mathrm{COI}} \tag{9.35}$$

where

$$P_{\mathrm{COI}} = \sum_{i=1}^{n_g} P_{mi} - \sum_{j=1}^{n_b} P_{lj} \quad \text{and} \quad \mathrm{p} = \mathrm{d}/\mathrm{d}t$$

Relative to COI, all generators have phase angles:

$$\theta_i = \delta_i - \delta_o \tag{9.36}$$

Hence, (9.31) incorporated with (9.35) and (9.36) gives

$$M_i(\mathrm{p}+D_i)\dot{\theta}_i + M_i(\mathrm{p}+D_i)\dot{\delta}_o = P_{mi} - P_{ei}$$

i.e.

$$M_i(\mathrm{p}+D_i)\dot{\theta}_i = P_{mi} - P_{ei} - \frac{M_i}{M_T}P_{\mathrm{COI}} \tag{9.37}$$

It is noted that the variables of COI satisfy the constraints

$$\sum_{i=1}^{n_g} M_i\dot{\theta}_i = \sum_{i=1}^{n_g} M_i\left(\dot{\delta}_i - \dot{\delta}_o\right) = \sum_{i=1}^{n_g} M_i\dot{\delta}_i - M_T\dot{\delta}_o = 0$$

9.3.2.2 Derivation of the TEF

The dynamics of the post-disturbance system are represented by (9.37). The energy function for this system can be derived as below:

(i) Neglecting damping and multiplying (9.37) by $\dot{\theta}_i$, get the sum for all generators in the system as

$$\sum_{i=1}^{n_g}\left[M_i\ddot{\theta}_i - P_{mi} + P_{ei} + \frac{M_i}{M_T}P_{\text{COI}}\right]\dot{\theta}_i \tag{9.38}$$

(ii) Equation (9.32) can be rewritten as

$$P_{ei} = E_i^2 G_{ii} + \sum_{\substack{j=1\\ \neq i}}^{n_g} E_i E_j \left[B_{ij}\sin\left(\delta_i - \delta_j\right) + G_{ij}\cos\left(\delta_i - \delta_j\right)\right]$$

$$P_{ei} = E_i^2 G_{ii} + \sum_{\substack{j=1\\ \neq i}}^{n_g} \left[\overline{B}_{ij}\sin\left(\delta_i - \delta_j\right) + \overline{G}_{ij}\cos\left(\delta_i - \delta_j\right)\right]$$

where

$$\overline{B}_{ij} = E_i E_j B_{ij} \quad \text{and} \quad \overline{G}_{ij} = E_i E_j G_{ij}$$

(iii) Substitute P_{ei} to get the sum in the form

$$\sum_{i=1}^{n_g}\left[M_i\ddot{\theta}_i - P_i + \sum_{\substack{j=1\\ \neq i}}^{n_g}\left\{\overline{B}_{ij}\sin\theta_{ij} + \overline{G}_{ij}\cos\theta_{ij}\right\}\right]\dot{\theta}_i \tag{9.39}$$

where

$$P_i = P_{mi} - E_i^2 G_{ij},\, \theta_{ij} = \theta_i - \theta_j = \delta_i - \delta_j$$

(iv) In the expression (9.39), as $\overline{B}_{ij} = \overline{B}_{ji}$ and $\overline{G}_{ij} = \overline{G}_{ji}$, it is seen that

$$\left.\begin{aligned}
\sum_{i=1}^{n_g}\sum_{\substack{j=1\\ \neq i}}^{n_g}\overline{B}_{ij}\sin\theta_{ij}\dot{\theta}_i &= \sum_{i=1}^{n_g-1}\sum_{j=i+1}^{n_g}\overline{B}_{ij}\sin\theta_{ij}\dot{\theta}_{ij}\\
\sum_{i=1}^{n_g}\sum_{\substack{j=1\\ \neq i}}^{n_g}\overline{G}_{ij}\cos\theta_{ij}\dot{\theta}_i &= \sum_{i=1}^{n_g-1}\sum_{j=i+1}^{n_g}\overline{G}_{ij}\cos\theta_{ij}\dot{\theta}_{ij}
\end{aligned}\right\} \tag{9.40}$$

(v) Substituting expression (9.40) into (9.39) and integrating the resulting expression with respect to time, from $t = t^s$ at which $\dot{\theta}(t^s) = 0$ and $\theta(t^s) = \theta_s$, the energy function V describing the total system transient energy for the post-disturbance system is given by

$$V = \frac{1}{2}\sum_{i=1}^{n_g} M_i\dot{\theta}_i^2 - \sum_{i=1}^{n_g} P_i(\theta_i - \theta_{is})$$
$$-\sum_{i=1}^{n_g-1}\sum_{j=i+1}^{n_g}\left[\overline{B}_{ij}\left(\cos\theta_{ij} - \cos\theta_{ij}^s\right) - \int_{\theta_i^s+\theta_j^s}^{\theta_i+\theta_j} \overline{G}_{ij}\cos\theta_{ij}\,\mathrm{d}(\theta_i + \theta_j)\right] \quad (9.41)$$

where

θ_{is} = the angle of bus i at the post-disturbance SEP.

The TEF (9.41) consists of the following four terms:

(i) $$½\sum_{i=1}^{n_g} M_i\dot{\theta}_i^2 = ½\sum_{i=1}^{n_g} M_i\left(\dot{\delta}_i - \dot{\delta}_o\right)^2$$
$$= ½\sum_{i=1}^{n_g} M_i\dot{\delta}_i^2 - \sum_{i=1}^{n_g} M_i\dot{\delta}_i\dot{\delta}_o + ½\sum_{i=1}^{n_g} M_i\dot{\delta}_o^2$$
$$= ½\sum_{i=1}^{n_g} M_i\dot{\delta}_i^2 - \left(M_T\dot{\delta}_o\right)\dot{\delta}_o + ½\,\dot{\delta}_o^2\sum_{i=1}^{n_g} M_i$$
$$= ½\sum_{i=1}^{n_g} M_i\dot{\delta}_i^2 - M_T\dot{\delta}_o^2 + ½\,M_T\dot{\delta}_o^2$$
$$= ½\sum_{i=1}^{n_g} M_i\dot{\delta}_i^2 - ½\,M_T\dot{\delta}_o^2$$

= total change in *KE* of all rotors in the COI frame of reference

This change equals the change in *KE* of all generator rotors minus the change in *PE* associated with the COI.

(ii) $$\sum_{i=1}^{n_g} P_i(\theta_i - \theta_{is}) = \sum_{i=1}^{n_g} P_i(\delta_i - \delta_{is}) - (\delta_o - \delta_{os})\sum_{i=1}^{n_g} P_i$$
= change in *PE* of all rotors relative to the COI

This change equals the change in *PE* of all generator rotors minus the change in PE associated with the COI.

(iii) $\sum_{i=1}^{n_g-1}\sum_{j=i+1}^{n_g} \overline{B}_{ij}\left(\cos\theta_{ij} - \cos\theta_{ijs}\right)$ = the change in stored magnetic energy of all branches. It is independent of the path of integration

(iv) $\sum_{i=1}^{n_g-1}\sum_{j=i+1}^{n_g}\int_{\theta_i^s+\theta_j^s}^{\theta_i+\theta_j}\overline{G}_{ij}\cos\theta_{ij}\,\mathrm{d}(\theta_i+\theta_j)=$ the change in dissipated energy of all branches. It depends on the path of θ_i.

The first term is called the kinetic energy, Γ_{ke}, and is a function of only the generator speeds. The sum of the second, third and fourth terms is called the potential energy 'Γ_{pe}' and is a function of only the generator angles.

Therefore, in a multi-machine power system, the energy function V describing the total system transient energy for the post-disturbance system is given by

$$V=\Gamma_{ke}-\Gamma_{pe} \tag{9.42}$$

where

$$\Gamma_{ke}=\tfrac{1}{2}\sum_{i=1}^{n_g}M_i\dot{\theta}_i^2$$

$$\Gamma_{pe}=\sum_{i=1}^{n_g}P_i(\theta_i-\theta_{is})+\sum_{i=1}^{n_g-1}\sum_{j=i+1}^{n_g}\left[\overline{B}_{ij}(\cos\theta_{ij}-\cos\theta_{ijs})-\int_{\theta_i^s+\theta_j^s}^{\theta_i+\theta_j}\overline{G}_{ij}\cos\theta_{ij}\mathrm{d}(\theta_i+\theta_j)\right]$$

To assess the system stability, both the critical energy function V_{cr} and system energy at the instant of fault clearing V_c are calculated. The difference, $\Delta V=V_{cr}-V_c$, is defined as stability index or stability margin, which is positive when the system is stable. Otherwise, the system is unstable.

To calculate V_c: The angles and speeds of all generators in the system at the instant of fault clearing are required. They can be obtained by running up the simulation in time-domain. V_{cr} is defined as the potential energy at the controlling unstable equilibrium point 'UEP' for a particular disturbance under study.

The integral term in (9.41) representing the dissipated energy is difficult to be evaluated as the system trajectory is unknown. So, a linear angle trajectory is assumed. It has been found that this assumption is acceptable for a first swing transient [5, 6]. It can be derived as below.

Assume $\theta_i(t)$ and $\theta_j(t)$ are the angular trajectories of machines i and j with respect to time, respectively. θ_{ic} and θ_{iu} denote the angles of ith generator at the clearance and UEP state. Thus, θ_{ic} and θ_{iu} represent the initial and final vectors of angular positions of the n_g generators. The linear angle trajectory between the initial state 'at $t=0$, $\theta_i=\theta_{ic}$' and the final state 'at $t=1$, $\theta_i=\theta_{iu}$' is expressed by

$$\theta_i=\theta_{ic}+(\theta_{iu}-\theta_{ic})t \quad 0\le t\le 1 \quad i=1,2,\ldots,n \tag{9.43}$$

Differentiating (9.43)

$$\left.\begin{aligned} d\theta_i &= (\theta_{iu} - \theta_{ic})dt \\ d\theta_j &= (\theta_{ju} - \theta_{jc})dt \end{aligned}\right\} \tag{9.44}$$

Add the two equations to give

$$d(\theta_i + \theta_j) = (\theta_{iu} - \theta_{ic} + \theta_{ju} - \theta_{jc})dt \tag{9.45}$$

Subtract $d\theta_j$ from $d\theta_i$ to give

$$d(\theta_i - \theta_j) = d\theta_{ij} = (\theta_{iu} - \theta_{ic} - \theta_{ju} + \theta_{jc})dt \tag{9.46}$$

Using (9.45) and (9.46) eliminates dt as below:

$$d(\theta_i + \theta_j) = \frac{\theta_{iu} - \theta_{ic} + \theta_{ju} - \theta_{jc}}{\theta_{iju} - \theta_{ijc}} d\theta_{ij} \tag{9.47}$$

Thus, substituting $d(\theta_i+\theta_j)$ from (9.47) into the expression representing the dissipated energy, $\int_{\theta_i^s+\theta_j^s}^{\theta_i+\theta_j} \overline{G}_{ij} \cos\theta_{ij} d(\theta_i+\theta_j)$ gives an expression that can be integrated with respect to θ_{ij} between any two points as

$$I_{ij} = \overline{G}_{ij} \frac{\theta_{iu} - \theta_{ic} + \theta_{ju} - \theta_{ja}}{\theta_{iju} - \theta_{ijc}} (\sin\theta_{iju} - \sin\theta_{ijc}) \tag{9.48}$$

Therefore, by using the dissipated energy expressed by (9.48) between the conditions at the clearance of the disturbance and the controlling UEP, and then substituting for V_{cr} and V_c from (9.41) the stability index is given by

$$\begin{aligned} \Delta V = &-\frac{1}{2}\sum_{i=1}^{n_g} M_i \dot{\theta}_{ic}^2 - \sum_{i=1}^{n_g} P_i(\theta_{iu} - \theta_{ic}) \\ &- \sum_{i=1}^{n_g-1}\sum_{j=i+1}^{n_g}\left[\overline{B}_{ij}(\cos\theta_{iju} - \cos\theta_{ijc}) - \overline{G}_{ij}\frac{\theta_{iu} - \theta_{ic} + \theta_{ju} - \theta_{jc}}{\theta_{iju} - \theta_{ijc}}(\sin\theta_{iju} - \sin\theta_{ijc})\right] \end{aligned} \tag{9.49}$$

where

$(\theta_c, \dot{\theta}_c) \triangleq$ the conditions at the clearance of disturbance
$(\theta_u, 0) \triangleq$ the conditions at controlling UEP

The main steps of transient stability assessment using TEF method for multi-machine power system are summarised in the flowchart shown in Figure 9.13.

9.3.2.3 Calculation of critical energy

The critical energy, V_{cr}, represents the boundary of stability region. It is the most difficult step to calculate V_{cr} when using the TEF method for stability assessment.

The calculation depends mainly on computing the UEP that may be made by one of the following approaches.

The closest UEP approach

At different initial values of bus angles, the steady state equations of the post-disturbance system are solved to determine all unstable equilibrium points (UEPs). They can be obtained by computing the set of generator's angles that satisfy:

$$f_i = P_{mi} - P_{ei} - \frac{M_i}{M_T} P_{\mathrm{COI}} = 0 \quad i = 1, 2, \ldots, n_g \tag{9.50}$$

For a multi-machine power system with n_g generators, there are 2^{ng-1} solutions. Each solution gives a value of potential energy. The chosen UEP is the one that results in the minimum potential energy. It is noted that the results are to some extent pessimistic and usually of little practical value as this approach implies the assumption of worst fault location. In addition, it is found that the trajectory of severely disturbed generators passes close to a UEP different from that having the minimum potential energy. This can mostly be avoided by applying the approach described next.

The controlling UEP approach

The system trajectories for all critically stable cases get close to UEPs (called controlling UEPs) that are closely related to the boundary of system separation. This approach is based on using the disturbed trajectory to determine its intersection with the post-disturbance principle singular surface, θ_{ss}. At this intersection, a direction vector $\boldsymbol{h}$ is formed and along this direction a one-dimensional minimisation problem can be solved to minimise

$$F(\theta) = \sum_{i=1}^{n_g} f_i^2(\theta) = \sum_{i=1}^{n_g} \left[P_{mi} - P_{ei} - \frac{M_i}{M_T} P_{\mathrm{COI}} \right]^2 \tag{9.51}$$

and obtain $\hat{\theta}_u$ that is considered as a starting point to apply a suitable numerical technique to achieve the controlling UEP. It is noticed that there are two aspects to determining the controlling UEP: (i) the effect of the different generators and (ii) the effect of the post-disturbance network, in particular, its energy-absorbing capacity.

These two aspects must be considered when determining the controlling UEP as the more severely disturbed generators may or may not lose synchronism with the rest of the system. It depends on whether the potential energy-absorbing capacity of the network is relevant to convert the kinetic energy at clearing the disturbance into potential energy.

The boundary of stability region based on controlling unstable equilibrium point (BCU) approach

If the starting point, $\hat{\theta}_u$, for the UEP is not sufficiently close to the exact UEP the convergence problem may take place, in particular, when the system is highly stressed or highly unstressed. In this case BCU approach can be used to overcome some of these problems. The algorithm for determining the stability boundary can be found in Chapter 3 of Reference 3. Moreover, the BCU approach is based on the relationship between the boundary of stability region of a power system and that of a reduced system [7, 8]. Some of the other work is concerned with determining the UEP for detailed generator models rather than the classic model [9].

At the desired calculated controlling UEP, the change in potential energy, ΔV_{PE}, can be obtained. It is preferred to be normalised with respect to the kinetic energy at the end of disturbance, ΔV_{PEn}, to give a reliable indication of the degree of stress on different generators.

If the disturbance is large enough, the post-disturbance trajectory approaches the controlling UEP that has in this case the lowest normalised potential energy index at the instant of clearing the disturbance. Thus,

$$\Delta V_{PEn} = \frac{\Delta V_{PE}}{V_{KE}} \tag{9.52}$$

$$\begin{aligned} \Delta V_{PE} &= V_{PEu} - V_{PEc} \\ &= -\sum_{i=1}^{n_g} P_i(\theta_{iu} - \theta_{ic}) - \sum_{i=1}^{n_g-1} \sum_{j=i+1}^{n_g} \Big[\overline{B}_{ij}(\cos\theta_{iju} - \cos\theta_{ijc}) \\ &\quad - \overline{G}_{ij} \frac{\theta_{iu} - \theta_{ic} + \theta_{ju} - \theta_{jc}}{\theta_{iju} - \theta_{ijc}} (\sin\theta_{iju} - \sin\theta_{ijc}) \Big] \end{aligned} \tag{9.53}$$

Based on the former explanation, the computational steps of stability analysis using the TEF for a multi-machine power system can be outlined as described below:

- Collect the input data, applying steady-state power flow analysis, the synchronous machine parameters and specification of disturbance. The initial values of machine internal bus voltages and rotor angles are calculated. The machines are represented by classical model.
- According to the specification of the disturbance and system topology, the system admittance matrix and its reduced form are built.
- The conditions at disturbance clearance are determined, and then the system admittance matrix and its reduced form at the end of disturbance are computed.
- The relevant mode of disturbance is determined by identifying the most affected generators.
- The UEP is calculated.
- The total energy at clearing time and the critical transient energy as well as the stability index can be computed to decide whether the system is stable.

The flowchart in Figure 9.13 presents the main steps outlined above.

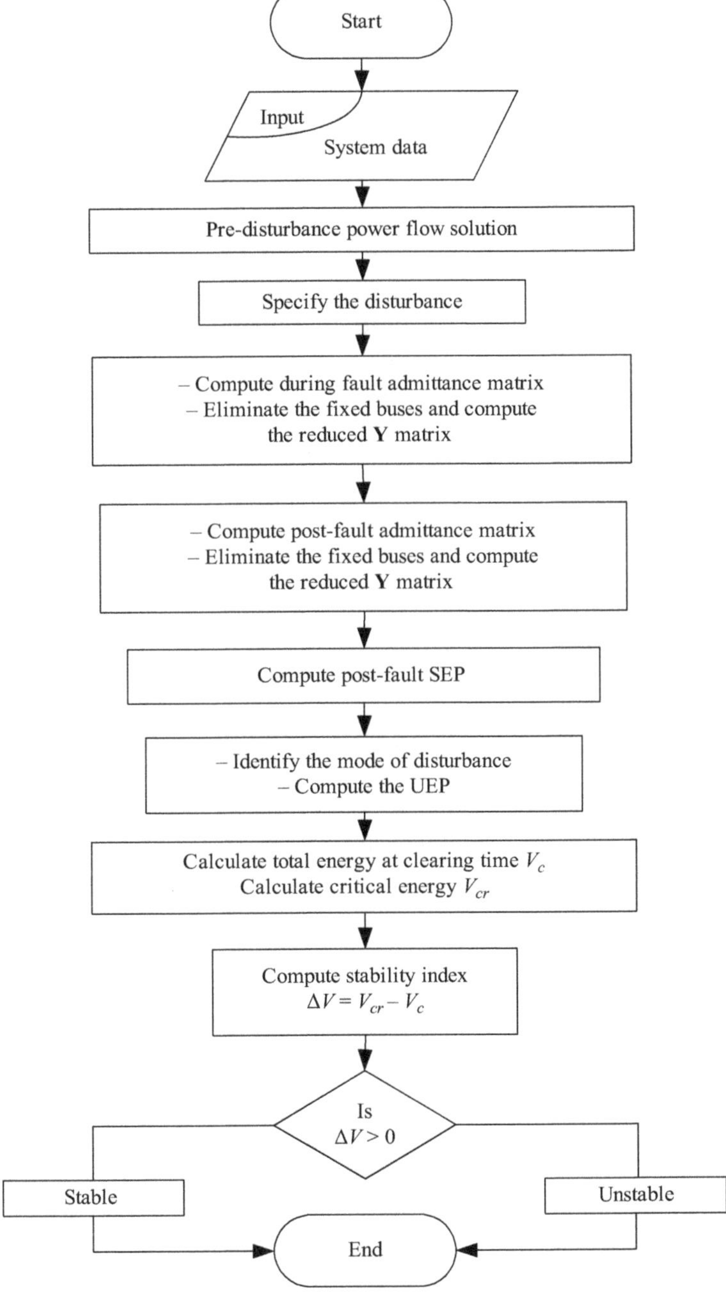

Figure 9.13 Flowchart of transient stability assessment using TEF

References

1. Momoh J.A., El-Hawary M.E. *Electric Systems, Dynamics, and Stability with Artificial Intelligence Applications*. New York, NY, US: Marcel Dekker; 2000
2. Fouad A.A., Vittal V. *Power System Transient Stability Analysis Using the Transient Energy Function Method*. Upper Saddle River, NJ, US: Prentice Hall; 1992
3. Chiang H.D. *Direct Methods for Stability Analysis of Electric Power Systems*. Hoboken, NJ, US: John Wiley & Sons; 2011
4. Sallam A.A. 'Power systems transient stability assessment using catastrophe theory'. *IEE Proceedings*. 1989;**136**(2) Pt C:108–14
5. Uyemura, K., Matsuki J., Yamada J., Tsuji T. 'Approximation of an energy function in transient stability analysis of power systems'. *Electrical Engineering in Japan*. 1972;**92**(4):96–100
6. Athay, T., Sherkat V.R., Podmore R., Virmani S., Puech C. 'Transient energy stability analysis'. System engineering for power. Emergency operating state control-Section IV. U.S. Department of Energy Publication No. CONF-790904-PL, 1979
7. Chiang H.D., Wu F.F., Varaiya P.P. 'A BCU method for direct analysis of power system transient stability'. *IEEE Transactions on Power Systems*. 1994;**9**(3):1194–208
8. Chu C.C., Chiang H.D. (eds.). 'Boundary properties of the BCU method for power system transient stability assessment'. *International Symposium on Circuits and Systems ISCAS 2010, IEEE*; Paris, France, May/Jun 2010. pp. 3453–6
9. Chen L., Min Y., Xu F., Wang K.P. 'A continuation-based method to compute the relevant unstable equilibrium points for power system transient stability analysis'. *IEEE Transactions on Power Systems*. 2009;**24**(1):165–72

Part IV

Stability enhancement and control

Chapter 10
Artificial intelligence techniques

Traditional analytic and time analysis approaches may not easily handle online real-time applications for large systems due to computational time requirements. In particular, the power systems being non-linear and time varying, application of traditional approaches to a power system for the purpose of identifying its parameters, controlling the operation to maintain stability and damping oscillations following disturbances is not suitable for online monitoring. They are more suitable for offline design and investigations.

Advent of artificial intelligence (AI) techniques based on logic mathematics has encouraged power system engineers, planners and designers to employ these techniques with the goal of reducing computation time and designing fast algorithms that are adequate for power system online applications. Many AI and computational intelligence techniques, such as artificial neural network (ANN), fuzzy logic (FL), neuro-FL (NFL), particle swarm optimisation (PSO), genetic algorithms, exist. The basics of ANN, FL and NFL as well as the adaptive neuro-fuzzy control (ANFC) are presented in this chapter as they are used, in addition to the time analysis techniques, for some applications (e.g. power system stabilisers and static var compensators) to power systems in the subsequent chapters.

10.1 Artificial neural networks

ANNs are biologically inspired computational models that consist of processing elements (called neurons) interconnected together to constitute the network structure. They are essentially non-linear function approximations that utilise process inputs to estimate process outputs. An important feature of the ANNs is the ability to adjust their connections through an adaptive learning process called Learning. Learning can be accomplished using a series of examples and patterns. Information obtained through learning is retained and represented by a set of connection weights within the neural network structure [1, 2].

A simple neuron model consists of two main parts: a linear combiner and a nonlinear activation function. Typically, the neuron has more than one input and can be mathematically modelled as shown in Figure 10.1.

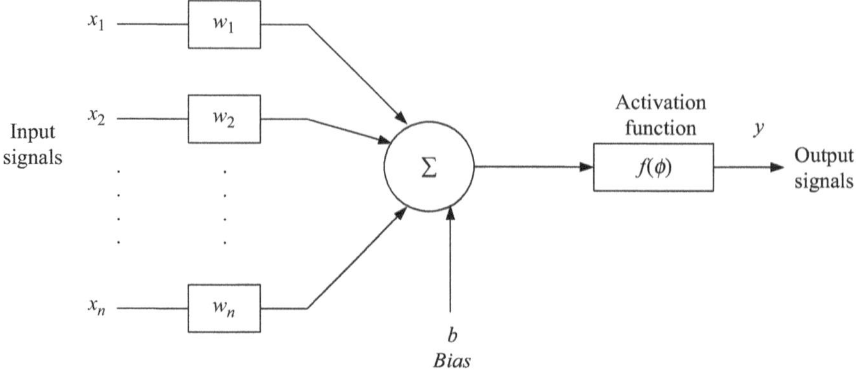

Figure 10.1 Simple model of an artificial neuron

The input signals, $x_1, x_2, \ldots, x_n$, are multiplied by weights $w_1, w_2, \ldots, w_n$ and added together to produce the net input to the activation function. The output signal of the neuron, y, can be expressed as

$$y = f\left(\sum_{k=1}^{n} w_k x_k + b\right) \tag{10.1}$$

It is worth mentioning that the weights are the most important coefficients in determining the output of the neural network. They are used to adjust the relative importance of the connections between the neurons, according to a modification rule.

It can be noted from (10.1) that the effect of the bias, b, is to increase or decrease the input to the activation function. The activation function is utilised to transform the activity level of the neuron into the output signal. Many activation functions such as a hard-limit, sigmoid, Gaussian and hyperbolic tangent functions have been used successfully to build neural networks [1–4]. The choice of the activation function relies on the applications where the neural network is used. The most common activation functions used in multi-layer networks are the sigmoid and hyperbolic tangent functions.

The outputs of sigmoid and hyperbolic tangent functions are described using (10.2) and (10.3), respectively.

$$f(x) = \frac{1}{1 + e^{-x}} \tag{10.2}$$

$$f(x) = \frac{e^x - e^{-x}}{e^x + e^{-x}} \tag{10.3}$$

10.2 Neural network topologies

The neurons themselves are not very powerful in terms of computation or representation. However, their interconnection allows one to encode relations between the variables and gives powerful processing capabilities. The way the neurons are connected within the neural network and the type of activation function used to construct the network yield to different network architectures. In general, three different types of network architectures can be identified.

10.2.1 Single-layer feed-forward architecture

A feed-forward network has a layered structure. A single-layer network (Figure 10.2) consists of multi-input and multi-output signals. The input signals are connected to each of the neurons in the network. The sum of the products of the weights and the inputs is calculated in each node. The input layer is not accounted for as no computation has taken place there.

Regardless of how many neurons the network has or what kind of activation function is chosen, the limitation of this type of network can only approximate a linear function. The common approach of approximating a non-linear function can be obtained by using a multi-layer perceptron.

10.2.2 Multi-layer feed-forward architecture

In this type of network, two or more single-layer networks are connected together to form one network. Each layer consists of neurons that receive their inputs from the neurons located in the layer directly before them and send their outputs to the neurons located in the subsequent layer. The layer whose output is the network output is called an output layer, while other layers are called hidden layers.

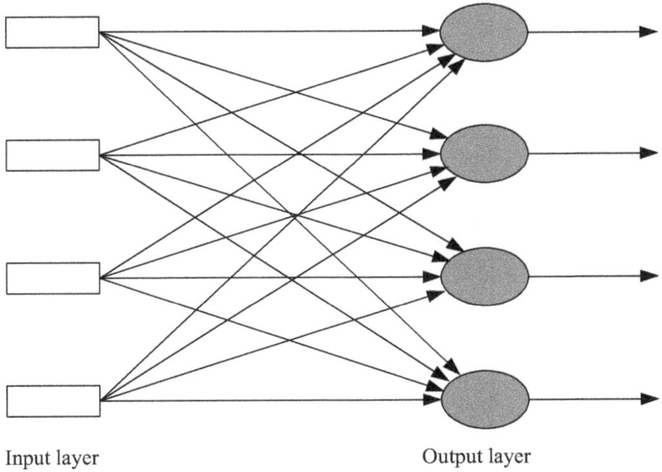

Figure 10.2 Single-layer feed-forward network

A multi-layer perceptron network 'often known as an MLP network' with one hidden and one output layer is shown in Figure 10.3.

10.2.3 Recurrent networks

A recurrent neural network (RNN) is another class of neural network that contains feedback connections between the outputs and inputs of the network. Using this kind of network structure will allow signal flow in both forward and backward directions, providing the network with a dynamic memory and is useful to mimic dynamic systems [5]. Compared with the MLP, RNN is more difficult to train due to the feedback connections. Configuration of an RNN is shown in Figure 10.4.

10.2.4 Back-propagation learning algorithm

As previously mentioned, by adjusting the weights of the neural network, the output of the network will be altered. The weight modifications can be achieved by applying a suitable learning algorithm, which leads the network to converge to the

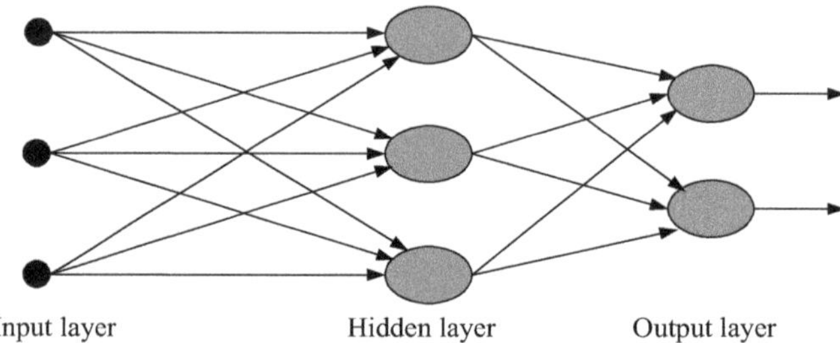

Figure 10.3 Multi-layer perceptron network

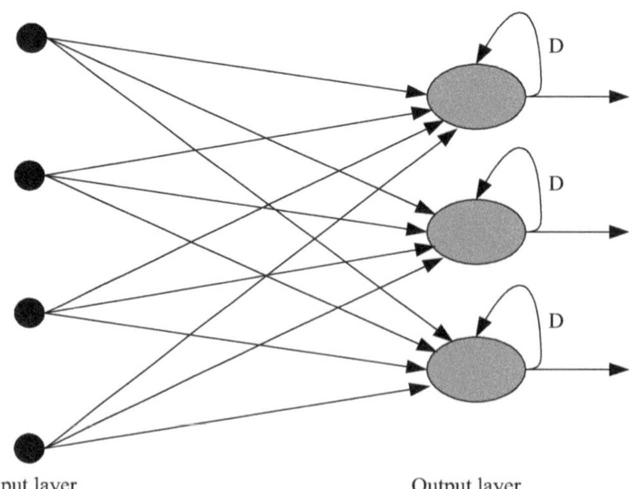

Figure 10.4 Recurrent neural network

desired value. Back-propagation is a learning algorithm, which has been widely utilised to train the multi-layer neural networks [6, 7]. The learning technique is based on the gradient method, which, using the derivatives of error, minimises the error between the actual and the desired outputs. By passing the derivative of the error from the output layer of the network back towards the input layer, the weights of the network can be adjusted in a suitable manner.

The error function E for a given training data set p can be described in the form of a well-known error function called the sum of squared errors:

$$E = \frac{1}{2} \sum_{p} \left(\sum_{k} (d_{pk} - y_{pk})^2 \right) \tag{10.4}$$

where d_{pk} is the desired output at instant time k for pattern p and y_{pk} is the actual output at instant time k for pattern p. The objective is to reduce the error function E to zero so that the output of the network is equal to the desired value. It is assumed that by minimising the error of each pattern individually, E will also be minimised. Therefore, notation p can be neglected assuming there is only one pattern to be considered.

A three-layer network (Figure 10.5) is considered as an example in order to explain the back-propagation algorithm. The output of neuron j 'located in the hidden layer' is given as

$$O_j = \varphi(\text{net}_j) \tag{10.5}$$

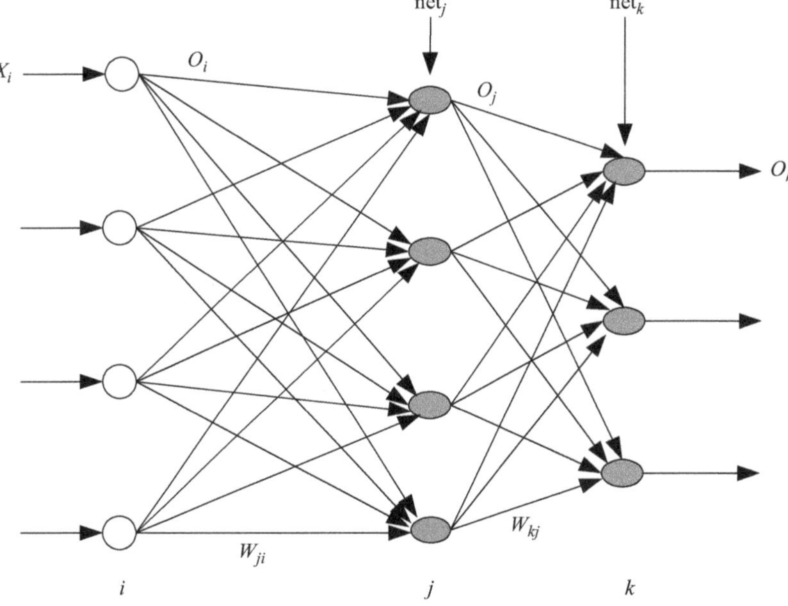

Figure 10.5 Three-layer neural network

where φ represents the activation function used for the output of neuron j. net_j is called the local field and it is given by

$$\text{net}_j = \sum_i W_{ji}\, O_i \tag{10.6}$$

where W_{ji} and O_i are the weight and the input signal to neuron j, respectively.

Similarly, the output of neuron k, which represents the output of the neural network, can be expressed as

$$O_k = \varphi(\text{net}_k) \tag{10.7}$$

The local field net_k is given by

$$\text{net}_k = \sum_j W_{kj}\, O_j \tag{10.8}$$

where W_{kj} is the weight related to neuron k and O_j is the output of neuron j.

The network (Figure 10.5) is capable of calculating the total error E for a given training set. Typically, the weights are the only parameters of the network that can be iteratively modified to make the error function as low as possible. The relation between the error function E and the weights of the network can be defined as a quadratic function as illustrated in Figure 10.6. If the slope is positive, the weights should be decreased by a small amount to lower the error. On the contrary, if the slope is negative, the weights of the network should be increased.

By applying the chain rule method, the partial derivative of the error function E with regard to the weights W_{kj}, leading to an output unit change, can be calculated as

$$\frac{\partial E}{\partial W_{kj}} = \frac{\partial E}{\partial O_k}\frac{\partial O_k}{\partial \text{net}_k}\frac{\partial \text{net}_k}{\partial W_{kj}} \tag{10.9}$$

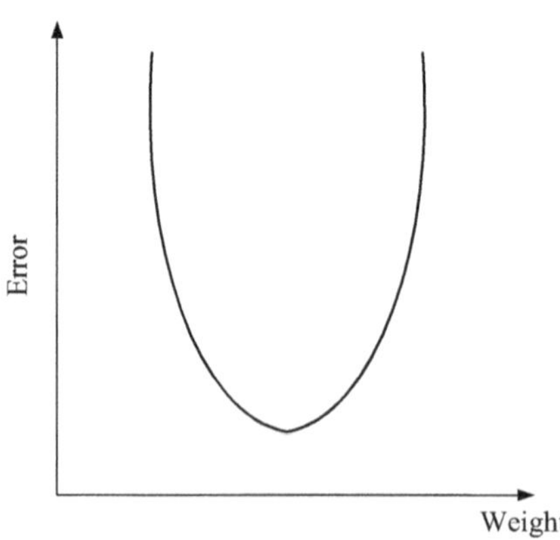

Figure 10.6 Error versus weights of network

The weights can be updated using the gradient descent method as follows:

$$W_{kj}(n+1) = W_{kj}(n) + \eta \frac{\partial E}{\partial W_{kj}} \tag{10.10}$$

where η is the learning rate of the back-propagation algorithm.

It can be noted from Figure 10.5 that changing W_{kj} will only affect the output of neuron k, while changing W_{ji} will affect the output of neurons j and k. Therefore, the change in E with regard to W_{ji} can be written as the sum of the changes to each of the output units. The adaptation of weights between hidden j, and input i, layers can be expressed as follows:

$$\frac{\partial E}{\partial W_{ji}} = \frac{\partial E}{\partial O_j} \frac{\partial O_j}{\partial \mathrm{net}_j} \frac{\partial \mathrm{net}_j}{\partial W_{ji}} \tag{10.11}$$

The term $\frac{\partial E}{\partial O_j}$ can be calculated as

$$\frac{\partial E}{\partial O_j} = \frac{\partial E}{\partial O_k} \frac{\partial O_k}{\partial \mathrm{net}_k} \frac{\partial \mathrm{net}_k}{\partial O_j} \tag{10.12}$$

Hence, the adaptation of the hidden layer weights can be written as

$$\frac{\partial E}{\partial W_{ji}} = \sum_k \frac{\partial E}{\partial O_k} \frac{\partial O_k}{\partial \mathrm{net}_k} \frac{\partial \mathrm{net}_k}{\partial O_j} \frac{\partial O_j}{\partial \mathrm{net}_j} \frac{\partial \mathrm{net}_j}{\partial W_{ji}} \tag{10.13}$$

$$W_{ji}(n+1) = W_{ji}(n) + \eta \frac{\partial E}{\partial W_{ji}} \tag{10.14}$$

The back-propagation algorithm requires a large number of training examples in order to provide an acceptable level of accuracy. It is also important to carefully select the learning rate to ensure the convergence of the network, as a large value of η might lead to network instability and a small value will cause a very slow convergence.

10.3 Fuzzy logic systems

In the real world, there are a lot of imprecise conditions that defy a simple, true or false statement as a description of their state. A computer system and its binary logic are incapable of adequately representing these vague (yet understandable) states and conditions. FL, which was developed in the mid-1960s by L.A. Zadeh, is a branch of mathematics that deals with vague and linguistic representations of data that mimic human understanding or intuition [8]. It expands the reach of traditional binary logic by allowing for the use of analog values as inputs and outputs in logic calculations. FL was developed based on the concept of Fuzzy Set Theory. It is considered a valuable tool, which can be used to solve highly complex problems where a mathematical model is too difficult or impossible to create.

The applications of FL can be found in many engineering and scientific works. FL has been successfully used in numerous applications, such as control systems engineering, image processing, power system engineering, industrial automation, robotics, consumer electronics, optimisation, medical diagnosis and treatment plans, as well as stock trading [9]. Regarding the power system, FL has also been useful in the application of parameter identification. A fuzzy identifier has been used to track the parameters of the power system and update the adaptive controller. Based on the knowledge of the plant, the input signals to the fuzzy system are fuzzified: a rule table is constructed and the output signals are finally defuzzified. Parameters of the fuzzy identifier are updated in real time by minimising a defined cost function using the gradient descent method. More description of fuzzy theory is presented in Section 10.3.1.

10.3.1 Fuzzy set theory

Fuzzy set can be defined by changing the usual definition of the characteristic function of a crisp set and introducing degree of membership. A fuzzy set A in a reference set X (called the universe of discourse) is defined by a mapping function (called the Membership Function, MF) that takes values in the range between 0 and 1, which can be mathematically written as $\mu_A: X \rightarrow [0, 1]$. The MF is a curve that defines how each point in the input space is mapped to a membership value between 0 and 1. The higher the membership X has in the fuzzy set A, the truer that X is A [10]. Many MFs, such as triangular, trapezoidal, bell and Gaussian, are used in FL; however, triangular and trapezoidal are the most common MFs. Unfortunately, there are no general rules or guidelines for selecting the appropriate shape of the MFs. The fact that trapezoidal and triangular shapes are the MFs most used in the literature is because they produce good results for most input variables in various applications.

FL has operators defined in a similar way to the classical Boolean logic. The *AND* operator can be evaluated, e.g. using min while the max operator represents the *OR* operator and the *NOT* is replaced by $1 - A$ in FL. Let X be a fuzzy set, and A and B two fuzzy sets with the MFs $\mu A(x)$ and $\mu B(x)$, respectively. Then the union, intersection and complement of fuzzy sets can be respectively defined as

$$\mu_A \cup \mu_B(x) = max(\mu_A(x), \mu_B(x)) \tag{10.15}$$

$$\mu_A \cap \mu_B(x) = min(\mu_A(x), \mu_B(x)) \tag{10.16}$$

$$\acute{\mu}_A(x) = 1 - \mu_A(x) \tag{10.17}$$

Some other definitions are also available in the literature. For instance, the intersection operator 'also known as the T-norm operator' could also be described as the algebraic product of two fuzzy sets:

$$\mu_A \cap \mu_B(x) = \mu_A(x) \cdot \mu_B(x) \tag{10.18}$$

The choice of the fuzzy operator 'in the end' depends on the expert knowledge and implementation feasibility.

10.3.2 Linguistic variables

As FL deals with events and situations with subjectively defined attributes, a proposition in FL does not have to be either true or false. For example, a room temperature can be described as cold, cool, comfortable, warm or hot, as opposed to only cold or hot. The descriptions mentioned previously are known as the linguistic variables in FL terminology. The range of possible values of a linguistic variable is called the universe of discourse. In the case of the temperature example, the universe of discourse can be within the interval [10°C, 35°C]. However, for simplicity, a common practice is to normalise or scale the values to be in the range of $[-1, +1]$ [9].

10.3.3 Fuzzy IF–THEN rules

A single fuzzy IF–THEN statement can be explained as follows:

$$IF(x \text{ is } A) \rightarrow THEN(y \text{ is } B)$$

where x and y are the input and output variables, respectively. A and B are the linguistic values defined by fuzzy sets on the ranges x and y. The *IF* part is called the antecedent while *THEN* part is called the consequent. The *IF–THEN* rule can be interpreted in such a way that if the antecedent is a fuzzy statement that is true to some degree of membership, then the consequent is also true to that same degree.

10.3.4 Structure of an FL system

The basic structure of an FL system is illustrated in Figure 10.7. It can be seen that in order to design an FL system, four steps are required to be considered:

(1) Fuzzification
Fuzzification is the process of mapping the input data into corresponding universes of discourses and converting the input into suitable linguistic values. The task of the fuzzification process can be summarised as follows:

- measure the values of input variables
- map the values of input variables to a corresponding universe of discourse
- convert the input data into appropriate linguist values

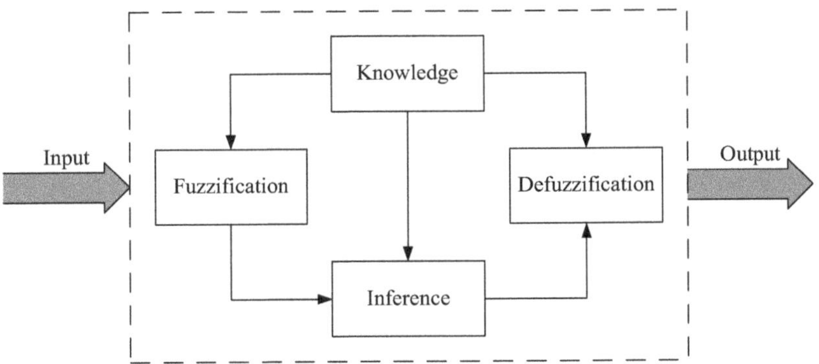

Figure 10.7 Structure of a fuzzy system

(2) Knowledge base
The purpose of the knowledge base or rule base step is to provide definitions that express the relation between the input and output fuzzy variables. It defines the control goals by means of a set of linguistic control rules. The rule base is often expressed in the form of *IF–THEN* rules.

(3) Fuzzy inference
The fuzzy inference is considered the core of any FL system. The fuzzy inference mechanism is a process by which the input values for each of the fuzzy variables in the antecedent are matched with all rules in the fuzzy rule base and an inferred fuzzy set is obtained. The membership values obtained in the fuzzification step are combined through a specific fuzzy operator to obtain the firing strength of each rule. Based on the firing strength, the consequent part of each qualified rule is produced.

There are two methods that are used in fuzzy inference known as Mamdani and Sugeno inference systems [9]. The difference between the two methods resides in the consequent part. Mamdani fuzzy inference expects the output MF to be fuzzy sets while Sugeno fuzzy inference method treats the consequent parts as either linear polynomials or constants in the form of single spikes. The output of each rule is weighted by the firing strength of the rule and the final output is the weighted average of all rule outputs.

(4) Defuzzification
The process by which a non-fuzzy (crisp) output is obtained from the fuzzy set is called defuzzification. Several defuzzification methods – centre of area (COA), mean of maximum (MOM), smallest of maximum (SOM) and largest of maximum (LOM) – are used in the defuzzification process of a fuzzy system. The two approaches most commonly used are the COA and MOM methods. The COA (also known as the centre of gravity, COG) calculates the centre of gravity of the distribution of the membership degrees under the curve. This method can be expressed in the discrete form as

$$COA = \frac{\sum_{k=1}^{n} x\mu_A(x)}{\sum_{k=1}^{n} \mu_A(x)} \tag{10.19}$$

where $\mu_A(x)$ is the MF of a fuzzy set A defined in the universe x and n is the number of quantisation levels of the output.

The MOM method calculates the output value by averaging only the part of the inferred fuzzy set whose MFs reach the maximum. The output of using this method can be described in the discrete form as

$$MOM = \sum_{i=1}^{l} \frac{x_i}{l} \tag{10.20}$$

where l is the number of elements, x_i, with membership equal to the maximum value. Figure 10.8 shows different types of defuzzification methods for a fuzzy function.

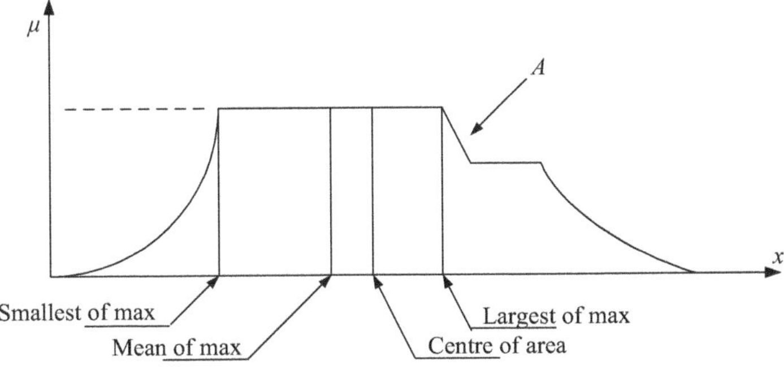

Figure 10.8 Defuzzification methods for a fuzzy function

10.4 Neuro-fuzzy systems

The neuro-fuzzy system is an AI approach, resulting from the merging of an FL system and a neural network structure. The basic idea of the integrated system (called neuro-fuzzy system) is to model an FL system by a neural network and apply the learning algorithms developed in the field of neural networks to adapt the parameters of the fuzzy system. The motive for combining FL with neutral networks is to take advantage of their strengths and overcome the shortcomings. In fact, FL and neural networks can be treated as complementary technologies. In a neuro-fuzzy system, the fuzzy system can be provided by an automatic tuning mechanism without altering its functionality. It has the advantage of tuning the rules of the fuzzy system using learning algorithms applied to neural networks. In return, the neural network can improve the transparency by having the rule-based fuzzy reasoning considered in its construction [11].

Many neuro-fuzzy network structures have been presented in the literature. Some of these networks are Fuzzy Adaptive Learning Control Network [12], Adaptive Neuro-Fuzzy Inference System (ANFIS) [13], Fuzzy Net (FUN) [14] and others [15]. However, one of the most well-known networks that has been reported in many publications and applied to various applications is the ANFIS.

Neuro-fuzzy control (NFC) has been widely used in many control system applications [16–20]. It represents a control approach where FL and ANNs are combined. An NFC can be defined as a multi-layer network that has the elements and functions of typical FL control systems, with additional capability to adjust its parameters through learning techniques [21].

A background about NFC and the online adaptation technique used to adjust the parameters of the NFC controller are described in Section 10.4.1.

10.4.1 Adaptive neuro-fuzzy inference system

ANFIS was first introduced by Takagi and Sugeno in 1985 and further developed by Jang [12]. The network is built to have the capability of ANNs in adapting and learning, together with the merit of approximate reasoning offered by FL.

Unlike neural networks, the weights of the connections between nodes located in one layer and the nodes in the subsequent layer are constant and have values of one.

There are two main ANFIS structures known as the first-order or zero-order Sugeno models. A typical rule in the first-order Sugeno model has the form of

$$IF\ input\ 1 = x_1\ and\ input\ 2 = x_2 \rightarrow THEN\ output\ is\ y = ax_1 + bx_2 + c$$

where $\{a,b,c\}$ is a parameter set.

For the zero-order Sugeno model, the output y is considered a constant and does not depend on the inputs to the network. The fuzzy *IF–THEN* rule for this type can be written as

$$IF\ input\ 1 = x_1\ and\ input\ 2 = x_2 \rightarrow THEN\ output\ is\ y = c$$

where $a = b = 0$.

The basic structure of a first-order ANFIS with two inputs and one output is depicted in Figure 10.9 [22].

The function of each layer can be described as follows:

- Layer 1: input membership layer

The first layer represents the MFs and contains adaptive nodes. The membership value specifying the degree to which an input value belongs to a fuzzy set is determined in this layer. The output of the nodes in this layer can be defined by

$$\left.\begin{aligned} O_{1,i} &= \mu_{Ai}(x_1) \quad \text{for } i = 1,2 \\ O_{1,i} &= \mu_{Bi-2}(x_2) \quad \text{for } i = 3,4 \end{aligned}\right\} \tag{10.21}$$

Assuming that the MFs are triangular functions, the output from node A_i can be given as

$$O_{1,i} = \mu_{Ai}(x_1) = \max\left(\min\left(\frac{x_1 - a}{b - a}, \frac{c - x_1}{c - b}\right), 0\right) \quad \text{for } i = 1,2 \tag{10.22}$$

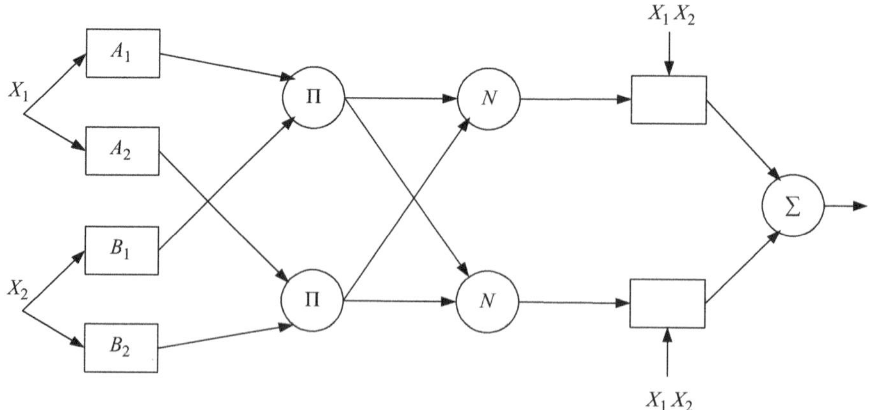

Figure 10.9 ANFIS structure with two inputs and one output

where x_1 is the input of node i, A_i is linguistic label associated with this node and $\{a, b, c\}$ is the parameter set of the triangular MF.

- Layer 2: firing strength layer

The output of each node in this layer is the product of all incoming signals and represents the firing strength of a rule. Each node in this layer is a fixed node, which performs the fuzzy *AND* operation using the algebraic product. The output of each node in this layer is given by

$$O_{2,i} = w_i = \mu_{Ai}(x_1)\mu_{Bi}(x_2) \quad \text{for } i = 1, 2 \tag{10.23}$$

- Layer 3: normalised firing strength layer

The nodes in this layer are fixed nodes and they calculate the normalised firing strength for each rule, which is given by

$$O_{3,i} = \overline{w}_i = \frac{w_i}{\sum_{i=1}^{n} w_i} \quad \text{for } i = 1, 2 \tag{10.24}$$

- Layer 4: consequent layer

The output of each node in this layer is adaptive and represents the weighted consequent part of the rule table. The output of each node can be expressed as

$$O_{4,i} = f_i = \overline{w}_i(p_i x_1 + q_i x_2 + r_i) \quad \text{for } i = 1, 2 \tag{10.25}$$

The parameter sets $\{p_i, q_i, r_i\}$ are called consequent parameters (CPs).

- Layer 5: defuzzification layer

This layer is the output layer and acts as a defuzzifier. The single node in this layer is a fixed node, which computes the overall output as the summation of all incoming signals. The output of this layer is given by

$$O_{5,i} = y = \sum_{i=1}^{n} f_i \tag{10.26}$$

Like any other neural network, a set of parameters in ANFIS is required to be updated in order for the network to be adaptive. These parameters are the MFs represented by Layer 1 and the CPs represented by Layer 4. The common adaptation technique is based on a gradient descent method [13].

10.4.2 Structure of the NFC

The structure of a typical NFC is shown in Figure 10.10. The two inputs to the controller are usually the error signal and change-of-error. The error signal $e(t)$ represents the difference between the actual output of the plant and a desired set-point while the change-of-error $\Delta e(t)$ is the difference between the error $e(t)$ and the previous error value $e(t-1)$. A negative sign of $e(t)$ means that the output of the

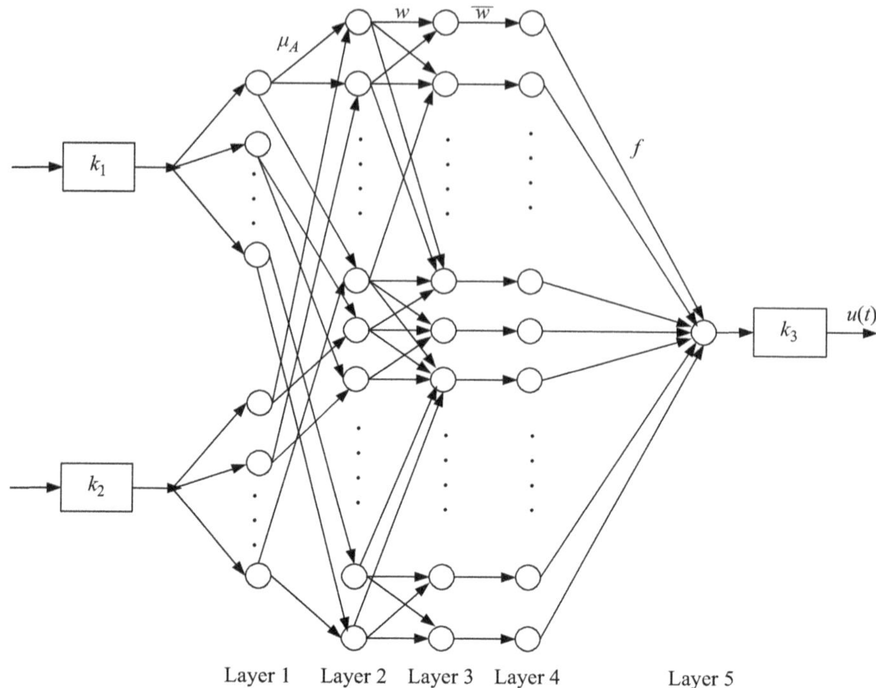

Figure 10.10 Architecture of an adaptive neuro-fuzzy controller

plant $y(t)$ has a value above the desired value y_d as $e(t) = y_d - y(t)$, while a positive sign of $e(t)$ indicates that output of the plant is below the desired value. Furthermore, a negative sign of $\Delta e(t)$ suggests that the plant output has increased when compared to its previous value $y(t-1)$ while a positive $\Delta e(t)$ means the opposite.

The input scaling factors, K_1 and K_2, are typically used to map the real input to the normalised input space in which the MFs are defined. In general, the normalisation range can be in the range of $[-1, +1]$ for the universe of discourse. It is noted that the input scaling factors influence the sensitivity of the NFC and affect its performance [9, 23, 24]. On the other hand, the output scaling factor K_3 is used to map the output of the fuzzy inference system to the real output. The value of K_3 should be appropriately selected so the output range of the NFC will not exceed a certain boundary, where a physical limitation is violated. It is clear that the output scaling factor has the most influence on system stability and oscillation tendency [9].

Considering there are seven MFs located in the first layer of the NFC and have triangular shapes, seven triangular MFs are associated with fuzzy linguistic sets used for each input to the controller. The input MFs of the NFC are depicted in Figure 10.11.

As shown in Figure 10.11, the centres of the MFs are distributed evenly along the normalised input space, which is a common technique used in most fuzzy control applications. Given that the peak value of a MF is equal to 1, the cross point

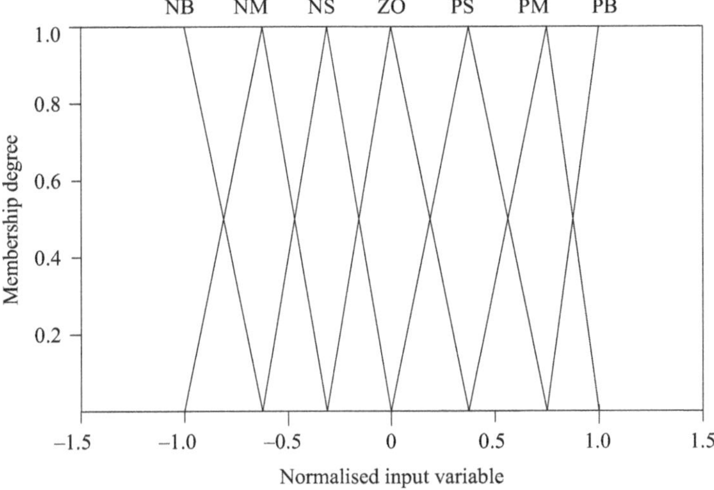

Figure 10.11 Triangular membership functions of the inputs

Table 10.1 Sugeno-type rule-base table with 49 rules

Δe	*e*						
	PB	**PM**	**PS**	**ZO**	**NS**	**NM**	**NB**
NB	0.0	−0.333	−0.666	−1.0	−1.0	−1.0	−1.0
NM	0.333	0.0	−0.333	−0.666	−1.0	−1.0	−1.0
NS	0.666	0.333	0.0	−0.333	−0.666	−0.666	−1.0
ZO	−1.0	0.666	0.333	0.0	−0.333	−0.666	−1.0
PS	−1.0	−1.0	0.666	0.333	0.0	−0.333	−0.666
PM	−1.0	−1.0	−1.0	0.666	0.333	0.0	−0.333
PB	−1.0	−1.0	−1.0	−1.0	0.666	0.333	0.0

PB, Big positive; PM, medium positive; PS, small positive; ZO, zero; NS, small negative; NM, medium negative; NB, Big negative.

between two MFs is at 0.5. The linguistic terms used for the MFs are big positive, medium positive, small positive, zero (ZO), small negative, medium negative and big negative. As the NFC has two inputs of which each has seven MFs, the rule-base table associated with the controller will contain 49 rules. The Sugeno-type rule-base table with 49 rules is summarised in Table 10.1.

10.4.3 Online adaptation technique

To make the NFC adaptive, two set of parameters are required to be adjusted. This can be achieved through network learning of which the MFs of linguistic terms and the CPs can be modified, using certain adaptation techniques [25]. One of these adaptation schemes is the back-propagation algorithm. The back-propagation algorithm can be employed to adjust the centres of the MFs and the CPs of the NFC. Given the cost function, described in (10.27), the centres of the MFs and CPs can be updated online using the gradient descent method [22, 26].

$$J_c(k) = \frac{1}{2} e_c[(k+1)]^2 \tag{10.27}$$

where $e_c(k+1)$ is the error signal between the estimated output, provided by the identifier, and the desired system output at time step $(k+1)$.

Assuming θ is an arbitrary parameter in the NFC, updating the MFs and CPs can be achieved using the gradient optimisation method given as [27]

$$\theta(k+1) = \theta(k) - \eta \frac{\partial J_c}{\partial \theta} \tag{10.28}$$

where

$$\nabla_\theta J_c(k) = \left[e_c(k+1) \frac{\partial e_c(k+1)}{\partial u(k)} \right] \left[\frac{\partial u(k)}{\partial \theta} \right] \tag{10.29}$$

$$\frac{\partial u(k)}{\partial \theta} = \sum_{O^* \epsilon S} \frac{\partial u(k)}{\partial O^*} \frac{\partial O^*}{\partial \theta} \tag{10.30}$$

where η and $u(k)$ are the learning rate and output of the NFC, respectively. S and O^* are the set of nodes whose outputs depend on θ and the output of nodes belonging to S, respectively.

For the output node, $\frac{\partial u(k)}{\partial O^*}$ is given as

$$\frac{\partial u(k)}{\partial O^*} = K_3 \tag{10.31}$$

For an internal node, $\frac{\partial u(k)}{\partial O^*}$ is given as

$$\frac{\partial u(k)}{\partial O_i^l} = K_3 \sum_{n=1}^{P} \frac{\partial u(k)}{\partial {O_n}^{l+1}} \frac{\partial {O_n}^{l+1}}{\partial O_i^l} \tag{10.32}$$

where O_i^l is the output of the ith node of the lth layer and P is the number of nodes in the $(l+1)$ layer. The updating of the membership centres and CPs of the NFC takes place in every sampling period.

The numbers of neurons (Figure 10.10) in the adaptive NFC are 14, 49, 49, 7 and 1 for layers L_1, L_2, L_3, L_4 and L_5, respectively. This means that 14 centre points in the MFs (seven for each input) and seven CPs need to be updated online. A network with this number of parameters to update is considered relatively complex and computationally expensive, especially for real-time applications [27].

The purpose of the adaptive controller is to be applied to an SVC device (explained in Chapters 13 and 14) in order to damp power system oscillations and enhance system stability. As the SVC is an electronic device, designed based on high-speed power electronic components, it is desirable to design a controller that possesses the characteristic of having a fast response time. Therefore, the objective is to design a simplified version of the ANFC, described in Section 10.4.2, that requires less computation time, and apply it to the SVC device. It is necessary for the simplified controller to provide similar performance as compared to the ANFC.

10.5 Adaptive simplified NFC

The essence of designing an NFC is to be able to transform an FL controller (FLC) and represent it in a neural network structure. This implies that it is imperative that the FLC is the main part taken into consideration when designing an NFC. If an FLC is constructed, an NFC can be designed and represented by a neural network, accordingly. Therefore, in order to develop a simplified version of an NFC, a simplified FLC is first required.

In general, it is always desirable to design the simplest control system that performs the expected task it is built for, as long as its accuracy is not compromised. In fact, in designing FLCs, interpretability and accuracy are the most important aspects to be considered [28].

Although, accuracy and interpretability represent contradictory objectives, an optimum fuzzy control system design should satisfy both criteria, to a certain degree. For instance, a complex FLC might successfully control a high-order non-linear system accurately; nonetheless, the drawback would be the difficulty in expressing the behaviour of the controller in an understandable way. On the contrary, a simple FLC can be easily understood but, its performance is not satisfactory. Therefore, a trade-off between the interpretability and accuracy should always be considered.

10.5.1 Simplification of the rule-base structure

There have been several publications proposing the design of simplified FL controllers using different simplification approaches. One of these techniques is reducing the size of the fuzzy rule tables [29–31]. The principle of designing the proposed adaptive simplified NFC (ASNFC) is based on the concept of reducing the fuzzy rule-base table (Table 10.1). It can be seen from Table 10.1 that the rule-base table can be viewed as a Toeplitz structure with zero diagonal line. Having such a structure provides the advantage of using the symmetrical property of the table to construct a one-dimensional fuzzy-rule table.

A new variable, called the signed distance, can be introduced to build a simplified FL controller (SFLC). This new variable represents the distance, d, to an actual state from the main diagonal line, called the switching line. The distance can be positive or negative, depending on the position of the actual state in the rule-base table, illustrated in Figure 10.12 [32].

As shown in Figure 10.12, a control action can be proportionally related to the perpendicular distance from any consequent in the table to the switching line. Three

Δe	e						
	PB	PM	PS	ZO	NS	NM	NB
NB	0.0	−0.333	−0.666	−1.0	−1.0	−1.0	−1.0
NM	0.333	0.0	−0.333	−0.666	−1.0	−1.0	−1.0
NS	0.666	0.333	0.0	−0.333	−0.666	−1.0	−1.0
ZO	1.0	0.666	0.333	0.0	−0.333	−0.666	−1.0
PS	1.0	1.0	0.666	0.333	0.0	−0.333	−0.666
PM	1.0	1.0	1.0	0.666	0.333	0.0	−0.333
PB	1.0	1.0	1.0	1.0	0.666	0.333	0.0

$-d$
Switching line
$+d$

Figure 10.12 The distance, d, between the switching line and actual state

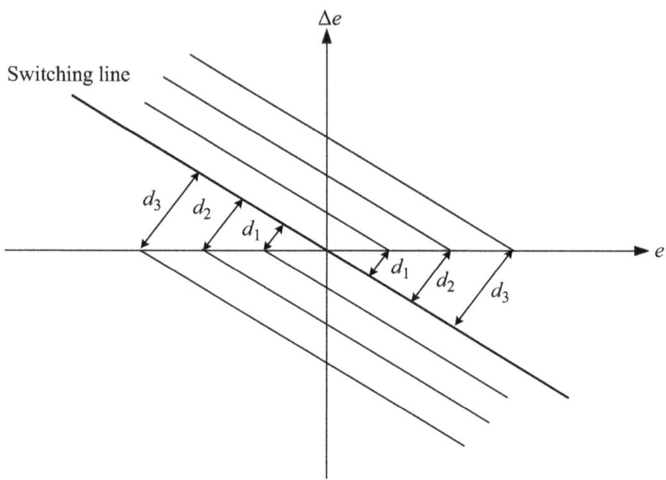

Figure 10.13 Illustration of d_1, d_2 and d_3 between the switching line and actual states in the phase-plane

distances, d_1, d_2 and d_3, can be obtained in both the upper half-plane and the lower half-plane. These distances will have negative signs if they are located in the upper half-plane and positive signs if they are in the lower half-plane. This can clearly be illustrated in Figure 10.13.

The switching line (Figure 10.13) can be represented by a general straight line equation given by

$$\mathrm{A}e + \mathrm{B}\Delta e + \mathrm{C} = 0 \tag{10.33}$$

where A, B and C are constants and are equal to $\mathrm{A} = \mathrm{B} = -1$, $\mathrm{C} = 0$.

The perpendicular distance between the switching line and a given point $P(e, \Delta e)$, located on the phase-plane, can be shown in Figure 10.14 and expressed as [33]:

$$d = \frac{|\mathrm{A}e + \mathrm{B}\Delta e + \mathrm{C}|}{\sqrt{\mathrm{A}^2 + \mathrm{B}^2}} \tag{10.34}$$

Substituting A, B and C into (10.34) yields

$$d = f_1(e, \Delta e) = \frac{|-(e + \Delta e)|}{\sqrt{2}} \tag{10.35}$$

Four symmetrical triangular MFs with 50 per cent overlap in the range of [0, +1] are chosen to be the input to the SFLC. This range is considered, as opposed to the range of [−1, +1] used in the typical FLC, as d_1 can be located in the upper half-plane or the lower half-plane (Figure 10.13), with negative or positive signs, respectively. Therefore, a calculation of only one distance from the switching line is required and a positive or negative sign can be associated, based on the location of the actual state in the phase-plane.

The values of d_1, d_2 and d_3 for the upper and lower half-plane represent the centres of the triangular MFs. The rule-base table is reduced to one dimension with the fuzzy linguistic terms, ZO, small (S), medium (M) and big (B) for the distance, *d*, and fuzzy singletons for the control signal. The control signal for any point in the phase-plane is given by [29]

$$u = K_3 S_{\mathrm{u}} u_{\mathrm{us}} \tag{10.36}$$

where K_3 is the output scaling factor, u_{us} is the unsigned control action and S_u is given by

$$S_{\mathrm{u}} = f_2(e, \Delta e) = \begin{cases} 1 & \text{when } e + \Delta e \geq 0 \\ -1 & \text{otherwise} \end{cases} \tag{10.37}$$

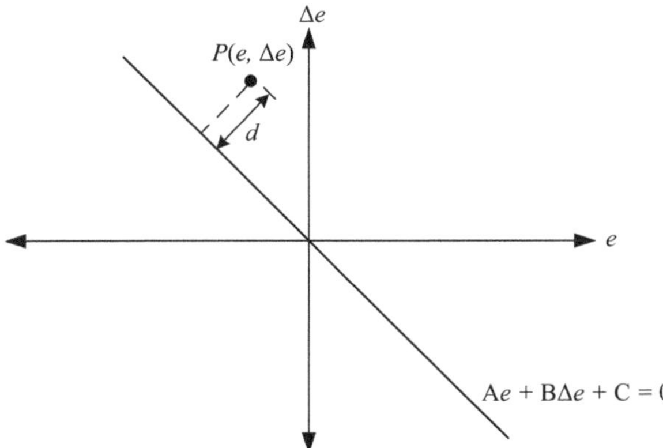

Figure 10.14 Perpendicular distance between point P(e, Δe) and the switching line in the phase-plane

Table 10.2 Reduced rule-base table

		d		
	ZO	***S***	***M***	***B***
u_{us}	0.0	0.33	0.66	1.0

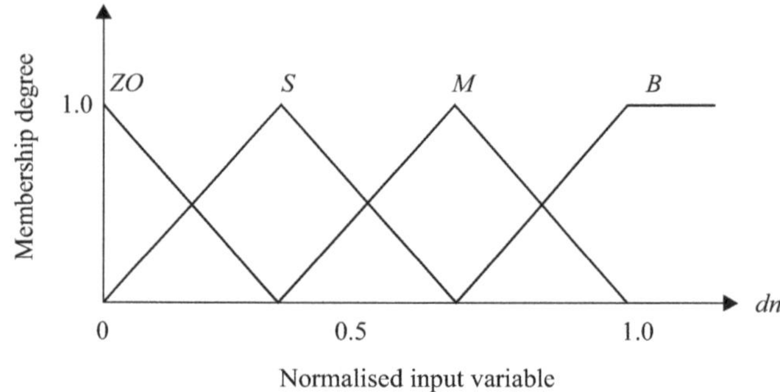

Figure 10.15 Input membership functions for the SFLC

The reduced rule-base table and input MFs are summarised in Table 10.2 and illustrated in Figure 10.15, respectively.

10.6 Control system design of the proposed ASNFC

The reduced rule-base table summarised in Table 10.2 is used for the design of the proposed ASNFC. As the objective is to design a simple NFC with a reduced number of parameters to update, a zero-order Sugeno-type fuzzy controller-based ANFIS is employed to construct the proposed controller. The overall structure of the proposed ASNFC is illustrated in Figure 10.16.

The ASNFC comprises an ANFIS network, with a reduced number of layers and nodes, and two function blocks, f_1 and f_2, given by (10.35) and (10.37), respectively. As shown in Figure 10.16, the input to the ANFIS network is the distance in the phase-plane, d, while the output is the unsigned control action, u_{us}. The cost function considered to update the centres of the MFs and the CPs of the proposed controller is defined in (10.38).

$$J(k) = \frac{1}{2} e_c(k+1)^2 = \frac{1}{2}\left[\Delta \hat{P}_{svc}(k+1) - \Delta P_d(k+1)\right]^2 \tag{10.38}$$

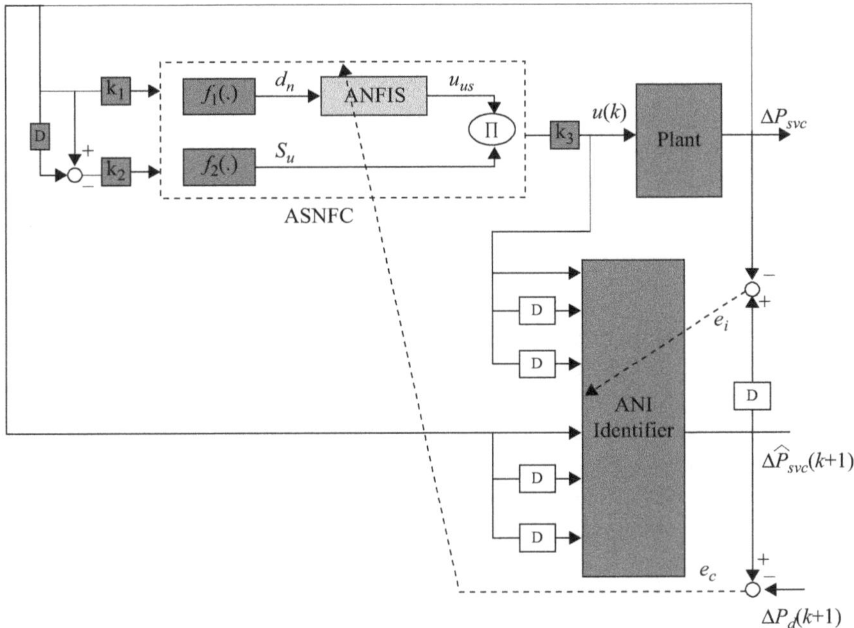

Figure 10.16 Overall control system structure of ASNFC

where $\Delta\hat{P}_{svc}(k+1)$ and $\Delta P_d(k+1)$ are the estimated power deviation and the desired value at time step $k+1$, respectively. As the desired value at time step $k+1$ is always zero, (10.38) can be written as

$$J(k) = \frac{1}{2}e_c(k+1)^2 = \frac{1}{2}\left[\Delta\hat{P}_{svc}(k+1)\right]^2 \tag{10.39}$$

The update of the MF centres and CPs is taking place at every sampling time, employing (10.28) and (10.29). Equation (10.29) can be expressed in terms of $\Delta\hat{P}_{svc}(k+1)$ and the control signal $u(k)$ as

$$\nabla_\theta J_c(k) = \left[\Delta\hat{P}_{svc}(k+1)\frac{\partial\Delta\hat{P}_{svc}(k+1)}{\partial u(k)}\right]\left[\frac{\partial u(k)}{\partial\theta}\right] \tag{10.40}$$

Three terms need to be calculated in (10.40). The terms $\Delta\hat{P}_{svc}(k+1)$ and $\frac{\partial\Delta\hat{P}_{svc}(k+1)}{\partial u(k)}$ are the estimated output and the Jacobian of the plant, respectively. They can be obtained from the neuro-identifier. The term $\frac{\partial u(k)}{\partial\theta}$ can be calculated using the back-propagation algorithm, as expressed in (10.30).

The new ANFIS network consists of four layers and has 4, 4, 4 and 1 neurons for layers 1, 2, 3 and 4, respectively. As the rule-base table is reduced to a one-dimensional rule-table instead of a two-dimensional rule-table, the second layer described in Section 10.4.1 will not be applicable in the proposed design. It is obvious that the number of layers and neurons of the new ANFIS network is significantly reduced, which yields to a simple type of ANFIS network. In addition, the control parameters required to be updated online are reduced from twenty-one (fourteen centre points in the MFs plus seven CPs) to eight (four centre points in the MFs plus four CPs). This will reduce the overall computation time of the controller.

References

1. Haykin S. *Neural Networks and Learning Machines*. 3rd edn. Upper Saddle River, NJ, US: Pearson Prentice Hall; 2008
2. Astrom K., Hugglund T. *Advanced PID Control*. Research Triangle Park, NC, US: International Society of Automation (ISA); 2006
3. Zhou Z., Shi W., Bao Y., Yang M. (eds.). 'A Gaussian function based chaotic neural network'. *Proceedings of the International Conference on Computer Application and System Modelling*, vol. 4; Taiyuan, China, Oct 2010. pp. 203–6
4. Karlik B., Olgac V. 'Performance analysis of various activation functions in generalised MLP architectures of neural networks'. *International Journal of Artificial Intelligence and Expert Systems*. 2010;**1**(4):111–22
5. Akpan V., Hassapis G. (eds.). 'Adaptive recurrent neural network training algorithm for nonlinear model identification using supervised learning'. *Proceedings of 2010 American Control Conference*; Baltimore, MD, US, Jun/Jul 2010. pp. 4937–42
6. Werbos P.J. 'Backpropagation through time: What I does and how to do it'. *Proceedings of IEEE*. 1990;**78**:1550–60
7. Werbos P.J. *Neural Network for Control*. Cambridge, MA, US: MIT Press; 1995
8. Zadeh L. 'Fuzzy sets'. *Information and Control*. 1965;**8**:338–53
9. Reznik L. *Fuzzy Controllers*. Oxford, UK: Newnes-Publishers; 1997
10. Wang L. *Adaptive Fuzzy Systems and Control: Design and Stability Analysis*. Upper Saddle River, NJ, US: Prentice Hall; 1994
11. Nurnberger A., Nauck D., Kruse R. 'Neuro-fuzzy control based on the NEFCON-model: recent developments'. *Soft Computing – A Fusion of Foundations, Methodologies and Applications*. 1999;**2**(4):168–82
12. Lin C.T., Lee C.G. 'Neural network based fuzzy logic control and decision system'. *IEEE Transactions on Computers*. 1991;**40**(12):1320–36
13. Jang J.-S.R. 'ANFIS: adaptive-network-based fuzzy inference system'. *IEEE Transactions on Systems, Man and Cybernetics*. 1993;**23**(3):665–85
14. Sulzberger S.M., Gurman N.N.T., Vestil S. (eds.). 'FUN: optimisation of fuzzy rule based systems using neural networks'. *IEEE International Conference on Neural Networks*; San Francisco, CA, US, Mar/Apr 1993. pp. 312–16
15. Abraham A. (eds.). 'Neuro-fuzzy systems: state-of-the-art modelling techniques'. *6th International Work Conference on Neural Networks (IWANN)*; Granada, Spain, Jun 2001. Germany: Springer Verlag; 2001. pp. 269–76

16. Barton Z. 'Robust control in a multi-machine power system using adaptive neuro-fuzzy stabilisers'. *IEE Proceedings on Generation, Transmission and Distribution.* 2004;**151**(2):261–7
17. Farrag M.E.A., Putrus G.A. 'Design of an adaptive neuro-fuzzy inference control system for the unified power-flow controller'. *IEEE Transactions on Power Delivery.* 2012;**27**(1):53–61
18. Munasinghe S.R., Kim M.S., Lee J.J. 'Adaptive neuro-fuzzy controller to regulate UTSG water level in nuclear power plants'. *IEEE Transactions on Nuclear Science.* 2005;**52**(1):421–9
19. Uddin M.N., Wen H. 'Development of a self-tuned neuro-fuzzy controller for induction motor drives'. *IEEE Transactions on Industry Applications.* 2007;**43**(4):1108–16
20. Wang J., Lee C. 'Self-adaptive recurrent neuro-fuzzy control of an autonomous underwater vehicle'. *IEEE Transactions on Robotic and Automation.* 2003;**19**(2):283–95
21. Jen Y. *Advanced Fuzzy System Design and Application*. Germany: Springer Publisher; 2003
22. Albakkar A., Malik O.P. (eds.). 'Intelligent FACTS controller based on ANFIS architecture'. *IEEE Power Engineering Society General Meeting*; Detroit, MI, US, Jul 2011. pp. 1–7
23. Jang P.R., Sun C., Mitzutani E. *Neuro-Fuzzy and Soft Computing – A computational Approach to Learning and Machine Intelligence:* Upper Saddle River, NJ, US: Prentice Hall; 1997
24. Abdelnour G.M., Chang C.H., Huang F.H., Cheung J.Y. 'Design of a fuzzy controller using input and output mapping factors'. *IEEE Transactions on Systems, Man and Cybernetics.* 1991;**21**(2):952–60
25. Ramirez-Gonzalez M., Malik O.P. 'Power system stabilizer design using an online adaptive neurofuzzy controller with adaptive input link weights'. *IEEE Transactions on Power Systems.* 2008;**23**(3):914–22
26. Lee S.J., Ouyang C.S. 'A neuro-fuzzy modeling with self-constructing rule generation and hybrid SVD-based learning'. *Transactions on Fuzzy Systems.* 2003;**11**(3):341–53
27. Yao W., Wen J.J., Wu Q.H. 'Wide-area damping controller for FACTS devices for inter-area oscillations considering communication time delays'. *IEEE Transactions on Power Systems.* 2014;**29**(1):318–29
28. Gacto M.J., Alcalá R., Herrera F. 'Interpretability of linguistic fuzzy rule-base systems: an overview of interpretability measures'. *Information Science – Applications*. 2011;**181**(20):4340–60
29. Ramirez-Gonzalez M., Malik O.P. (eds.). 'Simplified fuzzy logic controller and its application as a power system stabilizer'. *International Conference on Intelligent System Applications to Power Systems*; Curitiba, Nov 2009. pp. 1–6
30. Kaynak O., Jezernik K., Szeghegyi A. (eds.). 'Complexity reduction of rule based models: a survey'. *IEEE International Conference on Fuzzy Systems*; Honolulu, HI, US, 2002, vol. 2. pp. 1216–21

31. Viswanathan K., Oruganti R. 'Nonlinear function controller: a simple alternative to fuzzy logic controller for a power electronic converter'. *IEEE Transactions on Industrial Electronics.* 2005;**52**(5):1439–48
32. Choi B.J., Kwak S.W., Kim B.K. 'Design and stability analysis of single-input fuzzy logic controller'. *IEEE Transactions on Systems, Man and Cybernetics-Part-B Cybernetics.* 2000;**30**(2):303–9
33. Protter M.H., Protter P.E. *Calculus with Analytic Geometry.* Boston, MA, US: Jones and Bartlett Publishers; 1988

Chapter 11
Power system stabiliser

In an AC-interconnected power system, all synchronous generators rotate at the same speed, i.e. synchronous speed, under steady-state conditions. With all synchronous generators operating in synchronism, the power system is said to be stable. The ability of all generators on a system to maintain synchronism and to return to a stable operating point following a system disturbance leads to the concept of power system stability.

Power systems are often subjected to a variety of disturbances, such as sudden changes in load, short circuit faults on transmission lines and loss of transmission lines. This can result in the lack of balance between the mechanical input and the electrical output of a generating unit resulting in deviations in generator speed from the synchronous speed. This leads to individual generating units oscillating against each other. In other cases, particularly under heavily loaded conditions or loss of one or more transmission lines, the natural oscillation frequency of the system may not be adequately damped. In that case, even small disturbances, e.g. normal load fluctuations, can cause generating unit shaft oscillations of increasing magnitude resulting in angular instability.

In an interconnected power system, two distinct types of oscillations can exit simultaneously. In one type, called the local mode, a generator swings against the rest of the system with an oscillation frequency generally in the range of 0.8–2.0 Hz. In the other oscillation mode, called the inter-area mode, a number of generators in one part of the interconnected system (area 1) swing against machines in another part of the system (area 2) in a frequency range of 0.4–0.8 Hz. Depending on the system characteristics, there also can be a small overlap in the local and inter-area modes of oscillations.

Continuously acting automatic voltage regulators (AVRs) are employed on all synchronous generators. It is widely recognised that although AVRs are essential to maintain a proper voltage at the generator terminals and in the system, high gain fast-acting AVRs have the potential of introducing negative damping in the excitation control system [1, 2].

The local and inter-area modes of oscillations can be damped by introducing a supplementary signal through the synchronous generator excitation system. This was recognised in the 1950s. A lot of successful experience gained since then has shown that a supplementary control signal, properly derived from an appropriately selected feedback signal, acting through the AVR can significantly enhance damping of rotor oscillations. The device used to generate the supplementary control signal is called a power system stabiliser (PSS).

11.1 Conventional PSS

A schematic block diagram of the generator excitation system is shown in Figure 11.1. The AVR output is based on the voltage error difference between the generator terminal voltage reference set point and the actual terminal voltage magnitude. The large inductance of the field winding causes a delay in the machine flux and hence in the terminal voltage response. This delay, which can be interpreted as a phase lag, causes the un-damping effect and may cause oscillatory stability problems.

11.1.1 Configuration of common PSS

The negative damping effect of the high gain fast-acting AVRs and the time delay in the excitation circuit can be reduced by employing a PSS. The PSS output modulates the generator excitation so as to develop a torque in phase with the rotor speed deviations and adds damping to the characteristic electromechanical oscillations [3, 4]. Innumerable studies and tests on all types of utility-scale generators have proven that power system stability can be improved even beyond the classical steady-state stability limit and overall damping increased by using a properly tuned and tested PSS.

The most commonly used PSS (CPSS) provides phase compensation for the phase difference between the AVR input and the generator shaft speed (Figure 11.1) through adjustable lead-lag compensation functions (T1–T4) (Figure 11.2) over a dynamic frequency of interest, usually 0.4–2.0 Hz. The PSS gain, Ks, is determined to be the highest within the constraints of the PSS control loop stability. The high-frequency filters allow for the suppression of potentially unstable torsional oscillations or other sources of torsional noise. The wash out filter is a high-pass filter to remove any DC signals. It generally has a long time constant (T_w) of 5–10 s. It is common to have an output limiter to limit the PSS output from overwhelming the AVR forcing during transient conditions.

Commonly used PSS input signals are change in generator shaft speed $\Delta\omega$, electrical frequency deviation Δf, variation in electrical power output ΔP_e and

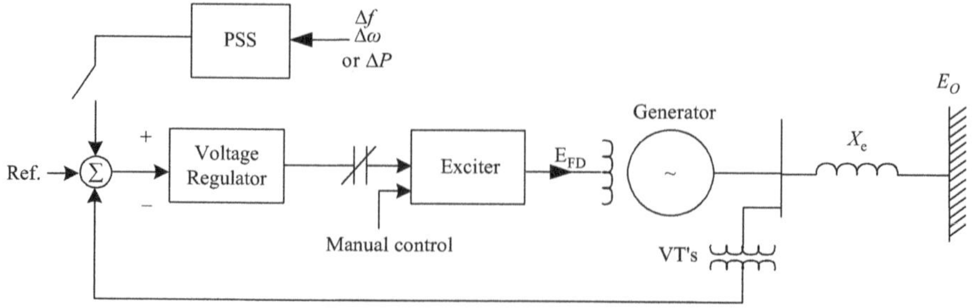

Figure 11.1 Generator excitation system block diagram

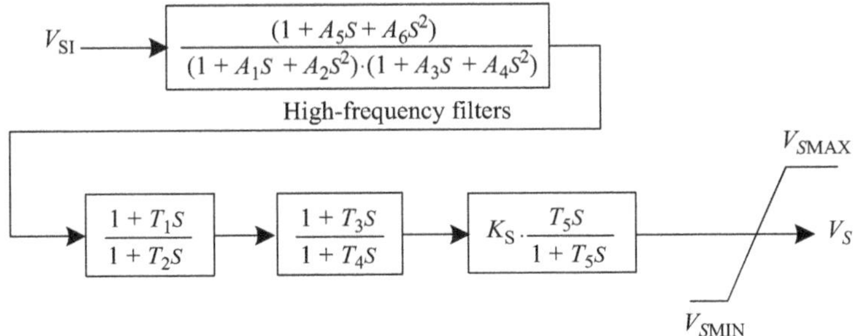

Figure 11.2 Power system stabiliser structure

accelerating power. In some cases even a combination of these is used. Depending on the feedback signal used, alternative forms of PSS have been developed [5].

11.1.2 PSS input signals

The early designs of PSSs employed a direct measurement of shaft speed [2]. As this requires the use of a torsional filter to attenuate torsional components, an additional phase lag is introduced that limits the allowable stabiliser gain. Placement of the speed pick-up transducers on the generating unit shaft requires special care so that undesirable frequencies are not picked up [6].

Another input signal that has been employed successfully is the terminal frequency. Frequency has been used directly. Also, in some cases the terminal voltage and current inputs were combined to generate a signal to approximate the generator shaft speed, called as 'compensated' frequency.

Frequency signal is more sensitive to inter-area mode of oscillations than the local mode. It can thus provide better damping of the inter-area mode [3]. Frequency signal also needs to be filtered for torsional components. In addition, changes in power system configuration or noise caused by large industrial loads may produce large frequency transients that can affect the generator field voltage [7].

It is simple to measure electric power, and it can be related to the generator speed through the torque equation:

$$\text{Accelerating torque} = \text{Input (mechanical) torque} - \text{Output (electrical) torque} \tag{11.1}$$

It can be written as

$$\frac{2H}{\omega}\frac{\mathrm{d}^2\delta}{\mathrm{d}t^2} = T_m - T_e \tag{11.2}$$

where

H is the inertia constant (s)
ω is the frequency (elec. rad/s)
T_m is the mechanical input torque (N-m) and
T_e is the electrical output torque (N-m)

Considering that the mechanical time constant is much larger than the electrical time constant, ignoring the variations in the mechanical torque, the shaft acceleration (which leads the speed by 90°) may be considered as a scaled version of electrical power. The stabilising signal, derived from the deviations in electrical power in combination with high-pass and low-pass filters, can thus provide pure damping torque. This power-based PSS has been used as the basis for a number of PSSs. Such a PSS can provide pure damping at only one frequency, and also, unwanted PSS output is produced whenever mechanical power changes. This can put severe limits on the gain and output of the PSS.

With a number of limitations manifest in deploying PSSs with any of the speed, frequency or power signals as input, efforts were made to directly measure the accelerating power of the generator [8–10]. As these methods involved significant complexity in the design, an indirect method of deriving the accelerating power, PSS2A of [5], was developed. The principle of this PSS is based on deriving the integral-of-accelerating power signal from shaft speed and electrical power signals by integrating and manipulating (11.1) into the form

$$\int \frac{\Delta P_a}{2H} \mathrm{d}t \rightarrow -\frac{\Delta P_e(s)}{2Hs} + G(s)\left[\frac{\Delta P_e(s)}{2Hs} + \Delta\omega\right] \tag{11.3}$$

where

P_a is the accelerating power
P_e is the electrical power and
$G(s)$ is the transfer function of a low-pass filter

The two input signals pass through high-pass filters and are processed in individual channels before being added to form one input signal to the stabiliser gain and lead/lag stage. This PSS does not require a torsional filter in the path involving the electrical power signal. However, as it employs two inputs, speed and active power, it is sensitive to the relationship between the two inputs and, therefore, is critical to match the two signal paths in terms of gain and filter constants.

11.1.3 Characteristics of common PSS

Successful experience has been gained with the injection of a supplementary feedback signal through the generator excitation system to enhance damping of generator rotor oscillations. As described in Section 11.1.2, various input signals have been used as input to a PSS that, in general, consists of a second-order phase lead/lag network with a gain. For proper damping action, appropriate PSS settings are determined by adjusting the lead, lag and gain of the stabiliser. The conventional PSS, adopted by most electric utilities, is designed offline using linear control theory and is based on a model of the power system with a fixed configuration linearised for one operating condition. It is simple in structure, has flexibility and is easy to implement. It has made significant contribution in enhancing the quality of electrical supply.

Power systems, in general, are complex and non-linear systems. Their parameters not only depend on the operating condition, but both their configuration and parameters can change with time. This may create discrepancies between the mathematical model and the physical conditions. Therefore, with the conventional linear control theory-based PSS it is difficult to realise the desired control performance over wide operating conditions of the power plant. To further improve the performance and stability of the power system, various other approaches using linear quadratic (LQ) optimal control, H-infinity, variable structure, rule-based and artificial intelligence (AI) technologies [3, 11–18] have been proposed in the literature to design a fixed parameter PSS. One common feature of all fixed parameter controllers is that the design is done offline. To yield satisfactory control performance, it is desirable to develop a controller that considers the non-linear nature of the plant and has the ability to adjust its parameters online according to the environment in which it is working, i.e. track the plant operating conditions.

The conventional stabiliser parameters have to be designed for each application. Its parameters, once designed, tuned and implemented, are fixed. They can be set to contribute optimal damping at only one oscillation frequency. A power system is subject to multi-modes of oscillations. As the parameters of a conventional PSS are tuned for one set of operating conditions, the selected parameters are a compromise between the local and inter-area mode oscillations. Therefore, the fixed parameter PSS generally cannot maintain the same quality of performance under all conditions of operation.

11.2 Adaptive control-based PSS

The common procedure in process control is to compare the actual measured values of the output with the desired values and the difference, the error, is fed as input to the process through a regulator and an actuator. Various criteria are available for the computation of the control to minimise the error. Using this technique, the desired control law is obtained as

$$u(t) = f[\boldsymbol{\theta}_s(t), \boldsymbol{y}(t), \boldsymbol{u}(t-T)] \tag{11.4}$$

where

$\boldsymbol{\theta}_s(t)$ is the system parameter vector
$\boldsymbol{y}(t)$ is the output vector $[y(t)\ y(t-T)\ldots]^t$
$\boldsymbol{u}(t-T)$ is the control vector $[u(t-T)\ \ u(t-2T)\ldots]^t$
t superscript denotes the transpose
T is the sampling period and
$f[.]$ denotes function

If the parameter vector is known, control to meet specific performance criterion can be computed directly. However, the dynamics of a complex non-linear system vary with time depending on the operating conditions, disturbances and so on.

An adaptive controller has the ability to modify its behaviour depending on the performance of the closed-loop system. The basic functions of the adaptive controller may be described as

- identification of unknown parameters, or measurement of a performance index
- decision of the control strategy
- online modification of the controller parameters

Depending on how these functions are synthesised, different types of adaptive controllers are obtained. Various adaptive control techniques have been proposed for excitation control since the mid-1970s. A brief review of the adaptive control techniques from the excitation control aspect is presented in this section.

Two distinct approaches – direct adaptive control and indirect adaptive control – can be used to control a plant adaptively. In the direct control, the parameters of the controller are directly adjusted to reduce some norm of the output error. In the indirect control, the parameters of the plant are estimated as the elements of a vector at any instant *k*, and the parameters' vector of the controller is adapted based on the estimated plant vector.

11.2.1 Direct adaptive control

A very common form of direct adaptive control is the model reference adaptive control (MRAC). The objective of an MRAC system is to update the controller parameters such that the closed-loop system maintains a performance specified by a reference model. It requires a suitable model, an adaptive mechanism and a controller.

The structure of a MRAC system is shown in Figure 11.3. In MRAC, the actual system performance is measured against a desired closed-loop performance specified by a reference model that is driven by the same input as the controlled system. The objective is to minimise the error, the difference between the actual system output and the reference model output. The 'adaptation mechanism' block in Figure 11.3 is used to update the parameters of the controller. Various methods are available to minimise the error function.

The most important feature in ensuring the success of MRAC is the selection of a proper reference model and its parameters. The selected parameters must be

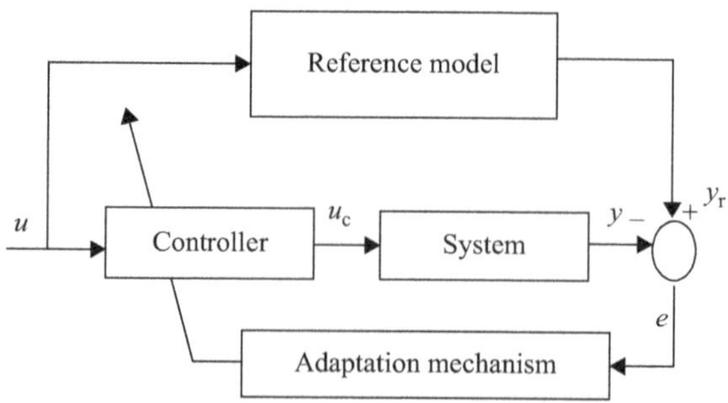

Figure 11.3 MRAC structure

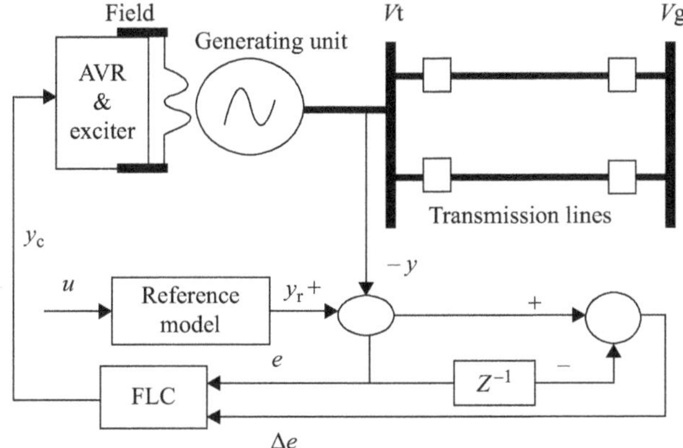

Figure 11.4 System configuration with MRAC-based APSS

such that the system is capable of following the reference model output and that the control signal remains within the physical control limits. A systematic method to determine a proper reference model for the plant is described in [19].

Application of an adaptive PSS (APSS) based on the MRAC principle is shown in Figure 11.4. A fuzzy logic controller (FLC) with self-learning capability is used to adapt the system performance to track the reference model. Two inputs, generator speed deviation and its derivative, and the supplementary control output, each have seven membership functions. The FLC uses the Mamdani-type fuzzy proportional derivative (PD) rule base [20]. Updating the centre points of the controller input membership functions, i.e. the weights of the fuzzy controller, using the steepest descent algorithm provides it with a self-learning capability. It can thus adapt the system performance to track the reference model.

Results of a number of studies show that this APSS provides good damping over a wide operating range and improves the performance of the system. An illustrative example showing the system response to a three phase to ground fault at the middle of one transmission line and successful reclosure with the self-learning MRAC-based FLC and a fixed centre FLC is given in Figure 11.5.

11.2.2 Indirect adaptive control

A general configuration of the indirect adaptive control as a self-tuning controller is shown in Figure 11.6. At each sampling instant, the input and output of the generating unit are sampled and a plant model to represent the dynamic behaviour of the generating unit at that instant in time is obtained by some online identification algorithm. It is expected that the model obtained at each sampling instant can track the system operating conditions.

The required control signal is computed based on the identified model. Various control techniques can be used to compute the control. All control algorithms assume that the identified model is the true mathematical description of the controlled system.

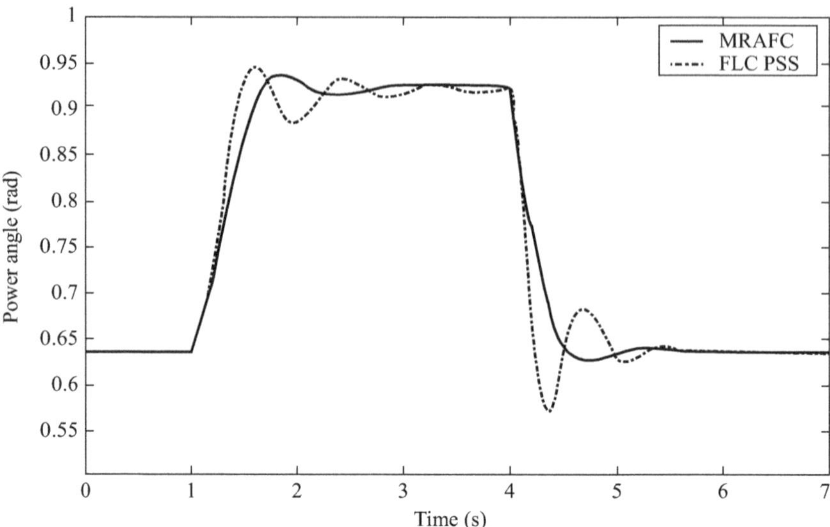

Figure 11.5 Three-phase to ground fault with APSS (MRAFC) and fixed FLC PSS ($P = 0.95$ pu, 0.9 pf lag)

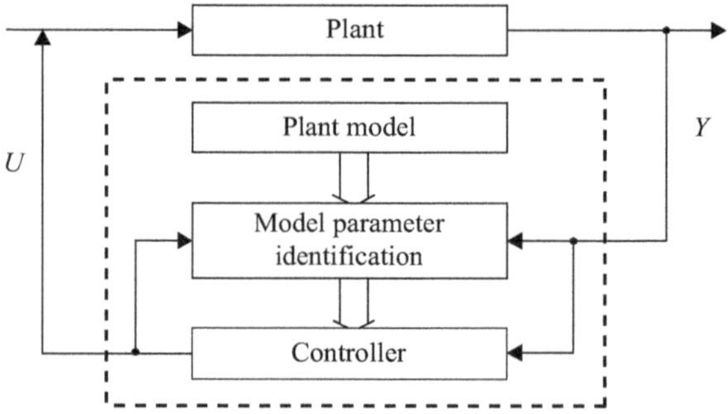

Figure 11.6 Block diagram of a self-tuning controller

In the analytical approach to the design of an adaptive controller, sampled data design techniques are used to compute the control. The indirect adaptive control procedure involves:

- Selection of a sampling frequency, f_s, about ten times the normal frequency of oscillation to be damped.
- Updating of the system model parameters (coefficients of system transfer function in the z-domain) each sampling interval T (=1/fs) using an identification technique suitable for real-time application. A number of identification

routines, in recursive form, e.g. recursive least squares (RLS), recursive extended least squares (RELS), can be used to determine the transfer function of the controlled plant in the discrete domain.

- Use the updated estimates of the parameters to compute the control output based on the control strategy chosen. Various control strategies, among them optimal, minimum variance (MV), pole-zero assignment, pole assignment and pole shift (PS), have been proposed.

11.2.2.1 System model

The generating unit is described by a discrete ARMAX model of the form

$$A(z^{-1})y(t) = B(z^{-1})u(t) + e(t) \tag{11.5}$$

where

$A(z^{-1})$ and $B(z^{-1})$ polynomials in the delay operator z^{-1} are of the form

$$A(z^{-1}) = 1 + a_1 z^{-1} + \cdots + a_i z^{-i} + \cdots + a_{n_a} z^{-n_a} \tag{11.6}$$

$$B(z^{-1}) = b_1 z^{-1} + \cdots + b_i z^{-i} + \cdots + b_{n_b} z^{-n_b} \tag{11.7}$$

$n_a \geq n_b$

The variables $y(t)$ and $u(t)$ are the system output and system input, respectively, and $e(t)$ is assumed to be a sequence of independent random variables with zero mean.

11.2.2.2 System parameter estimation

The control is computed based on the identified model parameters, a_i and b_i. Thus, to compute the control appropriate to the varying conditions the system parameters have to be estimated online. The correctness of the identification determines the preciseness of the identified model that tries to reflect the true system. For a time-varying system the tracking ability of the identification method is very important.

An online estimate of the system parameters is obtained by providing in the regulator a mathematical model having a desired structure describing the actual process. Such a model may be expressed as

$$\hat{y}(t) = g[\theta_m, \xi(t)] \tag{11.8}$$

where

$\hat{y}(t)$ is the predicted (estimated) value of the system output
θ_m is the model parameter vector and
$\xi(t)$ is the information known at the time of prediction

The model parameter vector may either be constant, θ_m, or be a function of time, $\theta_m(t)$. For the model to track the system dynamics, i.e. tune itself to the system, its parameters must be updated continuously at an interval that is consistent with the time constants of the system.

Several methods can be used to obtain an estimate for the model parameter vector, $\theta_m(t)$ [21]. A commonly used technique of achieving a continuous tracking of the system behaviour is the RLS parameter estimation technique. It minimises the square of the error between the actual system output and the model output, and the estimated parameter vector $\hat{\theta}_m(t)$ is given by

$$\hat{\theta}_m(t) = h\left[\hat{\theta}_m(t-T), P(t), \xi(t)\right] \tag{11.9}$$

where

$P(t)$ is the covariance matrix of the error of estimates. In general terms it contains the entire history of the process.

To enhance the ability of the identifier to track the operating conditions of the actual system, a forgetting factor is used to discount the importance of the older data. It can be chosen as a constant or a variable. A variable forgetting factor, employed to improve the tracking ability especially under large disturbances, is calculated online every sampling interval [22].

11.2.3 Indirect adaptive control strategies

Four control strategies that need explicit clarification are described below.

11.2.3.1 LQ control

In the LQ control algorithm the objective is to minimise a performance index [23]. The performance is chosen so that the system output error is minimised with respect to the system input. The LQ controller has the advantage that it will always result in a stable closed-loop system provided that the parameter estimates are exact. However, the achievement of this characteristic imposes heavy computational burden because it requires the solution of a matrix Riccati equation. Also, this controller is designed in the state space form and a common identification technique estimates the system parameters in the input/output form. Thus, an observer is required to convert the system parameters into a canonical form.

11.2.3.2 MV control

In this control strategy, the objective is to minimise the variance of the output [24]. Output error at the next sampling instant for zero control is predicted first. The control that will drive this predicted error to zero is then computed. Although this control strategy has nice properties, it has characteristics that make it difficult to use for excitation control.

In this strategy, the controller poles are obtained directly from the identified system zeros. The closed-loop system will be unstable if the dynamics of the sampled system are non-minimum phase, i.e. the system has a zero on or outside the unit circle in the z-domain. This might cause an unstable control computation if identified zeros are not cancelled exactly with the system zeros. When the cancellation of large parameter errors is not possible within one sample due to the limits on the control signal, the MV controller will produce an oscillatory response. The excitation signal is band limited, and the use of MV controller will result in excessive control and a poor control action. These problems associated with the MV controller can be avoided by using a pole-zero or pole-assigned (PA) controller.

11.2.3.3 Pole-zero and PA control

In the pole-zero assignment (PZA) controller the poles and zeros in the closed-loop are pre-specified by the designer [25]. Whereas, in the MV case all poles are shifted towards the centre of the unit circle in the z-domain, poles and zeros in the PZA case are shifted to locations that produce the desired closed-loop characteristics. This permits a trade-off between performance and control effort. Although this controller does not suffer from the problems of non-minimum phase and band limited output associated with the MV controller, the designer has to know the system characteristics to achieve the desired characteristics. In this respect, this algorithm can be compared to MRAC.

Pre-selection of the locations of poles and zeros is difficult for non-deterministic case and their poor choice may lead to unstable control computations.

In the PA controller only poles, instead of both poles and zeros, are assigned [26]. Otherwise, it is exactly the same as the PZA controller.

11.2.3.4 PS control

The PS controller is in essence the PA controller, but the closed-loop poles are obtained by shifting the open-loop poles radially towards the centre of the unit circle in the z-domain. Shifting the poles towards the centre is directly related to increased damping. This approach has the advantage of producing a stable controller. Detailed description of the PS control algorithm and its application as an APSS is given in Section 11.3.

11.3 PS control-based APSS

Extensive amount of work has been done to develop and implement an APSS based on the PS strategy. Such a PSS can adjust its parameters online according to the environment in which it works and can provide good damping over a wide range of operating conditions of the power system.

11.3.1 Self-adjusting PS control strategy

In the PS control strategy, in closed-loop (with PSS) the poles of the controlled system are shifted from their open-loop (without PSS) locations towards the centre in the z-plane by a factor less than one. This factor, called the 'pole shifting factor', is varied online to always produce maximum damping contribution without exceeding the control limits. To determine the desired control, such a system may be modelled by a linear low-order discrete model with time-varying parameters.

The parameters of the system model of a given structure, estimated as in Section 11.2.2.2, are used in the control algorithm to compute the updated control. A block diagram of the regulator is shown in Figure 11.6. Because the control is based on the estimated model parameter vector, $\hat{\theta}_m(t)$, (11.4) now becomes

$$u(t) = f\left[\hat{\theta}_m(t), y(t), U(t-T)\right] \tag{11.10}$$

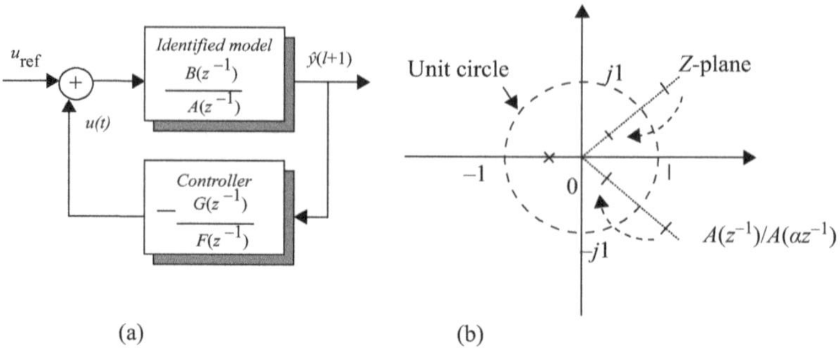

Figure 11.7 (a) Closed-loop system block diagram and (b) pole-shifting process

For the system modelled by (11.5), assume that the feedback loop has the form (cf. Figure 11.7(a))

$$\frac{u(t)}{y(t)} = -\frac{G(z^{-1})}{F(z^{-1})} \tag{11.11}$$

From (11.5) and (11.11) the closed loop characteristic polynomial $T(z^{-1})$ can be derived as

$$A(z^{-1})F(z^{-1}) + B(z^{-1})G(z^{-1}) = T(z^{-1}) \tag{11.12}$$

Unlike the pole-assignment algorithm in which $T(z^{-1})$ is prescribed [26], the PS algorithm makes $T(z^{-1})$ take the form of $A(z^{-1})$ but the pole locations are shifted by a factor α, i.e.

$$T(z^{-1}) = A(\alpha z^{-1}) \tag{11.13}$$

In the PS algorithm, α, a scalar, is the only parameter to be determined and its value reflects the stability of the closed-loop system. Supposing λ is the absolute value of the largest characteristic root of $A(z^{-1})$, then $\alpha\lambda$ is the largest characteristic root of $T(z^{-1})$. To guarantee the stability of the closed-loop system, α ought to satisfy the following inequality (stability constraint):

$$-\frac{1}{\lambda} < \alpha > \frac{1}{\lambda} \tag{11.14}$$

The PS process is presented schematically in Figure 11.7(b). It can be seen that once $T(z^{-1})$ is specified, $F(z^{-1})$ and $G(z^{-1})$ can be determined by (11.12), and thus the control signal $u(t)$ can be calculated from (11.11).

To consider the time domain performance of the controlled system, a performance index J is formed to measure the difference between the predicted system output, $\hat{y}(t+1)$, and its reference, $y_r(t+1)$:

$$J = E[\hat{y}(t+1) - y_r(t+1)]^2 \tag{11.15}$$

E is the expectation operator. $\hat{y}(t+1)$ is determined by system parameter polynomials $A(z^{-1})$, $B(z^{-1})$ and past $y(t)$ and $u(t)$ signal sequences. Considering that $u(t)$ is a function of the pole-shifting factor α, the performance index J becomes

$$\min \quad \alpha J = f\left[A(z^{-1}), B(z^{-1}), u(t), y(t), \alpha, y_r(t+1)\right] \tag{11.16}$$

The pole-shifting factor α is the only unknown variable in (11.16) and thus can be determined by minimising J.

Constraints:

When minimising $J(t+1,\ \alpha)$, it should be noted that α will be subject to the following constraints:

- The stabiliser must keep the closed-loop system stable. It implies that all roots of the closed-loop characteristic polynomial $A(z^{-1})$ must lie within the unit circle in the z-plane (cf. (11.14)).
- The control limit should be taken into account in the stabiliser design to avoid servo saturation or equipment damage. The optimal solution of α should also satisfy the following inequality (control constraint):

$$u_{\min} \leq u(t, \alpha) \leq u_{\max} \tag{11.17}$$

Pole patterns of $T(z^{-1})$ for a 50-ms three-phase to ground fault at the middle of one line of a double-circuit transmission line connecting a generator to a constant voltage bus (Figure 11.8) are shown in Figure 11.9. The pole pattern before the application of control is shown in Figure 11.9(a). As two poles map outside the unit circle, the closed-loop system is in an unstable state. The pole pattern after the PS control is applied is shown in Figure 11.9(b). As all the poles lie within the unit circle,

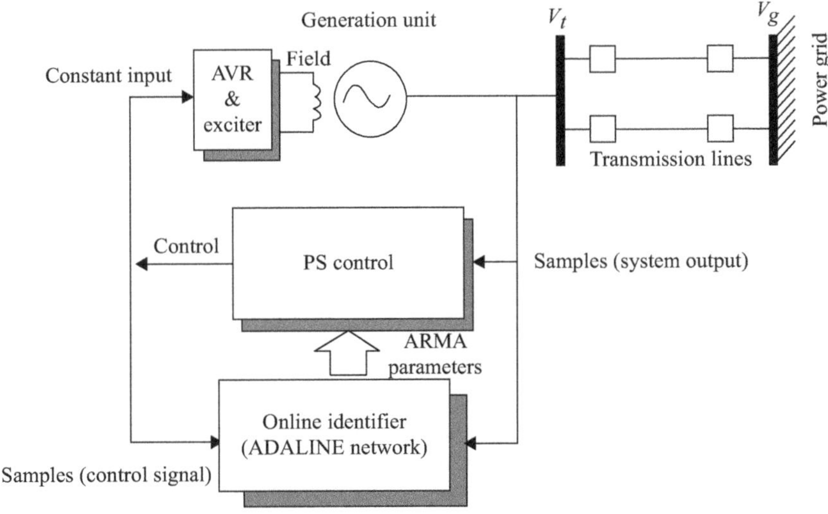

Figure 11.8 Power system with adaptive PSS

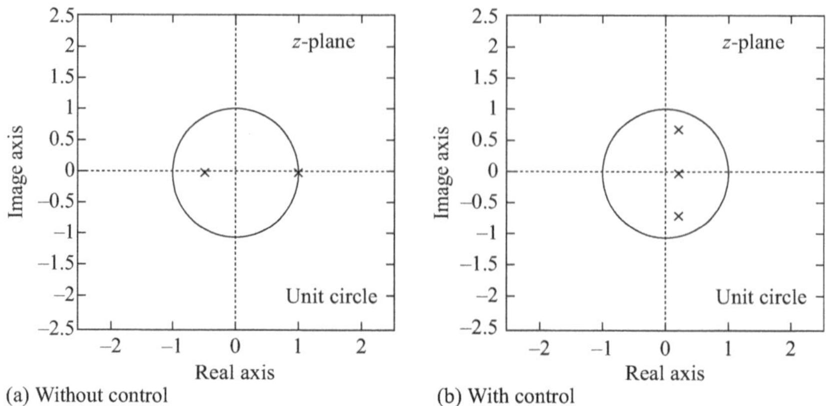

Figure 11.9 Pole patterns for $T(z^{-1})$ (a) with and (b) without pole-shift control

the closed-loop system is stable. It shows that the PS control assures the stability of the closed-loop system and also optimises the performance given by (11.16).

11.3.2 Performance studies with pole-shifting control PSS

Performance of the self-tuning adaptive controller based on the pole-shifting control algorithm has been investigated by conducting simulation studies on single machine [22, 27] and multi-machine system [28], on a single machine [29] and on a multi-machine physical model [30] in the laboratory, and on a 400-MW thermal machine under fully loaded conditions connected to the system [31].

The single machine power system consists of a synchronous generator connected to a constant voltage bus through two transmission lines (Figure 11.8). A non-linear seventh-order model is used to simulate the dynamic behaviour of this system. The differential equations used to simulate the synchronous generator and the parameters used in simulation studies are given in [22, 27]. The generator has an IEEE Standard 421.5, Type ST1A AVR and Exciter. An IEEE Standard 421.5, PSS1A Type CPSS [32] is used for comparative studies.

The system output is sampled at the rate of 20 Hz for parameter identification and control computation. Studies performed with various sampling rates show that the performance is practically the same for a sampling rate in the range of 20–100 Hz. Sampling frequencies above 100 Hz are of no practical benefit and the performance deteriorates for sampling rate under 20 Hz. A sampling rate of 20 Hz is chosen to make sure that there is enough time available for updating the parameters and control computation. In most studies, deviation of electrical power output is used as the input to the PSS. The control output is limited to 0.1 pu.

Results of a simulation study to demonstrate the effect of the APSS on the transient stability margin are summarised in Table 11.1. With the single machine infinite bus system initially operating at 0.95 pu power, 0.9 pf lag, a three phase to ground fault was applied near the sending end of one transmission line. It can be observed from Table 11.1 that the APSS provides the largest maximum clearance time.

Table 11.1 Transient stability margin results

	Without PSS	With CPSS	With APSS
Maximum clearing time (ms)	120	150	165

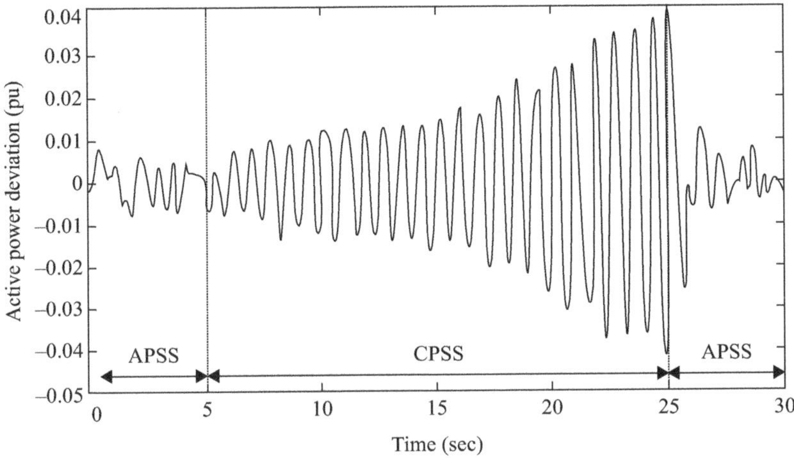

Figure 11.10 Dynamic stability improvement by the APSS

This APSS was implemented on a microprocessor and tested in real time on a physical model of a single-machine infinite bus system. With the system operating at a stable operating point, the APSS was applied and the torque reference increased gradually to the level, $P = 1.307$ pu, $pf = 0.95$ lead, $v_t = 0.950$ pu. At this load, the system was still stable with the APSS.

At 5 s (Figure 11.10) the APSS was replaced by the CPSS. After the switch over, the system began to oscillate and diverge, which means that the CPSS is unable to keep the system stable at this load level. At about 25 s, the APSS was switched back to control the unstable system and the system came under control very quickly as shown in Figure 11.10. This test demonstrates that the ASPSS can provide a larger dynamic stability margin than the CPSS. Also, more power can be transmitted with the help of the APSS if an overload operation is necessary under certain circumstances.

11.4 AI-based APSS

Various approaches using analytical and/or AI-based algorithms can be used to design an adaptive controller. It is also possible that the analytical and AI techniques be integrated such that some functions are performed using analytical approach while the others are performed using AI techniques. Successful implementation of purely AI and integrated approaches is illustrated by application as an APSS to improve damping and stability of an electric generating unit.

11.4.1 APSS with NN predictor and NN controller

Identification of the power plant model using an online recursive identification technique is a computationally extensive task. Neural networks (NNs) offer the alternative of a model-free method. An adaptive NN-based controller using indirect adaptive control method has been developed. It combines the advantages of NNs with the good performance of the adaptive control. In this controller, the learning ability of the NNs is employed in the adaptation process by training the NN in real-time each sampling period.

The controller consists of two sub-networks as shown in Figure 11.11. One network is an adaptive neuro-identifier (ANI) that identifies the power plant in terms of its internal weights and predicts the dynamic characteristics of the plant. It is based on the inputs and outputs of the plant and does not need the states of the plant.

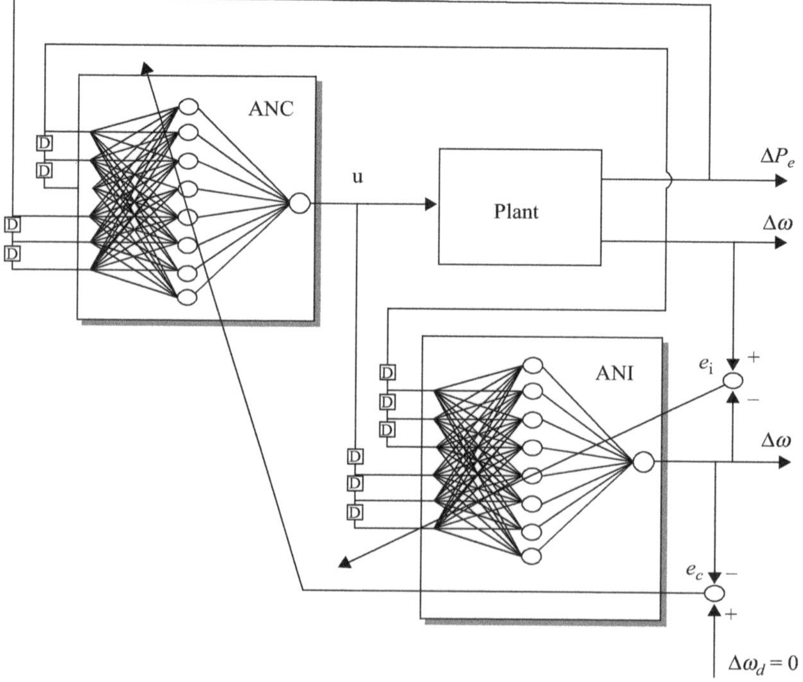

Figure 11.11 Controller structure for single-machine study

The second sub-network is an adaptive neuro-controller (ANC) that provides the necessary control action to damp the oscillations of the power plant.

The success of the control algorithm depends on the accuracy of the identifier in predicting the dynamic behaviour of the plant. The ANI and ANC are initially trained offline over a wide range of operating conditions and a wide spectrum of possible disturbances. After the offline training stage, the controller is hooked up in the system. Further updating of the weights of the ANI and ANC is done online for every sampling period. Online training enables the controller to track the plant variations as they occur and to provide control signal accordingly.

Employing a feed-forward multi-layer network in each of the two sub-networks, a NN-based APSS (NAPSS) has been built [33]. The two networks are trained further in each sampling period using an online version of the back-propagation algorithm. The errors used to train the ANI and ANC are both scalar, and the learning is done only once in each sampling period for each of the two sub-networks. This simplifies the training algorithm in terms of the computation time.

Performance of the adaptive network-based APSS was also tested on a five machine interconnected power system shown in Figure 11.12. Generating units in the five machine power system without infinite bus are modelled by fifth-order differential equations [34]. Results for a three-phase to ground fault on one circuit of the double-circuit transmission line between bus nos. 3 and 6 are shown in Figure 11.13.

The adaptive NN-based PSSs were installed on two generators and CPSSs were installed on the other three generators. It can be seen that both the local mode and the inter-area mode oscillations are damped effectively.

11.4.2 Adaptive network-based FLC

The characteristics of fuzzy logic and NNs complement each other in respect of their prospects and concepts. That offers the possibility of using a hybrid

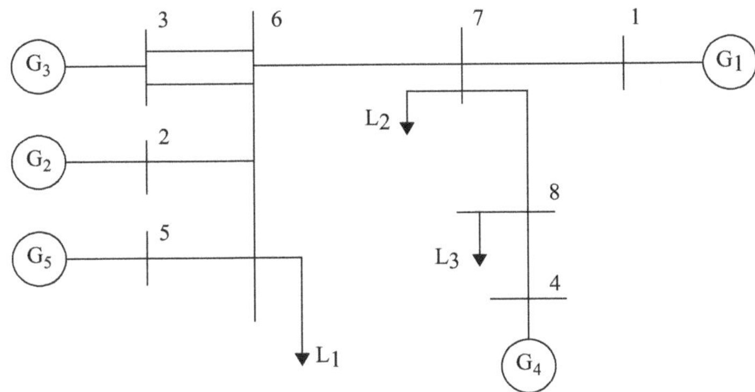

Figure 11.12 Five machine power system

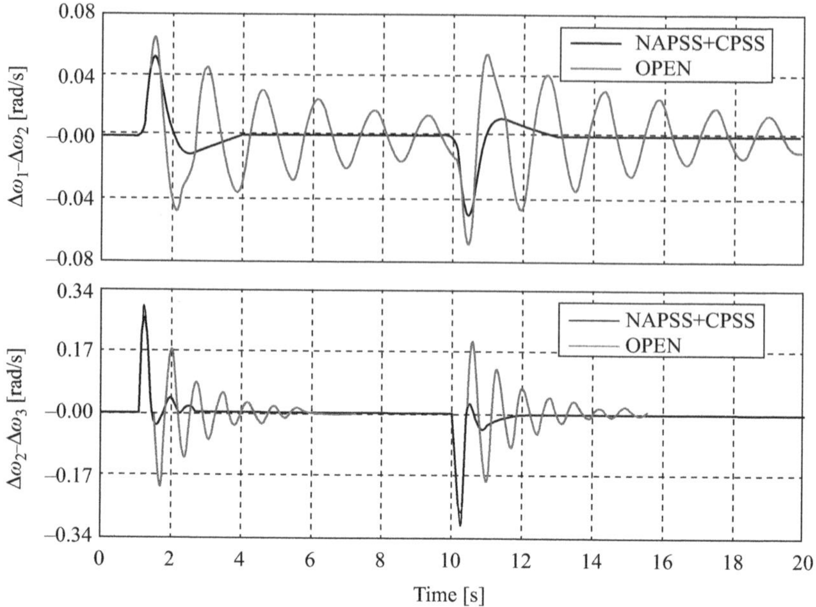

Figure 11.13 System response with NAPSS installed on generators G_1 and G_3 and CPSS on G_2, G_4 and G_5

neuro-fuzzy approach in the form of an adaptive network-based FLC whereby it is possible to take advantage of the positive features of both fuzzy logic and neural networks. Such a system can automatically find an appropriate set of rules and membership functions [35].

11.4.2.1 Architecture

In the neuro-fuzzy controller, the system is implemented in the framework of network architecture. Considering the functional form of the FLC (Figure 11.14), it becomes apparent that the FLC can be represented as a five-layer feed-forward network, in which each layer corresponds to one specific function with the node functions in each layer being of the same type. With this network representation of the fuzzy logic system, it is straightforward to apply the back-propagation or a similar method to adjust the parameters of the membership functions and inference rules.

In this network, the links between the nodes from one layer to the next layer only indicate the direction of flow of signals and part or all of the nodes contain the adjustable parameters. These parameters are specified by the learning algorithm and should be updated according to the given training data and a gradient-based learning procedure to achieve a desired input/output mapping. It can be used as an

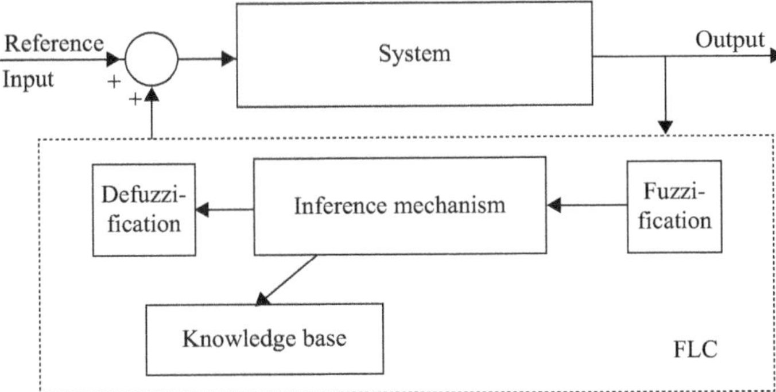

Figure 11.14 Basic structure of fuzzy logic controller

identifier for non-linear dynamic systems or as a non-linear controller with adjustable parameters.

11.4.2.2 Training and performance

Because the neuro-fuzzy controller has the property of learning, fuzzy rules and membership functions of the controller can be tuned automatically by the learning algorithm. Learning is based on the error in the controller output. Thus, it is necessary to know the error that can be evaluated by comparing the output of the neuro-fuzzy controller and a desired controller.

To train this controller as an adaptive network-based fuzzy PSS (ANF PSS), training data were obtained from a self-optimising pole-shifting APSS. Training was performed over a wide range of operating conditions of the generating unit including various types of disturbances. Based on earlier experience, seven linguistic variables for each input variable were used to get the desired performance.

Extensive simulation [36] and experimental studies with the ANF PSS show that it can provide good performance over a wide operating range and can significantly improve the dynamic performance of the system over that with a fixed parameter CPSS.

11.4.2.3 Self-learning ANF PSS

In the above case the ANF PSS was trained by data obtained from a desired controller. However, in a general situation, the desired controller may not be available. In that case, the neuro-fuzzy controller can be trained using a self-learning approach [37].

In the self-learning approach two neuro-fuzzy systems are used in a manner similar to Figure 11.11: one acting as the controller and the other acting as the predictor. The plant identifier can compute the derivative of the plant's output with respect to the plant's input by means of the back-propagation process illustrated by the line passing through the forward identifier and continuing back through the neuro-fuzzy controller that uses it to learn the control rule.

The self-learning ANF PSS was initially trained offline on a power system simulation model over a wide range of operating conditions and disturbances. Electric power deviation and its integral were used as the input to the stabiliser. The ANF PSS, with the parameters, membership functions and inference rules obtained from the offline training procedure, was implemented on a DSP mounted on a PC and its performance was evaluated on a physical model of a power system in the laboratory. A digital CPSS was also implemented in the same environment on the DSP board for comparative studies.

Out of the various tests, results for a 0.25-pu step decrease in the input torque reference applied at 1 s and removed at 9 s with the generator operating at 0.9 pu power, 0.85 pf lag and 1.10 pu V_t are shown in Figure 11.15. The ANF PSS provides a consistently good performance for either of the two disturbances.

Simulation studies on a single machine connected to a constant voltage bus and on a multi-machine power system [34] and experimental studies on a physical model of a power system have demonstrated the effectiveness of the ANN PSS in improving the performance of a power system over a wide operating range and a broad spectrum of disturbances.

11.4.2.4 Neuro-fuzzy controller architecture optimisation

Adaptive fuzzy systems offer a potential solution to the knowledge elicitation problem. The controller structure, expressed in terms of the number of membership

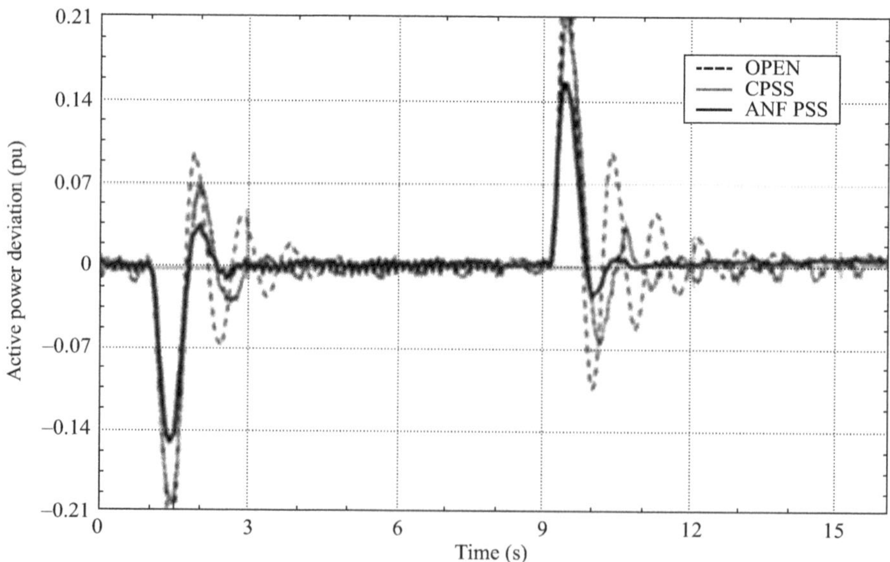

Figure 11.15 Comparison of ANF PSS and CPSS responses to a 0.25 pu step torque disturbance ($P = 0.9$ pu, 0.85 pf lag)

functions and the number of inference rules, is usually derived by trial and error. The number of inference rules has to be determined from the standpoint of overall learning capability and generalisation capability.

The above problem can be resolved by employing a genetic algorithm to determine the structure of the adaptive fuzzy controller. By employing both genetic algorithm and adaptive fuzzy controller, the inference rules parameters can be tuned and the number of membership functions can be optimised at the same time.

11.5 Amalgamated analytical and AI-based PSS

11.5.1 APSS with neuro identifier and PS control

A self-tuning APSS described above can improve the dynamic performance of the synchronous generator by allowing the parameters of the PSS to adjust as the operating conditions change. However, proper care needs to be taken in the design of the RLS algorithm for identification to make it stable, especially under large disturbances.

It is possible to make the identification more robust by using an NN for identifying the system model parameters. An analytical technique, such as the PS control, can be retained to compute the control signal. One approach, using a radial basis function (RBF) network for model parameter identification, is described below [38]. The APSS shown in Figure 11.6 now consists of an ANN identifier and the pole-shifting control algorithm described above.

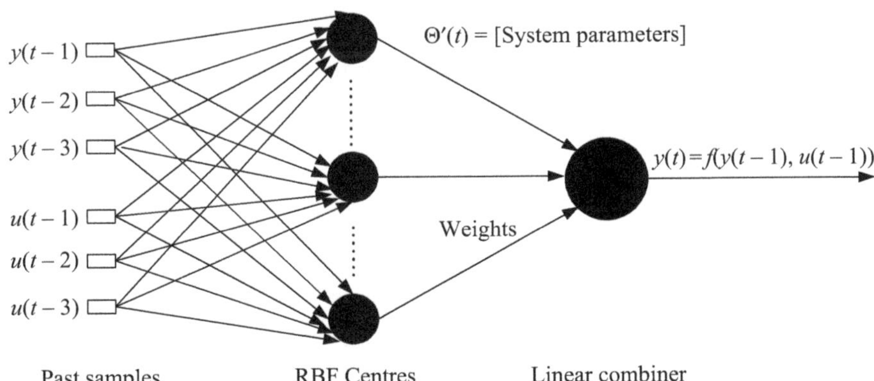

Figure 11.16 Radial basis function network model

The RBF network (Figure 11.16) is used to identify the system model parameters, a_i, b_i, (11.6) and (11.7). The network consists of three layers: input, hidden and output. The input vector is

$$V(t) = [\Delta P_e(t-T),\ \Delta P_e(t-2T),\ \Delta P_e(t-3T),\ u(t-T),\ u(t-2T),\ u(t-3T)] \tag{11.18}$$

Each of the six input variables is assigned to an individual node in the input layer and passes directly to the hidden layer without weights. The hidden nodes, called the RBF centres, calculate the Euclidean distance between the centres and the network input vector. The result is passed through a widely used Gaussian function characterised by a response that has a maximum value of 1 when the distance between the input vector and the centre is 0. Thus, a radial basis neuron acts as a detector that produces '1' whenever the input vector is identical to the centre (active neuron). The other neurons with centres quite different from the input vector will have outputs near 0 (non-active neurons).

The connections between the hidden neurons and the output node are linear weighted sums as described by the equation:

$$y = \sum_{i=1}^{nh} \theta^t \exp\left(-\frac{\|p - c_i\|^2}{\sigma^2}\right) \tag{11.19}$$

where

c_i, σ, θ^{t} and nh are the centres, widths, weights and the number of hidden layer neurons, respectively.

To make the proposed RBF identifier faster for online applications, the hidden layer is created as a competitive layer wherein the centre closest to the input vector becomes the winner and all the other non-active centres are deactivated. Also, the scalar weights are modified as a vector θ_t whose size equals the size of the input vector. The weight vector is given by

$$\theta'(t) = \left[a'_1 a'_2 a'_3 b'_1 b'_2 b'_3\right] \tag{11.20}$$

Linearising the output of the RBF, $y(t) = f[y(t-1), u(t-1)]$, by Taylor series expansion at each sampling instant, a one-to-one relationship between the weight vector θ' and the system model parameters $\hat{\theta}_{\mathrm{m}}(t)$, (11.9), can be obtained. These parameters are then used in computing the control signal.

The RBF identifier was first trained offline to choose appropriate centres using data collected at a number of operating points for various disturbances. The n-means clustering algorithm used for training yielded 15 centres for the RBF model. After the offline training, the weights (system parameters) were updated online to obtain the appropriate control signal using the pole-shifting controller. A 100-ms sampling period was chosen for digital implementation.

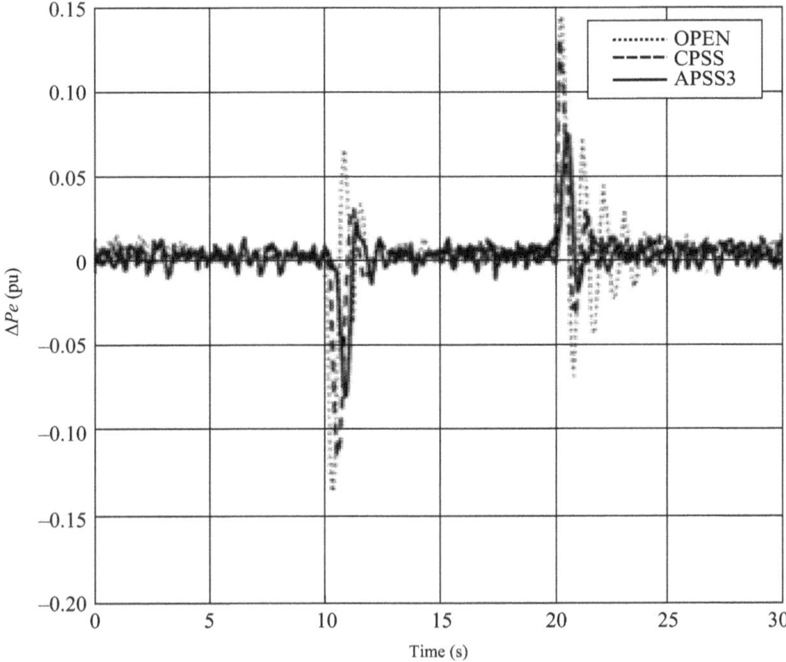

Figure 11.17 *ΔP_e response for 0.1 pu input torque reference step change with APSS*

© 2009 IEEE. Reprinted with permission from Malik O.P. 'Adaptive and artificial intelligence based PSS', *Proceedings, IEEE PES 2003 General Meeting*, Vol. 3, pp. 1792–7

Results of an experimental study for a 0.10-pu decrease in torque reference applied at 10 s and removed at 20 s, with the generator operating at 0.6 pu power, 0.92 pf lead and V_t of 0.99 pu are shown in Figure 11.17. It can be seen that the APSS can provide a well-damped response.

11.5.2 APSS with fuzzy logic identifier and PS controller

Takagi–Sugeno (TS) fuzzy systems have been successfully employed in the design of stabilisation control of non-linear systems.

A non-linear plant can be represented by a set of linear models interpolated by membership functions of a TS fuzzy model. Although the TS system identifier is a NARMAX model, at each sample an average linear discrete auto-regressive moving average (ARMA) model can be determined to identify the controlled plant according to the current active rules. This ARMA model can be used to determine the control signal by the pole-shifting control strategy. Using this approach, a self-tuning adaptive controller has been developed and applied as a PSS [39].

The proposed single-input single-output TS model used for the identification of dynamic systems is composed of fuzzy rules, the consequent part of which provides the rule output at time k based on the past inputs and past outputs with fuzzy sets designed in the universe of discourse. The consequent part of the rule then identifies the parameters of a desired order discrete model of the plant. Two parallel online learning procedures, one each for the identification of premise and consequent parameters, are used to track the plant in real time [40].

In the proposed TS system for generating unit identification, two input signals, the past control input, $u(k-1)$, and the past generator speed output, $y(k-1)$, are used to identify a third-order model of the plant. The output at sample k is the estimated generator speed output, $\hat{y}(k)$. The TS system is trained by using the steepest descent algorithm for the premise parameters and RLS algorithm for the consequent parameters using the error of the system output and the estimated TS output. Initially a set of three equally spaced membership functions, over the normalised universe, are used for the inputs of the system.

The response of the system with the TS system-based identifier and PS controller-based APSS has been studied for various disturbances at different operating conditions. One illustrative result for a three-phase to ground fault is shown in Figure 11.18.

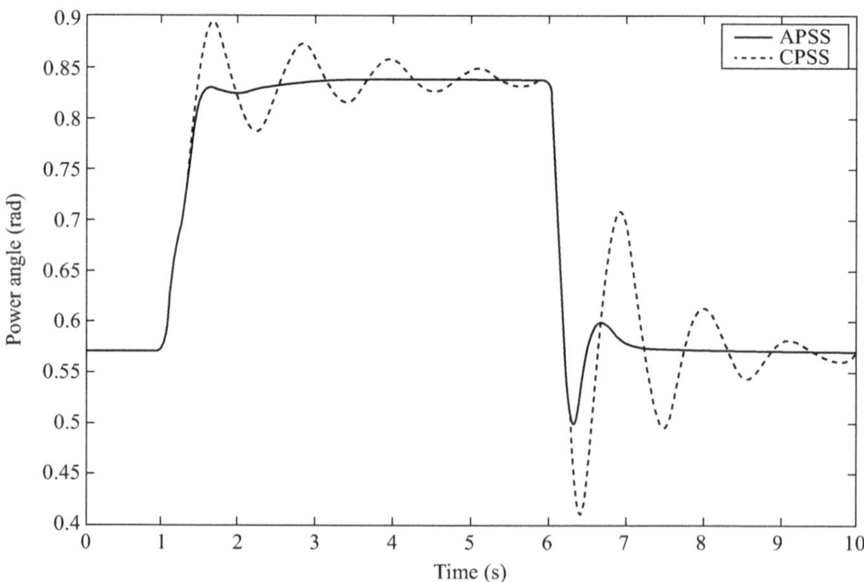

Figure 11.18 Three phase to ground fault at the middle of one transmission line and successful reclosure ($P = 0.95$ pu, 0.9 pf lag)

11.5.3 APSS with RLS identifier and fuzzy logic control

FLCs have attracted considerable attention as candidates for novel computational systems because of the advantages they offer over the conventional computational systems. They have been successfully applied to the control of non-linear dynamic systems, especially in the field of adaptive control, by making use of online training.

A self-learning adaptive FLC has been developed. Only the inputs and outputs of the plant are measured and there is no need to determine the states of the plant. Using online training by the steepest descent method and the identified system model, the adaptive FLC is able to track the plant variations as they occur and compute the control.

In the proposed controller, a discrete model of the plant is first identified using the RLS parameter identification method. This allows a continuous tracking of the system behaviour.

The control learning is based on the prediction of the identified model. The identified model output is used as input to the Mamdani-type PD controller [20]. The centre points of the controller inputs are updated [40] by treating them exactly the same as the weights of an NN and by using the steepest descent algorithm with chain rule.

The proposed adaptive FLC has been applied as an adaptive fuzzy PSS (AFPSS) [41]. For the AFPSS, the generating unit is identified as a third-order model. The controller has two input signals, the generator speed deviation and its derivative, with an initial set of seven equally spaced membership functions over the normalised universe of discourse. The output, the supplementary control signal, also having seven membership functions, is added to the AVR summing junction. A number of simulation studies have been performed for various disturbances at different operating conditions. An illustrative result for a 0.05-pu increase in torque and return to initial condition, shown in Figure 11.19, demonstrates the performance of this AFPSS.

11.6 APSS based on recurrent adaptive control

For nonlinear systems with a general form, unless the reference model is well defined, the traditional MRAC may cause system oscillations. This problem arises because the connection between the current system state and the controller parameters is ignored by the MRAC. The focus of recurrent adaptive control (RAC) is on optimising a certain objective function in which the lost connection is picked up.

Development of RAC is inspired on observing the similarity between the adaptive control system and the recurrent NNs (RNNs) [42]. Because of that, a modified version of the back propagation through time (BPTT) [43], a learning algorithm of RNNs, can be exploited in RAC. A new control algorithm for RAC, named recursive gradient (RG), which improves the performance of the original and truncated BPTT algorithms, has been developed and an APSS has been developed based on the RG algorithm.

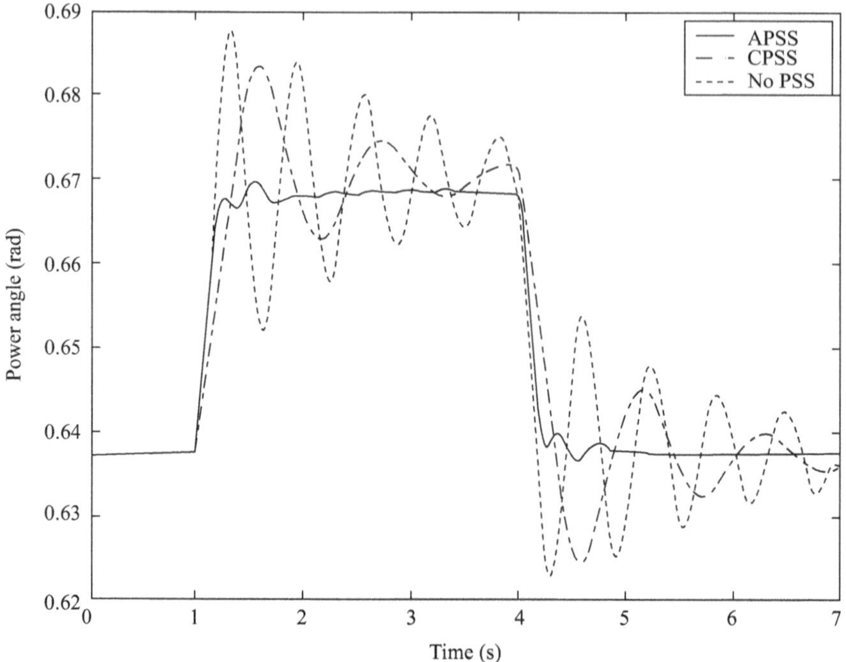

Figure 11.19 Response to a 0.05 pu step increase in torque and return to initial condition ($P' = 0.95$ pu, 0.9 pf lag)

The system in Figure 11.3 can also be expressed by the non-linear equations

$$\begin{cases} X(k+1) = F(X(k), U(k)) \\ U(k) = G(X(k), \theta) \end{cases} \tag{11.21}$$

where

$X(k)$, $X(k+1) \in R_p$, $U(k) \in R_q$ and $\theta \in R_r$. p, q and r are the number of system states, system inputs and controller parameters, respectively. At each discrete time k, the controller parameters θ are updated online to minimise a predefined objective function $J(X(k+1))$, (11.22), that is used to evaluate the control performance.

$$\min_{\theta} \quad J(X(k+1)) \tag{11.22}$$

In many cases, the performance index $J(X(k+1))$ is not directly related to the state but the output of the system. The output of the system, $Y(k)$, can be a subset of the state $X(k)$ or, in general way, a function of the state. Many non-linear optimisation methods can be utilised to minimise the performance index $J(X(k+1))$.

One of the most popular local optimisation methods is the gradient descent algorithm [44] with one-step or multi-step optimisation control. In these schemes it is also assumed that the current state $X(k)$ is independent of the controller parameter θ and, therefore,

$$\left.\frac{\partial X(k)}{\partial \theta}\right|_{\theta=\theta(k)} = 0 \tag{11.23}$$

In this approach, the feedback loop, which makes the adaptive control system to be a recurrent system, is ignored. However, the real control system is recurrent.

By using the rule given by (11.24):

$$\theta(k+1) = \theta(k) - \alpha \cdot \left.\frac{\partial X(k+1)}{\partial \theta}\right|_{\theta=\theta(k)} \cdot \frac{\partial J(X(k+1))}{\partial X(k+1)} \tag{11.24}$$

where

α is the step size and

$$\begin{aligned}\left.\frac{\partial X(k+1)}{\partial \theta}\right|_{\theta=\theta(k)} = {} & \left.\frac{\partial X(k)}{\partial \theta}\right|_{\theta=\theta(k)} \left[\frac{\partial X(k+1)}{\partial X(k)} + \frac{\partial U(k)}{\partial X(k)}\frac{\partial X(k+1)}{\partial U(k)}\right] \\ & + \left.\frac{\partial U(k)}{\partial \theta}\right|_{\theta=\theta(k)} \frac{\partial X(k+1)}{\partial U(k)}\end{aligned} \tag{11.25}$$

to update the controller parameters in RAC, the assumption (11.23) can be removed.

Control algorithm (11.24) can be solved using the BPTT or the truncated BPTT algorithm. Both versions of BPTT control algorithm require extensive computation. This problem can be overcome using the RG algorithm in which the controller parameters θ are updated by the following rule:

$$\theta(k+1) = \theta(k) - \alpha \cdot \sum_{m=0}^{k}\left[\lambda^m \cdot \left.\frac{\partial X(k+1)}{\partial \theta}\right|_{\theta=\theta(k), X(k-m)}\right] \cdot \frac{\partial J(X(k+1))}{\partial X(k+1)} \tag{11.26}$$

where $1 > \lambda > 0$ and α is the step size.

Although the RG control algorithm has been developed for RAC control applications, it can also be used to train RNNs.

To design an APSS based on the RAC, a model that tracks the dynamics of the synchronous machine has to be built first. A second-order operating condition dependent (OC-dependent) ARMA model [45] has been used in this application. By rewriting the A and B coefficients in the widely used model (2) as functions of the operating conditions represented by the easily measured active power output, P_e, and reactive power, Q_e, the OC-dependent ARMA model is given by

$$\begin{aligned}\Delta\hat{\omega}\ (k+1) = {} & a_1(P_e, Q_e)\Delta\omega(k) + a_2(P_e, Q_e)\Delta\omega(k-1) \\ & + b_1(P_e, Q_e)(u(k) - u(k-1))\end{aligned} \tag{11.27}$$

where

$$a_1(P_e, Q) = \sum_{i=1}^{N} a_{i1}\rho_i(P_e, Q_e)$$

$$a_2(P_e, Q_e) = \sum_{i=1}^{N} a_{i2}\rho_i(P_e, Q_e)$$

$$b_1(P_e, Q_e) = \sum_{i=1}^{N} b_{i1}\rho_i(P_e, Q_e)$$

The operating region vector is $\phi = [P_e\ Q_e]t$ and the region function $\rho i(\phi)$ is commonly chosen as a normalised Gaussian function.

In this model, one set of parameters can work for various operating conditions without updating [36]. The model (11.27) can be realised by N local model networks (LNNs) [46] that are a general form of RBF networks [47]. Each local model can be interpreted as a good approximation to the desired function in a region defined by the region function.

The synchronous generator with the OC-dependent linear controller can be written as

$$\begin{cases} X(k+1) = AX(k) + Bu(k) \\ u(k) = \mathrm{H}(X^T(k)\boldsymbol{\Phi}_c\theta_{LC}) \end{cases} \tag{11.28}$$

where

$$A = \begin{bmatrix} a_1(P_e, Q_e) & a_2(P_e, Q_e) & -b_1(P_e, Q_e) \\ 1 & 0 & 0 \\ 0 & 1 & 0 \end{bmatrix}$$

$$B = [b_1(P_e, Q_e) \quad 0 \quad 1]^T$$

$$\boldsymbol{\Phi}_\mathrm{c} = \begin{bmatrix} \rho_1(\phi) & 0 & 0 & & \rho_N(\phi) & 0 & 0 \\ 0 & \rho_1(\phi) & 0 & \cdots & 0 & \rho_N(\phi) & 0 \\ 0 & 0 & \rho_1(\phi) & & 0 & 0 & \rho_N(\phi) \end{bmatrix}$$

$$\theta_{\mathrm{LC}} = [g_{11}, g_{12}, h_{11}, \ldots, g_{N1}, g_{N2}, h_{N1}]^T$$

and H function involves hard constraints.

Using the objective function:

$$\begin{aligned} J(X(k+1)) &= \frac{1}{2}\left(\Delta\omega^2(k+1) + \beta u^2(k)\right) \\ &= \frac{1}{2}X^T(k+1)QX(k+1) \end{aligned} \tag{11.29}$$

where

β is the weight on energy expense in the objective function and

$$Q = \begin{bmatrix} 1 & 0 & 0 \\ 0 & 0 & 0 \\ 0 & 0 & \beta \end{bmatrix}$$

the RG control algorithm for the OC-dependent linear PSS, given in (11.30), is obtained as

$$\theta(k+1) = \theta(k) - \alpha' \cdot D'(k) \cdot Q \cdot X(k+1) \tag{11.30}$$

where

$$D'(k) = (1-\lambda) \cdot R(k) + \lambda D'(k-1)T(k)$$

$$R(k) = \begin{cases} \Phi_c{}^T X(k) B^T & u_{\min} \le u(k) \le u_{\max} \\ 0 & u_{\min} > u(k) \; or \; u(k) > u_{\max} \end{cases}$$

$$T(k) = \begin{cases} A + \Phi_C \theta_{LC}(k) B^T & u_{\min} \le u(k) \le u_{\max} \\ 0 & u_{\min} > u(k) \text{ or } u(k) > u_{\max} \end{cases}$$

$D'(-1) = 0$ (initial condition).

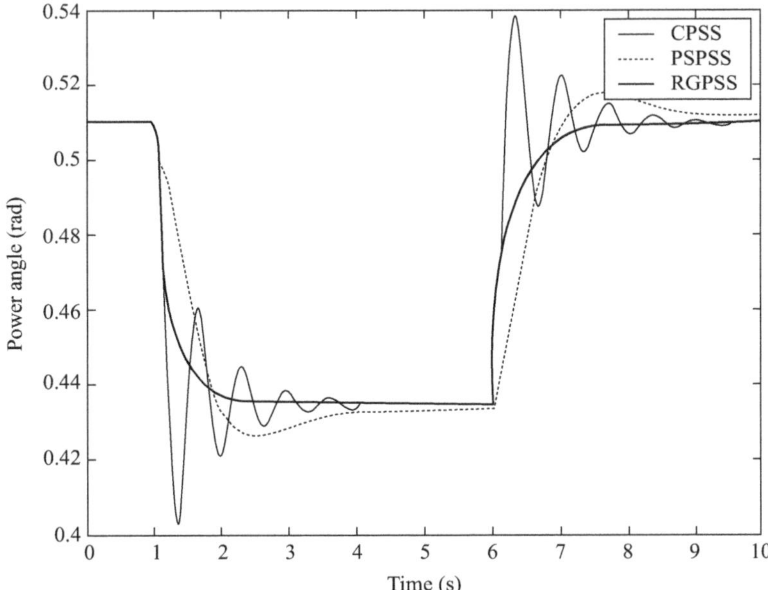

Figure 11.20 Response to a 50 ms three-phase short circuit fault at the middle of one transmission line at 1 s ($P_e = 1.0$ pu, pf = 0.85 lag)

© 2009 IEEE. Reprinted with permission from Zhao P., Malik O.P. Design of an adaptive PSS based on Recursive Adaptive Control theory. *IEEE Transactions on Energy Conversion.* 2009;**24**(4):884–92

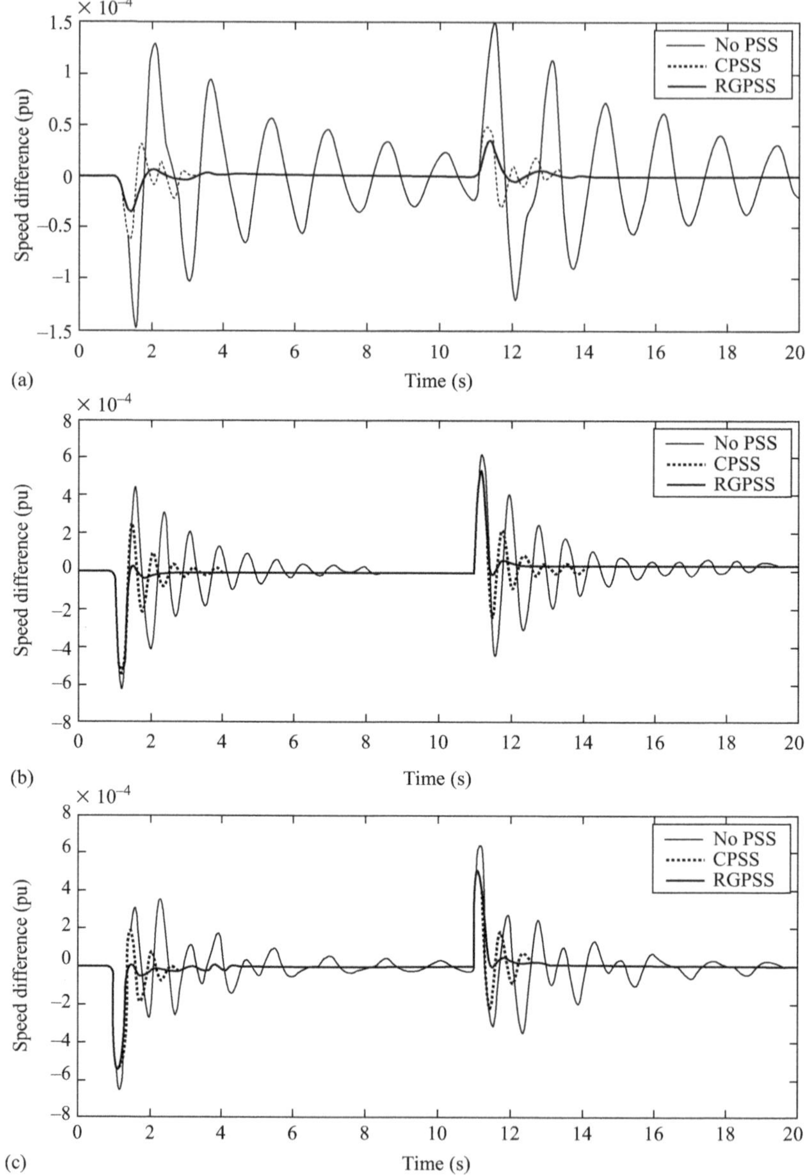

Figure 11.21 System response to 0.10 pu step increase in the mechanical torque change on the generator G_3 in the five-machine power system. PSSs are installed on G_1, G_2 and G_3 (a) Speed different between the generator G_1 and G_2 (b) Speed different between the generator G_2 and G_3 (c) Speed different between the generator G_1 and G_3

The OC-dependent ARMA model was first trained offline on a single machine constant voltage bus system described in Section 11.3.2 with white noise signal as the input. Based on this sufficiently trained model, the APSS is trained offline by the RG control algorithm over a wide range of operating conditions varying from 0.1 pu to 1.0 pu power and 0.6 lag to 0.8 lead power factor.

Performance of the synchronous machine operating at 1.0 pu power, 0.85 pf lag without PSS, with a CPSS and with the PSS based on the RG algorithm for a short-circuit at the middle of one line at 1 s, opening of the line after 50 mms and line reconnection at 6 s is shown in Figure 11.20. This APSS was also tested on the five machine power system (Figure 11.12). Response for a 0.1-pu increase in input torque reference of G3 at 1 s is shown in Figure 11.21. The system returns to its initial conditions at 11 s.

It can be seen that the multi-modal oscillations are also damped out more effectively than with a CPSS.

11.7 Concluding remarks

PSSs based on the control algorithms described above have been studied extensively in simulation. They have also been implemented and tested in real time on physical models in the laboratory with very encouraging results. The pole-shifting control algorithm-based APSS has also been tested on a multi-machine physical model [30], on a 400-MW thermal machine under fully loaded conditions connected to the system [31], and is now in regular service in a hydro power station after extensive testing in the field [48]. These studies have shown clearly the advantages of the advanced control techniques and intelligent systems.

Very satisfactory adaptive controllers can be developed and implemented using a number of approaches, i.e. purely analytical, purely AI techniques or by amalgamating the analytical and AI approaches. Which approach to use depends on the expertise of the designer and the developer of the controller, and the confidence that they or the client have in a particular technology.

References

1. De Mello F.P., Concordia C. 'Concepts of synchronous machine stability as affected by excitation control'. *IEEE Transactions on Power Apparatus and Systems.* 1969;**PAS-88**:316–29
2. Dandeno P.L., Karas A.N., McClymont K.R., Watson W. 'Effect of high-speed rectifier excitation systems on generator stability limits'. *IEEE Transactions on Power Apparatus and Systems.* 1968;**PAS-87**(1):190–201
3. Larsen E.V., Swann D.A. 'Applying power system stabilizers, Parts I, II and III'. *IEEE Transactions on Power Apparatus and Systems.* 1981;**PAS-100**(6): 3017–46
4. Kundur P., Klien M., Rogers G.J., Zywno M.S. 'Application of power system stabilizers for enhancement of overall system stability'. *IEEE Transactions on Power Systems.* 1989;**4**(2):614–26

5. IEEE Standard 421.5-2005. IEEE Recommended Practice for Excitation System Models for Power System Stability Studies, Apr 2006
6. Watson W., Coultes M.E. 'Static exciter stabilizing signals on large generators – mechanical problems'. *IEEE Transactions on Power Apparatus and Systems*. 1973;**92**(1):204–11
7. Keay F.W., South W.H. 'Design of a power system stabilizer sensing frequency deviations'. *IEEE Transactions on Power Apparatus and Systems*. 1971;**90**(2):707–13
8. Bayne J.P., Lee D.C., Watson W. 'A power system stabilizer for thermal units based on derivation of accelerating power'. *IEEE Transactions on Power Apparatus and Systems*. 1977;**96**(6):1777–83
9. deMello F.P., Hannett L.N., Underill J.M. 'Practical approaches to supplementary stabilizing from accelerating power'. *IEEE Transactions on Power Apparatus and Systems*. 1978;**97**(6):1515–22
10. Lee D.C., Beaulieu R.E., Service J.R.R. 'A power system stabilizer using speed and electrical power inputs – design and field experience'. *IEEE Transactions on Power Apparatus and Systems*. 1981;**100**(9):4151–57
11. Kundur P., Lee D.C., Zein el-Din H.M. 'Power system stabilizers for thermal units: analytical techniques and on-site validation'. *IEEE Transactions on Power Apparatus and Systems*. 1981;**100**(1):81–95
12. El-Metwally M.M., Rao N.D., Malik O.P. 'Experimental results on the implementation of an optimal control for synchronous machines'. *IEEE Transactions on Power Apparatus and Systems*. 1975;**94**(4):1192–200
13. Chen S., Malik O.P. 'H∞ optimisation based power system stabiliser design'. *IEE Proceedings-Generation, Transmission and Distribution*. 1995;**142**(2): 179–84
14. Chan W.C., Hsu Y.Y. 'An optimal variable structure stabilizer for power system stabilization'. *IEEE Transactions on Power Apparatus and Systems*. 1983;**102**(6):1738–46
15. Hiyama T. 'Application of rule-based stabilising controller to electrical power system'. *IEE Proceedings C*, 1989;**136**(3):175–81
16. Zadeh L.A., Fu K.S., Tanaka K., Shimura M. 'Calculus of fuzzy restriction' in Zadeh L.A. (ed.). *Fuzzy Sets and Their Applications to Cognitive and Decision Processes*. New York, NY, US: Academic Press; 1975. pp. 1–40
17. El-Metwally K.A. Hancock G.C., Malik O.P. 'Implementation of a fuzzy logic PSS using a micro-controller and experimental test results'. *IEEE Transactions on Energy Conversion*. 1996;**11**(1):91–6
18. Zhang Y., Malik O.P., Chen G.P. 'Artificial neural network power system stabilizers in multi-machine power system environment'. *IEEE Transactions on Energy Conversion*. 1995;**10**(1):147–55
19. Abdelazim T., Malik O.P. 'Power system stabilizer based on model reference adaptive fuzzy control'. *Electric Power Components and Systems*. 2005;**33**(9): 985–98
20. Mamdani M. 'Application of fuzzy algorithm for control of simple dynamic plant'. *Proceedings of the Institution of Electrical Engineers, IEE*. 1974; **121**(12):1585–88

21. Eykhoff, P. *System Identification*. London: John-Wiley Press; 1974
22. Cheng S.J., Chow Y.S., Malik O.P., Hope G.S. 'An adaptive synchronous machine stabilizer'. *IEEE Transactions on Power Systems.* 1986;**1**(3):101–7
23. Anderson B.D.O., Moore J.B. *Linear Optimal Control.* Upper Saddle River, NJ, US: Prentice Hall; 1971
24. Astrom K.J., Borisson U., Ljung L., Wittenmark B. 'Theory and application of adaptive control – A survey'. *Automatica*. 1983;**19**(5):471–86
25. Wellstead P.E., Edmunds J.M., Prager D., Zanka P. 'Self-tuning pole/zero assignment regulators'. *International Journal of Control.* 1979;**30**(1):1–26
26. Wellstead P.E., Prager D., Zanker P. 'Pole-assignment self-tuning regulator'. *Proceedings of the Institution of Electrical Engineers, IEE.* 1979;**126**(8): 781–7
27. Malik O.P., Chen G.P., Hope G.S., Qin Y.H., Yu G.Y. 'Adaptive self-optimizing pole-shifting control algorithm'. *IEE Proceedings-D.* 1992;**139**(5): 429–38
28. Cheng S.J., Malik O.P., Hope G.S. 'Damping of multi-modal oscillations in power systems using a dual-rate adaptive stabilizer'. *IEEE Transactions on Power Systems.* 1988;**3**(1):101–8
29. Chen G.P., Malik O.P., Hancock G.C. 'Implementation and experimental studies of an adaptive self-optimizing power system stabilizer'. *Control Engineering Practice.* 1994;**2**(6):969–77
30. Malik O.P., Stroev V.A., Shtrobel V.A., Hancock G.C., Beim R.S. 'Experimental studies with power system stabilizers on a physical model of a multi-machine power system'. *IEEE Transactions on Power Systems.* 1996;**11**(2): 807–12
31. Malik O.P., Mao C.X., Prakash K.S., Hope G.S., Hancock G.C. 'Tests with a microcomputer based adaptive synchronous machine stabilizer on a 400 MW thermal unit'. *IEEE Transactions on Energy Conversion.* 1993;**8**(1):6–12
32. IEEE Standard 421.5. IEEE Recommended Practice for Excitation Systems for Power System stability Studies. 1992
33. Shamsollahi P., Malik O.P. 'An adaptive power system stabilizer using on-line trained neural networks'. *IEEE Transactions on Energy Conversion.* 1997;**12**(4):382–7
34. Hariri A., Malik O.P. 'A self-learning adaptive-network-based fuzzy logic power system stabilizer in a multi-machine power system'. *Engineering Intelligent Systems*. 2001;**9**(3):129–36
35. Jang J.S.R. 'Adaptive-network-based fuzzy inference system'. *IEEE Transactions on Systems, Man and Cybernetics*. 1993;**23**(3):665–85
36. Hariri A., Malik O.P. 'A fuzzy logic based power system stabilizer with learning ability'. *IEEE Transactions on Energy Conversion.* 1996;**11**(4):721–27
37. Jang J.S.R. 'Self-learning fuzzy controllers based on temporal back-propagation'. *IEEE Transactions on Neural Networks.* 1992;**3**(5):714–23
38. Ramakrishna G., Malik O.P. (eds.). 'Adaptive control of power systems using radial basis function network and predictive control calculation'. *Conference Proceedings, IEEE Power Engineering Society Summer Meeting*; Edmonton, AB, Canada, Jul 1999. pp. 989–94

39. Abdelazim T., Malik O.P. 'Fuzzy logic based identifier and pole-shifting controller for PSS application'. *Proceedings of Power Engineering Society General Meeting, 2003, IEEE*; Toronto, Canada, Jul 2003. pp. 1680–5
40. Adams J.M., Rattan K.S. 'Backpropagation learning for a fuzzy controller with partitioned membership functions'. *Proceedings of Annual Meeting of the North American Fuzzy Information Processing Society, NAFIPS*, 2002. pp. 172–7
41. Abdelazim T., Malik O.P. (eds.). 'An adaptive power system stabilizer using on-line self-learning fuzzy system'. *Proceedings of Power Engineering Society General Meeting, 2003, IEEE*; Toronto, Canada, Jul 2003. pp. 1715–20
42. Seidl D.R., Lorenz R.D. (eds.). 'A structure by which a recurrent neural network can approximate a nonlinear dynamic system'. *Proceedings of the International Joint Conference on Neural Networks, IJCNN-91*; Seattle, WA, US, Jul 1991, vol. 2. pp. 709–14
43. Williams R.J., Zipser D. 'Gradient-based learning algorithms for recurrent networks and their computational complexity' in Chauvin Y., Rumelhart D.E. (eds.). *Backpropagation: Theory, Architecture, and Applications*. Hillsdale, NJ, US: Lawrence Erlbaum; 1995, ch. 13. pp. 422–86
44. Whitaker H.P., Yamron J., Kezer A. 'Design of model-reference-adaptive control systems for aircraft'. Report R-164, Instrumentation Laboratory, MIT, Cambridge, MA, US, 1958
45. Zhao P., Malik O.P. (eds.). 'Operating condition dependent ARMA model for PSS application'. *Power Engineering Society General Meeting, 2004, IEEE*; Denver, CO, US, Jul 2004. pp. 1749–54
46. Johansen T.A., Foss B.A. 'Constructing NARMAX models using ARMAX models'. *International Journal of Control.* 1993;**58**(5):1125–53
47. Ramakrishna G., Malik O.P. (eds.). 'RBF identifier and pole- shifting controller for PSS application'. *Electric Machines and Drives, 1999. International Conference IEMD '99*; Seattle, WA, US, May 1999. pp. 589–91
48. Eichmann A., Kohler A., Malik O.P., Taborda J.,(eds.). 'A prototype self-tuning adaptive power system stabilizer for damping of active power swings'. *Proceedings of Power Engineering Society Summer Meeting, 2000. IEEE*; Seattle, WA, US, Jul 2000, vol. 1. pp. 122–6
49. Malik O.P. 'Adaptive and artificial intelligence based PSS', *Proceedings, IEEE PES 2003 General Meeting*, vol. 3. pp. 1792–7
50. Shamsollahi P., Malik O.P. 'Application of neural adaptive power system stabilizer in a multi-machine power system', *IEEE Transactions on Energy Conversion.* 1999;**14**(3):731–6
51. Zhao P., Malik O.P. 'Design of an adaptive PSS based on Recursive Adaptive Control theory'. *IEEE Transactions on Energy Conversion.* 2009;**24**(4):884–92

Chapter 12
Series compensation

The transmission network connects scattered clusters of loads to different sets of generating units in the power system. Such networks are often required to be compensated for reactive power flow to maintain proper voltage levels. Compensation can be provided by inserting elements that are capable of delivering or absorbing reactive power resulting in changes of reactive power flow in the transmission network. Thus, the reactive power flow can be controlled to make an optimal use of the transmission system in a manner that the voltage along the transmission lines and at the network nodes is kept within desired values as well as increasing the power transfer capacity of the transmission lines. Consequently, the system stability may be improved.

The compensation highly pertains to the transmission system and is provided by installing capacitors and/or reactors at different locations in the transmission network. Capacitors are installed either in series with the transmission lines called 'series capacitive compensation' or in shunt at specific points, commonly, near the loads called 'shunt capacitive compensation'. Reactors can be used as shunt reactive compensators connected to the transmission lines or the transmission nodes at locations determined by the study and its objectives.

This chapter focuses on the series compensation of transmission network and its benefits, in particular, the system stability improvement. This entails explanation and discussion of some definitions and some basic concepts as below.

12.1 Definitions of transmission line parameters

Parameters ***ABCD***:

As explained in Chapter 4, the voltage and current at the sending end, $\boldsymbol{V}_S$ and $\boldsymbol{I}_S$, respectively, can be deduced from (4.23) by substituting $x = l$ (the total length of line), thus

$$\left.\begin{aligned} \boldsymbol{V}_S &= \boldsymbol{A}\boldsymbol{V}_R + \boldsymbol{B}\boldsymbol{I}_R \\ \boldsymbol{I}_S &= \boldsymbol{C}\boldsymbol{V}_R + \boldsymbol{D}\boldsymbol{I}_R \end{aligned}\right\} \tag{12.1}$$

and in matrix form, (12.1) can be written as

$$\begin{bmatrix} \boldsymbol{V}_S \\ \boldsymbol{I}_S \end{bmatrix} = \begin{bmatrix} \boldsymbol{A} & \boldsymbol{B} \\ \boldsymbol{C} & \boldsymbol{D} \end{bmatrix} \begin{bmatrix} \boldsymbol{V}_R \\ \boldsymbol{I}_R \end{bmatrix} \tag{12.2}$$

where

$$A = \cosh\gamma l, \quad B = Z_C \sinh\gamma l, \quad C = (1/Z_C)\sinh\gamma l, \quad D = A = \cosh\gamma l \tag{12.3}$$

and

$$Z_C = \sqrt{\frac{z}{y}} = \sqrt{\frac{r_L + jx_L}{g_c + jb_c}} \tag{12.4}$$

$$\gamma = \sqrt{zy} = \alpha + j\beta \tag{12.5}$$

The transmission line is described by complex parameters, ***ABCD***, which relate the voltage and current at the sending end to those at the receiving end. The parameters A and D are dimensionless, B has dimensions of Ω and C has dimensions of ℧.

The characteristic impedance, Z_C:

It is the square root of the ratio of the line series impedance per unit length to the line shunt admittance per unit length. It is given by (12.4). The reciprocal of Z_C is defined as the characteristic admittance, Y_C.

The propagation constant, γ:

It is the square root of the product of the line series impedance per unit length and the line shunt admittance per unit length as given by (12.5). Thus, γ is a complex number. The real part, α, is defined as the attenuation constant and the imaginary part, β, is called 'the phase constant'.

The natural power, P_n:

It is the power transmitted by the line that is terminated by its characteristic impedance. Natural power is often used to indicate the nominal capability of the line, sometimes called the 'characteristic impedance loading' or 'surge impedance loading, *SIL*' of the transmission line. It is defined as

$$P_n \triangleq SIL = \frac{V^2}{Z_C^*} \tag{12.6}$$

where V is the line-to-line voltage and P_n is a three-phase power. It includes both active and reactive power. If V is the line-to-neutral voltage, the value of P_n is per phase.

The total line angle, θ:

$$\theta = \mathrm{Im}(\gamma l) = \beta l = \frac{2\pi l}{\lambda}\ \text{rad} \tag{12.7}$$

where λ is the wavelength of the line. For fixed line parameters, the line angle is a constant.

Based on the definitions of the transmission line parameters given above, some concepts are discussed for two cases in the next two forthcoming sections.

The first case is the lossless case where the real part of both z and y are assumed to be zero. The second is a real case in which the resistance of the system is considered [1].

12.2 Compensation of lossless transmission line

A lossless transmission line uniformly compensated through its length is considered. Although this case is not real and cannot be seen in practice, it gives a good understanding of reactive compensation impact through simplified mathematics.

From (12.4) and (12.5), the characteristic impedance and propagation constant are

$$\boldsymbol{Z}_C = \sqrt{\frac{x_L}{b_c}} = R_C \tag{12.8}$$

$$\gamma = j\beta l = jl\sqrt{x_L b_c} \tag{12.9}$$

It is noted that Z_C is a real number, R_C, and γ becomes purely imaginary. Consequently, the sinh function is purely imaginary. Accordingly and from (12.3), the following relations can be written:

$$\left.\begin{aligned} \operatorname{Im}\boldsymbol{B} &= R_C \operatorname{Im}(\sinh\gamma l) \\ \operatorname{Im}\boldsymbol{C} &= \frac{1}{R_C}\operatorname{Im}(\sinh\gamma l) = \frac{\operatorname{Im}\boldsymbol{B}}{R_C^2} \end{aligned}\right\} \tag{12.10}$$

Thus,

$$\frac{\operatorname{Im}\boldsymbol{B}}{\operatorname{Im}\boldsymbol{C}} = R_C^2 = \frac{x_L}{b_c} \tag{12.11}$$

The line angle as defined by (12.7) is computed for lossless line as

$$\theta = \beta l = l\operatorname{Im}\left(\sqrt{zy}\right) = l\sqrt{x_L b_c} \tag{12.12}$$

From (12.6), the three-phase natural power is

$$P_n = \frac{V^2}{R_C}\text{MW} \tag{12.13}$$

where V is the line-to-line voltage in kV and R_C is the characteristic impedance in ohms. It is noted that this relation does not include reactive power and the voltage profile across the line when operating at its natural loading is flat.

12.2.1 Determination of amount of series compensation

It has been shown that it is worthy to operate the line at its natural loading. Satisfying this condition for all lines in the system is not practical as the line loadings depend on the load, generation and system configuration. So it is better to attempt to change the natural loading, i.e. *SIL* of each line, by using series compensation to conform to the flow on that line where the inductive reactance of the line is varied by varying the amount of series compensation. This can be illustrated by the relations below that determine the compensated line parameters as a ratio to the uncompensated case. The subscript o is added to denote the uncompensated case.

The characteristic impedance ratio is

$$\frac{R_C}{R_{Co}} = \sqrt{\frac{x_L}{b_{co}} \frac{b_{co}}{x_{Lo}}} = \sqrt{\frac{x_L}{x_{Lo}}} \tag{12.14}$$

The ratio of line angle is computed as

$$\frac{\theta}{\theta_o} = \sqrt{\frac{x_L b_{co}}{x_{Lo} b_{co}}} = \sqrt{\frac{x_L}{x_{Lo}}} \tag{12.15}$$

and the ratio of natural power is given by

$$\frac{P_n}{P_{no}} = \frac{R_{Co}}{R_C} = \sqrt{\frac{x_{Lo}}{x_L}} \tag{12.16}$$

It is to be noted that in (12.14) through (12.16) the capacitive susceptance, b_{co}, is unchanged for series compensated line. In addition, the subscript C as an upper case letter refers to the characteristic impedance while the subscript c as a lower case letter refers to the capacitive susceptance.

The total reactance of series compensation, X_C, is commonly expressed as a percentage of the total line inductive reactance, which is called the 'degree of series compensation'. As it is assumed that the lossless line is uniformly compensated, the net line reactance per unit length is written as

$$x_L = x_{Lo} - \frac{X_C}{l} \tag{12.17}$$

and the degree of series compensation, k, is defined as

$$k = \frac{X_C}{\text{Im}\,\boldsymbol{B}_o} \tag{12.18}$$

Another definition of the degree of series compensation, which is referred to as the nominal degree of series compensation, k_{nom}, is expressed as

$$k_{nom} = \frac{X_C}{x_{Lo} l} \tag{12.19}$$

Both of the above definitions may be used in accordance with (12.18) and (12.19). In terms of k_{nom} and using (12.17) the ratios of the characteristic impedance, line angle and natural power (12.14–12.16) can be rewritten as

$$\frac{R_C}{R_{Co}} = \frac{\theta}{\theta_o} = \sqrt{\frac{x_L}{x_{Lo}}} = \sqrt{1 - k_{nom}} \tag{12.20}$$

$$\frac{P_n}{P_{no}} = \frac{1}{\sqrt{1 - k_{nom}}} \tag{12.21}$$

Therefore, increasing the degree of series compensation increases the natural power and decreases the characteristic impedance and line angle as well.

The increase of natural power results from the amount of reactive power supplied by the series compensation. To determine this amount, it is supposed that the series capacitors are added uniformly along the length of the lossless line in order to increase the natural power to $\mathcal{K}$ times its uncompensated value, $\mathcal{K} > 1$. Thus, (12.21) gives

$$\frac{1}{\sqrt{1 - k_{nom}}} = \mathcal{K} \text{ and then}$$

$$k_{nom} = 1 - \frac{1}{\mathcal{K}^2} \tag{12.22}$$

Incorporating (12.19) and (12.22) obtains

$$X_C = \left(1 - \frac{1}{\mathcal{K}^2}\right) l x_{Lo} \tag{12.23}$$

Hence, the three-phase reactive power supplied by series compensation, ΔQ_C, can be computed as

$$\Delta Q_C = 3X_C(\mathcal{K} I_{no})^2 \tag{12.24}$$

Substituting X_C from (12.23) to (12.24) gives

$$\Delta Q_C = (\mathcal{K}^2 - 1) l P_{no} \sqrt{x_{Lo} b_{co}} \tag{12.25}$$

The reactive power can be expressed as a ratio to the uncompensated power as

$$\frac{\Delta Q_C}{P_{no}} = (\mathcal{K}^2 - 1)\, l\, \sqrt{x_{Lo} b_{co}} \tag{12.26}$$

From (12.20) and (12.22) the ratio of line angle is

$$\frac{\theta}{\theta_o} = \frac{1}{\mathcal{K}} \tag{12.27}$$

Example 12.1 Typical data of an overhead transmission line are given as: nominal voltage 345 kV, $r_L = 0.037$ Ω/km, $X_L = 0.367$ Ω/km, $b_c = 4.518$ μs/km, and the total length of the line is 160 km. The line is bundled conductors and r_L, x_L, b_c are per phase values.

Considering the line as a lossless line (neglecting the resistance), find the electric line parameters. If the natural power, SIL, is required to be 1.5 times the nominal value by using series compensation, calculate the degree of compensation and the amount of reactive compensation.

Solution:

The characteristic impedance $\boldsymbol{Z}_C = \sqrt{\frac{x_L}{b_c}} = R_C = 10^3\sqrt{\frac{0.367}{4.518}} = 285\,\Omega$

The propagation constant $\gamma = j\beta = j\sqrt{x_L b_c}$ where $\beta = 10^{-3}\sqrt{0.367 \times 4.518} =$ $0.00129\ \text{rad/km} = 0.074°/\text{km}$

The line angle $\theta = \beta l = 0.00129 \times 160 = 0.2064\ \text{rad} = 11.83°$

$$P_n \triangleq SIL = \frac{V^2}{R_C} = 417.63\ \text{MW (three phase)}$$

When the line is series compensated to increase P_n to be 1.5×417.63 MW, the required degree of series compensation 'k_{nom}' can be obtained using (12.22). Thus,

$$k_{nom} = 1 - (4/9) = 55.6\%$$

Using (12.25), the amount of reactive compensation, $\Delta Q_C = 107.55$ MVAR and using (12.27), the line angle $= 0.2064/1.5 = 0.1376\ \text{rad} = 7.89°$

12.2.2 Transient stability improvement for lossless compensated line

The series compensation supplies the system a reactive power that increases the natural power of the transmission line, which in turn increases its power transfer capacity. This results in an improvement of the system transient stability. The amount of reactive compensation can be determined by examining the method of compensation as well as the system operation when subjected to different faults.

Usually, a three-phase fault is considered as the most severe fault. The analysis below illustrates this concept through studying a simple uncompensated and compensated power system (one machine to an infinite bus).

For uncompensated system, the voltages at the two ends of the transmission line, sending and receiving, are $\boldsymbol{V}_S$ and $\boldsymbol{V}_R$, respectively. The receiving end voltage, $\boldsymbol{V}_R$, at the infinite bus is taken as a reference. The total reactance between the sending and receiving ends is denoted by X_L and the angle between $\boldsymbol{V}_S$ and $\boldsymbol{V}_R$ is denoted by δ. The equivalent circuit is shown in Figure 12.1.

$$\text{It is seen that } \boldsymbol{V}_S = V_S e^{j\delta} \quad \text{and} \quad \boldsymbol{V}_R = V_R\,\angle 0 \tag{12.28}$$

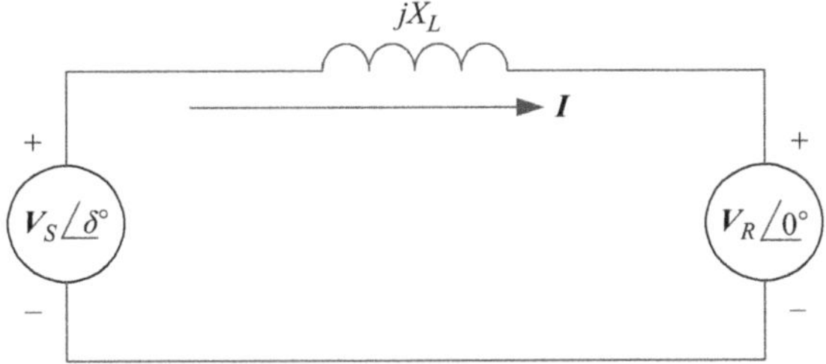

Figure 12.1 Equivalent circuit of a simple power system

The current flow from sending to receiving end, $\boldsymbol{I}$, is given by

$$\boldsymbol{I} = \frac{\boldsymbol{V}_S - \boldsymbol{V}_R}{jX_L} = \frac{V_S \sin\delta}{X_L} - j\frac{V_S \cos\delta - V_R}{X_L} \tag{12.29}$$

The complex power phasor at both the sending and the receiving ends can be computed as

$$\left.\begin{aligned} P_S + jQ_S &= \boldsymbol{V}_S \boldsymbol{I}^* \\ &= \left(\frac{V_S V_R \sin\delta}{X_L}\right) + j\left(\frac{V_S^2 - V_S V_R \cos\delta}{X_L}\right) \end{aligned}\right\} \tag{12.30}$$

$$\left.\begin{aligned} P_R + jQ_R &= \boldsymbol{V}_R \boldsymbol{I}^* \\ &= \left(\frac{V_S V_R \sin\delta}{X_L}\right) - j\left(\frac{V_R^2 - V_S V_R \cos\delta}{X_L}\right) \end{aligned}\right\} \tag{12.31}$$

For compensated system:
The voltages at sending and receiving end are obtained by the same (12.28). Assuming the reactance of the series capacitors is X_C, the relations to calculate the power at the two ends are the same as above but replacing the reactance X_L by X. From (12.17) and (12.19), it can be found that

$$X = X_L(1 - k_{nom}) \tag{12.32}$$

To calculate the reactive power of the capacitors, Q_C, using (12.29), first, it is seen that

$$I^2 = \frac{1}{X^2}\left(V_S^2 + V_R^2 - 2V_S V_R \cos\delta\right) \tag{12.33}$$

Using (12.32) the value (X_C/X^2) can be computed as

$$\frac{X_C}{X} = \frac{X_C}{X_L(1 - k_{nom})} = \frac{k_{nom}}{(1 - k_{nom})} \tag{12.34}$$

Hence,

$$\frac{X_C}{X^2} = \frac{1}{X}\frac{X_C}{X} = \frac{1}{X}\frac{k_{nom}}{1 - k_{nom}} = \frac{k_{nom}}{X_L(1 - k_{nom})^2} \quad (12.35)$$

Thus, from (12.33) and (12.35)

$$Q_C = I^2 X_C = \frac{k_{nom}}{X_L(1 - k_{nom})^2}\left(V_S^2 + V_R^2 - 2V_S V_R \cos\delta\right) \quad (12.36)$$

Example 12.2 The parameters of the equivalent circuit (Figure 12.1) have the per unit values as

$$V_S = 1.2\angle 11^\circ, \quad V_R = 1.0\angle 0^\circ, \quad X_L = 0.275$$

Find the power at both ends when the system is uncompensated as well as compensated at 50 per cent degree of compensation.

Solution:

It is noted that the subscript o is added to the parameters in case of uncompensated system. Using (12.30) and (12.31), the powers at the two ends of the transmission line are

$$P_{So} + jQ_{So} = \frac{1.2 \sin 11^\circ}{0.275} + j\frac{1.44 - 1.2\cos 11^\circ}{0.275} = 0.833 + j0.945 \text{ pu}$$

$$P_{Ro} + jQ_{Ro} = \frac{1.2 \sin 11^\circ}{0.275} + j\frac{1 - 1.2\cos 11^\circ}{0.275} = 0.833 + j0.6545 \text{ pu}$$

It is seen that P_{So} and P_{Ro} are equal as the system is lossless while Q_{So} and Q_{Ro} differ by the reactive power absorbed by the system reactance.

In case of compensation, $X = X_L(1 - k_{nom}) = 0.275(1 - 0.5) = 0.1375$ pu

Similarly, $P_S + jQ_S = 1.665 + j1.891$ and $P_R + jQ_R = 1.665 + j1.309$ pu

Therefore, the power transfer capacity of the line at the same power angle, δ, is doubled by the series compensation. It means that by referring to equal area criterion, the area between the P–δ curve and the input power line is increased resulting in more marginal stability and security.

As computed by (12.36) the reactive power of the capacitors is

$$Q_C = \frac{k_{nom}}{X_L(1 - k_{nom})^2}\left(V_S^2 + V_R^2 - 2V_S V_R \cos\delta\right)$$

$$= \frac{0.5}{0.275(1 - 0.5)^2}(1.44 + 1 - 2.4\cos 11^\circ) = 0.582 \text{ pu}$$

Example 12.3 For the system in Example 8.1-i, find the change of δ versus time at fault clearing time 0.08 s when the system is uncompensated and compensated with 30 per cent and 50 per cent, degree of compensation.

Solution:

Using the initial values of system parameters obtained in Example 8.1, and applying PSAT/MATLAB® toolbox to solve the swing equation, the variation of power angle versus time is shown in Figure 12.2. It is seen that the stability of the system is improved by compensation.

For high-voltage lines, the resistance is small compared to the reactance. In addition, for overhead lines, the real part of the line admittance is zero. Thus, these lines may be approximated with a little error as lossless lines. For long transmission lines, further detailed analysis to describe the transmission lines 'in a more realistic manner' is given in Section 12.3.

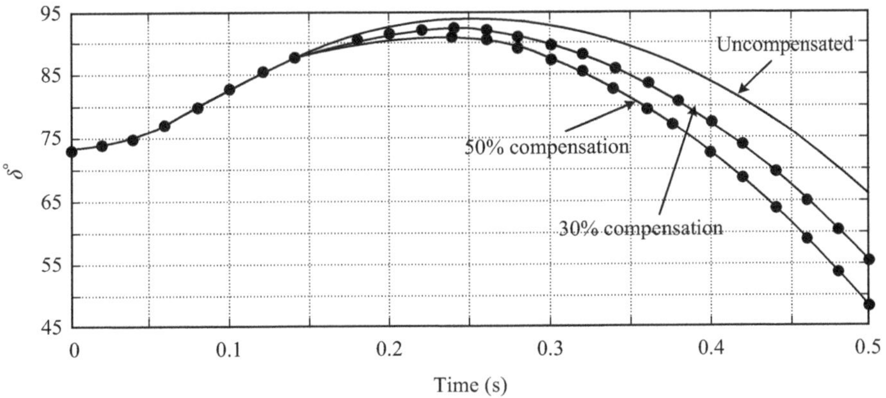

Figure 12.2 Power angle variation versus time

12.3 Long transmission lines

As explained in Chapter 4, Section 4.2.1, voltage and current, as phasors, $\boldsymbol{V}_s$ and $\boldsymbol{I}_s$, at a point distance s from the receiving end, are calculated using (4.22). For convenience they are written again as

$$\left.\begin{aligned} \boldsymbol{V}_s &= \frac{\boldsymbol{V}_R + Z_c\boldsymbol{I}_R}{2}\mathrm{e}^{\gamma s} + \frac{\boldsymbol{V}_R - \boldsymbol{Z}_c\boldsymbol{I}_R}{2}\mathrm{e}^{-\gamma s} \\ \boldsymbol{I}_s &= \frac{(\boldsymbol{V}_R/\boldsymbol{Z}_C) + \boldsymbol{I}_R}{2}\mathrm{e}^{\gamma s} - \frac{(\boldsymbol{V}_R/\boldsymbol{Z}_C) - \boldsymbol{I}_R}{2}\mathrm{e}^{-\gamma s} \end{aligned}\right\} \tag{12.37}$$

The first term in each equation is called the 'incident component' and the second term is the 'reflected component'. If the line is terminated at the receiving end by a constant impedance load $\boldsymbol{Z}_{Ld}$, then the voltage and current at the receiving end must satisfy the relation:

$$\boldsymbol{V}_R = \boldsymbol{Z}_{Ld}\boldsymbol{I}_R \tag{12.38}$$

Therefore, (12.37) can be expressed as a function of $\boldsymbol{Z}_{Ld}$ as below:

$$\left.\begin{aligned} \boldsymbol{V}_s &= \frac{\boldsymbol{V}_R}{2}\left(1+\frac{\boldsymbol{Z}_C}{\boldsymbol{Z}_{Ld}}\right)e^{\gamma s} + \frac{\boldsymbol{V}_R}{2}\left(1-\frac{\boldsymbol{Z}_C}{\boldsymbol{Z}_{Ld}}\right)e^{-\gamma s} \\ \boldsymbol{I}_s &= \frac{\boldsymbol{V}_R}{2\boldsymbol{Z}_C}\left(1+\frac{\boldsymbol{Z}_C}{\boldsymbol{Z}_{Ld}}\right)e^{\gamma s} - \frac{\boldsymbol{V}_R}{2\boldsymbol{Z}_C}\left(1-\frac{\boldsymbol{Z}_C}{\boldsymbol{Z}_{Ld}}\right)e^{-\gamma s} \end{aligned}\right\} \tag{12.39}$$

A complex constant '$\boldsymbol{\Lambda}$' is defined as the 'reflection coefficient' and written as

$$\boldsymbol{\Lambda} = \frac{\boldsymbol{Z}_{Ld} - \boldsymbol{Z}_C}{\boldsymbol{Z}_{Ld} + \boldsymbol{Z}_C} \tag{12.40}$$

Assuming the impedance ratio $\boldsymbol{Z}_{CLd} = \boldsymbol{Z}_C/\boldsymbol{Z}_{Ld}$, (12.40) can be written in the form

$$\left.\begin{aligned} \boldsymbol{V}_s &= \frac{\boldsymbol{V}_R}{2}(1+\boldsymbol{Z}_{CLd})e^{\gamma s} + \frac{\boldsymbol{V}_R}{2}(1-\boldsymbol{Z}_{CLd})e^{-\gamma s} \\ \boldsymbol{I}_s &= \frac{\boldsymbol{V}_R}{2\boldsymbol{Z}_C}(1+\boldsymbol{Z}_{CLd})e^{\gamma s} - \frac{\boldsymbol{V}_R}{2\boldsymbol{Z}_C}(1-\boldsymbol{Z}_{CLd})e^{-\gamma s} \end{aligned}\right\} \tag{12.41}$$

where

$$\boldsymbol{Z}_{CLd} = \frac{\boldsymbol{Z}_C}{\boldsymbol{Z}_{Ld}} = \frac{Z_C e^{j\theta_C}}{Z_{Ld} e^{j\theta_{Ld}}} = Z_{CLd} e^{j(\theta_C - \theta_{Ld})} \tag{12.42}$$

By examining the case at which $\boldsymbol{Z}_C = \boldsymbol{Z}_{Ld}$, i.e. the line is terminated at its receiving end by impedance equal to the line characteristic impedance, it is found that (12.38) becomes $\boldsymbol{V}_R = \boldsymbol{Z}_C\boldsymbol{I}_R$, and the voltage and current along the line as a function of the distance x can be obtained from (12.41) as

$$\left.\begin{aligned} \boldsymbol{V}_x &= \boldsymbol{V}_R e^{\gamma x} \\ \boldsymbol{I}_x &= \frac{\boldsymbol{V}_R}{\boldsymbol{Z}_C} e^{\gamma x} \end{aligned}\right\} \tag{12.43}$$

It is noted that both the reflection component and the reflection coefficient equal zero, and the driving point impedance at any distance x from the receiving end equals $\boldsymbol{Z}_C$.

In addition, the voltage and current magnitudes increase from the receiving end towards the sending end by the amount of attenuation constant, and the phase angles move linearly in the leading direction. On the other hand, the attenuation constant is usually small; consequently, the voltage magnitude is almost constant along the line and is exactly constant for the lossless line.

As defined by (12.6), the *SIL* is given by

$$SIL = \frac{V^2}{\boldsymbol{Z}_C^*} = \frac{V^2(R_C + jX_C)}{|Z_C|^2} \tag{12.44}$$

The parameter *SIL* provides a useful measure of the line nominal capability as well as the condition at which the line absorbs or delivers reactive power. As the reflected component of the voltage or current vanishes when the line is terminated by its characteristic impedance, the reactive power generated by the distributed line susceptance is exactly the reactive I^2X losses in the line. Therefore, at loadings greater than *SIL* and to maintain the voltage within normal limits it is necessary to supply VArs to the line. Conversely, the line generates excess VArs at loadings less than *SIL*.

It is concluded that to guarantee the network voltages at the different nodes to be nearly constant, the lines terminated by these nodes should be loaded with *SIL*.

The active and reactive power phasor power at the receiving end can be calculated as below.

$$\boldsymbol{S}_R = \boldsymbol{V}_R \boldsymbol{I}_R^* \; VA/\text{phase} \tag{12.45}$$

Substituting: $\boldsymbol{I}_R = (\boldsymbol{V}_S - \boldsymbol{A}\boldsymbol{V}_R)/\boldsymbol{B}$, $\boldsymbol{V}_R = V_R e^{j0}$, $\boldsymbol{V}_S = V_S e^{j\delta}$

$$\boldsymbol{A} = Ae^{j\mu}, \quad \boldsymbol{B} = Be^{j\zeta}$$

Then,

$$\boldsymbol{S}_R = \boldsymbol{V}_R \boldsymbol{I}_R^* = P_R + jQ_R = \left[\frac{V_S V_R}{B}\cos(\delta - \zeta) - \frac{AV_R^2}{B}\cos(\mu - \zeta)\right] - j\left[\frac{V_S V_R}{B}\sin(\delta - \zeta) - \frac{AV_R^2}{B}\cos(\mu - \zeta)\right] \tag{12.46}$$

The angle $\mu \ll \zeta$ and for lossless line $\zeta = 90°$, V_S and V_R are phase voltages in kV, thus

$$P_R = \frac{V_S V_R}{X}\sin\delta \tag{12.47}$$

The maximum active power at receiving end, P_{Rmax}, can be computed by using (12.46) and substituting $\delta° = \zeta°$ to obtain

$$P_{Rmax} = \frac{V_S V_R}{B} - \frac{AV_R^2}{B}\cos(\mu - \zeta) \; \text{MW/phase} \tag{12.48}$$

12.3.1 Series compensation for long transmission lines

In practice, series compensation is carried out by inserting capacitor banks in series with the transmission line located somewhere along the line or at its two ends. So, the compensation is not uniformly distributed along the line as it is assumed in the

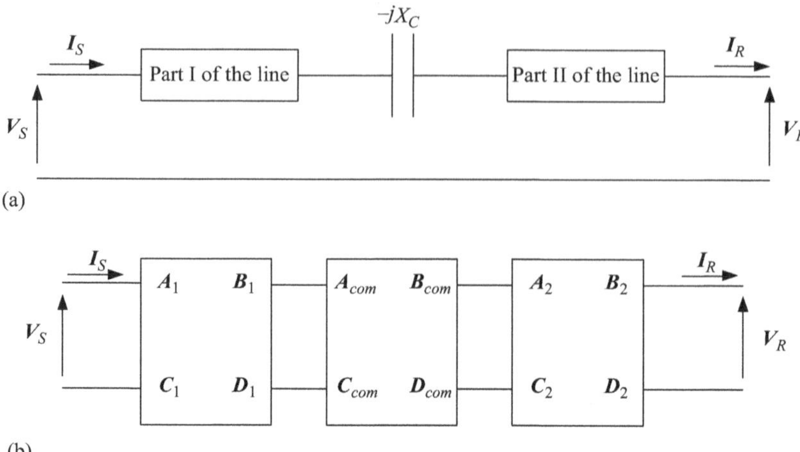

Figure 12.3 (a) Compensated transmission line and (b) representation of compensated line

former analysis. To describe the line in this case, the simplest way is to derive the line equations in terms of the equivalent $\boldsymbol{ABCD}$ parameters as below.

Consider the compensated line shown in Figure 12.3(a), where the series capacitors are located at a specific distance from the receiving end dividing the line into two parts described by $\boldsymbol{A}_1\boldsymbol{B}_1\boldsymbol{C}_1\boldsymbol{D}_1$ and $\boldsymbol{A}_2\boldsymbol{B}_2\boldsymbol{C}_2\boldsymbol{D}_2$ parameters. The system can be represented as shown in Figure 12.3(b). The compensation can be described by $\boldsymbol{A}_{com}\boldsymbol{B}_{com}\boldsymbol{C}_{com}\boldsymbol{D}_{com}$ where $\boldsymbol{A}_{com} = \boldsymbol{D}_{com} = 1$, $\boldsymbol{C}_{com} = 0$, $\boldsymbol{B}_{com} = -jX_C$ and the relation of $\boldsymbol{V}_S$ and $\boldsymbol{I}_S$ as a function of $\boldsymbol{V}_R$ and $\boldsymbol{I}_R$ is

$$\begin{bmatrix} \boldsymbol{V}_S \\ \boldsymbol{I}_S \end{bmatrix} = \begin{bmatrix} \boldsymbol{A}_1 & \boldsymbol{B}_1 \\ \boldsymbol{C}_1 & \boldsymbol{D}_1 \end{bmatrix} \begin{bmatrix} 1 & -jX_C \\ 0 & 1 \end{bmatrix} \begin{bmatrix} \boldsymbol{A}_2 & \boldsymbol{B}_2 \\ \boldsymbol{C}_2 & \boldsymbol{D}_2 \end{bmatrix} \begin{bmatrix} \boldsymbol{V}_R \\ \boldsymbol{I}_R \end{bmatrix}$$
$$= \begin{bmatrix} \boldsymbol{A}_1\boldsymbol{A}_2 - jX_C\boldsymbol{A}_1\boldsymbol{C}_2 + \boldsymbol{B}_1\boldsymbol{C}_2 & \boldsymbol{A}_1\boldsymbol{B}_2 - jX_C\boldsymbol{D}_2\boldsymbol{A}_1 + \boldsymbol{B}_1\boldsymbol{D}_2 \\ \boldsymbol{C}_1\boldsymbol{A}_2 - jX_C\boldsymbol{C}_1\boldsymbol{C}_2 + \boldsymbol{D}_1\boldsymbol{C}_2 & \boldsymbol{C}_1\boldsymbol{B}_2 - jX_C\boldsymbol{C}_1\boldsymbol{D}_2 + \boldsymbol{D}_1\boldsymbol{D}_2 \end{bmatrix} \begin{bmatrix} \boldsymbol{V}_R \\ \boldsymbol{I}_R \end{bmatrix} \quad (12.49)$$

If the line is uncompensated and comprises of two parts in cascade described by $\boldsymbol{A}_1\boldsymbol{B}_1\boldsymbol{C}_1\boldsymbol{D}_1$ and $\boldsymbol{A}_2\boldsymbol{B}_2\boldsymbol{C}_2\boldsymbol{D}_2$, respectively, then the parameters of the total line are

$$\boldsymbol{A}_{\mathrm{o}} = \boldsymbol{A}_1\boldsymbol{A}_2 + \boldsymbol{B}_1\boldsymbol{C}_2\boldsymbol{B}_{\mathrm{o}} = \boldsymbol{A}_1\boldsymbol{B}_2 + \boldsymbol{B}_1\boldsymbol{D}_2$$
$$\boldsymbol{C}_{\mathrm{o}} = \boldsymbol{C}_1\boldsymbol{A}_2 + \boldsymbol{D}_1\boldsymbol{C}_2\boldsymbol{D}_{\mathrm{o}} = \boldsymbol{C}_1\boldsymbol{B}_2 + \boldsymbol{D}_1\boldsymbol{D}_2$$

It is noted that the subscript o denotes the uncompensated case. Equation (12.49) can be rewritten in the form

$$\begin{bmatrix} \boldsymbol{V}_S \\ \boldsymbol{I}_S \end{bmatrix} = \begin{bmatrix} \boldsymbol{A}_{eq} & \boldsymbol{B}_{eq} \\ \boldsymbol{C}_{eq} & \boldsymbol{D}_{eq} \end{bmatrix} \begin{bmatrix} \boldsymbol{V}_R \\ \boldsymbol{I}_R \end{bmatrix} \quad (12.50)$$

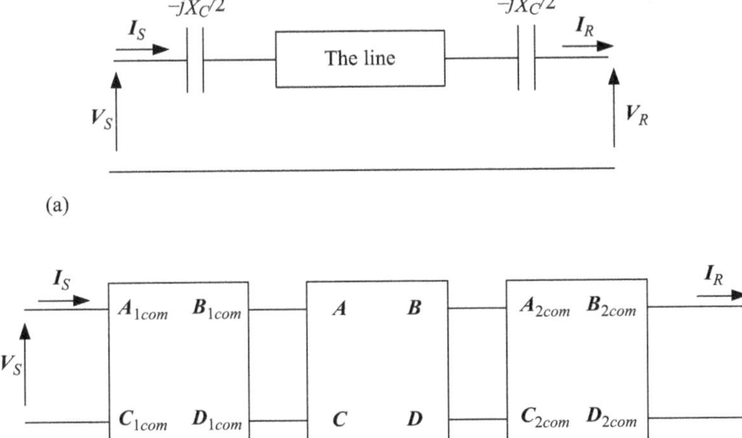

Figure 12.4 (a) Transmission line with series compensation at both ends and (b) equivalent representation

where

$$\boldsymbol{A}_{eq} = \boldsymbol{A}_{\mathrm{o}} - jX_C\boldsymbol{A}_1\boldsymbol{C}_2\boldsymbol{B}_{eq} = \boldsymbol{B}_{\mathrm{o}} - jX_C\boldsymbol{A}_1\boldsymbol{D}_2 = \boldsymbol{B}_{\mathrm{o}} + \Delta\boldsymbol{B}$$
$$\boldsymbol{C}_{eq} = \boldsymbol{C}_{\mathrm{o}} - jX_C\boldsymbol{C}_1\boldsymbol{C}_2\boldsymbol{D}_{eq} = \boldsymbol{D}_{\mathrm{o}} - jX_C\boldsymbol{C}_1\boldsymbol{D}_2$$

The same procedure can be applied to a line compensated at the sending and receiving end by equal capacitance ($-jX_C/2$) (Figure 12.4).

The equivalent ***ABCD*** parameters are given by

$$\begin{bmatrix} \boldsymbol{A}_{eq} & \boldsymbol{B}_{eq} \\ \boldsymbol{C}_{eq} & \boldsymbol{D}_{eq} \end{bmatrix} = \begin{bmatrix} 1 & -jX_C/2 \\ 0 & 1 \end{bmatrix} \begin{bmatrix} \boldsymbol{A} & \boldsymbol{B} \\ \boldsymbol{C} & \boldsymbol{D} \end{bmatrix} \begin{bmatrix} 1 & -jX_C/2 \\ 0 & 1 \end{bmatrix}$$
$$= \begin{bmatrix} \boldsymbol{A} - j\boldsymbol{C}X_C/2 & \boldsymbol{B} - \boldsymbol{C}\dfrac{X_C^2}{4} - j(\boldsymbol{A}+\boldsymbol{D})\dfrac{X_C}{2} \\ \boldsymbol{C} & \boldsymbol{D} - j\boldsymbol{C}X_C/2 \end{bmatrix} \tag{12.51}$$

It is to be noted that the location of the series capacitors along the transmission line, e.g. lumped at midpoint, equally divided at line ends or two equal capacitors at equal distances, affects the value of equivalent ***ABCD*** parameters, which in turn changes the transmission power transfer [2, 3]. From (12.51) the value of parameter ***B*** becomes

$$\boldsymbol{B}_{eq} = \boldsymbol{B} + \Delta\boldsymbol{B} \quad \text{where } \Delta\boldsymbol{B} = -\boldsymbol{C}\frac{X_C^2}{4} - j(\boldsymbol{A}+\boldsymbol{D})\frac{X_C}{2}$$

It is seen that $\Delta\boldsymbol{B}$ represents the change of ***B*** due to compensation. It can be considered an index to indicate the impact of compensation on the line reactance, i.e. the effectiveness of compensation [4]. As shown in (12.50) and (12.51) the value of $\Delta \boldsymbol{B}$ differs, and thus the effectiveness of series compensation varies according to its location.

Example 12.4 A single machine represented by an internal voltage source, $E = 1.4\angle 17°$ behind a reactance is connected to an infinite bus through a transformer and two identical parallel transmission lines as shown in Figure 12.5. The data of the transmission line are given below.

$r = 0.028$ Ω/km, $x_L = 0.325$ Ω/km, $b_c = 5.2$ μs/km, $l = 160$ km, the nominal voltage = 500 kV. Find the transmission network parameters '***ABCD***' that relate the machine voltage and current to the voltage and current at the receiving end 'infinite bus' for two cases: (i) uncompensated transmission lines and (ii) each line is compensated by series capacitors with 50 per cent degree of compensation located at the midpoint of the line.

If a fault occurs at point F (Figure 12.5) and cleared at power angle $\delta = 50°$ by isolating the faulty line, examine the transient stability of the system to indicate how much the stability is enhanced.

Solution:

For each transmission line

$$Z_C = \sqrt{\frac{z}{y}} = 250\,\Omega$$

$$\gamma = \sqrt{zy} = \alpha + j\beta \cong j\sqrt{xy}\left(1 - j\frac{r}{2x}\right)$$

$$\alpha = 0.000057 \text{ nepers/km}, \ \beta = 0.0013 \text{ rad/km}$$

$$X_L = 0.325 \times 160/250 = 0.208 \text{ pu of } Z_C$$

$$B_c = 5.2 \times 10^{-6} \times 160 \times 250 = 0.208 \text{ pu of } Z_C$$

$$\beta l = 0.208 \text{ rad} = 11.9°$$

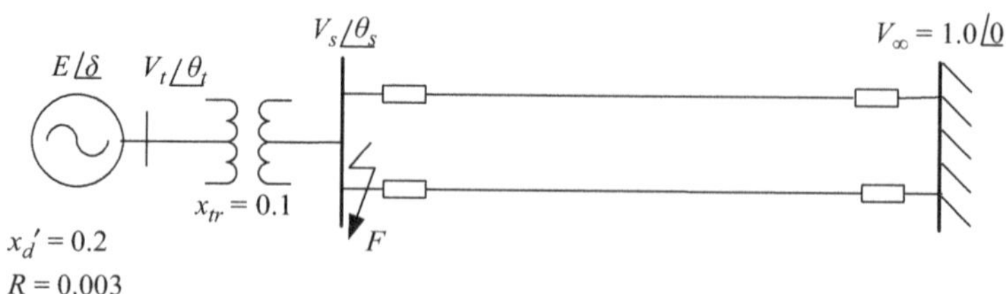

Figure 12.5 System of Example 12.4

As $\alpha \ll \beta$ it is assumed that $\gamma = j\beta$ to simplify the calculations without loss of generality.

Applying (12.3), the ***ABCD*** parameters of the line are

$$\boldsymbol{A}_L = 0.978\angle 0^\circ, \quad \boldsymbol{B}_L = j0.206 \text{ pu}, \quad \boldsymbol{C}_L = j0.206 \text{ pu}, \quad \boldsymbol{D}_L = 0.978\angle 0^\circ \quad (12.52)$$

For uncompensated transmission lines:

When one of the two lines is isolated* (Figure 12.6), the system ***ABCD* parameters are

$$\boldsymbol{A} = \boldsymbol{A}_L + j0.3\boldsymbol{C}_L = 0.916\angle 0^\circ$$
$$\boldsymbol{B} = \boldsymbol{B}_L + j0.3\boldsymbol{D}_L = j0.499 \text{ pu}$$
$$\boldsymbol{C} = \boldsymbol{C}_L = j0.206 \text{ pu}$$
$$\boldsymbol{D} = \boldsymbol{D}_L = 0.978\angle 0^\circ$$

Using (12.46), the active power transfer capacity at the receiving end is obtained as

$$P_R = (1.4/0.499)\sin 17^\circ = 0.82 \text{ pu/phase}$$

When the two lines are connected to the system in parallel* (Figure 12.7), the system ***ABCD* parameters are calculated as below

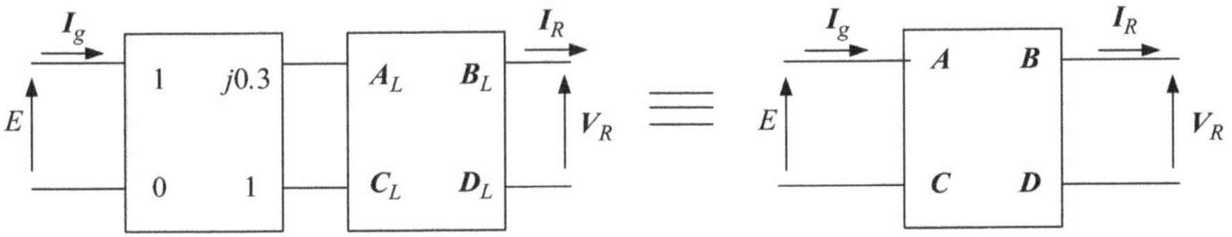

*Figure 12.6 Equivalent **ABCD** parameters of the system 'from machine voltage source to the infinite bus' with one line in operation*

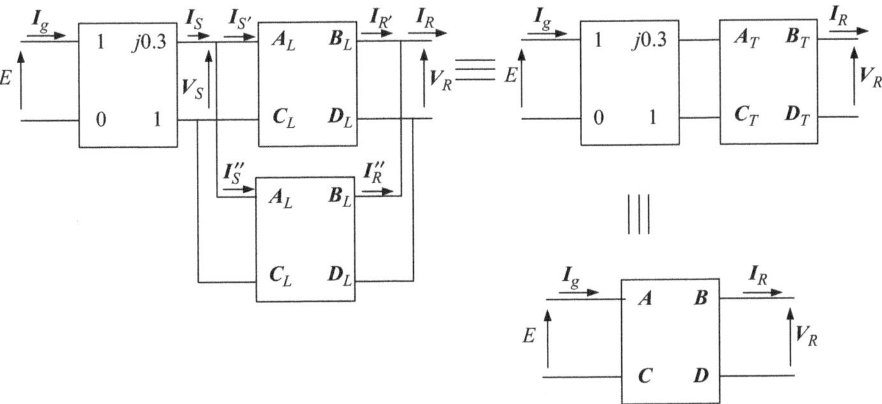

*Figure 12.7 Equivalent **ABCD** parameters of the system (from machine voltage source to the infinite bus) with two parallel lines in operation*

$\boldsymbol{ABCD}$ parameters are obtained by two steps: *the first* is to get the equivalent $\boldsymbol{A_T B_T C_T D_T}$ parameters for the two parallel lines:

$$\boldsymbol{I}_S = \boldsymbol{I}'_S + \boldsymbol{I}''_S \quad \text{and} \quad \boldsymbol{I}_R = \boldsymbol{I}'_R + \boldsymbol{I}''_R \tag{12.53}$$

$$\left.\begin{aligned} \boldsymbol{V}_S &= \boldsymbol{A}_L \boldsymbol{V}_R + \boldsymbol{B}_L \boldsymbol{I}'_R \\ \boldsymbol{V}_S &= \boldsymbol{A}_L \boldsymbol{V}_R + \boldsymbol{B}_L \boldsymbol{I}''_R \end{aligned}\right\} \tag{12.54}$$

Summation of the two equations, (12.54) gives

$$\boldsymbol{V}_S = \boldsymbol{A}_L \boldsymbol{V}_R + \frac{\boldsymbol{B}_L}{2} \boldsymbol{I}_R \tag{12.55}$$

Similarly,

$$\left.\begin{aligned} \boldsymbol{I}'_S &= \boldsymbol{C}_L \boldsymbol{V}_R + \boldsymbol{D}_L \boldsymbol{I}'_R \\ \boldsymbol{I}''_S &= \boldsymbol{C}_L \boldsymbol{V}_R + \boldsymbol{D}_L \boldsymbol{I}''_R \end{aligned}\right\} \tag{12.56}$$

Hence,

$$\boldsymbol{I}_S = 2\boldsymbol{C}_L \boldsymbol{V}_R + \boldsymbol{D}_L \boldsymbol{I}_R \tag{12.57}$$

From (12.55) and (12.57), it is observed that

$$\boldsymbol{A}_T = \boldsymbol{A}_L, \quad \boldsymbol{B}_T = \frac{\boldsymbol{B}_L}{2}, \quad \boldsymbol{C}_T = 2\boldsymbol{C}_L, \quad \boldsymbol{D}_T = \boldsymbol{D}_L \tag{12.58}$$

The second step is to get the system $\boldsymbol{ABCD}$ *parameters as below.*

Using (12.52) and (12.58) to obtain

$$\boldsymbol{A} = \boldsymbol{A}_T + j0.3\boldsymbol{C}_T = 0.854\angle 0^\circ, \quad \boldsymbol{B} = \boldsymbol{B}_T + j0.3\boldsymbol{D}_T = j0.396 \text{ pu}$$

$$\boldsymbol{C} = \boldsymbol{C}_T = j0.412 \text{ pu}, \quad \boldsymbol{D} = \boldsymbol{D}_T = 0.854\angle 0^\circ$$

The active power transfer capacity at the receiving end is

$$P_R = (1.4/0.396)\sin 17^\circ = 1.034 \text{ pu/phase}$$

For compensated transmission lines:

The capacitors are located at the midpoint of the transmission line. So, the line is divided into two equal parts of length 80 km and equal $\boldsymbol{A}_{L1}\boldsymbol{B}_{L1}\boldsymbol{C}_{L1}\boldsymbol{D}_{L1}$ parameters. Thus,

$$\boldsymbol{A}_{L1} = 0.995\angle 0^\circ, \quad \boldsymbol{B}_{L1} = j0.194 \text{ pu}, \quad \boldsymbol{C}_{L1} = j0.104 \text{ pu}, \quad \boldsymbol{D}_{L1} = 0.995\angle 0^\circ$$

**When one of the two lines is isolated:*

$$X_C = 50\% X_L = 0.104 \text{ pu}$$

Applying the parameters as written in (12.50) and taking into account that $\boldsymbol{A_o B_o C_o D_o}$ are the parameters of the uncompensated line given by (12.52), obtain $\boldsymbol{A_L B_L C_L D_L}$ for the compensated line as

$$\left.\begin{aligned} \boldsymbol{A}_L &= 0.978 - j0.104 \times 0.995 \times j0.104 = 0.989\angle 0^\circ \\ \boldsymbol{B}_L &= j0.206 - j0.104 \times 0.995 \times 0.995 = j0.103 \text{ pu} \\ \boldsymbol{C}_L &= j0.206 - j0.104 \times j0.104 \times j0.104 = j0.207 \text{ pu} \\ \boldsymbol{D}_L &= 0.978 - j0.104 \times j0.104 \times 0.995 = 0.989\angle 0^\circ \end{aligned}\right\} \quad (12.59)$$

Consequently, the system $\boldsymbol{ABCD}$ parameters are

$$\boldsymbol{A} = 0.989 + j0.3 \times j0.207 = 0.927\angle 0^\circ$$

$$\boldsymbol{B} = j0.103 + j0.3 \times 0.989 = j0.4 \text{ pu}$$

$$\boldsymbol{C} = \boldsymbol{C}_L = j0.207 \text{ pu}$$

$$\boldsymbol{D} = \boldsymbol{D}_L = 0.989\angle 0^\circ$$

Accordingly, the active power transfer capacity at receiving end is

$$P_R = (1.4/0.4)\sin 17^\circ = 1.023 \text{ pu/phase}$$

**When the two compensated lines are connected to the system in parallel:*

The same procedure is implemented as in the case of uncompensated lines but taking into account the values of parameters with compensation. Using (12.59), it is observed that

$$\boldsymbol{A}_T = \boldsymbol{A}_L = 0.989\angle 0^\circ, \quad \boldsymbol{B}_T = \boldsymbol{B}_L/2 = j0.052 \text{ pu}$$

$$\boldsymbol{C}_T = 2\boldsymbol{C}_L = j0.414 \text{ pu}, \quad \boldsymbol{D}_T = \boldsymbol{D}_L = 0.989\angle 0^\circ$$

Therefore, the system $\boldsymbol{ABCD}$ parameters are

$$\boldsymbol{A} = 0.989 + j0.3 \times j0.414 = 0.865\angle 0^\circ$$

$$\boldsymbol{B} = j0.052 + j0.3 \times 0.989 = j0.349 \text{ pu}$$

$$\boldsymbol{C} = j0.414 \text{ pu}$$

$$\boldsymbol{D} = 0.989\angle 0^\circ$$

The active power transfer capacity at receiving end is

$$P_R = (1.4/0.349)\sin\ 17^\circ = 1.173 \text{ pu/phase}$$

Examination of transient stability

It is of interest to evaluate the benefits that may be gained from the series compensation. As explained in Chapter 9, the equal area criterion can be used to check the system stability and to know how much the system is secure. Assuming that the active power received is 0.8 pu, the system analysis when the lines are either uncompensated or compensated is given below.

For the system with uncompensated lines:

Before the fault: $0.8 = (1.4/0.396)\sin\delta_o$ and $\delta_o = 13.1°$

After the fault: $0.8 = (1.4/0.499)\sin\delta_m$ and $\delta_m = 163.4°$

The area between the input power line and P versus δ curve during the fault

$A_1 = 0.8(50 - 13.1)\pi/180 = 0.515$ pu, where the electric power is zero

The area between the input power line and P versus δ curve after the fault:

$$\begin{aligned} A_2 &= (EV_R/B)(\cos\delta_c - \cos\delta_m) - 0.8(\delta_m - \delta_c)\pi/180 \\ &= (1.4/0.499)(0.643 + 0.958) - 1.58 = 2.92 \text{ pu} \end{aligned}$$

For the system with compensated lines:

Before the fault: $0.8 = (1.4/0.349)\sin\delta_o$ and $\delta_o = 11.5°$

After the fault: $0.8 = (1.4/0.4)\sin\delta_m$ and $\delta_m = 166.8°$

$$A_1 = 0.8(50 - 11.5)\pi/180 = 0.537 \text{ pu}$$

$$A_2 = (1.4/0.4)(0.643 + 0.974) - 0.8(166.8 - 50)\pi/180 = 4.03 \text{ pu}$$

It is concluded that:

- As in Table 12.1, the active power transfer capacity of the system at a specific power angle is increased by compensating the lines with series capacitors.
- The marginal stability increases with line compensation and the system is getting more secure. This can be noted by the ratio, A_2/A_1 (Table 12.2), where the increase due to compensation is about 32.27 per cent.

Table 12.1 Summary of results

System state	**Connected lines**	$A\angle 0°$	B **(pu)**	C **(pu)**	$D\angle 0°$	P_R **(pu/ph)**
Uncompensated	One line	0.916	*j*0.499	*j*0.206	0.978	0.82
	Two lines	0.854	*j*0.396	*j*0.412	0.854	1.034
Compensated	One line	0.927	*j*0.400	*j*0.207	0.989	1.023
	Two lines	0.865	*j*0.349	*j*0.414	0.989	1.173

Table 12.2 Results of transient stability study

System state	$\mathcal{A}_1$ (pu)	$\mathcal{A}_2$ (pu)	$\mathcal{A}_2/\mathcal{A}_1$
Uncompensated	0.515	2.92	5.67
Compensated	0.537	4.03	7.50

- The power transfer from the sending to the receiving end of the transmission line is approximately inversely proportional to the series reactance of the line, (12.47). Its maximum value (steady-state stability limit) is obtained at δ equals 90°. So, decreasing the series reactance by series capacitive compensation results in an increase of stability limit and improves system stability.

12.4 Enhancement of multi-machine power system transient stability

One of the major reasons to apply series compensation to the transmission lines in power systems is to improve the system transient stability. This can clearly be shown by the example below.

Example 12.5 As in Example 8.2, the nine-bus test system (Figure 12.8) is subjected to a three-phase short circuit at bus no. 7 for durations of 0.08 s and 0.20 s. Investigate the system transient stability when the system is uncompensated,

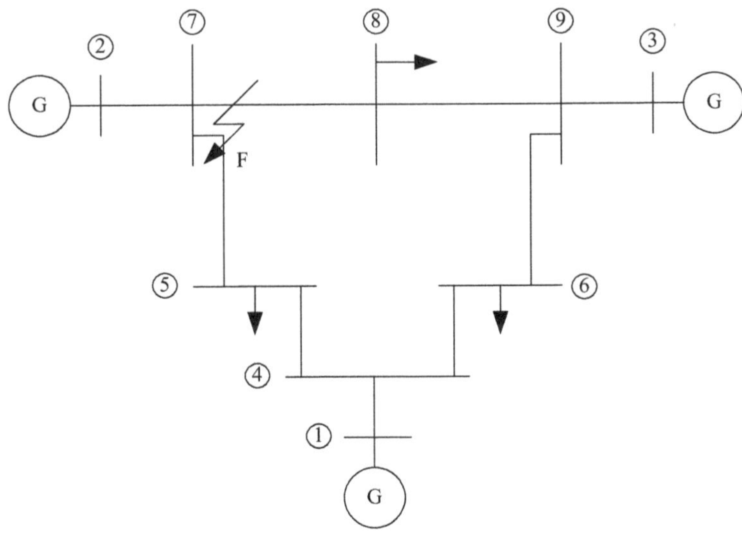

Figure 12.8 Nine-bus test system

30 per cent series capacitive compensation, 50 per cent series capacitive compensation and 90 per cent series capacitive compensation for all lines in the system. The system data are given in Appendix II.

Solution:

At the same initial operating conditions as calculated in Example 8.2, the second-order Runge–Kutta method and PSAT/MATLAB® toolbox are used for the transient analysis. The variation of rotor angle and speed of each generator are plotted as below (Figures 12.9–12.14). For all figures, it is to be noted that the solid curves represent the uncompensated case, the curves crossed by black circles are for 30 per cent compensation, the doted curves for 50 per cent compensation and the curves crossed by white circles represent 90 per cent compensation.

At fault duration 0.08 s

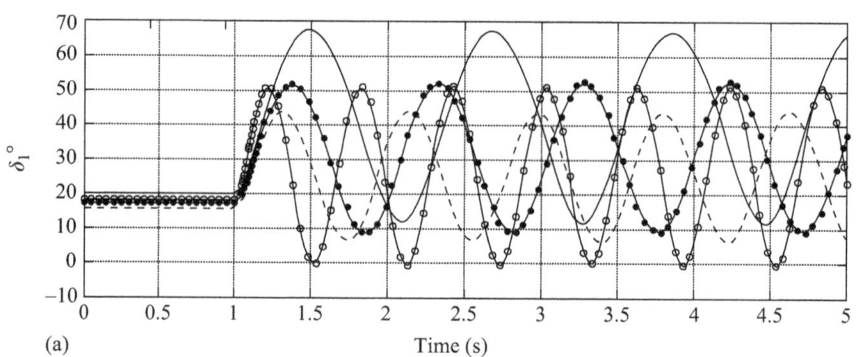

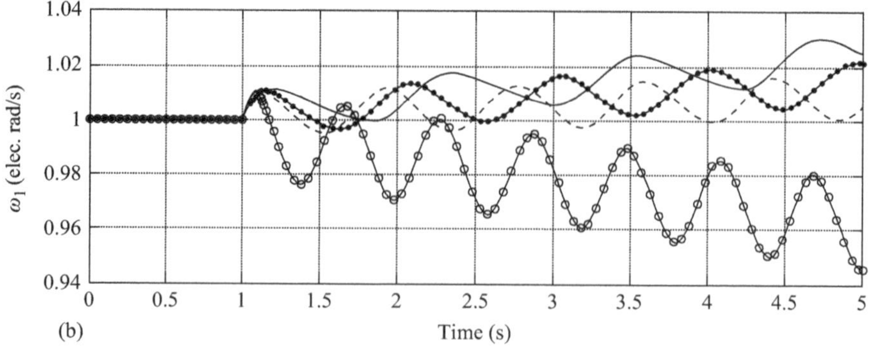

Figure 12.9 (a) δ_1° versus time (s) and (b) ω_1 (elec. rad/s) versus time (s)

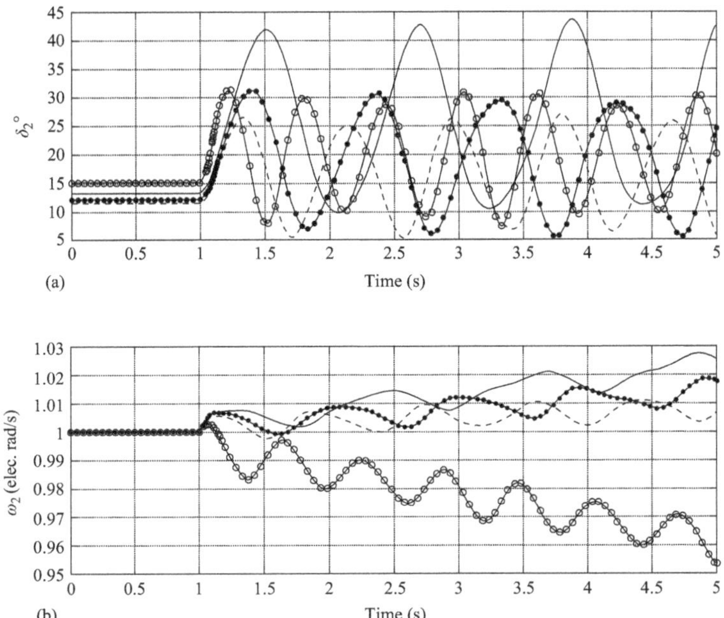

Figure 12.10 *(a) δ_2° versus time (s) and (b) ω_2 (elec. rad/s) versus time (s)*

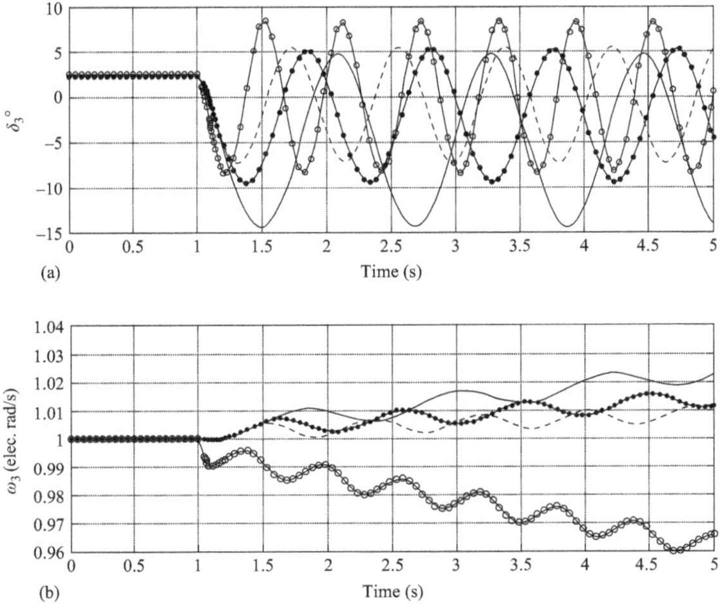

Figure 12.11 *(a) δ_3° versus time (s) and (b) ω_3 (elec. rad/s) versus time (s)*

At fault duration 0.20 s

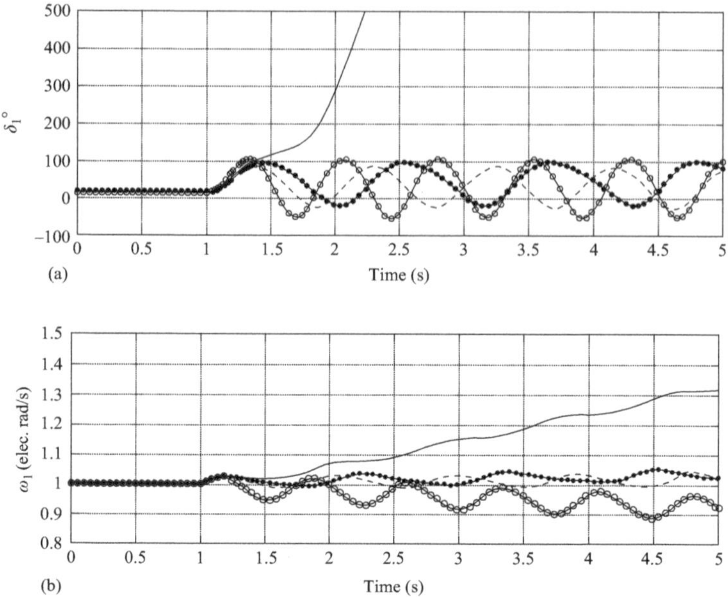

Figure 12.12 (a) δ_1° versus time (s) and (b) ω_1 (elec. rad/s) versus time (s)

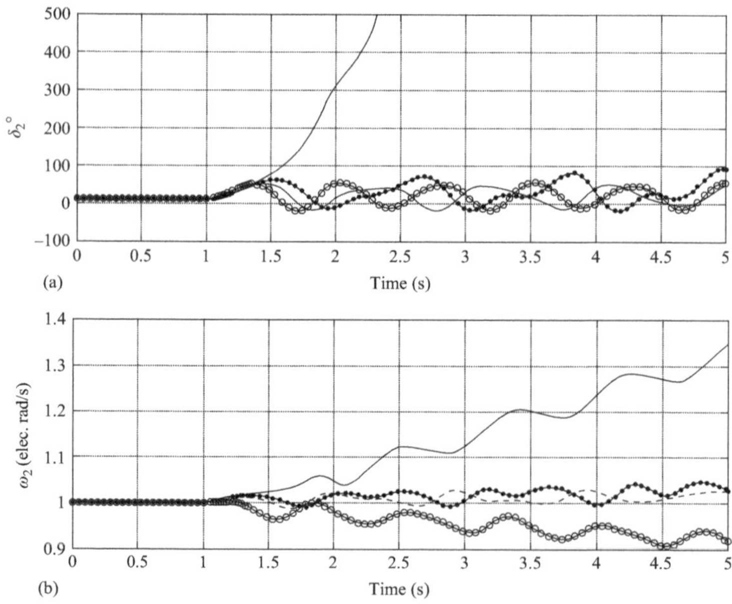

Figure 12.13 (a) δ_2° versus time (s) and (b) ω_2 (elec. rad/s) versus time (s)

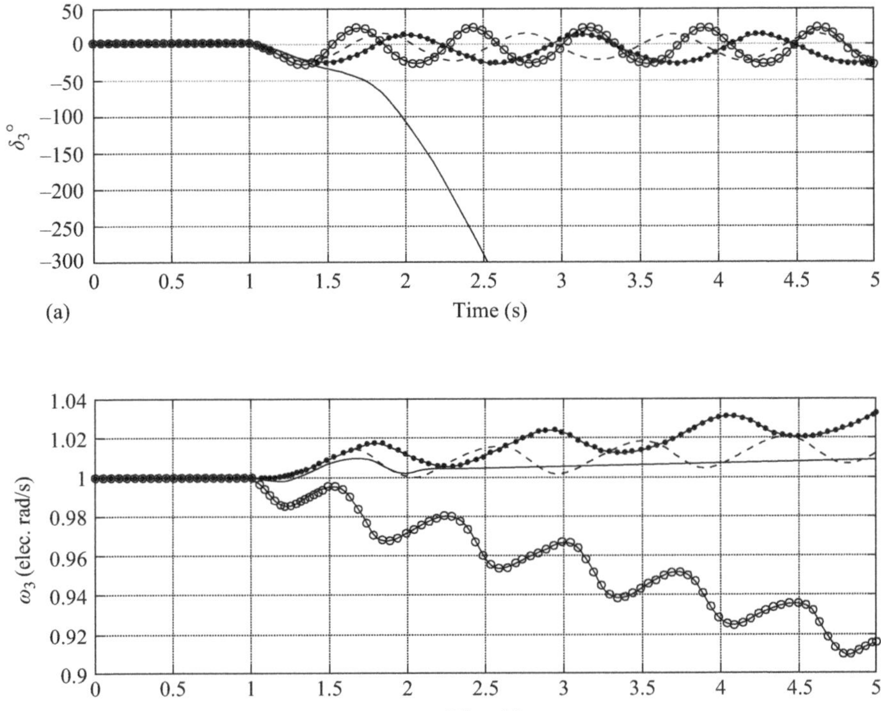

Figure 12.14 *(a) δ_3° versus time (s) and (b) ω_3 (elec. rad/s) versus time (s)*

12.5 Investigation of transmission power transfer capacity

It seems that increasing the degree of series compensation increases the improvement of system performance and system stability with no limit to the degree of compensation. Actually, this is not true as the power transfer capacity depends on various parameters that are not guaranteed to be simultaneously at the desired values giving the maximum power transfer. This can be realised from Example 12.5 where the improvement of system stability at 90 per cent compensation is less than that at 30 per cent. This concept is further discussed below.

As shown in Figure 12.15, a transmission line connects the node i as a sending end to the node j as a receiving end. The line is represented by a series impedance $Z_L = R + jX_L$ and compensated by series capacitors of reactance X_C. So, the net series reactance X is given by $X = X_L - X_C = X_L(1 - k_{nom})$, where k_{nom} is the degree of compensation.

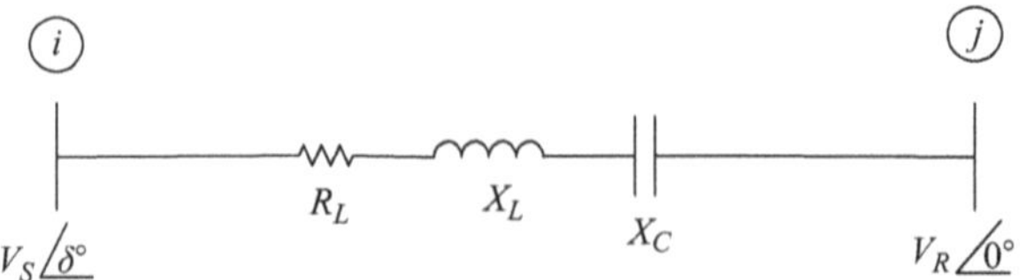

Figure 12.15 Series compensated line between node i and node j

The net series impedance is $\boldsymbol{Z} = R + jX$. The voltage at the receiving end, $\boldsymbol{V}_R$, is taken as a reference while the voltage at the sending end is defined as $\boldsymbol{V}_S$ with angle δ. Thus,

$$\boldsymbol{V}_S = V_S e^{j\delta} = V_S(\cos\delta + j\sin\delta) \tag{12.60}$$

Hence, the line current is given by

$$\boldsymbol{I} = \frac{\boldsymbol{V}_S - \boldsymbol{V}_R}{\boldsymbol{Z}} = \frac{V_R}{Z^2}\left\{\left[R\left(\frac{V_S}{V_R}\cos\delta - 1\right) + X\left(\frac{V_S}{V_R}\sin\delta\right)\right]\right.$$
$$\left. + j\left[R\left(\frac{V_S}{V_R}\sin\delta\right) - X\left(\frac{V_S}{V_R}\cos\delta - 1\right)\right]\right\} \tag{12.61}$$

The apparent power at both sending and receiving ends can be computed as

$$P_S + jQ_S = \boldsymbol{V}_S\boldsymbol{I}^* = \frac{V_S V_R}{Z^2}\left[R\frac{V_S}{V_R} - R\cos\delta + X\sin\delta\right.$$
$$\left. + j\left(X\frac{V_S}{V_R} - X\cos\delta - R\sin\delta\right)\right] \tag{12.62}$$

and

$$P_R + jQ_R = \boldsymbol{V}_R\boldsymbol{I}^* = \frac{V_R^2}{Z^2}\left\{\left[R\left(\frac{V_S}{V_R}\cos\delta - 1\right) + X\left(\frac{V_S}{V_R}\sin\delta\right)\right]\right.$$
$$\left. - j\left[R\left(\frac{V_S}{V_R}\sin\delta\right) - X\left(\frac{V_S}{V_R}\cos\delta - 1\right)\right]\right\} \tag{12.63}$$

Equation (12.63) can be rewritten as a function of the line impedance 'ϕ' as

$$P_R + jQ_R = \frac{V_R}{Z}\{[V_S\cos(\phi - \delta) - V_R\cos\phi] + j[V_S\sin(\phi - \delta) - V_R\sin\phi]\} \tag{12.64}$$

where

$$\phi = \tan^{-1}\frac{X}{R} = \tan^{-1}\frac{X_L(1 - k_{nom})}{R} \tag{12.65}$$

The real part of both (12.63) and (12.64) represents the value of P_R, i.e.

$$\begin{aligned} P_R &= \frac{V_R}{Z}[V_S\cos(\phi-\delta) - V_R\cos\phi] \\ &= \frac{V_R^2}{Z^2}\left[R\left(\frac{V_S}{V_R}\cos\delta - 1\right) + X\left(\frac{V_S}{V_R}\sin\delta\right)\right] \end{aligned} \tag{12.66}$$

Therefore, P_R is a function of the parameters; the ratio V_S/V_R, the line impedance and the power angle δ. It is impractical to decide the values of these parameters that maximise the power transfer at the receiving end. This is due to the fact that it may need a large value of δ to close to $\varnothing$ (12.66) or a voltage ratio out of the regulated value. In addition, for a multi-machine system that comprises many lines connecting different types of buses, the angles and magnitudes of bus voltages as well as the power flow into the lines are determined by load flow calculations depending on the system topology. So, no guarantee to achieve the maximum power transfers for each line when using load flow results. Thus, based on these facts it can be concluded that by continuous increase of the degree of compensation that affects both V_S/V_R and δ it is not guaranteed to obtain the maximum power transfer. It may be increased with increasing the degree of compensation until a certain limit, reach a plateau with no benefit and the gain may decrease beyond this limit. The same conclusion is applied to system stability improvement as increasing the transmission power transfer leads to an enlarged area between the input power line and the power versus angle curve, Equal Area Criterion, i.e. more marginal stability. This can be observed from Example 12.5, where the stability improvement of 50 per cent compensated system is larger than that of 90 per cent compensation. However, several studies have been done with a goal of achieving the optimal improvement of both power capability and system stability [5–10] as well as the schemes of applying and controlling the series compensation [11, 12].

12.6 Improvement of small signal stability

It has been shown that the transient stability of the power system is enhanced when applying series capacitors. An additional benefit may be obtained as well where the small signal stability is improved [13]. This can be demonstrated by the example below.

Example 12.6 Examine the small signal stability of the system given in Example 7.1 and shown in Figure 12.16 in terms of perturbed values of flux linkages and currents. Find the effect of degree of compensation of 30 per cent, 50 per cent and 90 per cent on the system performance.

Solution:

The exciting system is on manual control, constant E_{fd}. The linear flux linkage state space generator model, Chapter 7, Section 7.4, with the effect of governor included

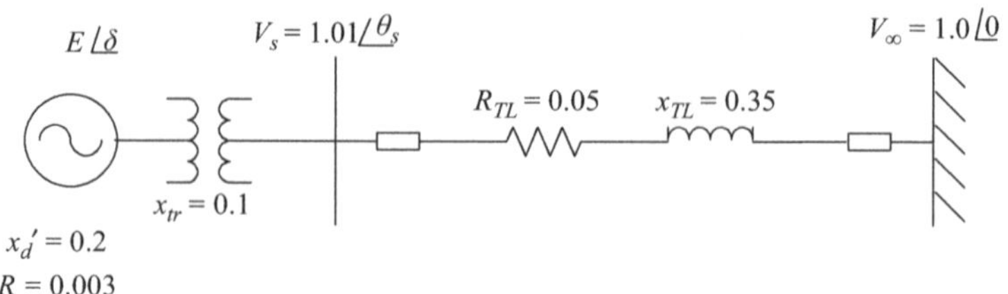

Figure 12.16 System for Example 11.6

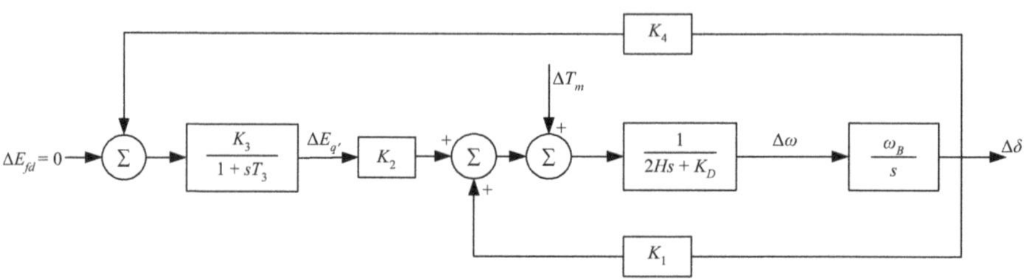

Figure 12.17 Representation of a single-machine infinite bus with linear generator model

in the mechanical torque is shown as a block diagram in Figure 12.17 [1]. The model is characterised by the following assumptions: (i) amortisseur effects and stator resistance are neglected, (ii) terms related to the derivative of stator flux linkages are neglected, (iii) balanced load conditions and (iv) constant speed in the speed voltage $\omega\psi$ terms. The change of both T_m and E_{fd} depends on the prime-mover and excitation controls. So, with constant mechanical input torque, $\Delta T_m = 0$; with constant exciter output voltage, $\Delta E_{fd} = 0$.

The model constants are related to the synchronous machine and transmission line parameters [1] as

$$K_1 = \frac{V_\infty}{R_E^2 + (X_q + X_E)(X_d' + X_E)}\Big\{E_{qo}\big[R_E \sin\delta_o + (x_d' + X_E)\cos\delta_o\big] + I_{qo}(X_q - X_d')(X_q + X_E)\sin\delta_o - R_E\cos\delta_o\Big\} \tag{12.67}$$

$$K_2 = \frac{1}{R_E^2 + (X_q + X_E)(X_d' + X_E)}\Big\{R_E E_{qo} + I_{qo}\Big[R_E^2 + (X_q + X_E)^2\Big]\Big\} \tag{12.68}$$

$$K_3 = \frac{1}{1 + K_A(X_d - X_d')(X_q + X_E)} \tag{12.69}$$

$$K_4 = V_\infty K_A(X_d - X_d')\big[(X_q + X_E)\sin\delta_o - R_E\cos\delta_o\big] \tag{12.70}$$

where

$$K_A = \frac{1}{R_E^2 + (X_q + X_E)(X_d' + X_E)} \tag{12.71}$$

$$R_E = 0.05, \quad X_E = 0.45, \quad \text{H} = 3.5\ \text{MW.s/MVA}$$

By using PSAT/MATLAB® toolbox the constants associated with block diagram (Figure 12.17) for uncompensated and compensated systems are calculated as summarised in Table 12.3. It is to be noted that this toolbox uses the concept of participation factors (PFs) that can be interpreted as the sensitivity of a given eigenvalue with respect to a corresponding entry of the system dynamic matrix and that the entries of the system matrix are a linear combination of PFs and eigenvalues. PFs have been introduced to analyse the connection between a system variable and some modes and to quantify the corresponding participation degree of a system variable in a mode and vice versa. Therefore, PFs can be used to detect the states most involved in a mode evolution. It is evident that once the modes of interest have been identified, PFs might help to obtain a reduced order model of the system that still preserves the dynamic of interests, selective modal analysis. More details about PFs can be found in [14].

In addition, the incremental saturation associated with perturbed values of flux linkages and currents is taken into account. Values of L_d and L_q are modified [15] to be L_{ds} and L_{qs} that given by $L_{sd} = K_{sd}L_d$ and $L_{qs} = K_{sd}L_q$, where

$$K_{sd} = \frac{1}{1 + B_{sat}A_{sat}\mathrm{e}^{B_{sat}(\psi_{to} - \psi_{T1})}}$$

The numerical values used in the program are $A_{sat} = 0.031$, $B_{sat} = 6.93$, $\Psi_{T1} = 0.8$

For uncompensated system:

State matrix A is obtained as

$$A = \begin{bmatrix} 0 & -0.2051 & -0.1771 \\ 376.9911 & 0 & 0 \\ 0 & -5.5497 & -11.6616 \end{bmatrix}$$

Eigen values are

$$\lambda_1 = -1.114, \quad \lambda_2 = -1.114, \quad \lambda_3 = -9.434$$

Table 12.3 Values of the constants associated with block diagram

System state		K_1	K_2	K_3	K_4	T_3
Uncompensated		1.436	1.240	0.228	2.089	0.086
Compensated	30%	1.679	1.384	0.209	2.341	0.079
	50%	1.866	1.511	0.196	2.535	0.074
	90%	4.551	0.000	0.062	0.000	0.024

Eigen vector matrix V:

$$V = \begin{bmatrix} 0.0027 - j0.0181 & 0.0027 + j0.0181 & 0.0093 \\ -0.9184 & -0.9184 & -0.3724 \\ 0.3232 - j0.2274 & 0.3232 + j0.2274 & 0.9280 \end{bmatrix}$$

Participation matrix P is:

$$P = \begin{bmatrix} 0.5845 + j0.0197 & 0.5845 - j0.0197 & -0.1690 \\ 0.5845 + j0.0197 & 0.5845 - j0.0197 & -0.1690 \\ -0.1690 - j0.0394 & -0.1690 - j0.0394 & 1.338 \end{bmatrix}$$

For compensated system at 30 per cent degree of compensation:

Eigen values:

$$\lambda_1 = -1.175, \quad \lambda_2 = -1.175, \quad \lambda_3 = -10.328$$

Eigen vector matrix V is:

$$V = \begin{bmatrix} 0.0029 - j0.0195 & 0.0029 + j0.0195 & 0.0097 \\ -0.9142 & -0.9142 & -0.3537 \\ 0.3317 - j0.2321 & 0.3317 + j0.2321 & 0.9353 \end{bmatrix}$$

Participation matrix P is:

$$P = \begin{bmatrix} 0.5817 + j0.0199 & 0.5817 - j0.0199 & -0.1635 \\ 0.5817 + j0.0199 & 0.5817 - j0.0199 & -0.1635 \\ -0.1635 - j0.0398 & -0.1635 + j0.0398 & 1.3269 \end{bmatrix}$$

For compensated system at 50 per cent degree of compensation:

Eigen values:

$$\lambda_1 = -1.221, \quad \lambda_2 = -1.221, \quad \lambda_3 = -11.133$$

Eigen vector matrix:

$$V = \begin{bmatrix} -0.0030 + j0.0205 & 0.0027 + j0.0181 & 0.0101 \\ 0.9119 & 0.9119 & -0.3409 \\ -0.3380 + j0.2318 & -0.3380 - j0.2318 & 0.9401 \end{bmatrix}$$

Participation matrix:

$$P = \begin{bmatrix} 0.5799 + j0.0214 & 0.5799 - j0.0214 & -0.1598 \\ 0.5799 + j0.0214 & 0.5799 - j0.0214 & -0.1598 \\ -0.1598 - j0.0429 & -0.1598 + j0.0429 & 1.3196 \end{bmatrix}$$

For compensated system at 90 per cent degree of compensation:

Eigen values:

$$\lambda_1 = -0.000, \quad \lambda_2 = -0.000, \quad \lambda_3 = -42.499$$

Eigen vector matrix:

$$V = \begin{bmatrix} -0.0000 - j0.0415 & 0.0027 + j0.0415 & 0.000 \\ -0.9991 & -0.9991 & -0.000 \\ 0 & 0 & 1.000 \end{bmatrix}$$

Participation matrix:

$$P = \begin{bmatrix} 0.5000 & 0.5000 & 0 \\ 0.5000 & 0.5000 & 0 \\ 0 & 0 & 1.0000 \end{bmatrix}$$

Summary of results: The eigenvalues and some coefficients are tabulated in Table 12.4.

It is seen that the magnitude of the negative real part of eigenvalues become larger in case of compensation, particularly, at 50 per cent degree of compensation. It is noted that the increase of marginal stability does not continue as the degree of compensation increases where at 90 per cent compensation the system becomes critically stable: two eigenvalues are zero.

It is to be noted that, in the above analysis, the series capacitors as passive compensators consist of fixed or switchable susceptances. They help in modifying transmission system reactance, increasing power transfer and stabilisation and controlling the reactive power. The behaviour of voltage and reactive power on an uncompensated transmission system is influenced by system impedance, variable active and reactive load characteristics as a function of supply voltage and loading on synchronous machines. Therefore, series capacitors as series compensation device must be equipped with controllers to effectively allow continuous control of reactive power that flows through the system and improving the performance 'steady and dynamic' of the overall power system. However, series compensation

Table 12.4 Eigenvalues and coefficients

System state	Eigenvalues			K_S	K_{srf}	K_{drf}	ω_n	E_P
	λ_1	λ_2	λ_3					
Uncompensated	−1.114	−1.114	−9.434	0.846	1.016	13.573	7.398	0.131
30% compensation	−1.175	−1.175	−10.328	1.001	1.196	14.383	8.024	0.128
50% compensation	−1.221	−1.221	−11.133	1.116	1.327	14.976	8.452	0.127
90% compensation	−0.000	−0.000	−42.499	4.551	4.551	0.000	15.656	0.000

K_S: Steady-state synchronising torque coefficient
K_{srf}: Synchronising torque coefficient at rotor oscillating frequency
K_{drf}: Damping coefficient at rotor oscillating frequency
ω_n: Un-damped natural frequency of the oscillatory mode
E_P: Damping ratio of the oscillatory mode

is implemented not only with series capacitors but also with current or voltage source devices. Some configuration of such devices is explained in Chapter 14.

12.7 Sub-synchronous resonance

It has been shown that applying series capacitors to long transmission lines improves the system stability as well as increases the power transfer capacity. On the other hand, series capacitors may cause self-excited oscillations at low frequencies because of the low X/R ratio or sub-synchronous frequencies due to induction generator effect.

The problem of self-excited torsional frequency due to torsional interaction is a serious problem since it may cause damage to the turbine-generator shaft [16–19]. The problem of oscillations at frequencies lower than the synchronous frequency is termed as 'sub-synchronous resonance' (SSR).

The formal definition of SSR as reported in [20] is SSR is an electric power system condition where the electric network exchanges energy with a turbine-generator at one or more of the natural frequencies of the combined system below the synchronous frequency of the system.

Therefore, the stability limit and hence the operational capability of long transmission lines are greatly improved by the use of series capacitor compensation, but unfortunate experiences of SSR with generator shaft torques have demonstrated the need for caution in the use of such series compensation [21]. Resonance can occur between the natural frequencies of oscillation inherent in the rotating masses of synchronous generators and prime movers coupled by shafts that are elastic, and the natural frequencies of the electric system to which the generator is connected.

Sudden change of torque to the main turbine-generator coupling, produced by a transient variation of electric power, can excite torsional natural resonant frequencies. When series capacitors are used to compensate the reactance of the transmission system the torsional natural resonant oscillations in the turbine-generator shafts may be excited by the power system natural frequency. Self-sustaining torsional oscillations can shorten shaft life.

The effects of long transmission line systems connecting generation at a remote point to the main power system have been studied. It is necessary to calculate the electrical natural frequencies, ENFs, to be able to locate the zone of torsional interaction, which is around the points of peak of resonance (the points of coincidence of ENFs with mechanical natural frequencies). Simply, for series capacitor compensated transmission lines their series LC combinations have natural frequencies ω_n that are defined by the equation:

$$\omega_n = \sqrt{\frac{1}{LC}} = \omega_B \sqrt{\frac{X_C}{X_L}} \tag{12.72}$$

where ω_B is the system base frequency, and X_L and X_C are the inductive and capacitive reactances, respectively. It is crucial to know which one of these sub-synchronous frequencies may interact with one of the natural torsional modes of the turbine-generator shaft. For a multi-machine power system, the problem of identifying the electric natural frequencies is difficult as the transmission network topology is more complicated.

12.7.1 The mechanical system

In modelling the system for SSR analysis, it is useful to consider the entire system as multi-subsystems connected to each other through the transmission network. In the mechanical system, it is sufficient to model the turbine-generator components, e.g. it is not important to model the boiler for SSR analysis. Torsional modes of the shaft oscillation can be obtained from the turbine-generator manufacturer or calculated by expressing the turbine-generator model as a spring-mass model whose parameters are known. The matrices expressing the dynamic equation of motion include both the inertia matrix and the velocity damping matrix, and when coupled by the stiffness matrix to the applied torque vector permit the derivation of eigenvalues representing the mechanical modes of natural oscillations as well as the derivation of the eigenvectors of the mode shape. The system is assumed to be represented by the spring-mass model shown in Figure 12.18 with a rotating exciter and assumes an unforced and un-damped mechanical system [22, 23]. It is to be noted that in the case of using a static exciter 'as it is common these days' the masses are reduced to only four: M_1, M_2, M_3 and M_4.

It can be described by a set of second-order differential equations as

$$\frac{1}{\omega_B}\boldsymbol{H}\ddot{\boldsymbol{\delta}} + \boldsymbol{K}\boldsymbol{\delta} = 0 \tag{12.73}$$

where all quantities are in pu, except time in seconds and δ in radians, and

$\boldsymbol{H}$: moment of inertia matrix $= \text{diag}[2H_1, 2H_2, \ldots, 2H_5]$, δ: angular displacement matrix.

$\boldsymbol{K}$: shaft-stiffness matrix

$$= \begin{bmatrix} K_{12} & -K_{12} & & & \\ -K_{12} & K_{12}+K_{23} & -K_{23} & & \\ & -K_{23} & K_{23}+K_{34} & -K_{34} & \\ & & -K_{34} & K_{34}+K_{45} & -K_{45} \\ & & & -K_{45} & K_{45} \end{bmatrix}$$

K_{ij}: spring constant of shaft between mass i and mass j

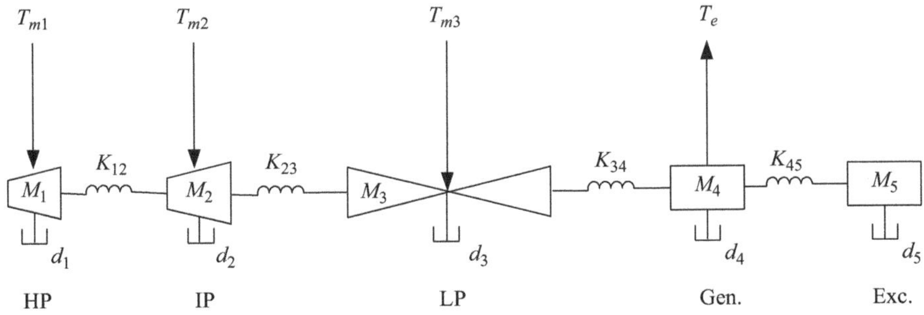

Figure 12.18 Spring-mass model for the shaft of a steam turbine unit

At resonance, all masses oscillate at the same frequency, ω_m, such that

$$\delta_i = \hat{\delta}_i \sin(\omega_m t + \alpha) \tag{12.74}$$

Substituting (12.74) into (12.73) gives

$$\frac{\omega_m^2}{\omega_B} H\hat{\delta} - K\hat{\delta} = 0 \tag{12.75}$$

Assuming, $M = H^{-1}K$ and $\lambda_m = \frac{\omega_m^2}{\omega_B}$, (12.75) can be rewritten as

$$M\hat{\delta} = \lambda_m \hat{\delta} \tag{12.76}$$

Therefore, by definition the eigenvalues of the matrix M, 'λ_m' provide the mechanical natural frequencies, where

$$\omega_m = \sqrt{\lambda_m \omega_B} \tag{12.77}$$

To each λ_m corresponds an eigenvector Q_m that gives the mode shape of oscillations.

Example 12.7 Typical data of the turbine shaft are given as

The stiffness: $K_{12} = 29.437$, $K_{23} = 62.241$, $K_{34} = 73.906$, $K_{45} = 5.306$

Turbine inertia constant(s): HP-stage = 0.0649, IP-stage = 0.2552, LP-stage = 1.539

Generator rotor inertia constant = 0.98 s, Exciter rotor inertia constant = 0.0298 s

Find the mechanical natural frequencies and the mode shapes.

Solution:

$$H = \mathrm{diag}[0.1298 \quad 0.5104 \quad 3.078 \quad 1.96 \quad 0.0596]$$

$$K = \begin{bmatrix} 29.437 & -29.437 & 0 & 0 & 0 \\ -29.437 & 91.678 & -62.241 & 0 & 0 \\ 0 & -62.241 & 136.147 & -73.906 & 0 \\ 0 & 0 & -73.906 & 79.212 & -5.306 \\ 0 & 0 & 0 & -5.306 & 5.306 \end{bmatrix}$$

$$H^{-1} = \begin{bmatrix} 7.7042 & 0 & 0 & 0 & 0 \\ 0 & 1.9592 & 0 & 0 & 0 \\ 0 & 0 & 0.3249 & 0 & 0 \\ 0 & 0 & 0 & 0.5102 & 0 \\ 0 & 0 & 0 & 0 & 16.7785 \end{bmatrix}$$

$$M = \begin{bmatrix} 226.7874 & -226.7874 & 0 & 0 & 0 \\ -57.6744 & 179.6199 & -121.9455 & 0 & 0 \\ 0 & -20.2212 & 44.2323 & -24.0110 & 0 \\ 0 & 0 & -37.7071 & 40.4143 & -2.7071 \\ 0 & 0 & 0 & -89.0268 & 89.0268 \end{bmatrix}$$

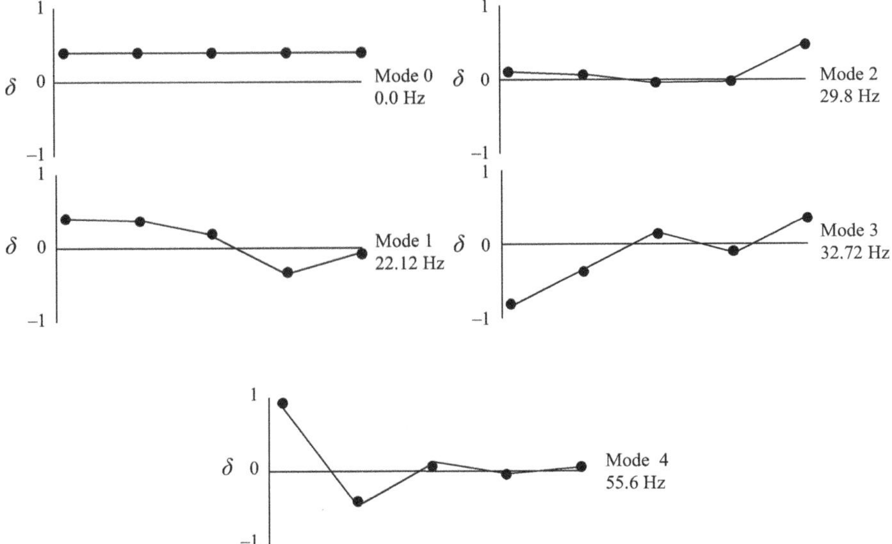

Figure 12.19 Mode shapes of the mechanical system of Example 12.7

Using MATLAB, the eigenvalues in ascending order are −0.0000, 51.2133, 93.1688, 112.0765, 323.6221.

Thus, the corresponding mechanical natural frequencies (12.77) are

$$f_{mo} = 0, \quad f_{m1} = 22.12, \quad f_{m2} = 29.8, \quad f_{m3} = 32.72, \quad f_{m4} = 55.6, \text{cps}$$

$\omega_{mo} = 0$, $\omega_{m1} = 138.91$, $\omega_{m2} = 187.36$, $\omega_{m3} = 205.5$, $\omega_{m4} = 349.2$ rad/s and the eigenvector matrix, Q_m, is:

$$Q_m = \begin{bmatrix} 0.4472 & 0.4299 & 0.1200 & -0.8118 & 0.9193 \\ 0.4472 & 0.3328 & 0.0707 & -0.4106 & -0.3925 \\ 0.4472 & 0.1471 & -0.0066 & 0.1565 & 0.0287 \\ 0.4472 & -0.3230 & -0.0460 & -0.0964 & -0.0038 \\ 0.4472 & -0.7606 & 0.9892 & 0.3724 & 0.0015 \end{bmatrix}$$

So, the mode shapes of the mechanical system are as shown in Figure 12.19.

12.7.2 The electrical network

For a simple power system, for example, a generator connected to an infinite bus, it is easy to model the line by its series impedance and then applying (12.72) calculates the ENFs. On the other hand, to study the generator dynamic performance, the linearised generator model in *d–q* axes used for stability analysis can be used for SSR analysis as well, Chapter 7. In case of a multi-machine power system, the generalised form of a ring, which can be a part or whole of the power system, is

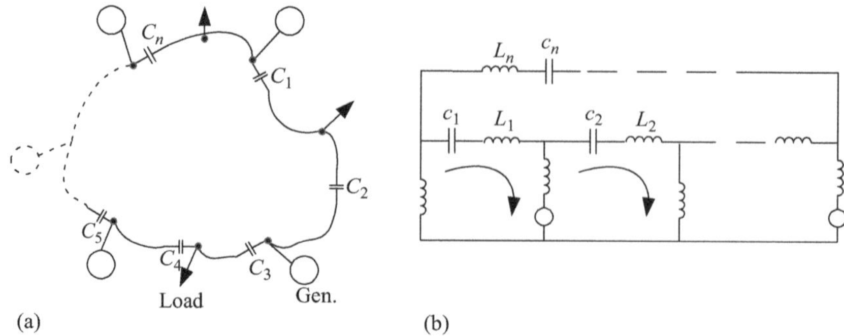

Figure 12.20 Generalised form of a ring system: (a) General loop configuration for an interconnected system and (b) the equivalent L-C loops of the system

shown in Figure 12.20(a). The possibility of series compensation in every interconnecting transmission line is considered but of course such compensation can be set at zero for as many lines as desired or at any required value whose resonance effect is to be determined.

The ENFs can be calculated by the following steps [24]:

(i) calculation of the single-phase equivalent circuit as shown in Figure 12.20(b).
(ii) calculation of the inductance matrix L, which is given by

$$L = \begin{bmatrix} L_{11} & L_{12} & \cdots & L_{1n} \\ L_{21} & L_{22} & \cdots & \cdots \\ \cdots & \cdots & L_{33} & \cdots \\ \vdots & \vdots & \ddots & \vdots \\ L_{n1} & \cdots & \cdots & L_{nn} \end{bmatrix} \tag{12.78}$$

where
L_{ij} is the summation of inductances, which are included in the loop i
L_{ij} is the inductance of the mutual link between loops i and j

(iii) calculation of the capacitance matrix C, assuming $(1/c_i) = S_i L_{ij}$, i and $j = 1, 2, \ldots, n$

where
c_i = capacitance inserted in loop i
S_i = degree of compensation for that loop
L_{ij} = inductance between the two directly linked buses i and j

Then the capacitive reactance matrix F for Figure 12.20(b) is

$$F = \begin{bmatrix} 1/c_1 & & & & 1/c_1 \\ & 1/c_2 & & 0 & 1/c_2 \\ & & 1/c_3 & & 1/c_3 \\ & & 0 & \ddots & \vdots \\ 1/c_1 & 1/c_2 & \cdots & \cdots & \cdots & 1/c_n \end{bmatrix} \tag{12.79}$$

Thus, the capacitance matrix is

$$C = F^{-1} \tag{12.80}$$

(iv) the ENF's of a network are those that arise from its configuration, without applied emf [25]. They can be found as the square roots of the reciprocal of the eigenvalues of the V matrix where

$$V = LC \tag{12.81}$$

and those values are viewed from the stator side. Alternatively, they can be viewed from the rotor side

$$f_r = f_n \pm f_s \tag{12.82}$$

where f_r is the rotor frequency, f_n is the natural frequency, f_s is the synchronous frequency, positive sign for super-synchronous frequency and negative sign for sub-synchronous frequency.

Steps (i)–(iv) can be applied for two or more coupled rings, but, of course, as the network gets more complex, the representation also gets more difficult. Another technique, called the 'state space approach' can be used as it is suitable for digital computer solution [21]. However, several computer programs, e.g. PSCAD, EMTP, are available to analyse the power system in steady state, transient state or 'in general' the dynamic performance of the power system.

Example 12.8 Reactance diagram of the nine-bus test system is shown in Figure 12.21. Find the ENFs when all lines in the system are compensated by series capacitors at 50 per cent degree of compensation. Investigate the torsional interaction for one of the system generators that has the turbine-shaft system given in Example 12.7.

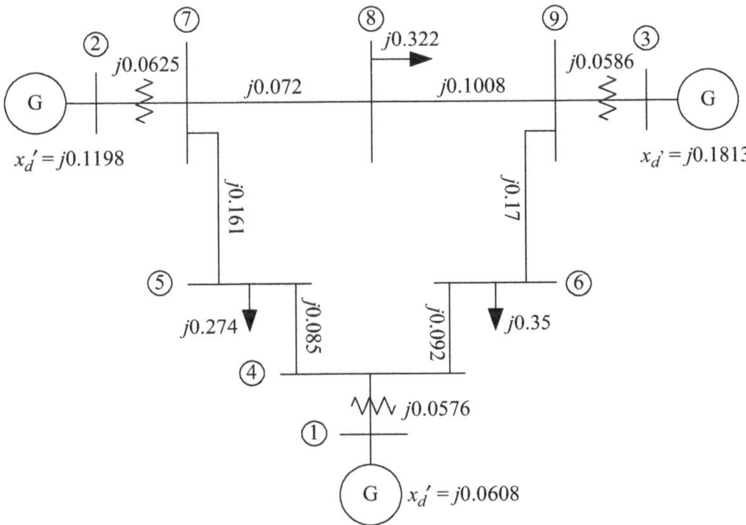

Figure 12.21 Reactance diagram of nine-bus test system

Solution:
The equivalent loop configuration is shown in Figure 12.22.

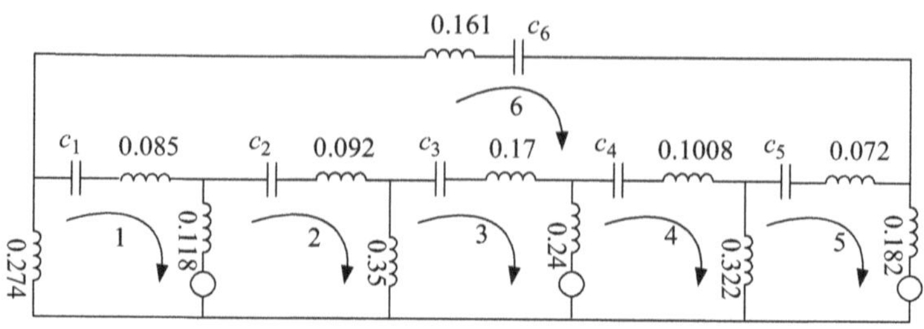

Figure 12.22 Equivalent loop configuration

The inductance matrix is:

$$L = \begin{bmatrix} 0.477 & 0.118 & 0.000 & 0.000 & 0.000 & 0.085 \\ 0.118 & 0.543 & 0.350 & 0.000 & 0.000 & 0.092 \\ 0.000 & 0.350 & 0.76 & 0.240 & 0.000 & 0.170 \\ 0.000 & 0.000 & 0.240 & 0.660 & 0.322 & 0.101 \\ 0.000 & 0.000 & 0.000 & 0.322 & 0.576 & 0.072 \\ 0.085 & 0.092 & 0.170 & 0.101 & 0.072 & 0.681 \end{bmatrix}$$

The capacitive reactance matrix is:

$$F = \begin{bmatrix} 0.0425 & & & & & 0.0425 \\ & 0.0460 & & & & 0.0460 \\ & & 0.0850 & & & 0.0850 \\ & & & 0.0504 & & 0.0504 \\ & & & & 0.0360 & 0.0360 \\ 0.0425 & 0.0460 & 0.0850 & 0.0504 & 0.0360 & 0.0805 \end{bmatrix}$$

The capacitance matrix is:

$$C = \begin{bmatrix} 17.9522 & -5.5772 & -5.5772 & -5.5662 & -5.5772 & 5.5772 \\ -5.5772 & 16.1619 & -5.5772 & -5.5662 & -5.5772 & 5.5772 \\ -5.5772 & -5.5772 & 6.1875 & -5.5662 & -5.5772 & 5.5772 \\ -5.5662 & -5.5662 & -5.5662 & 14.2468 & -5.5662 & 5.5662 \\ -5.5772 & -5.5772 & -5.5772 & -5.5662 & 22.2005 & 5.5772 \\ 5.5772 & 5.5772 & 5.5772 & 5.5662 & 5.5772 & -5.5772 \end{bmatrix}$$

Then,

$$
¥ = \mathrm{L\,C} = \begin{bmatrix} 8.3791 & 0.2792 & -2.8444 & -2.8388 & -2.8444 & 2.8444 \\ -2.3490 & 6.6789 & -1.0078 & -5.1153 & -5.1255 & 5.1255 \\ -6.5785 & 1.0302 & 2.3627 & -1.8130 & -6.5785 & 6.5785 \\ -6.2448 & -6.2448 & -3.4213 & 6.8369 & 2.6996 & 6.2448 \\ -4.6032 & -4.6032 & -4.6032 & 1.7821 & 11.3968 & 4.6032 \\ 2.8991 & 2.8991 & 2.8991 & 2.8973 & 2.8991 & -0.8991 \end{bmatrix}
$$

Using MATLAB, the eigenvalues are obtained as −11.8538, 19.3129, 11.1760, 2.8122, 5.6125, 7.6955

The corresponding ENFs 'omitting the negative value' are 13.18, 17.2, 20.6, 22.9, 34.4 Hz.

The mechanical modes of oscillations and ENFs are plotted in Figure 12.23. The horizontal lines represent the mechanical modes of oscillation. The black-filled circles represent the ENFs that are close to one of the mechanical modes while the white circles are those apart from mechanical modes.

Therefore, at modes #1, 2 and 3, the torsional natural resonant oscillations in the turbine generator shafts may be excited by the power system natural frequency causing shaft damage.

The modes of oscillations may move upwards or downwards for other turbine shafts in the system, with different values of inertia constants and different shaft

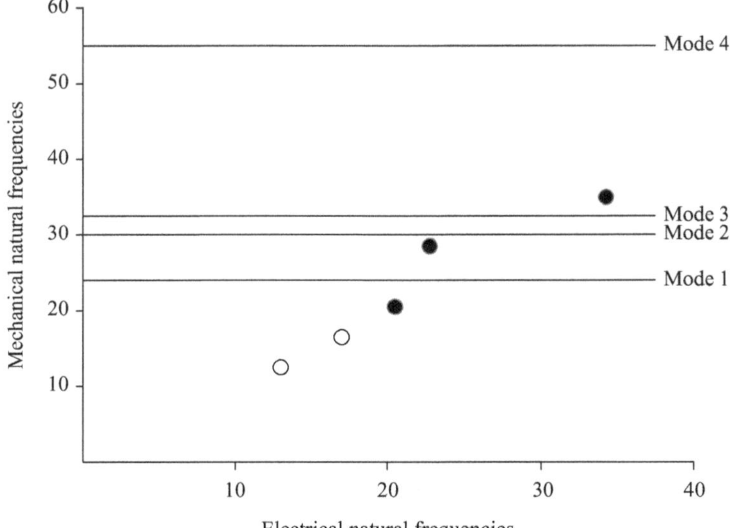

Figure 12.23 Mechanical modes of oscillations and electrical natural frequencies (50 per cent degree of compensation for all lines in the system)

stiffness, giving different modes. The inductance matrix changes with load while the capacitance matrix depends on the degree of compensation. Thus, changing the loads and/or degree of compensation yields a change of ENFs.

References

1. Anderson P.M., Farmer R.G. *The Series Compensation of Power Systems.* Encinitas, CA, US: PBLSH; 1996
2. Sallam A.A., Khafaga A.M. (eds.). 'Optimal parameters of series capacitors in compensated power systems'. *International Power Engineering Conference IPEC'93*; Singapore, Mar 1993
3. Sallam A.A., Khafaga A.M. (eds.). 'Optimal series compensation in power systems by using complex method'. *Proceedings of 3rd International Symposium of Electricity Distribution and Energy Management ISEDEM'93*; Singapore, Oct 1993. pp. 215–19
4. Belur S., Kumar A., Parthasarathy K., Prabhakara F.S., Khincha H.P. 'Effectiveness of series capacitors in long distance transmission lines'. *IEEE Transactions on Power Apparatus and Systems.* 1970;**89**(5):941–51
5. Leonidaki E.A., Georgiadis P., Hatziargyriou N.D. 'Decision trees for determination of optimal location and rate of series compensation to increase power system loading margin'. *IEEE Transactions on Power Systems.* 2003;**21**(3): 1303–10
6. Hedin R., Jalali S., Weiss S., Cope L., Johnson B., Mah D. *et al.* (eds.). 'Improving system stability using an advanced series compensation scheme to damp power swings'. *Sixth International Conference on AC and DC Power Transmission*, IET Conf. Publ. No. **423**, Apr/May 1996. pp. 311–14
7. Chen X.R., Pahalawaththa N.C., Annakkage U.D., Kumble C.S. 'Controlled series compensation for improving the stability of multi-machine power systems'. *IEE Proceedings – Generations, Transmission and Distribution.* 1995;**142** (4):361–66
8. Crary S.B., Saline L.E. 'Location of series capacitors in high-voltage transmission systems'. *IEEE Transactions on Power Apparatus and Systems, Part III Transactions of the AIEE.* 1953;**72**(2):1140–51
9. Kosterev D.N., Mittalstadt W.A., Mohler R.R., Kolodziej W.J. 'An application study for sizing and rating controlled and conventional series compensation'. *IEEE Transactions on Power Delivery.* 1996;**11**(2):1105–11
10. de Oliveira S.E.M., Gardos I., Fonseca E.P. 'Representation of series capacitors in electric power system stability studies'. *IEEE Transactions on Power Systems.* 1991;**6**(3):1119–25
11. Yu-Jen L. (ed.). 'Power systems transient stability preventive control incorporating network series compensation with the aid of if-then rules extracted from a multi-layer perceptron artificial neural network'. *4th International Conference on Electric Utility Deregulation and Restructuring and Power Technologies (DRPT), 2011*; Weihai, China, Jul 2011. US: IEEE; 2011. pp. 31–8

12. Fernandes A.A. (ed.). 'Series compensation using variable structure and Lyapunov function controls for stabilization multi-machine power system'. *16th IEEE International Conference on Control Applications, Part of IEEE Multi-Conference on Systems and Control*; Singapore, Oct 2007. pp. 1091–6
13. Lie T.T., Li G.J., Shrestha G.B., Lo K.L. (eds.). 'Coordinated decentralized optimal control of inter-area oscillations in power systems'. *International Conference on Energy Management and Power Delivery (EMPD'98)*; Singapore, Mar 1998, vol. 1. pp. 97–102
14. Garofalo F., Iannelli L., Vsca F. (eds.). 'Participation factors and their connections to residues and relative gain array'. *IFAC, 15th Triennial World Congress*; Barcelona, Spain, 2002, vol. 15, part 1. pp. 180–5
15. Kundur P. *Power System Stability and Control*. New York, NY, US: McGraw-Hill, Inc.; 1994. chapter 12
16. Kumar R., Harada A., Merkle M., Miri A.M. (eds.). 'Investigation of the influence of series compensation in AC transmission systems on bus connected parallel generating units with respect to sub-synchronous resonance (SSR)'. *IEEE Bologna Power Tech Conference*; Bologna, Italy, Jun 2003. pp. 23–6
17. de Oliveira A.L.P., Moraes M. (eds.). 'Sub-synchronous resonance analysis after Barra do Peixe 230 kV fixed series compensations installation at 230 kV MatoGrosso transmission system (Brazil)'. *Transmission and Distribution Conference and Exposition*; Latin America, 2008. pp. 1–6
18. de Oliveira A.L.P. (eds.). 'The main aspects of fixed series compensation dimensioning at Brazilian 230 kV transmission system'. *Transmission and Distribution Conference and Exposition*. Latin America; 2008. pp. 1–9
19. Jowder F.A.L. 'Influence of mode of operation of the SSSC on the small disturbance and transient stability of a radial power system'. *IEEE Transactions on Power Systems*. 2005;**20**(2):935–42
20. IEEE SSR Working Group. 'Proposed terms and definitions for sub-synchronous resonance'. *IEEE Symposium on Countermeasures for Sub-synchronous Resonance*, IEEE Pub. 81TH0086-9-PWR, 1981. pp. 92–7
21. Anderson P.M., Agrawal B.L., Van Ness J.E. *Sub-Synchronous Resonance in Power Systems*. New York, NY, US: IEEE Press; 1990
22. Fouad A.A., Khu K.T. 'Damping of torsional oscillations in power systems with series-compensated lines'. *IEEE Transactions on Power Apparatus and Systems*. 1978;**97**(3):744–53
23. Fouad A.A., Khu K.T. 'Sub-synchronous resonance zones in the IEEE "BENCHMARK" power system'. *IEEE Transactions on Power Apparatus and Systems*. 1978;**97**(3):754–62
24. Sallam A.A., Dineley J.L. (eds.). 'Sub-synchronous problems in an integrated power system'. *19th Universities Power Engineering Conference*; Aberdeen, UK, Apr 1984
25. Kimbark E.W. 'How to improve system stability without risking sub-synchronous resonance'. *IEEE Transactions on Power Apparatus and Systems*. Sept/Oct 1977;**96**(5):1608–18

Chapter 13
Shunt compensation

Improvement of AC power systems' performance is an important issue for power system planning and operation engineers. Application of series compensation of transmission system to achieve this goal is explained in Chapter 12. Another method is to compensate the transmission system by shunt compensators. In both methods the line reactance is controlled to modify the natural electrical characteristics of AC power systems. Consequently, the reactive power that flows through the system can effectively be controlled improving the system performance, in particular, increasing power transfer capacity, controlling steady state and dynamic voltage, controlling reactive power of dynamic loads, damping of power system oscillations and improving system stability [1–3]. For instance, if the shunt compensator injects reactive power near the load, the transmission line current can be reduced resulting in reduction of power losses, improved voltage regulation at load terminals and increased power transfer capacity.

Shunt compensation is applied by using shunt capacitors and shunt reactors that are permanently connected to the network or switched on and off according to operating conditions. Shunt capacitors help increase the system load ability [4] and reduce the voltage drop in the line by improving the power factor. Shunt reactors are used to limit voltage rise under both open line and light load conditions. Principles illustrating the impact of shunt compensation on transmission system parameters as well as its benefits are discussed in the next sections.

13.1 Shunt compensation of lossless transmission lines

For simplicity of analysis consider the line is lossless and is uniformly compensated throughout its length. The line parameters are affected as given below.

13.1.1 Shunt-compensated line parameters

Based on the definitions given in Chapter 12, Section 12.1, the parameters characteristic impedance, line angle and natural power of shunt compensated line can be given as a ratio of the uncompensated line, denoted by subscript *o* as

The characteristic impedance ratio:

$$\frac{\boldsymbol{Z}_C}{\boldsymbol{Z}_{Co}} = \frac{R_C}{R_{Co}} = \sqrt{\frac{x_{Lo}\, b_{Co}}{b_C\, x_{Lo}}} = \sqrt{\frac{b_{Co}}{b_C}} \tag{13.1}$$

The ratio of line angle:

$$\frac{\theta}{\theta_o} = \sqrt{\frac{x_{Lo} b_C}{x_{Lo} b_{Co}}} = \sqrt{\frac{b_C}{b_{Co}}} \tag{13.2}$$

and the ratio of natural power:

$$\frac{P_n}{P_{no}} = \frac{R_{Co}}{R_C} = \sqrt{\frac{b_C}{b_{Co}}} = \sqrt{\frac{b_{Co} + \Delta b_C}{b_{Co}}} = \sqrt{1 + \frac{\Delta b_C}{b_{Co}}} \tag{13.3}$$

where $\Delta b_C \triangleq$ change of b_C due to shunt compensation

Thus, the degree of shunt compensation '$\mathscr{d}$' is defined as

$$\mathscr{d} = \frac{\Delta b_C}{b_{Co}} = \left(\frac{P_n}{P_{no}}\right)^2 - 1 \tag{13.4}$$

It is found that by adding capacitive shunt compensation to the line, the line angle is increased: the characteristic impedance decreases while the natural power increases. It means that more power can be carried by the line while maintaining a flow that corresponds to the characteristic impedance loading. On the other hand, if the shunt compensation is inductive the relations above can be applied with a negative Δb_C.

The amount of reactive power delivered by the shunt capacitive compensator 'ΔQ_C' to increase the natural power from P_{no} to P_n can be computed as below.

Assuming that the natural power, P_n, is required to be $\mathbb{R}$ times the original value, P_{no}, (13.3) and (13.4) give

$$\frac{P_n}{P_{no}} = \mathbb{R} = \sqrt{1 + d} \tag{13.5}$$

and

$$\mathscr{d} = \mathbb{R}^2 - 1 \tag{13.6}$$

It is shown that the amount of shunt susceptance added is ($\mathbb{R}^2 - 1$) times the original value. Thus, the reactive power supplied by the added shunt compensation 'ΔQ_C' is given by

$$\Delta Q_C = \Delta b_C \ell V^2 = (\mathbb{R}^2 - 1) b_{Co} \ell V^2 \tag{13.7}$$

Also, ΔQ_C as a ratio of P_{no} is obtained by

$$\frac{\Delta Q_C}{P_{no}} = \left[(\mathbb{R}^2 - 1) b_{Co} \ell V^2\right] / \left[V^2 \sqrt{(b_{Co}/x_{Lo})}\right] = \ell(\mathbb{R}^2 - 1)\sqrt{b_{Co} x_{Lo}} \tag{13.8}$$

From (13.2) and (13.3) the ratio of line angle is

$$\frac{\theta}{\theta_o} = \mathbb{R} \tag{13.9}$$

Example 13.1 Repeat Example 12.1 with using shunt compensation rather than series compensation.

Solution:

The original values of $\boldsymbol{Z}_C$, β and θ are 285Ω, 0.00129 rad/km and −11.83°, respectively. They are the same as calculated in Example 11.1. Also, *SIL* = 417.63 MW 'three-phase'.

When the line is shunt compensated to increase P_n to 1.5 × 417.63 MW, the required degree of shunt compensation '$\mathscr{d}$' can be obtained using (12.6). Thus,

$$\mathscr{d} = \mathbb{R}^2 - 1 = (1.5)^2 - 1 = 1.25$$

Using (13.7) the amount of reactive compensation is

$$\Delta Q_C = 1.25 \times 160 \times 4.518 \times 10^{-6} \times 345^2 = 107.55 \text{ MVA}$$

By (13.9) the line angle = 0.2064 × 1.5 = 0.3096 rad = 17.75°

Comparing the results obtained in Example 13.1 with the results of Example 12.1, it is noted that (i) the reactive power added to the system to get a given natural power is the same whether added by series or shunt compensation and (ii) the line angle for shunt compensation case is $\mathbb{R}^2$ 'equals 2.25' times larger than the series compensation case.

13.1.2 Transient stability enhancement for shunt-compensated lossless lines

Considering the system shown in Figure 13.1, it comprises an ideal shunt compensator with unlimited capacity to control the voltage and fast response at the midpoint of the transmission line. The shunt compensation is assumed to be capable of holding the voltage at midpoint constant at all times.

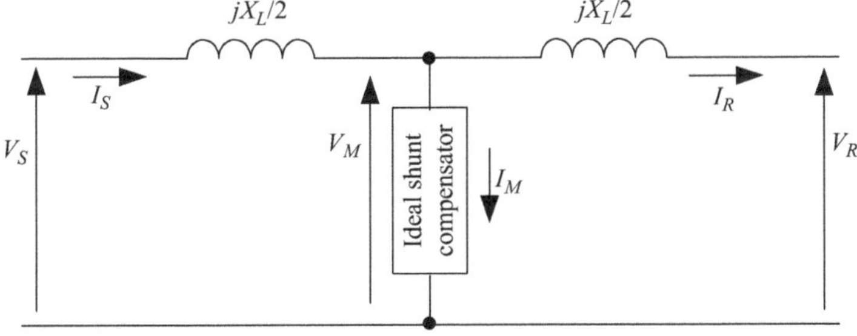

Figure 13.1 Shunt-compensated transmission line

As shown in Figure 13.1 the voltages can be expressed as

$\boldsymbol{V}_S = V_S e^{\theta_S}$, $\boldsymbol{V}_M = V_M e^{\theta_R}$ and $\boldsymbol{V}_R = V_R e^{\theta_R}$, where the voltage angles are referred to an arbitrary reference.

The currents ($\boldsymbol{I}_S$ and $\boldsymbol{I}_R$) and powers (P_S, Q_S and P_R, Q_R) at the sending and receiving ends, respectively, are computed using the relations below:

$$\left.\begin{aligned} \boldsymbol{I}_S &= \frac{\boldsymbol{V}_S - \boldsymbol{V}_M}{jX_L/2} = \frac{2(V_S \sin\theta_S - V_M \sin\theta_M)}{X_L} + j\frac{2(V_M \cos\theta_M - V_S \cos\theta_S)}{X_L} \\ \boldsymbol{I}_R &= \frac{\boldsymbol{V}_M - \boldsymbol{V}_R}{jX_L/2} = \frac{2(V_M \sin\theta_M - V_R \sin\theta_R)}{X_L} + j\frac{2(V_R \cos\theta_R - V_M \cos\theta_M)}{X_L} \end{aligned}\right\} \tag{13.10}$$

$$\left.\begin{aligned} P_S + jQ_S &= \boldsymbol{V}_S \boldsymbol{I}_S^* = \frac{2V_S V_M}{X_L}\sin(\theta_S - \theta_M) + j\frac{2}{X_L}\left(V_S^2 - V_S V_M \cos(\theta_S - \theta_M)\right) \\ P_R + jQ_R &= \boldsymbol{V}_R \boldsymbol{I}_R^* = \frac{2V_M V_R}{X_L}\sin(\theta_M - \theta_R) - j\frac{2}{X_L}\left(V_R^2 - V_M V_R \cos(\theta_M - \theta_R)\right) \end{aligned}\right\} \tag{13.11}$$

The reactive power of the compensator, Q_{comp}, can be defined as the difference between the generated reactive power at the sending end, Q_S, and the summation of reactive power losses, Q_{loss}, and reactive power at receiving end, Q_R. Using the current directions shown in Figure 13.1, it is considered a positive term as input to the compensator. Thus,

$$Q_{comp} = Q_S - \sum (Q_{loss} + Q_R) \tag{13.12}$$

Q_{loss} is calculated as the reactive power losses in the two sections of the line, i.e.

$$Q_{loss} = Q_{loss(S)} + Q_{loss(R)} \tag{13.13}$$

where

$$Q_{loss(S)} = \boldsymbol{I}_S^2 (X_L/2) \quad \text{and} \quad Q_{loss(R)} = \boldsymbol{I}_R^2 (X_L/2)$$

Using (13.10) gives

$$\left.\begin{aligned} Q_{loss(S)} &= \frac{2V_S^2}{X_L} + \frac{2V_M^2}{X_L} - \frac{4V_S V_M}{X_L}\cos(\theta_S - \theta_M) \\ Q_{loss(R)} &= \frac{2V_M^2}{X_L} + \frac{2V_R^2}{X_L} - \frac{4V_M V_R}{X_L}\cos(\theta_M - \theta_R) \end{aligned}\right\} \tag{13.14}$$

and Q_S is given using (13.11). Hence,

$$Q_{comp} = \frac{2V_SV_M}{X_L}\cos(\theta_S - \theta_M) + \frac{2V_MV_R}{X_L}\cos(\theta_M - \theta_R) - \frac{4V_M^2}{X_L} \qquad (13.15)$$

Therefore, by specifying the desired value of voltage at the point at which the compensator is connected to the system, the required reactive power supplied by the compensator can be determined using (13.15) as well as the active power at the sending and receiving ends is obtained using (13.11). It is found that the power transfer capability of the transmission line is increased when using shunt compensator, which in turn enhances both the system stability and system security.

Example 13.2 The parameters of the equivalent circuit (Figure 13.1) have the per unit values:

$$V_S = 1.0\angle 14.15°, \quad V_R = 0.9\angle 0°, \quad X_L = 0.275$$

Find the power at both ends when the system is uncompensated delivering a power of 0.8 pu and when midpoint shunt is compensated to hold the magnitude of V_R at $1.0\angle 0°$. Examine the system stability.

Solution:

For uncompensated system, the apparent power at sending and receiving ends (12.30) and (12.31) are

$$P_{So} + jQ_{So} = 0.80 + j0.463 \text{ pu} \quad \text{and} \quad P_{Ro} + jQ_{Ro} = 0.8 + j0.228 \text{ pu}$$

The relation of P_S versus δ is given by

$P_S = 3.27 \sin\delta$ (plotted in Figure 13.2)

In case of shunt compensation, taking the voltage at receiving end, $\boldsymbol{V}_R$, as a reference the angles and voltages in the set of (13.11) through (13.15) are $\theta_R = 0°$, $\theta_S \triangleq \delta = 14.15°$, $V_S = V_R = V$ 'to simplify the relations'. Consequently, $\theta_M = (\delta/2)$ and the set of relations becomes

$$\left.\begin{aligned} P_S + jQ_S &= \frac{2VV_M}{X_L}\sin(\delta/2) + j\frac{2}{X_L}\left(V^2 - VV_M\cos(\delta/2)\right) \\ P_R + jQ_R &= \frac{2V_MV}{X_L}\sin(\delta/2) - j\frac{2}{X_L}\left(V^2 - VV_M\cos(\delta/2)\right) \end{aligned}\right\} \qquad (13.16)$$

$$Q_{loss(S)} = Q_{loss(R)} = \frac{2V^2}{X_L} + \frac{2V_M^2}{X_L} - \frac{4VV_M}{X_L}\cos(\delta/2) \qquad (13.17)$$

$$Q_{comp} = \frac{2VV_M}{X_L}\cos(\delta/2) + \frac{2VV_M}{X_L}\cos(\delta/2) - \frac{4V_M^2}{X_L} \qquad (13.18)$$

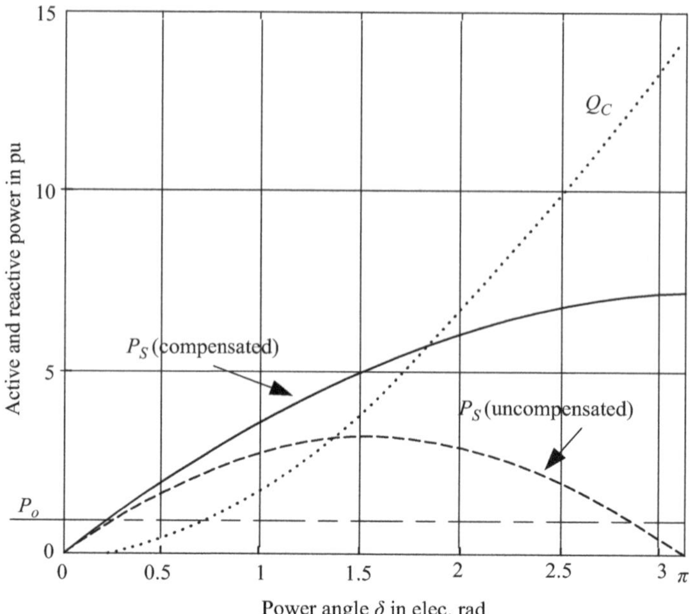

Figure 13.2 Variation of Q_C, P_S for compensated and uncompensated systems

To plot the variables in the set of relations versus the angle δ, V_M is assumed to equal V as a further simplification to generally examine the system stability. If the system voltages are $V_S = V_M = V_R = V$ pu, the set of relations, again, becomes of the form:

$$\left.\begin{aligned} P_S + jQ_S &= \frac{2V^2}{X_L}\sin(\delta/2) + j\frac{2V^2}{X_L}(1 - \cos(\delta/2)) \\ P_R + jQ_R &= \frac{2V^2}{X_L}\sin(\delta/2) - j\frac{2V^2}{X_L}(1 - \cos(\delta/2)) \end{aligned}\right\} \tag{13.19}$$

$$Q_{loss(S)} = Q_{loss(R)} = \frac{4V^2}{X_L}(1 - \cos(\delta/2)) \tag{13.20}$$

$$Q_{comp} = \frac{4V^2}{X_L}(\cos(\delta/2) - 1) \tag{13.21}$$

Phasor diagram of the system in this special case is shown in Figure 13.3.

The compensation reactive power, Q_C, is the negative of the absorbed reactive power, i.e. $Q_C = -Q_{comp}$

It is found from (13.19) that $P_S = (2/0.275)\sin 14.15 = 0.896$ pu and the relation of P_S versus δ for the compensated system is given by

$$P_S = 7.27\sin(\delta/2) \text{ (plotted in Figure 13.2)}$$

$$Q_{loss(S)} = Q_{loss(R)} = 0.11 \text{ pu}, \quad Q_M = -0.11 \text{ pu} \quad \text{and} \quad Q_C = 0.11 \text{ pu}$$

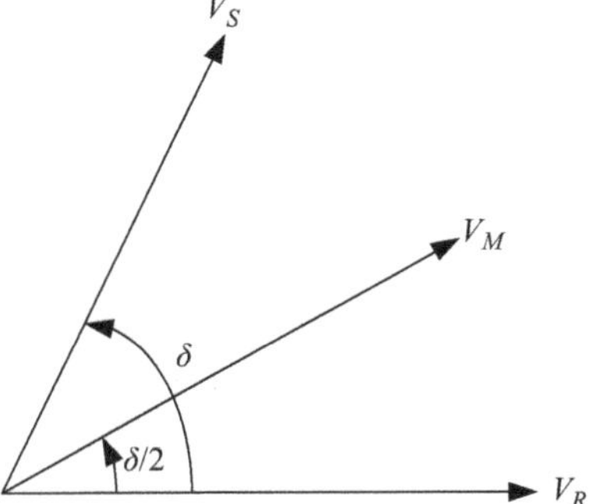

Figure 13.3 System phasor diagram with shunt compensation

As depicted in Figure 13.2, the area under P_S versus δ curve for the compensated case is much larger than that of the un-compensated case. Thus, the system has more stability margin.

13.2 Long transmission lines

It is appropriate to describe the long transmission line by ***ABCD*** parameters. It is found that the values of these parameters depend on the location of shunt compensation. In case of locating the compensators at a point somewhere along the line the equivalent ***ABCD*** parameters (Figure 13.4) can be calculated using the relations below.

$$\left.\begin{aligned} \boldsymbol{A}_{eq} &= \boldsymbol{A}_1\boldsymbol{A}_2 + \boldsymbol{B}_1\boldsymbol{C}_2 - jX_C\boldsymbol{A}_2\boldsymbol{B}_1 \\ \boldsymbol{B}_{eq} &= \boldsymbol{A}_1\boldsymbol{B}_2 + \boldsymbol{B}_1\boldsymbol{D}_2 - jX_C\boldsymbol{B}_1\boldsymbol{B}_2 \\ \boldsymbol{C}_{eq} &= \boldsymbol{A}_2\boldsymbol{C}_1 + \boldsymbol{C}_2\boldsymbol{D}_1 - jX_C\boldsymbol{A}_2\boldsymbol{D}_1 \\ \boldsymbol{D}_{eq} &= \boldsymbol{B}_2\boldsymbol{C}_1 + \boldsymbol{D}_1\boldsymbol{D}_2 - jX_C\boldsymbol{D}_2\boldsymbol{D}_1 \end{aligned}\right\} \tag{13.22}$$

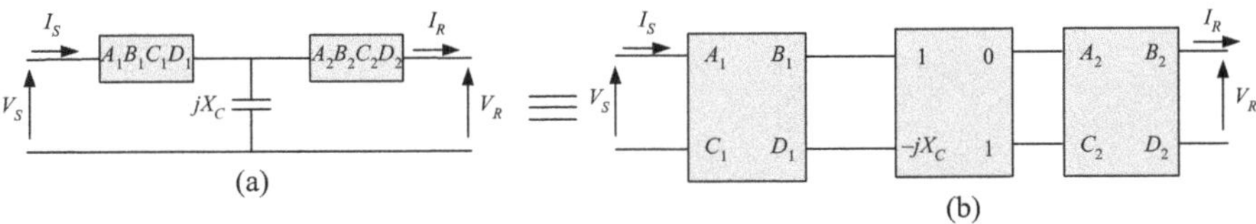

Figure 13.4 (a) Transmission line with shunt compensation at midpoint and (b) equivalent representation

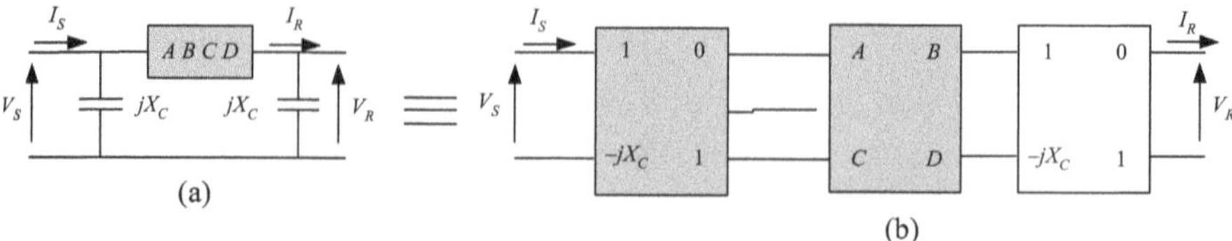

Figure 13.5 (a) Transmission line with shunt compensation at both the sending and receiving ends and (b) equivalent representation

Similarly, when the line is shunt compensated at both sending and receiving ends (Figure 13.5) ***ABCD*** parameters are

$$\left.\begin{aligned} \boldsymbol{A}_{eq} &= \boldsymbol{A} - jX_{C2}\boldsymbol{B} \\ \boldsymbol{B}_{eq} &= \boldsymbol{B} \\ \boldsymbol{C}_{eq} &= \boldsymbol{C} - jX_{C1}\boldsymbol{A} - jX_{C2}\boldsymbol{D} - X_{C1}X_{C2}\boldsymbol{B} \\ \boldsymbol{D}_{eq} &= \boldsymbol{D} - jX_{C1}\boldsymbol{B} \end{aligned}\right\} \tag{13.23}$$

Example 13.3 The system shown in Figure 13.6 comprises a transmission line with the data given in Example 12.4. The internal machine and receiving end voltages are given as $1.2\angle 17^\circ$ pu and $0.9\angle 0^\circ$ pu, respectively. Find the power transfer capacity. If the system is shunt capacitive compensated at the receiving end, calculate the power transfer capacity and the compensated reactive power necessary to raise the receiving end voltage up to $1\angle 0^\circ$.

Solution:

For uncompensated system the system ***ABCD*** parameters as calculated in Example 12.4 are

$$\begin{aligned} \boldsymbol{A} &= \boldsymbol{A}_L + j0.3\boldsymbol{C}_L = 0.916\angle 0^\circ \\ \boldsymbol{B} &= \boldsymbol{B}_L + j0.3\boldsymbol{D}_L = j0.499 \text{ pu} \\ \boldsymbol{C} &= \boldsymbol{C}_L = j0.206 \text{ pu} \\ \boldsymbol{D} &= \boldsymbol{D}_L = 0.978 \end{aligned}$$

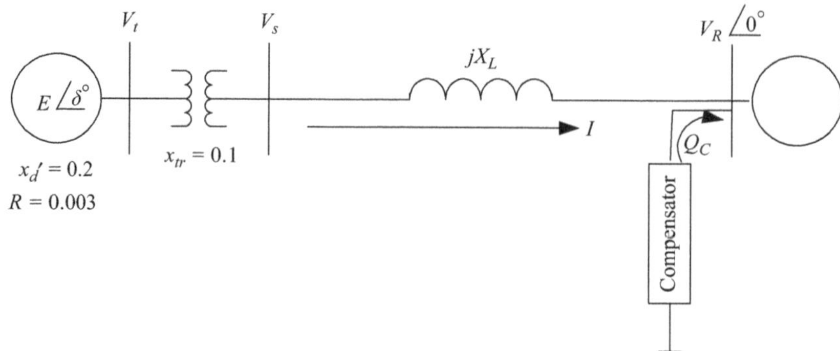

Figure 13.6 System studied in Example 12.3

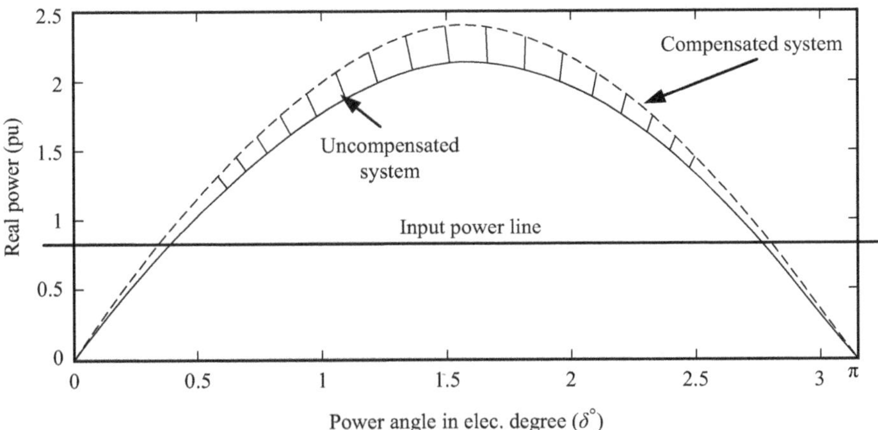

Figure 13.7 Power–angle curves for compensated and uncompensated systems

$\boldsymbol{E} = Ee^{j\delta} = 1.2e^{j17^\circ}$ and $\boldsymbol{V}_R = V_R e^{j0^\circ} = 0.9\angle 0^\circ$ ($\boldsymbol{V}_R$ is taken as a reference)

$$P_S + jQ_S = \boldsymbol{E}\boldsymbol{I}^* = \frac{EV_R}{X}\sin\delta + j\frac{(E^2 - EV_R\cos\delta)}{X} \tag{13.24}$$

$$P_R + jQ_R = \boldsymbol{V}_R\boldsymbol{I}^* = \frac{EV_R}{X}\sin\delta - j\frac{(V_R^2 - EV_R\cos\delta)}{X} \tag{13.25}$$

$$Q_{loss} = \frac{1}{X}\left(E^2 + V_R^2 - 2EV_R\cos\delta\right) \tag{13.26}$$

Thus, the following parameters can be calculated as:

$$P_R = (1.2 \times 0.9/0.499)\sin 17^\circ = 2.16\sin 17^\circ = 0.63 \text{ pu}$$
$$Q_R = (1/0.499)(1.2 \times 0.9\cos 17^\circ - 0.81) = 0.446 \text{ pu}$$
$$Q_{loss} = (1/0.499)(1.44 + 0.81 - 2 \times 1.2 \times 0.9\cos 17^\circ) = 0.37 \text{ pu}$$

For compensated system the compensation is required to increase the voltage at receiving end, $V_R = 1\angle 0^\circ$. In this case it is found that

$$P_R = (1.2 \times 1.0/0.499)\sin 17^\circ = 2.4\sin 17^\circ = 0.70 \text{ pu}$$
$$Q_R = (1/0.499)(1.2 \times 1.0\cos 17^\circ - 1.0) = 0.296 \text{ pu}$$
$$Q_{loss} = (1/0.499)(1.44 + 1.0 - 2 \times 1.2 \times 1.0\cos 17^\circ) = 0.29 \text{ pu}$$

It is seen that the reactive power flow into the line at the receiving end is reduced from 0.446 to 0.296. So, the difference 'Q_C' must be provided by the shunt compensation, i.e. $Q_C = 0.15$ pu.

The reactive power losses are reduced by a percentage of 27.6.

P versus δ curves for both the uncompensated and compensated cases are shown in Figure 13.7. It is found that the area under P–δ curve for the compensated system is increased by the hatched area. Therefore, the compensation provides the system more marginal stability.

13.3 Static var compensators

It has been shown that the shunt capacitive compensation improves power system stability by controlling both steady-state voltage and reactive power. It supplies partially or fully the load with the required reactive power. Thus, it increases the power system load-ability and reduces both the line current and the voltage drop in the line. In the case of system operation at light loads or under open line conditions the voltage may increase. So, shunt inductive compensation is needed to limit the voltage rise by continuously controlling the reactive power, which in turn will reduce the transmission losses and increase the transmission capacity of active power. Consequently, the shunt compensation is used to compensate for the effects of stressed and light load conditions of the transmission system. It should be a combination of fast controlled capacitances and inductances.

The static var compensator (SVC) is defined as a shunt device connected at a proper location in the transmission system [5]. It is built up by static components (capacitors and inductors), which may be controlled very fast by semiconductors, thyristors [6].

SVCs have different configurations as explained in Chapter 14. The basic configuration 'FC-TCR' is depicted in Figure 13.8. It is a combination of fixed capacitor, FC, branch and thyristor-controlled reactor (TCR), branch as well as a filter for low-order harmonics for each of the three phases [7].

A continuous range of reactive power consumption is obtained by using thyristor firing angle, α, control. However, odd harmonic current components during the control process are generated in the reactor current. Full conduction is obtained with α and equals 90°. The fundamental component of the reactor current is reduced by increasing α, i.e. the inductance is increased and the reactive power absorbed by the reactor is reduced. It is to be noted that the change in the reactor current may only take place at discrete points of time, i.e. adjustments cannot be made more frequently than once per half-cycle. Static compensators of TCR type are characterised by the ability to perform continuous control, maximum delay of one half-cycle and

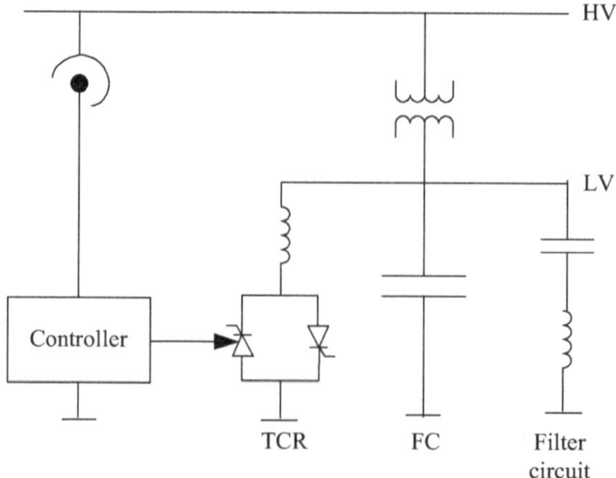

Figure 13.8 FC-TCR configurations

no transients. The main drawbacks of this configuration are the generation of low-frequency harmonic current components and higher losses when working in the inductive region [8, 9]. However, SVCs have a wide range of applications [10–12].

Using Fourier transform, the fundamental component of the reactor current, I_1, in terms of the firing angle α, is given by

$$\begin{aligned} I_1 &= \frac{V_{rms}}{\omega L}\frac{(2\pi - 2\alpha + \sin(2\alpha))}{\pi} \\ &= \frac{V_{rms}}{X_L}\frac{\sigma - \sin\sigma}{\pi} \end{aligned} \tag{13.27}$$

where

$\sigma \triangleq$ the conduction angle $= 2(\pi - \sin\sigma)$

The amplitude of each harmonic component is defined by

$$I_h = \frac{4V_{rms}}{\pi X_L}\left[\frac{\sin(h+1)\alpha}{2(h+1)} + \frac{\sin(h-1)\alpha}{2(h-1)} - \cos(\alpha)\frac{\sin(h\alpha)}{h}\right] \tag{13.28}$$

where

V_{rms} is rms value of the voltage applied to the compensator
h is harmonic order
α is thyristor firing angle

To eliminate low-frequency current harmonics, 3rd, 5th, 7th, etc., delta connections for triplet harmonics and passive filters for the others may be used as shown in Figure 13.9. The fixed capacitors may be connected in series with current limiting reactors.

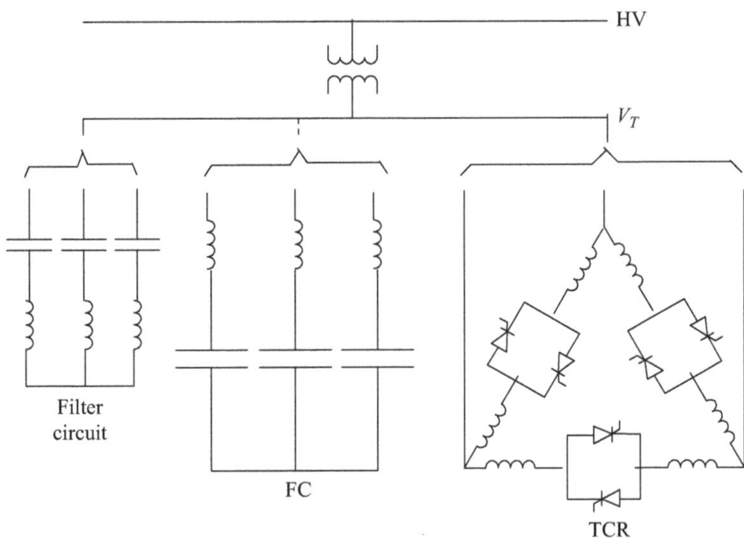

Figure 13.9 TCR-delta-connection with FC and tuned filter for harmonic elimination

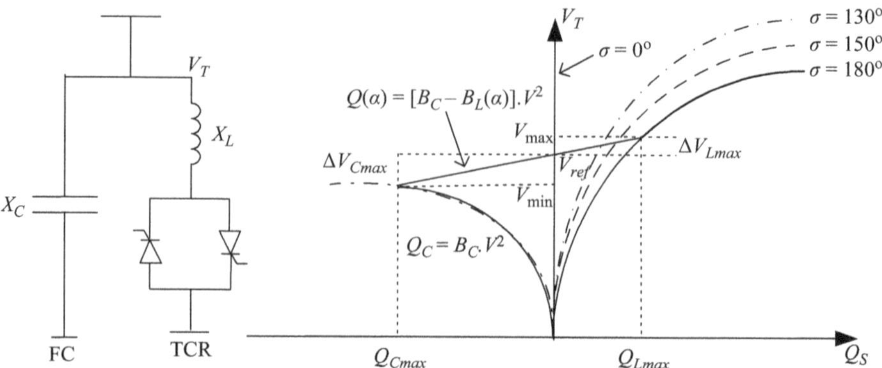

Figure 13.10 V_T–Q_S characteristic of FC-TCR compensator

13.3.1 Characteristics of FC-TCR compensators

The amount of reactive power interchanged with the system, Q_S, depends on the applied voltage, V_T. The steady-state $Q_S - V_T$ characteristic of FC-TCR compensator is shown in Figure 13.10, indicating the amount of reactive power generated or absorbed (Q_C or Q_L, respectively) by the compensator as a function of the applied voltage. At the rated voltage the characteristic is linear and limited by the rated power of both the capacitor and reactor. Between the voltage limits of the linear part, V_{max} and V_{min}, the control range is defined. When the rated voltage is applied to the compensator there is no interchange of reactive power between the compensator and the power system. This rated voltage is taken as a reference voltage, V_{ref}, for the control process. The slope of the linear characteristic reflects a change in voltage with compensator rating and, therefore, can be considered a slope reactance resulting in the SVC response to the voltage variation. The slope reactance, X_{SL}, is given by

$$X_{SL} = \frac{\Delta V^2_{Cmax}}{Q_{Cmax}} = \frac{\Delta V^2_{Lmax}}{Q_{Lmax}} \tag{13.29}$$

Out of the control range, the $V_T - Q_S$ characteristic is non-linear. Assuming the compensator is ideal, the change of Q_S versus V_T out of the control range can be considered a linear relationship as an approximation.

13.3.2 Modelling of FC-TCR compensators

For power flow analysis and stability study of power systems, a model of the SVC is required to express its performance by mathematical relations that introduce accurate representation [12]. Assuming an ideal SVC, the steady-state V–I characteristic is as shown in Figure 13.11.

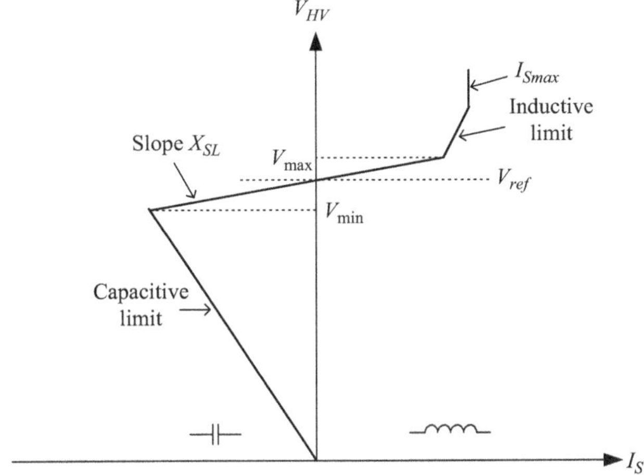

Figure 13.11 Steady-state V–I characteristic

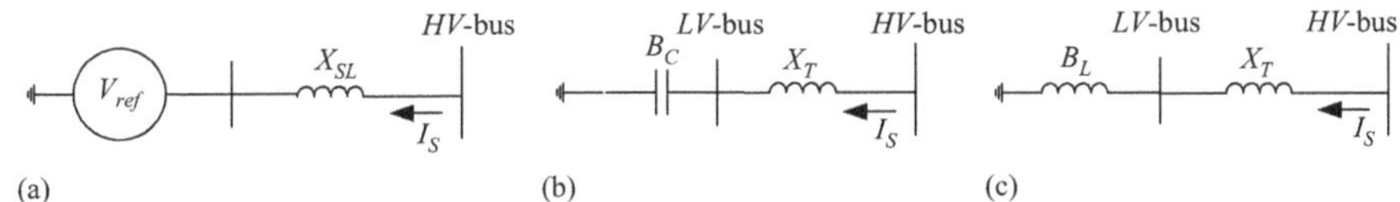

Figure 13.12 SVC modes of operation: (a) normal mode; (b) capacitive mode; and (c) inductive mode

For power flow analysis: referring to the schematic diagram (Figure 13.8) and *V–I* characteristic (Figure 13.11), it is seen that the SVC has three operation modes:

(i) The normal mode of operation: it is in the control range and the SVC is equivalent to a voltage source, V_{ref}, behind the slope reactance, X_{SL}, connected to *HV*-bus (Figure 13.12(a)).
(ii) The capacitive mode of operation: when the SVC operation reaches the capacitive limit it becomes a fixed capacitive susceptance, B_C, connected to the *LV*-bus, which is connected to *HV*-bus through the transformer leakage reactance, X_T (Figure 13.12(b)).
(iii) The inductive mode of operation: in this mode the SVC is represented by fixed inductive susceptance, B_L, when it reaches the inductive limit (Figure 13.12(c)). The SVC current is limited to I_{Smax}.

In the control range where $I_{min} < I_{SVC} < I_{max}$ and $V_{min} < V < V_{max}$, the SVC is represented as *PV*-node at an auxiliary or phantom bus with $P = 0$ and $V = V_{ref}$. The slope reactance is added between the auxiliary bus and *HV*-bus (node of coupling to the system). The *HV*-bus is a *PQ* bus with $P = 0$ and $Q = 0$ (Figure 13.13(a)). If the SVC transformer is represented, the reactance from the *HV*-bus to the auxiliary bus is a portion of the transformer leakage reactance. The *LV*-bus is a *PV* bus remotely regulating the auxiliary bus voltage to equal V_{ref} (Figure 13.13(b)).

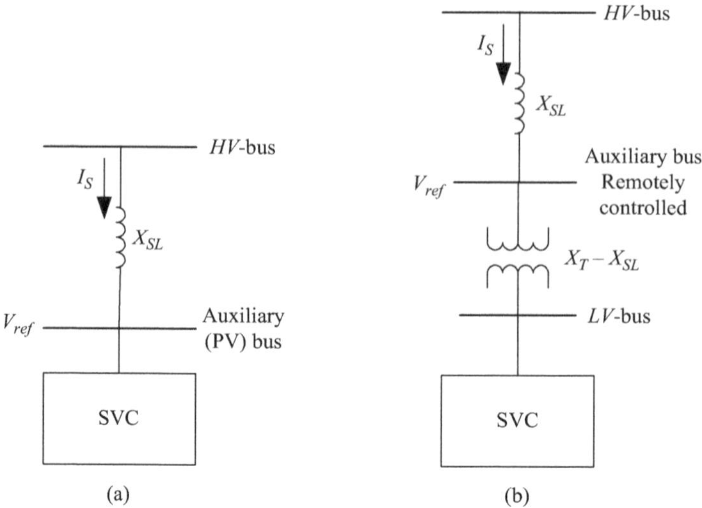

Figure 13.13 SVC representations in load flow analysis: (a) SVC as a node at the auxiliary bus and (b) SVC connected to a regulated auxiliary bus through transformer

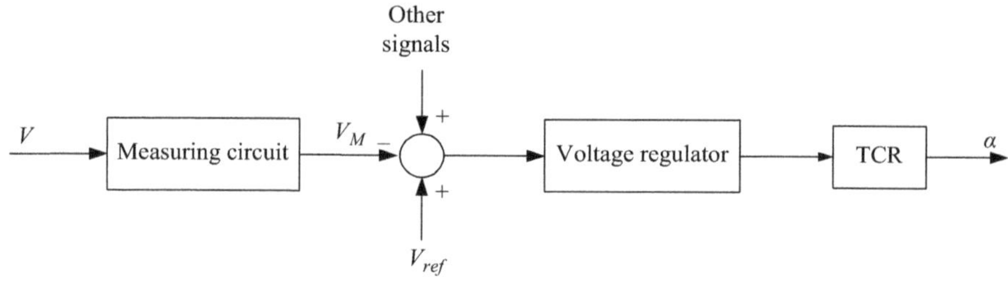

Figure 13.14 FC-TCR functional block diagram

In the case of operating the SVC outside the control range, it is represented as a shunt element with susceptance B, defined as

$$B = (1/X_C) \quad \text{if } V < V_{\min} \quad \text{and} \quad B = (1/X_L) \quad \text{if } V > V_{\max} \tag{13.30}$$

For stability study: The general approach to the modelling of an FC-TCR-type SVC is represented by the functional block diagram shown in Figure 13.14. It includes a measuring circuit block, voltage regulator block and TCR block. The measuring circuit block comprises instrument transformers, A/D converters and rectifiers. It contains a transport delay and very small time constants. Thus, the measuring circuit may be represented by a simple time constant and a unit gain. The voltage regulator block represents a proportional type regulator that could be with lead/lag circuit (integral type may be used as well). The TCR block represents the variation of thyristor firing angle. Accordingly, the simplified model of an FC-TCR compensator is shown in Figure 13.15. A voltage control signal can be modulated by a fast power oscillation damping (POD) control in case of severe system stability problems after system faults.

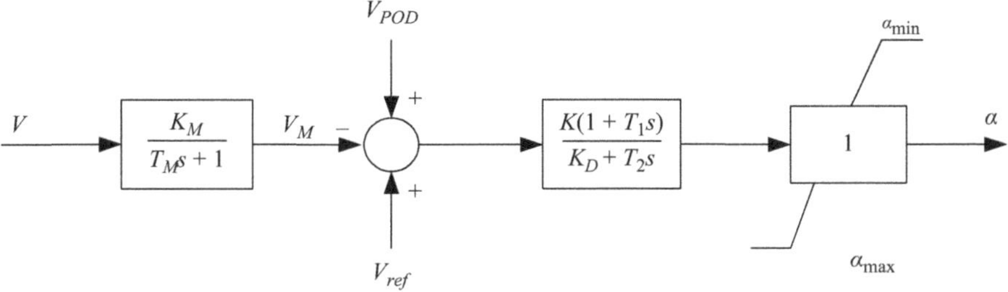

Figure 13.15 Simplified model of an FC-TCR-type SVC

The SVC-regulator model takes into account the firing angle α as an output assuming a balanced fundamental frequency operation. Thus, the model can be developed with respect to a sinusoidal voltage. It can be represented by two differential equations of v_M and α deduced from Figure 13.15 and an algebraic equation to calculate the reactive power 'Q' [13]. So, the representation can be written as below:

$$\left.\begin{aligned} \dot{v}_M &= \frac{(K_M V - v_M)}{T_M} \\ \dot{\alpha} &= \frac{\left(-K_D\alpha + K\dfrac{T_1}{T_2 T_M}(v_M - K_M V) + K\left(V_{ref} + V_{POD} - v_M\right)\right)}{T_2} \\ Q &= -\frac{2\alpha - \sin 2\alpha - \pi\left(2 - \dfrac{X_L}{X_C}\right)}{\pi X_L} V^2 = -b_{SVC}(\alpha) V^2 \end{aligned}\right\} \tag{13.31}$$

The thyristor firing angle α is allowed to vary within upper and lower limits. The SVC state variables are initialised after the power flow solution. To impose the desired voltages at the compensated bus, a PV generator bus with zero active power should be used in the power flow solution. After the power flow solution the PV bus is removed and the SVC equations are used. During the state variable initialisation a check for SVC limits is performed.

Example 13.4 The system transient stability has been evaluated in Example 8.1 for a nine-bus test system, when subjected to a three-phase short circuit at bus no. 7 'at the beginning of line 7-5' for durations of 0.08 s and 0.20 s. The fault is cleared by isolating line #7-5. It is required to investigate the system stability improvement using FC-TCR compensator. The system data are given in Appendix II.

Solution:

SAT/MATLAB® toolbox and typical control system parameters of the model in Figure 13.15 and Table 13.1 are used to check the transient stability when using FC-TCR shunt compensator at different locations. It is found that the best results are obtained when connecting the compensator at bus no. 7. The change of power angle and angular speed of each generator for fault durations of 0.08 s and 0.2 s are shown in Figures 13.16–13.18 and Figures 13.19–13.21, respectively.

Table 13.1 Control system parameters

Variable	Description	Unit	Value
S	Power rating	MVA	100
V	Voltage rating	KV	230
f	Frequency rating	Hz	60
T_2	Regulator time constant	S	10
K	Regulator gain	pu/pu	100
V_{ref}	Reference voltage	pu	1
α_{fmax}	Maximum firing angle	Rad	1
α_{fmin}	Minimum firing angle	Rad	−1
K_D	Integral deviation	pu	0.001
T_1	Transient regulator time constant	S	0
K_M	Measure gain	pu/pu	1
T_M	Measure time delay	S	0.01
x_L	Reactance (inductive)	pu	0.2
x_C	Reactance (capacitive)	pu	0.1

Fault duration 0.08 s

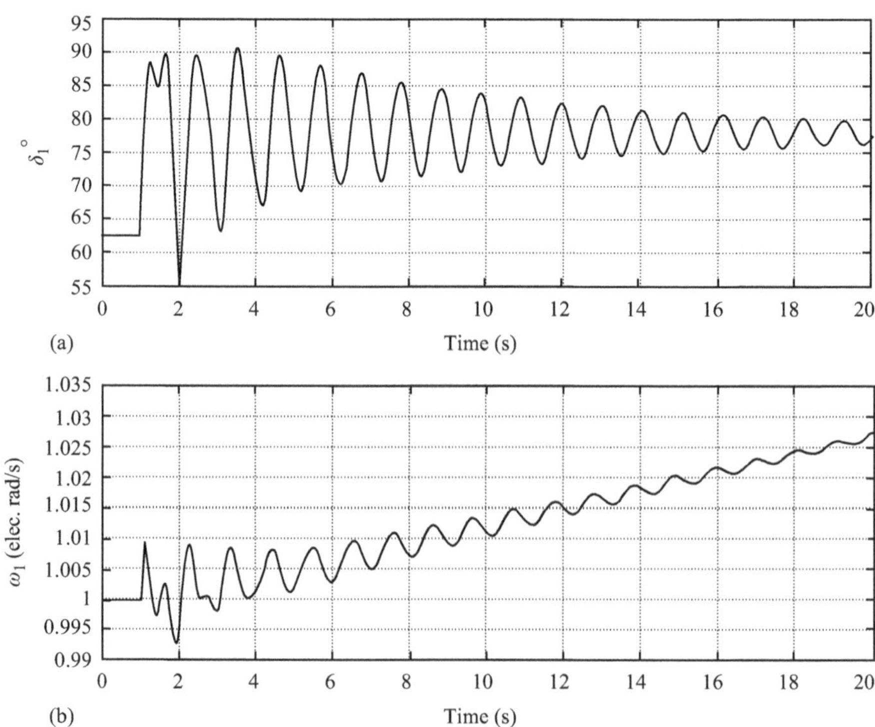

Figure 13.16 Generator 1: (a) δ versus time and (b) ω versus time, fault duration = 0.08 s

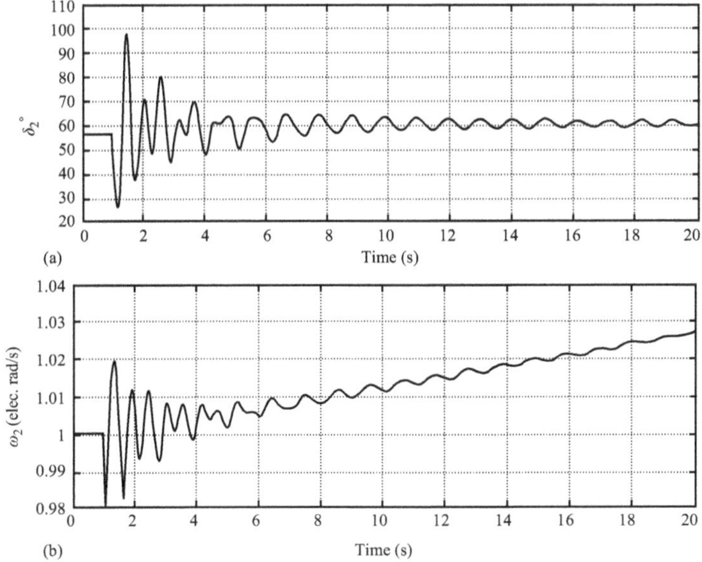

Figure 13.17 *Generator 2: (a) δ versus time and (b) ω versus time, fault duration = 0.08 s*

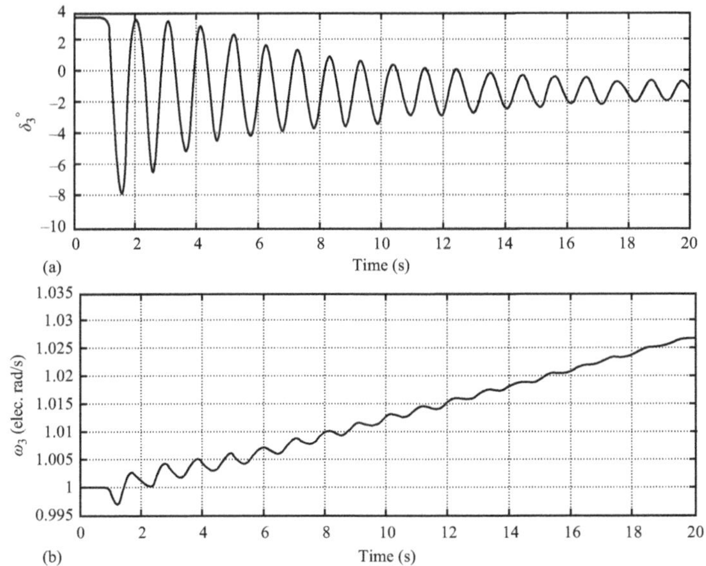

Figure 13.18 *Generator 3: (a) δ versus time and (b) ω versus time, fault duration = 0.08 s*

Fault duration 0.2 s

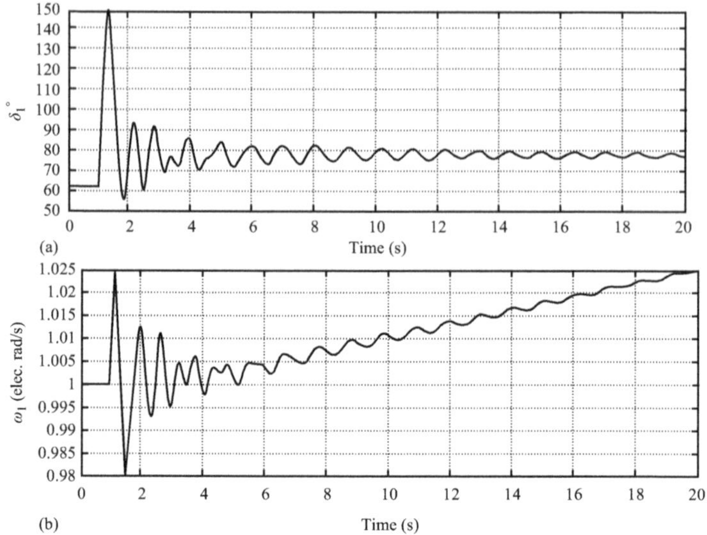

Figure 13.19 Generator 1: (a) δ versus time and (b) ω versus time, fault duration = 0.2 s

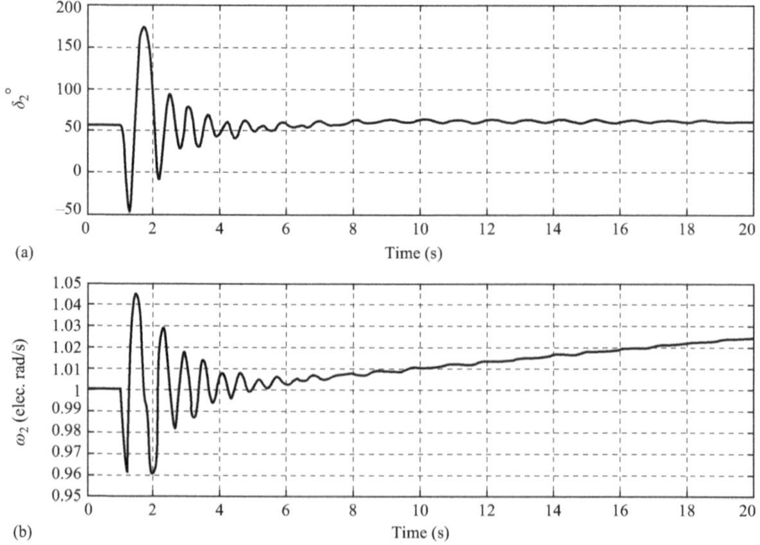

Figure 13.20 Generator 2: (a) δ versus time and (b) ω versus time, fault duration = 0.2 s

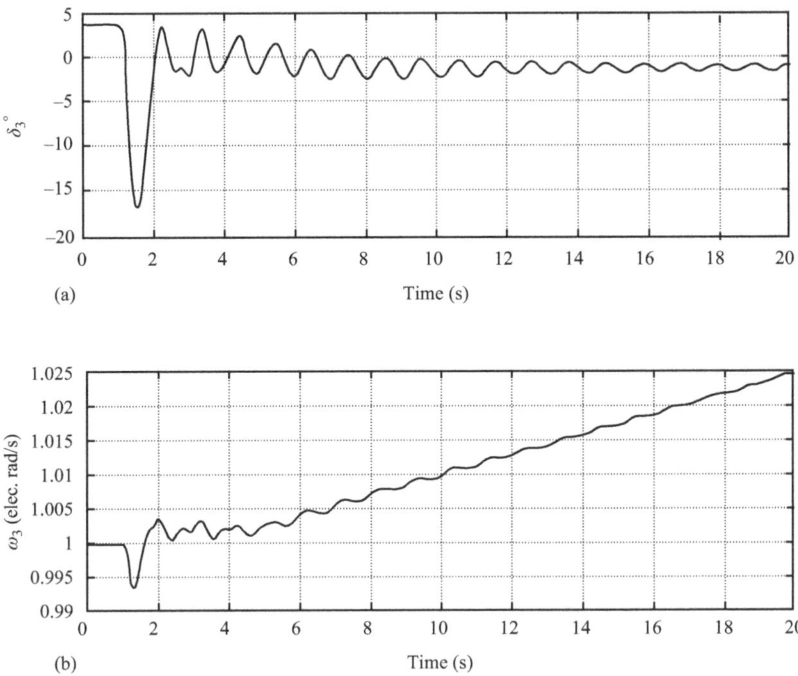

Figure 13.21 Generator 3: (a) δ versus time and (b) ω versus time, fault duration = 0.2 s

13.4 Static synchronous compensator (STATCOM)

A static compensator, STATCOM, acts as a solid-state synchronous voltage source in analogy with a synchronous machine generating a balanced set of three sinusoidal voltages at the fundamental frequency with controllable amplitude and phase angle. This device, however, has no inertia [14].

Principle of operation: The static compensator consists of a voltage source converter, a coupling transformer and controls. In this application the DC energy source device can be replaced by a DC capacitor, so that the steady-state power exchange between the static compensator and the AC system can only be reactive, as illustrated in Figure 13.22, where I_q is the converter output current, in quadrature with the converter voltage V_i. The magnitude of the converter voltage, and thus the reactive power output of the converter, is controllable. If V_i is greater than the terminal voltage 'V_{HV}' the static compensator will supply reactive power to the AC system. If V_i is smaller than V_{HV}, the static compensator absorbs reactive power.

STATCOM model: The STATCOM is modelled as a current injection model. The STATCOM current is always kept in quadrature in relation to the bus voltage so

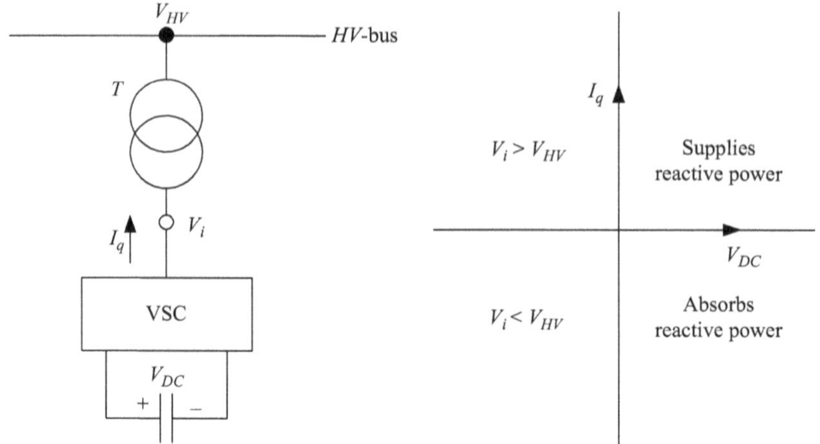

Figure 13.22 Static compensator comprising voltage source converter (VSC), coupling transformer T and control

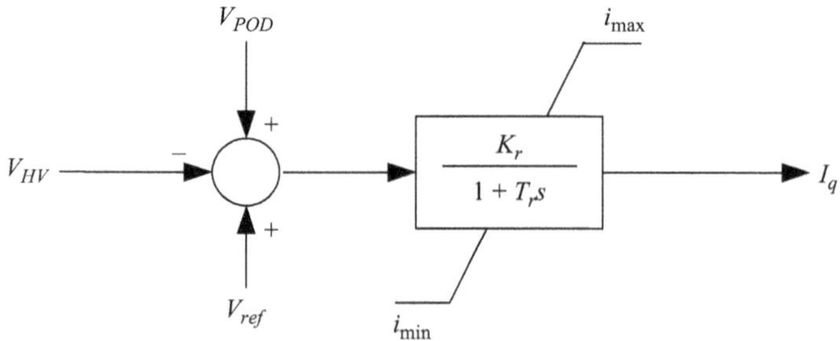

Figure 13.23 Block diagram of STATCOM simplified model

that only reactive power is exchanged between the a.c. system and the STATCOM. The dynamic model is shown in Figure 13.23 where it can be seen that the STATCOM assumes a single time constant regulator role like the SVC. The differential equation and the reactive power injected at the STATCOM node are given, respectively, by

$$\left.\begin{aligned} \dot{I}_q &= \frac{\left[K_r\left(V_{ref} + V_{POD} - V_{HV}\right) - I_q\right]}{T_r} \\ Q &= I_q V_{HV} \end{aligned}\right\} \tag{13.32}$$

where

K_r: regulator gain
V_{ref}: reference voltage of the STATCOM regulator.
V_{POD}: additional stabilising signal, which is the output of the power oscillation damper
T_r: regulator time constant

Example 13.5 Repeat Example 13.4 using STATCOM instead of SVC.

Solution:

A typical data of control parameters (Figure 13.23) is summarised in Table 13.2.

Using PSAT/MATLAB® toolbox and connecting the STATCOM as a shunt compensator at different locations in the system, the best results are obtained with the compensator connected in shunt at bus no. 8. The results, variation of δ and ω with time, are shown in Figures 13.24–13.26 for fault duration 0.08 s and Figures 13.27–13.29 for fault duration 0.2 s.

Table 13.2 Control system parameters

Variable	Description	Value	Unit
S	Power rating	100	MVA
V	Voltage rating	13.8	KV
f	Frequency rating	60	Hz
T_r	Regulator time constant	50	S
K_r	Regulator gain	0.1	pu/pu
I_{max}	Maximum current	0.2	pu
I_{min}	Minimum current	−0.2	pu

Fault duration 0.08 s

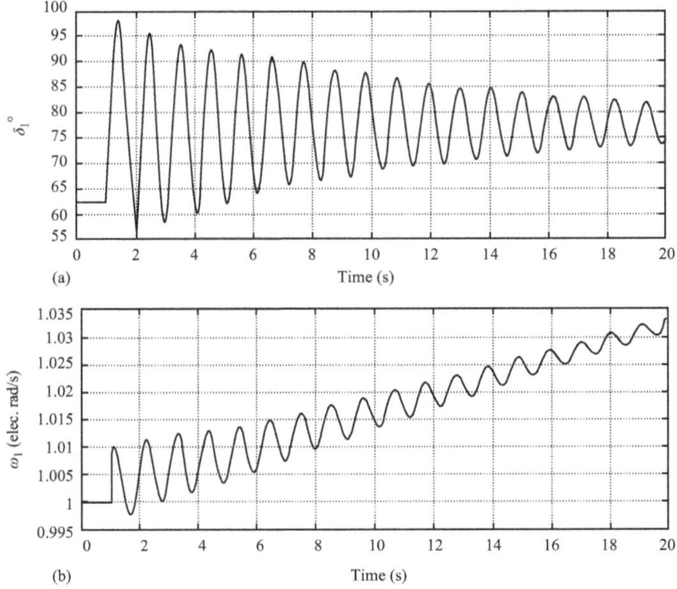

Figure 13.24 Generator 1: (a) δ versus time and (b) ω versus time, fault duration = 0.08 s

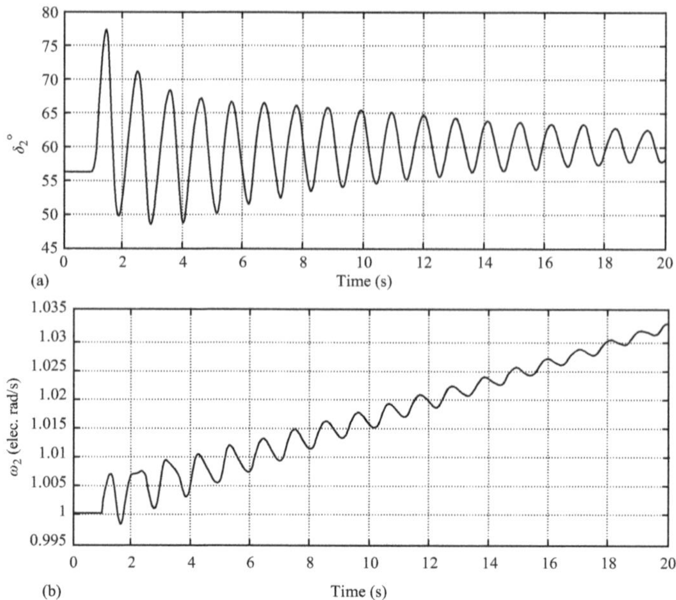

Figure 13.25 *Generator 2: (a) δ versus time and (b) ω versus time, fault duration = 0.08 s*

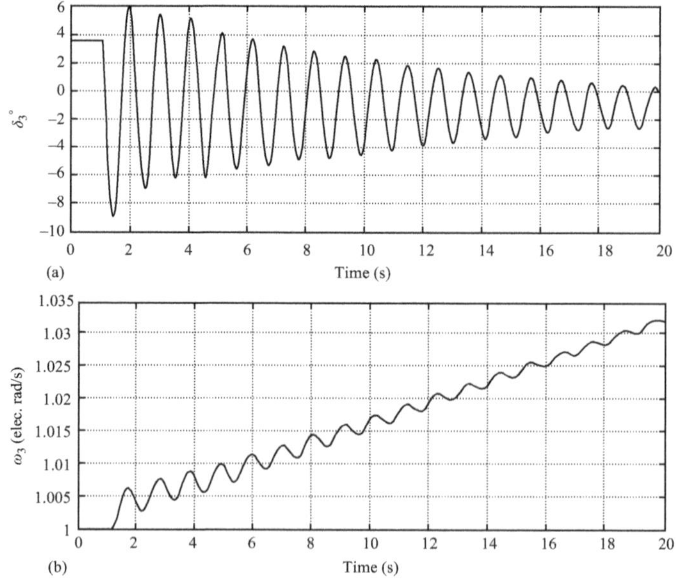

Figure 13.26 *Generator 3: (a) δ versus time and (b) ω versus time, fault duration = 0.08 s*

Fault duration 0.2 s

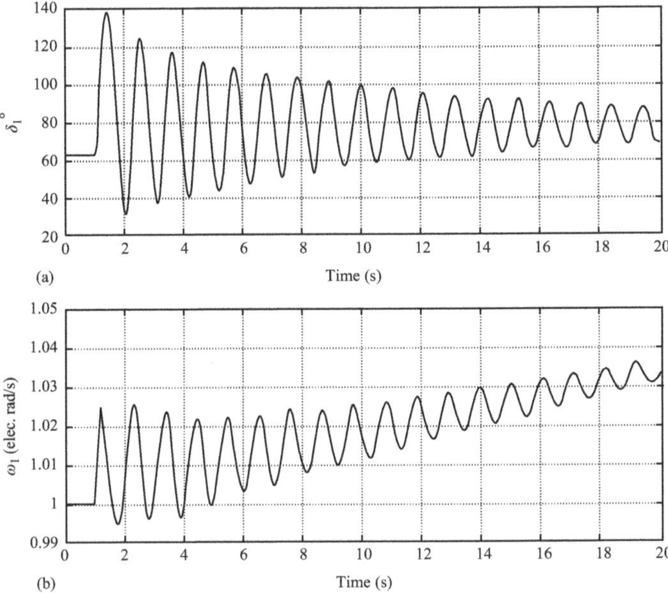

Figure 13.27 Generator 1: (a) δ versus time and (b) ω versus time, fault duration = 0.2 s

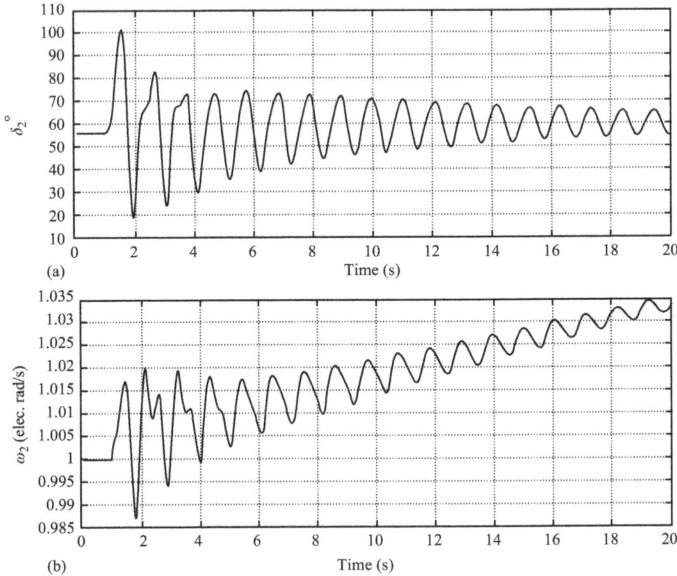

Figure 13.28 Generator 2: (a) δ versus time and (b) ω versus time, fault duration = 0.2 s

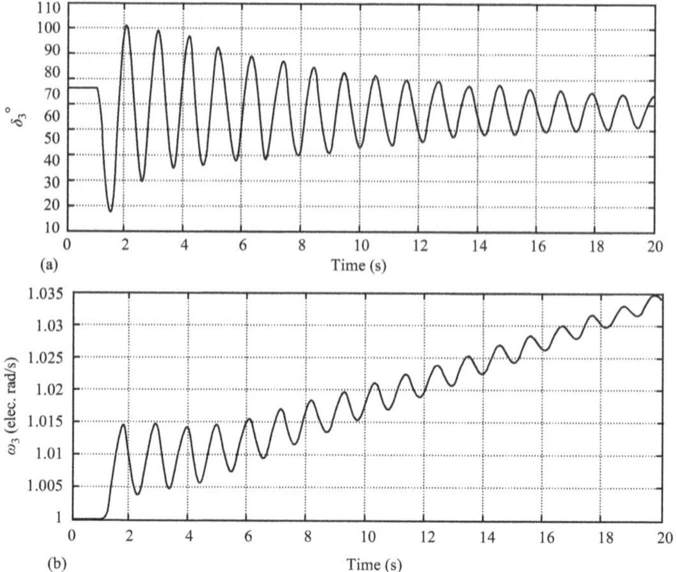

Figure 13.29 Generator 3: (a) δ versus time and (b) ω versus time, fault duration = 0.2 s

It is to be noted that the system without compensation is unstable when subjected to a fault of duration 0.2 s (Example 8.2). From the results obtained above, Examples 13.4 and 13.5, it is concluded that:

- The system is stabilised when using either SVC or STATCOM compensators in the case of fault duration of 0.2 s.
- For the stable state at which the fault duration is 0.08 s, damping of the oscillations is increased when using shunt compensation.
- The preferred location as well as the effectiveness of compensation depends on the type of compensation. It is seen that the STATCOM is connected at bus no. 8 while the SVC is connected at bus no. 7.

In power system stability studies, a combination of both series and shunt compensation may be applied simultaneously to enhance the system stability, increase power transfer capability and modify the voltage profile at different buses. Studies should be performed to determine the cost of each compensator before finalising the design.

13.5 Application of ASNFC to shunt-compensated power systems

In the preceding sections the analytic time analysis has been used to study the dynamic performance of the power system. On the other hand, the adaptive simplified neuro-fuzzy controller, ASNFC, has been applied for designing the PSS as a

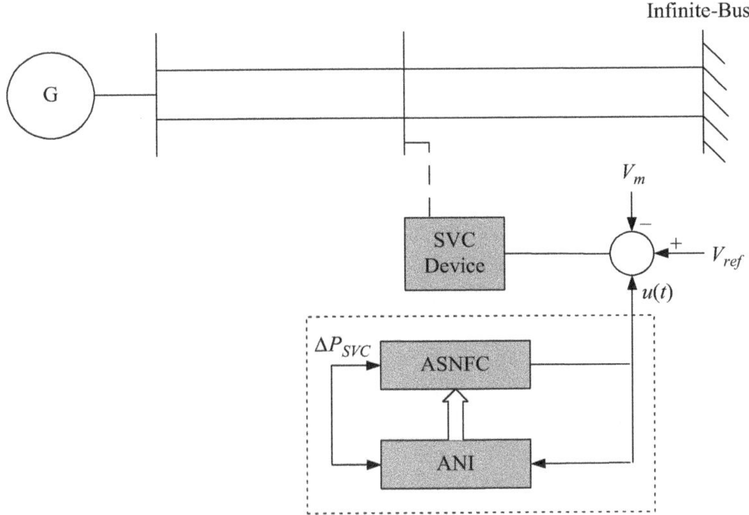

Figure 13.30 ASNFC for an SVC device in an SMIB system

supplementary controller of the excitation system to damp the generator oscillations as explained in Chapter 10, Section 10.5. In this section, the application of a proposed ASNFC as an artificial intelligent controller is used to control the SVC operation as a shunt compensator. The system model considered for studying the performance of the proposed adaptive controller is depicted in Figure 13.30. The simulation study has been performed on a single machine connected to an infinite bus, SMIB, through a long transmission line, with the SVC at the middle of the line. As shown in Figure 13.30, the SVC is used rather than the ideal shunt compensator. It is modelled as a susceptance that varies within a limit, depending on the control provided by the regulator of the SVC [15, 16].

The general structure of ASNFC system given in Chapter 10, Section 10.6, is applied to the SVC to damp power system oscillations. The control structure has two steps: estimation of the system parameters online, using artificial neural-identifier, ANI, and calculation of the controller parameters using the gradient descent method. If the system parameters vary, the identifier will provide an estimate of these parameters and the adaptation mechanism will subsequently tune the controller parameters. Inputs to the ASNFC are the power deviation at the SVC bus, $\Delta P_{SVC}(k)$, and its derivative, $\Delta \dot{P}_{svc}$, while the output of the controller is $u(k)$. The input scaling blocks, K_1 and K_2, map the real input to the normalised input space in which the membership functions are defined. The output block K_3 is used to map the output of the fuzzy inference system to the real output.

13.5.1 Simulation studies

An SVC device connected at the middle of an SMIB system is used to test the performance of the proposed ASNFC. An adaptive neuro-identifier is used to track the behaviour of the plant online and update the ASNFC. The proposed controller is designed based on a zero-order Sugeno-type fuzzy controller with one input to the ANFIS network.

The performance of the ASNFC is compared with the ANFC and SVC conventional power system stabiliser (SCPSS). The SCPSS was carefully tuned for the best possible performance at the nominal load of 0.70 pu power and 0.85 pf *lag*. The parameters of the SVC were then kept unchanged for the test performed. A sampling frequency of 25 Hz is used in the simulation and the learning rate is set to be 0.08. The absolute limits for the control output are limited to ± 0.1 pu and the absolute limits for the SVC output are limited to ± 0.15 pu. The performance of the proposed controller has been verified for a three-phase to ground short circuit [15].

The parameters of (i) the seventh-order model differential equations used to simulate the generating unit, (ii) the transmission line and (iii) the transfer functions of the exciter, the governor, the SVC device, the SCPSS and the generator conventional power system stabiliser (GCPSS) are provided in Appendix V [17, 18].

13.5.2 Three-phase to ground short circuit test

The GCPSS is applied to the generator during the three-phase to ground short circuit test. The behaviour of the system with the proposed ASNFC is investigated in this test. The generator speed deviation response, under the normal operating condition, to a three-phase to ground short circuit at the middle of one of the transmission lines connecting the generator to the middle bus is illustrated in Figure 13.31. The fault occurs at 1.0 s and is cleared 80 ms later by disconnection

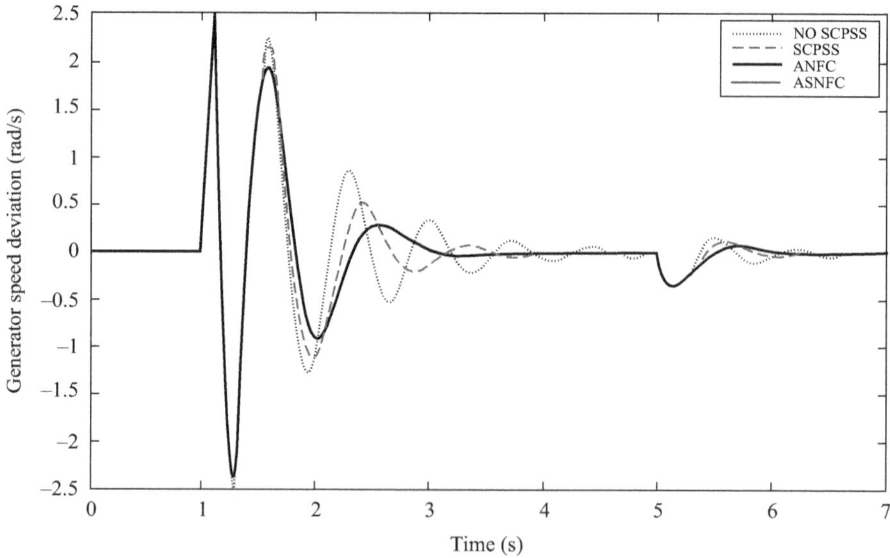

Figure 13.31 Generator speed deviation in response to a three-phase fault at the middle of a transmission line connecting the generator to the middle bus with a successful re-closure

of the faulted line and successful re-closure at 5.0 s. The results show that the ASNFC reduces the speed deviation after the fault and helps the system to reach the new operating point quickly. More investigation and further explanation are provided in [19].

References

1. Happ H.H., Wirgau K.A. 'Static and dynamic VAR compensation in system planning'. *IEEE Transactions on Power Apparatus and Systems*. Sept/Oct 1978;**97**(5):1564–78
2. Diaz U.A.R., Hermádez J.H.T. (eds.). 'Reactive shunt compensation planning by optimal power flows and linear sensitivities'. *2009 Electronics, Robotics and Automotive Mechanics Conference*; Cuernavaca, Mexico, Sept 2009. pp. 326–31
3. Mahdavian M., Shahgholian G., Shafaghi P., Bayati-Poudeh M. (eds.). 'Effect of static shunt compensation on power system dynamic performance'. *IEEE International Symposium on Industrial Electronics (ISIE), 2011*; Gdansk, Poland, Jun 2011. pp. 1029–32
4. Edris A.A. 'Controllable VAR compensator: a potential solution to load-ability problem of low capacity power systems'. *IEEE Transactions on Power Systems*. 1987;**2**(3):561–7
5. Goh S.H., Saha T.K., Dong Z.Y. (eds.). 'Optimal reactive power allocation for power transfer capability assessment'. *IEEE PES General Meeting*; Montreal, Canada, Jun 2006. pp. 1–7
6. Sahadat M.N., Al-Masood N., Hossain M.S., Rashid G., Chowdhury A.H. (eds.). 'Real power transfer capability enhancement of transmission lines using SVC'. *Power and Energy Engineering Conference (APEEC), 2011 Asia Pacific*; Wuhan, China, Mar 2011. pp. 104–7
7. Tyll H.K., Schettler F. (eds.). 'Historical overview on dynamic reactive power compensation solutions from the begin of AC power transmission towards present applications'. *Power Systems Conference and Exposition PSCE'09*; Seattle, WA, US, Mar 2009. pp. 1–7
8. Dixon J., Rodriguez J. 'Reactive power compensation technologies: state-of-art review'. *IEEE/JPROC*. 2005;**93**(2):2144–64
9. Hauth R.L., Miske S.A., Nozari F. 'The role and benefits of static VAR systems in high voltage power system applications'. *IEEE Transactions on Power Apparatus and Systems*. 1982;**101**(10):3761–70
10. Lajoie L.G., Larsen E.V. 'Hydro-Quebec multiple SVC application control stability study'. *IEEE Transactions on Power Delivery*. 1990;**5**(3):1543–51
11. Bronfeld J.D. (ed.). 'Utility application of static VAR compensation'. *Southern Tier Technical Conference, 1987, Proceedings of the 1987 IEEE*; Binghamton, NY, US, Apr 1987. IEEE; 1987. pp. 53–63
12. Jayabarathi R., Sindhu M.R., Devarajan N., Nambiar T.N.P. (eds.). 'Development of a laboratory model of hybrid static compensator'. *Power India*

Conference; New Delhi, India, Apr 2006. IEEE: Curran Associates; 2007. pp. 377–82

13. Talebi N., Ehsan M., Bathaee S.M.T. (eds.). 'Effects of SVC and TCSC control strategies on static voltage collapse phenomena'. *IEEE Proceedings, Southeast Conference*; Greensboro, NC, US, Mar 2004. pp. 161–8
14. Tan Y.L. 'Analysis of line compensation by shunt connected FACTS controllers: A comparison between SVC and STATCOM'. *Power Engineering Review, IEEE*. 1999;**19**(8):57–8
15. Albakkar A. 'Adaptive simplified neuro-fuzzy controller as supplementary stabilizer for SVC'. PhD Thesis. Alberta, Canada: University of Calgary; 2014
16. Albakkar A., Malik O.P. (eds.). 'Adaptive neuro-fuzzy controller based on simplified ANFIS network'. *IEEE Power Engineering Society General Meeting*. San Diego, CA, US, Jul 2012. pp. 1–6
17. Gokaraj R. 'Beyond gain-type scheduling controllers: new tools of identification and control for adaptive PSS'. PhD Dissertation. Alberta, Canada: Department of Electrical and Computer Engineering, University of Calgary; May 2000
18. IEEE Excitation System Model Working Group. 'Excitation system models for power system stability studies'. IEEE Standard 421.5. IEEE, 1992
19. Albakkar A.M. 'Adaptive simplified neuro-fuzzy system as a supplementary controller for an SVC device'. PhD Thesis. Alberta, Canada: University of Calgary; Sept 2014

Chapter 14
Compensation devices

14.1 Introduction

With the increase in demand and limitations on increasing power transfer capacity, power systems become more stressed and may be exposed to the risk of losing stability following a disturbance. One of the solutions to reducing this risk is to optimise the utilisation of power system components by maximising their performance. The transmission network, as one of the major components in power systems, attracts the power engineers to enhance its performance in both steady-state and transient conditions. As described in Chapters 12 and 13, the performance of a transmission network can be improved by using either series or shunt compensation or a combination of both. The basic configurations of compensation methods are described as well.

New compensation devices based on power electronics, which are very efficient for better utilisation of the existing transmission networks without sacrificing the desired stability margin, have been developed. The transmission network equipped with such devices is called 'flexible AC transmission system, FACTS'. Different configurations of FACTS devices, such as controllers of network compensation based on what is described in Chapters 12 and 13, are described in this chapter [1].

Voltage instability refers to system voltage collapse, which makes the system voltage decay to a level from which the system is unable to recover. Voltage collapse occurs when a system is loaded beyond its maximum load-carrying limit. The consequence of voltage collapse may lead to a partial or full power interruption in the system [2]. The only way to save the system from voltage collapse is to reduce the reactive power load or add additional reactive power prior to reaching the point of voltage collapse. Introducing sources of reactive power, i.e. shunt capacitors and/or FACTS controllers at appropriate location(s) in the system, is the most effective way for electric utilities to improve voltage stability of the system. Recent development and the use of FACTS controllers in power transmission system have led to many applications of these controllers not only to improve the voltage stability of the existing power network but also to provide operating flexibility to the power system [3].

14.2 Flexible AC transmission system

FACTS devices have been defined by IEEE as 'alternating current transmission system incorporating power electronic based and other static controllers to enhance controllability and increase power transfer capability' [4]. There are six well-known FACTS devices utilised by the utilities for this purpose. These FACTS devices are thyristor-controlled series capacitor (TCSC), synchronous series compensator (SSSC), static var compensator (SVC), static synchronous compensator (STATCOM), phase-shifting transformer (PST) and static and unified power flow controller (UPFC) [5]. Each of these devices has its unique characteristics and limitations. From the utility's perspective, the objective is to achieve voltage stability with the help of the most beneficial FACTS device.

14.2.1 Thyristor-controlled series capacitor

14.2.1.1 Principle of operation

Basic configuration of a TCSC comprises controlled reactors in parallel with sections of a capacitor bank as shown in Figure 14.1. This combination allows smooth control of the fundamental frequency capacitive reactance over a wide range. The capacitor bank for each phase is mounted on a platform to ensure full insulation to ground. The valve contains a string of series-connected high-power thyristors. The inductor is of the air-core type. A metal-oxide varistor is connected across the capacitor to prevent over-voltages.

The characteristic of the TCSC main circuit depends on the relative reactances of the capacitor bank, $X_C = 1/\omega_n C$, and the thyristor branch, $X_V = \omega_n L$, where ω_n is the fundamental angular frequency in rad/s, C is the capacitance of the capacitor bank in F and L is the inductance of the parallel reactor in H.

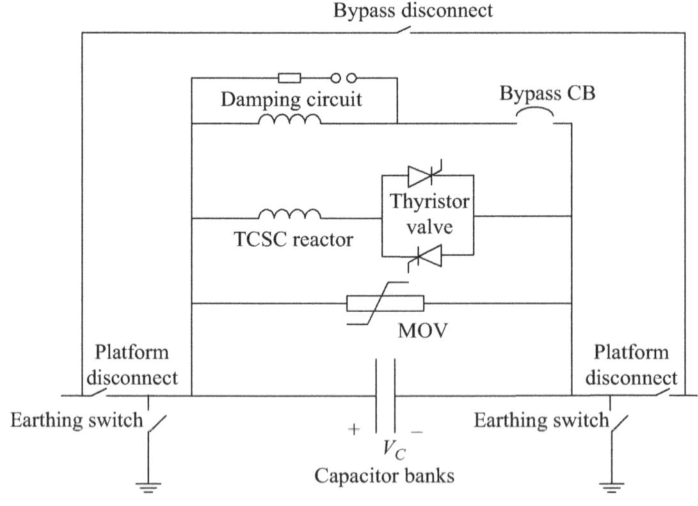

Figure 14.1 Basic structure of TCSC

The TCSC can operate in several different modes with varying values of apparent reactance, X_{app}. In this context, X_{app} is defined simply as the imaginary part of the quotient given below, in which the phasors represent the fundamental value of the capacitor voltage, V_{C1}, and the line current, I_{L1}, at rated frequency.

$$X_{app} = \mathrm{Im}\left(\frac{\boldsymbol{V}_{C1}}{\boldsymbol{I}_{L1}}\right) \tag{14.1}$$

The characteristic of the TCSC depends on the relative reactances of the capacitor bank and thyristor branch. The resonance frequency, ω_r, at which the capacitive reactance, X_C, equals the inductive reactance, X_L, is expressed as [6]

$$\left.\begin{aligned} X_C &= -\frac{1}{\omega_n C} \\ X_L &= \omega_n L \end{aligned}\right\} \tag{14.2}$$

Thus,

$$\omega_r = \frac{1}{\sqrt{\mathrm{LC}}} = \omega_n \sqrt{\frac{-X_C}{X_L}} \tag{14.3}$$

It is also practical to define a boost factor, K_B, as the quotient of the apparent and physical reactance, X_C, of the TCSC:

$$K_B = \frac{X_{app}}{X_C} \tag{14.4}$$

Blocking mode: When the thyristor valve is not triggered and the thyristors remain non-conducting the TCSC will operate in blocking mode. Line current passes through the capacitor bank only. The capacitor phasor voltage, V_C, is given in terms of the line phasor current, I_L. In this mode the TCSC performs in the same way as a fixed series capacitor with a boost factor equal to 1.

Bypass mode: If the thyristor valve is triggered continuously it will remain conducting all the time and the TCSC will behave like a parallel connection of the series capacitor bank and the inductor of the thyristor valve branch.

In this mode the capacitor voltage at a given line current is much lower than in the blocking mode. The bypass mode is therefore used to reduce the capacitor stress during faults.

Capacitive boost mode: If a trigger pulse is supplied to the thyristor with forward voltage just before the capacitor voltage crosses the zero line, a capacitor discharge current pulse will circulate through the parallel inductive branch. The discharge current pulse is added to the line current through the capacitor bank and causes a capacitor voltage, which is added to the voltage caused by the line current (Figure 14.2). The capacitor peak voltage will thus be increased in proportion to the charge passing through the thyristor branch. The fundamental voltage also increases almost in proportion to the charge.

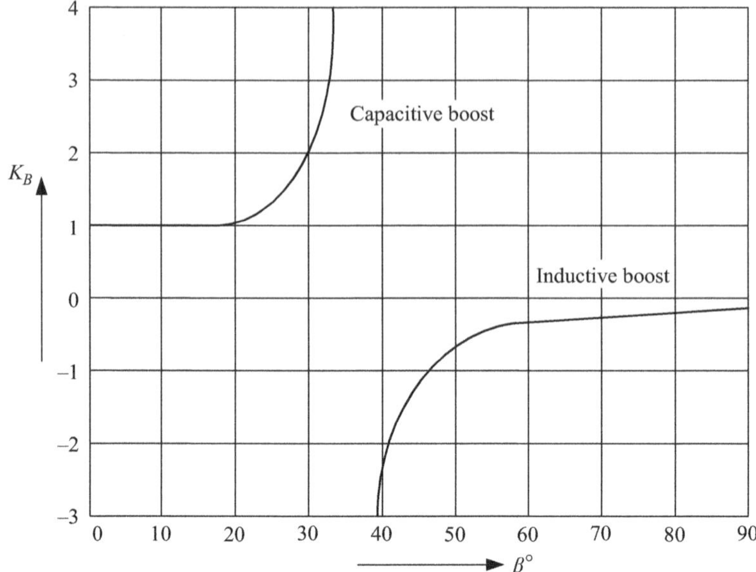

Figure 14.2 Boost factor, K_B, versus conduction angle, β, for a TCSC

The TCSC has the means to control the angle of conduction, β, as well as to synchronise the triggering of the thyristors with the line current.

14.2.1.2 Application of TCSC for damping electromechanical oscillations

The basic power flow equation shows that modulating the voltage and reactance influences the flow of active power through the transmission line. In principle, a TCSC is capable of fast control of the active power through a transmission line. The possible control of transmittable power points to this device as being used to damp electromechanical oscillations in the power system. Features of this damping effect are:

- The effectiveness of the TCSC for controlling power swings increases with higher levels of power transfer.
- The damping effect of a TCSC on an intertie is unaffected by the location of the TCSC.
- The damping effect is insensitive to the load characteristic.
- When a TCSC is designed to damp inter-area modes, it does not excite any local modes.

14.2.2 Static synchronous series compensator

A voltage source converter (VSC) can be used in series in a power transmission system. Such a device is referred to as a static SSSC.

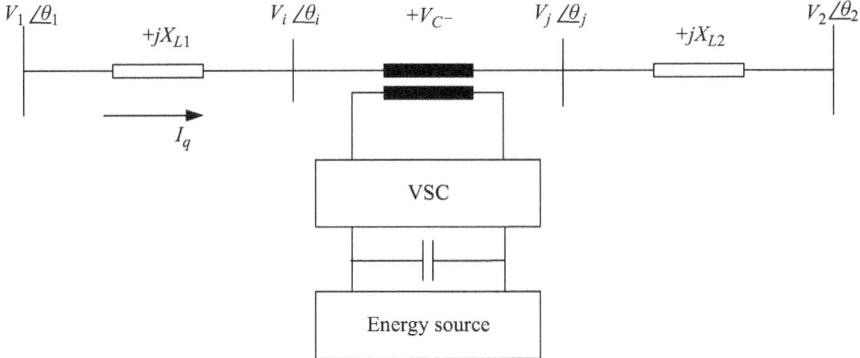

Figure 14.3 Basic configuration of a static synchronous series compensator (SSSC)

14.2.2.1 Principle of operation

A VSC connected in series with a transmission line through a transformer is shown in Figure 14.3. A source of energy is needed to provide the DC voltage across the capacitor and make up for the losses of the VSC.

In principle, an SSSC is capable of interchanging active and reactive power with the power system. However, if only reactive power compensation is intended, the size of the energy source could be quite small. The injected voltage can be controlled in terms of magnitude and phase if there is a sufficiently large energy source. With reactive power compensation, only the magnitude of the voltage is controllable as the vector of the inserted voltage is perpendicular to the line current.

In this case the series injected voltage can either lead or lag the line current by 90°. This means that the SSSC can be smoothly controlled at any value leading or lagging within the operating range of the VSC. Thus, an SSSC can behave in a similar way to a controllable series capacitor and a controllable series reactor. The basic difference is that the voltage injected by an SSSC is not related to the line current and can be independently controlled. This important characteristic means that the SSSC can be used with great effect for both low and high loading [7].

14.2.2.2 Applications

The general application of a controllable series capacitor applies also to the SSSC, dynamic power flow control and voltage plus angle stability enhancement. The fact that an SSSC can induce both capacitive and inductive voltage on a line widens the operating region of the device. For power flow control, an SSSC can be used to both increase and reduce the flow. In the stability area it offers more potential for damping electromechanical oscillations. However, the inclusion of a high-voltage transformer in the scheme means that, compared with controllable series capacitors, it is at a cost disadvantage. The transformer also reduces the performance of the SSSC due to an extra reactance being introduced. This shortcoming may be overcome in the future by introducing transformerless SSSCs. The scheme also calls for a protective device that bypasses the SSSC in the event of high fault currents on the line.

14.2.3 Static var compensator

Over the years SVCs of many different designs have been built. Nevertheless, the majority of them have similar controllable elements. The most common ones are

- thyristor-controlled reactor (TCR)
- thyristor-switched capacitor (TSC)
- thyristor-switched reactor (TSR)
- mechanically switched capacitor (MSC)

14.2.3.1 Principle of operation

In the case of the TCR a fixed reactor, typically an air-core type, is connected in series with a bidirectional thyristor valve. The fundamental frequency current is varied by phase control of the thyristor valve. A TSC comprises a capacitor in series with a bidirectional thyristor valve and a damping reactor. The function of the thyristor switch is to connect or disconnect the capacitor for an integral number of half-cycles of the applied voltage. The capacitor is not phase-controlled, being simply on or off. The reactor in the TSC circuit serves to limit current under abnormal conditions as well as to tune the circuit to a desired frequency [8].

The impedances of the reactors, capacitors and power transformer define the operating range of the SVC. The corresponding *V–I* diagram has two different operating regions. Inside the control range, voltage is controllable with an accuracy set by the slope. Outside the control range the characteristic is that of a capacitive reactance for low voltages and that of a constant current for high voltages. The low-voltage performance can be easily improved by adding an extra TSC bank (for use under low-voltage conditions only) [9].

The TSR is a TCR without phase control of the current, being switched in or out like a TSC. The advantage of this device over the TCR is that no harmonic currents are generated. The MSC is a tuned branch comprising a capacitor bank and a reactor. It is designed to be switched no more than a few times a day as the switching is performed by circuit-breakers. The purpose of the MSC is to meet steady-state reactive power demand [7].

14.2.3.2 SVC configurations

Controlled reactive power compensation is usually achieved in electric power systems by means of the SVC configurations shown in Figure 14.4.

A further variation of SVC configuration is achieved by using multi-level converters that have less harmonic generation and higher voltage capability because of serial connection of bridges or semiconductors. The arrangement of three-level converters is the most popular arrangement (Figure 14.5). Due to reduced harmonic interaction with the surrounding system multi-level converter-based SVCs

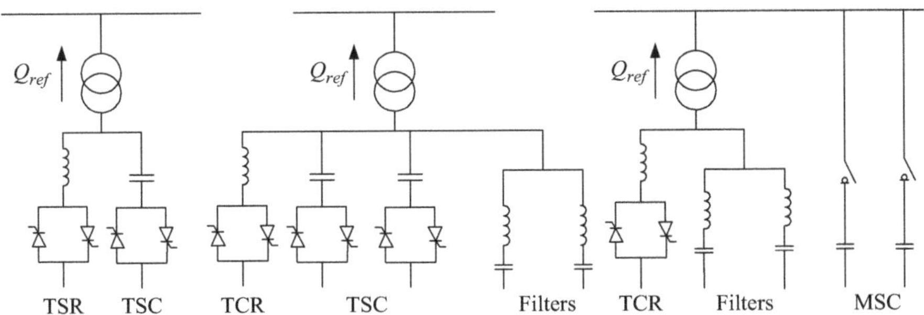

Figure 14.4 SVC configurations used to control reactive power compensation in electric power systems

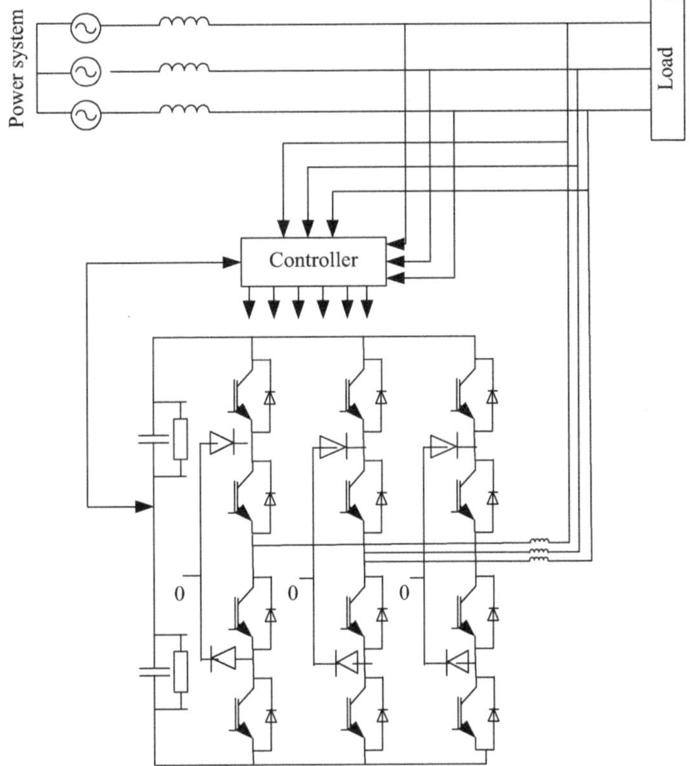

Figure 14.5 Three-level converter-based SVC

need fewer components and are easier to integrate in power systems than other types of static compensators. In addition, power losses of a multi-level converter are considerably lower than those of other SVCs of the same power rating but still slightly higher than those of the thyristor-based SVCs.

14.2.3.3 SVC applications

SVCs are installed to perform the following functions:

- dynamic voltage stabilisation: increased power transfer capability, reduced voltage variation
- synchronous stability improvements: increased transient stability, improved power system damping
- dynamic load balancing
- steady-state voltage support

Typically, SVCs are rated such that they are able to vary the system voltage by at least ±5%. This means that the dynamic operating range is normally about 10–20 per cent of the short-circuit power at the point of common connection. Three different locations are suitable for the SVC. One is close to major load centres, such as large urban areas, another is in critical substations, normally in remote grid locations, and the third is at the in feeds to large industrial or traction loads.

The two most popular configurations of this type of shunt controller are the fixed capacitor (FC) with a TCR and the TSC with TCR. Among these two setups, the second, TSC-TCR, minimises standby losses; however, from a steady-state point of view, this is equivalent to the FC-TCR. In this chapter, the FC-TCR structure is used for the analysis of the SVC, which is shown in Figure 14.6.

The TCR consists of a fixed reactor of inductance L and a bidirectional thyristor valve fired symmetrically in an angle control range of 90–180°, with respect to the SVC voltage.

Assuming controller voltage equal to the bus voltage and performing a Fourier series analysis on the inductor current wave form, at fundamental frequency the TCR, can be considered to act like variable inductance given by [9]:

$$X_V = X_L \frac{\pi}{2(\pi - \alpha) + \sin 2\alpha} \tag{14.5}$$

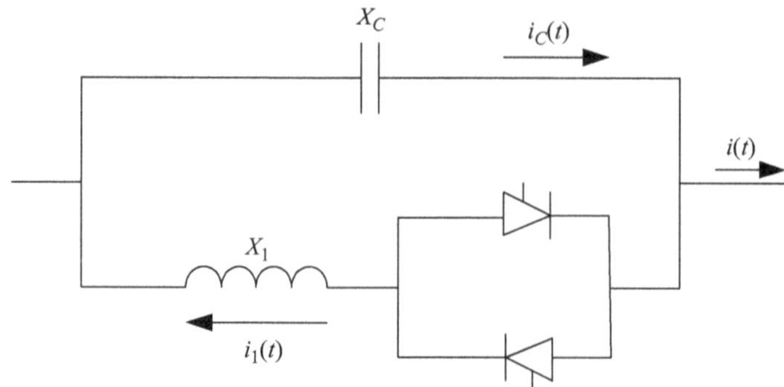

Figure 14.6 Equivalent FC-TCR circuit of SVC

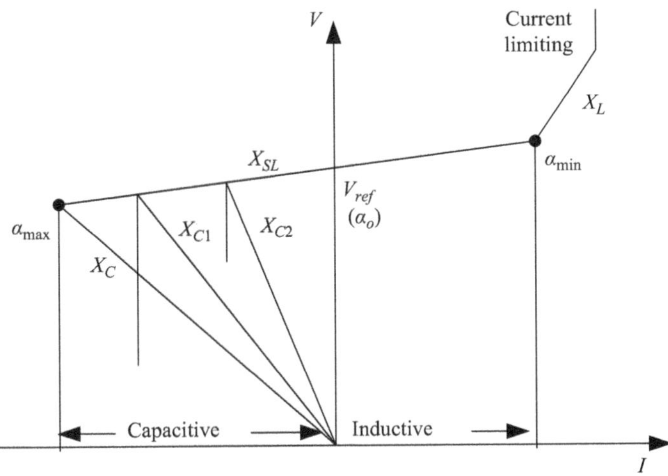

Figure 14.7 Typical steady state V–I characteristic of a SVC

where X_L is the reactance caused by the fundamental frequency without thyristor control and α is the firing angle. Hence, the total equivalent impedance of the controller can be represented as

$$X_{eq} = X_C \frac{\frac{\pi}{k}}{\sin 2\alpha - 2\alpha + \pi\left(2 - \frac{1}{k}\right)} \tag{14.6}$$

where $k = X_C/X_L$. The limits of the controller are given by the firing angle limits, which are fixed by design.

The typical steady-state control law of an SVC used here is depicted in Figure 14.7 and may be represented by the following voltage–current characteristic:

$$V = V_{ref} + X_{SL} I \tag{14.7}$$

where V and I stand for the total controller rms voltage and current magnitudes, respectively, and V_{ref} represents a reference voltage. Typical values for the slope X_{SL} are in the range of 2–5 per cent, with respect to the SVC base; this is needed to avoid hitting limits for small variations of the bus voltage. A typical value for the controlled voltage range is ±5 per cent about V_{ref} [7]. At the firing angle limits, the SVC is transformed into a fixed reactance. Of course, changing the reactance of the FC banks (from X_C to X_{C1} or to X_{C2}) will change the capacitive region accordingly.

14.2.4 Static synchronous compensator

The static compensator is based on a solid-state synchronous voltage source in analogy with a synchronous machine generating balanced set of (three) sinusoidal voltages at the fundamental frequency with controllable amplitude and phase angle. This device, however, has no inertia [10].

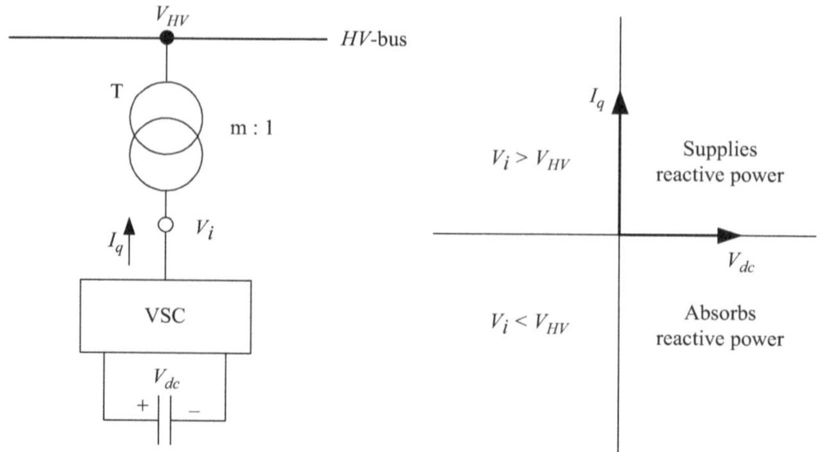

Figure 14.8 Static compensator, comprising VSC, coupling transformer T, and control

14.2.4.1 Principle of operation

A static compensator consists of a VSC, a coupling transformer and controls. In this application the DC energy source device can be replaced by a DC capacitor, so that the steady-state power exchange between the static compensator and the AC system can only be reactive, as illustrated in Figure 14.8. I_q is the converter output current, perpendicular to the converter voltage V_i. The magnitude of the converter voltage, and thus the reactive output of the converter, is controllable. If V_i is greater than the terminal voltage, V_t, the static compensator will supply reactive power to the AC system. If V_i is smaller than V_t, the static compensator absorbs reactive power.

The AC circuit is considered in steady state, whereas the DC circuit is described by the following differential equation, in terms of the voltage V_{dc} on the capacitor [11]. The power injection at the AC bus has the form

$$\left.\begin{aligned} P &= V^2G - kV_{dc}VG\cos(\theta-\alpha) - kV_{dc}VB\sin(\theta-\alpha) \\ Q &= -V^2B + kV_{dc}VB\cos(\theta-\alpha) - kV_{dc}VG\sin(\theta-\alpha) \end{aligned}\right\} \tag{14.8}$$

where $k = \sqrt{3/8}\,\mathrm{m}$

14.2.4.2 Applications

The functions performed by STATCOMs are [12–14]

- dynamic voltage stabilisation: increased power transfer capability, reduced voltage variations
- synchronous stability improvements: increased transient stability, improved power system damping, damping of SSR
- dynamic load balancing
- power quality improvement
- steady-state voltage support

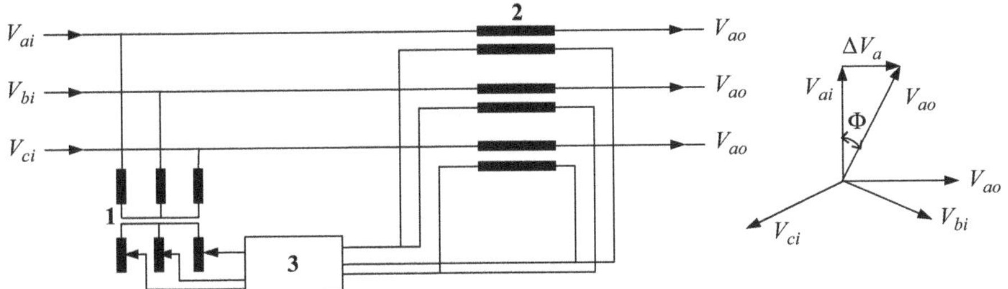

Figure 14.9 Phase shifter with quadrature voltage injection

14.2.5 Phase-shifting transformer

Phase angle regulating transformers, phase shifters, are used to control the flow of electric power over transmission lines. Both the magnitude and the direction of the power flow can be controlled by varying the phase shift across the series transformer [15] (Figure 14.9).

14.2.5.1 Principle of operation

The phase shift is obtained by extracting the line-to-ground voltage of one phase and injecting a portion of it in series with another phase. This is accomplished by using two transformers: the regulating 'or magnetising' transformer, which is connected in shunt, and the series transformer.

The star–star and star–delta connections used are such that the series voltage being injected is in quadrature with the line-to-ground voltage.

A portion of the line voltage is selected by the switching network and inserted in series with the line voltage. The added voltage is in quadrature with the line voltage, e.g. the added voltage on phase 'a' (ΔV_a) is perpendicular to V_{bc}. The angle of a phase shifter is normally adjusted by on-load tap-changing (LTC) devices. The series voltage can be varied by the LTC in steps determined by the taps on the regulating winding. Progress in the field of high-power electronics has made it possible for thyristors to be used in the switching network.

14.2.6 Unified power flow controller

The UPFC consists of two switching converters operated from a common DC link as shown in Figure 14.10. At times it is also referred to as a combination of STATCOM (shunt) and SSSC (series) compensators.

14.2.6.1 Principle of operation

In Figure 14.10 converter 2 performs the main function of the UPFC by injecting, through a series transformer, an a.c. voltage with controllable magnitude and phase angle in series with the transmission line. The basic function of converter 1 is to supply or absorb the real power demanded by converter 2 at the common d.c. link.

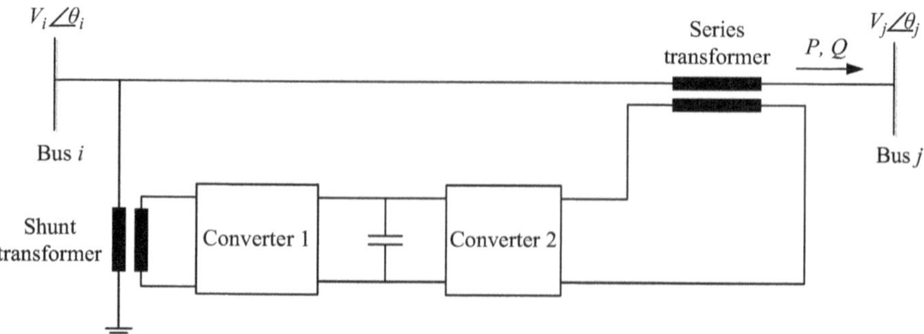

Figure 14.10 Basic circuit arrangement of the unified power flow controller (UPFC)

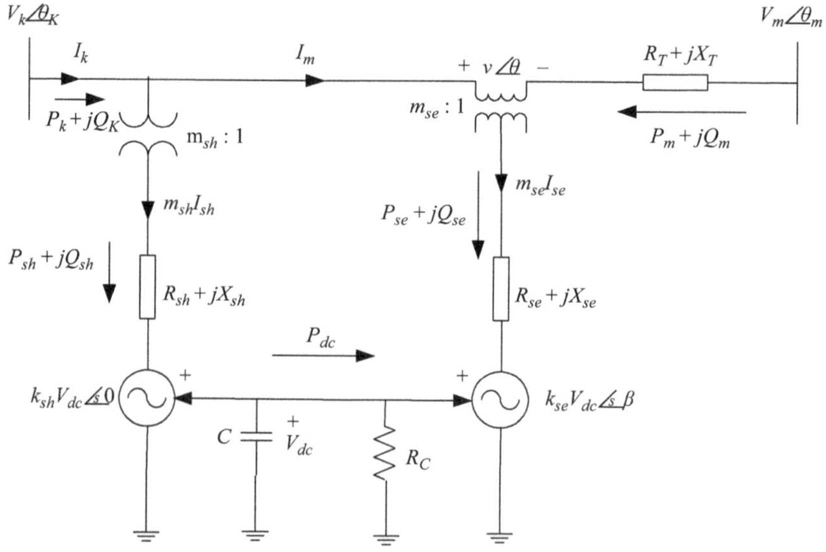

Figure 14.11 UPFC model

It can also generate or absorb controllable reactive power and provide independent shunt reactive compensation for the line. Converter 2 supplies or absorbs the required reactive power locally and exchanges the active power as a result of the series injection voltage [1].

The UPFC model is shown in Figure 14.11. According to this figure, some parameters can be adjusted for keeping voltage level and power flow of the network. The model can be obtained from the models of the associated shunt compensator, e.g. STATCOM, and series compensator, e.g. SSSC, where the d.c.

voltage V_{dc} is common for the two inverters. The related formulae of the power flow are described below [16–18]:

$$\left.\begin{aligned} P_k &= P_{sh} + \mathrm{Re}(\boldsymbol{V}_k \boldsymbol{I}_m^*) \\ Q_k &= Q_{sh} + \mathrm{Im}(\boldsymbol{V}_k \boldsymbol{I}_m^k) \end{aligned}\right\} \tag{14.9}$$

$$\left.\begin{aligned} P_m &= -\mathrm{Re}(\boldsymbol{V}_m \boldsymbol{I}_m^*) \\ Q_m &= -\mathrm{Im}(\boldsymbol{V}_m \boldsymbol{I}_m^*) \end{aligned}\right\} \tag{14.10}$$

The power P_{sh} and Q_{sh} absorbed by the shunt side are

$$\left.\begin{aligned} P_{sh} &= V^2 G_{sh} - k_{sh} V_{dc} V_k G_{sh} \cos(\theta_k - \alpha) - k_{sh} V_{dc} V_k B_{sh} \sin(\theta_k - \alpha) \\ Q_{sh} &= V^2 B_{sh} - k_{sh} V_{dc} V_k B_{sh} \cos(\theta_k - \alpha) - k_{sh} V_{dc} V_k G_{sh} \sin(\theta_k - \alpha) \end{aligned}\right\} \tag{14.11}$$

The current, $\boldsymbol{I}_m$, and the voltage, $\boldsymbol{V}$, produced due to series compensation are given by

$$\left.\begin{aligned} \dot{\boldsymbol{I}}_m &= \frac{(1 - \boldsymbol{a}_1)(\boldsymbol{V}_m - \boldsymbol{V}) - \boldsymbol{a}_2 \boldsymbol{V}_1}{R_T + jX_T} \\ \dot{\boldsymbol{V}} &= \boldsymbol{a}_1(\boldsymbol{V}_m - \boldsymbol{V}) + \boldsymbol{a}_2 \boldsymbol{V}_1 \end{aligned}\right\} \tag{14.12}$$

where

$$\boldsymbol{V}_1 = k_{se} V_{dc} e^{j\beta}$$

$$\boldsymbol{a}_1 = -\frac{R_{se} + jX_{se}}{(R_T - R_{se}) + j(X_T - X_{se})}$$

$$\boldsymbol{a}_2 = -\frac{R_{sh} + jX_{sh}}{(R_T - R_{sh}) + j(X_T - X_{sh})}$$

and

$$k_{sh} = \sqrt{3/8 m_{sh}} \quad \text{and} \quad k_{se} = \sqrt{3/8 m_{se}}$$

The DC circuit is modelled by the following differential equation:

$$\dot{\boldsymbol{V}}_{dc} = \frac{P_{sh}}{C\boldsymbol{V}_{dc}} + \frac{\mathrm{Re}(VI_m^*)}{C\boldsymbol{V}_{dc}} - \frac{\boldsymbol{V}_{dc}}{R_C C} - \frac{R_{sh}(P_{sh}^1 + Q_{sh}^1)}{C\boldsymbol{V}_{dc}\boldsymbol{V}_k^2} - \frac{R_{se} I_m^1}{C\boldsymbol{V}_{dc}} \tag{14.13}$$

14.2.6.2 Applications

A UPFC can regulate the active and reactive power simultaneously. In general, it has three control variables and can be operated in different modes. The shunt-connected converter regulates the voltage of bus i in Figure 14.10, and the series-connected converter regulates the active and reactive power or active power and the voltage at the series-connected node. In principle, a UPFC is able to perform the functions of the other FACTS devices, which have been described, namely voltage support, power flow control and improved stability [19–24].

References

1. Hingorani N., Gyugyi L. *Understanding FACTS: Concepts and Technology of Flexible AC Transmission Systems*. Piscataway, NJ, US: Wiley-IEEE Press; 2000
2. Dobson I., Chiang H.D. 'Towards a theory of voltage collapse in electric power systems'. *Systems & Control Letters*. 1989;**13**:253–62
3. Canizares C.A., Alvarado F.L. 'Point of collapse and continuation methods for large AD/DC systems'. *IEEE Transactions on Power Systems*. 1993;**7**(1):1–8
4. IEEE-PES and CIGRE. 'Facts overview'. IEEE Cat. #95TP108, 1995
5. Cañizares C.A. (ed.). 'Power flow and transient stability models of FACTS controllers for voltage and angle stability studies'. *Proceedings of the 2000 IEEE/PES Winter Meeting*; Singapore, Jan 2000. pp. 1–8
6. Acharya N., Sode-Yome A., Mithulananthan N. 'Comparison of shunt capacitor, SVC and STATCOM in static voltage stability margin enhancement'. *International Journal of Electrical Engineering Education, UMIST*. 2004; **41**(3):1–6
7. Sode-Yome A., Mithulananthan N., Lee K.Y. (eds.). 'Static voltage stability margin enhancement using STATCOM, TCSC and SSSC'. *IEEE/PES Transmission and Distribution Conference & Exhibition, Asia and Pacific*; Dalian, China, 2005. pp. 1–6
8. Canizares C.A., Faur Z.T. 'Analysis SVC and TCSC controllers in voltage collapse'. *IEEE Transactions on Power Systems*. 1999;**14**(1):158–65
9. Boonpirom N., Paitoonwattanakij K. (eds.). 'Static voltage stability enhancement using FACTS'. *The 7th International Power Engineering Conference IPEC/IEEE*; Singapore, Nov/Dec 2005, vol. 2. pp. 711–15
10. Tyll H.K., Schettler F. (eds.). 'Historical overview on dynamic reactive power compensation solutions from the begin of AC power transmission towards present applications'. IEEE/PES Power Systems Conference and Exposition, 2009, PSCE'09; Seattle, Washington, US, Mar 2009. pp. 1–7
11. Natesan R., Radman G. (eds.). 'Effects of STATCOM, SSSC and UPFC on voltage stability'. *Proceedings of the System Theory Thirty – Sixth Southeastern Symposium*; Atlanta, GA, US, 2004. pp. 546–50
12. Talebi N., Ehsan M., Bathaee S.M.T. (eds.). 'Effects of SVC and TCSC control strategies on static voltage collapse phenomena'. *IEEE Proceedings, SoutheastCon;* Greensboro, NC, US, Mar 2004. pp. 161–8
13. Kazemi A., Vahidinasab V., Mosallanejad A. (eds.). 'Study of STATCOM and UPFC controllers for voltage stability evaluated by saddle-node bifurcation analysis'. *First International Power and Energy Conference PECon, IEEE*; Putrajaya, Malaysia, Nov 2006. pp. 191–5
14. Verboomen J., Hertem D.V., Schavemaker P.H., Kling W.L., Belmans R. (eds.). 'Phase shifting transformers: principles and applications'. *International Conference on Future Power Systems*, 2005; Amsterdam, Holland, Nov 2005. pp. 1–6

15. Mathur R., Varma R. *Thyristor-Based FACTS Controllers for Electrical Transmission Systems*. NJ, US: Wiley-IEEE Press; 2002
16. Sun H., Luo C. (eds.). 'A novel method of power flow analysis with unified power flow controller (UPFC)'. *PES Winter Meeting, IEEE;* Singapore, Jan 2000, vol. 4. pp. 2800–5
17. Kawkabani B., Pannatier Y., Simond J.J. (eds.). 'Modeling and transient simulation of unified power flow controllers (UPFC) in power system studies'. *IEEE Power Tech Conference 2007*; Lausanne, Jul 2007. pp. 1–5
18. Shu-jun Y., Xiao-yan S., Yu-xin Y., Zhi Y. (eds.). 'Research on dynamic characteristics of unified power flow controller (UPFC)'. *Electric Utility Deregulation and Restructuring and Power Technologies (DRPT), 2011, 4th Int. Conference on*; Weihai, Shandong, China, Jul 2011. pp. 490–3
19. Ande S., Kothari M.L. (eds.). 'Optimization of unified power flow controllers (UPFC) using GEA'. *IEEE Power Engineering Conference, IPEC* 2007; Singapore, Dec 2007. pp. 53–8
20. Saied E.M., El-Shibini M.A. (eds.). 'Fast reliable unified power flow controller (UPFC) algorithm'. *7th International Conference on IET, AC-DC Power Transmission, 2001*; Nov 2001. London: IET; 2001. pp. 321–6
21. Sedraoui K., Al-haddad K., Chandra A. (eds.). 'Versatile control strategy of the unified power flow controller (UPFC)'. *IEEE, Electrical and Computer Engineering, 2000 Canadian Conference on*; Halifax, NS, Canada, May 2000, vol. 1. pp. 142–7
22. Balakrishnan F.G., Sreedharan S.K., Michael J. (eds.). 'Transient stability improvement in power system using unified power flow controller (UPFC)'. *4th International Conference on Computing, Communications, and Networking Technologies (ICCNT) 2013; Tiruchengode, India, Jul 2013*. Piscataway, NJ, US: IEEE; 2013. pp. 1–6
23. Sen K.K., Stacey E.J. 'UPFC-unified power flow controller: theory, modeling, and applications'. *IEEE Transactions on Power Delivery.* 1998;**13**(4):1453–60
24. Sharma N.K., Jagtap P.P. (eds.). 'Modeling and application of unified power flow controller (UPFC)'. *3rd International Conference on Energy Trends in Engineering and Technology (ICETET), 2010*; Goa, India, Nov 2010. pp. 350–5

Chapter 15
Recent technologies

Modern power systems have been growing in size and complexity. They are characterised by long distance bulk power transmission and wide area interconnections. Such networks have a chance to produce transmission congestion due to load increase (active and reactive power), particularly at peak periods, and also un-damped low-frequency power swings. This may cause severe problems such as reduction of power transfer capability of transmission lines, increased line losses, loss of generator synchronism. The system becomes stressed and has the risk of losing stability following a disturbance. Considerable progress has been made to overcome such problems by (i) controlling the active power of both the generators and loads; (ii) controlling the reactive power using compensators, SSSC, SVC, STATCOM, UPFC, etc.; and (iii) using fast-response excitation control and governor control on the generating units [1, 2].

The interest in applying new technologies in electric power systems is directly related to the expectation of improved performance, stability and efficiency. Some of the recently developed technologies, energy storage systems and phasor measurement devices, are presented in this chapter. Examples of the actual implementation, in particular those from the perspective of power system stability, and the trends in current research are discussed. Possible applications of energy storage in utility systems include transmission enhancement, power oscillation damping (POD), dynamic voltage stability, tie-line control, short-term spinning reserve, load levelling, reducing the need for under-frequency load shedding, allowing less stringent time limits for circuit breaker reclosing, sub-synchronous resonance damping and power quality improvement.

15.1 Energy storage systems

Electrical energy cannot be stored directly. It is possible to convert electrical energy to another form that can be stored. The stored energy then can be converted back to electricity when desired. There are a wide variety of possible forms in which the energy can be stored. Common examples include chemical energy (batteries), kinetic energy (flywheels or compressed air), gravitational potential energy (pumped hydroelectric) and energy in the form of electrical (capacitors) and magnetic field. These energy storage methods act as loads while energy is being stored and sources of electricity when the energy is returned to the system.

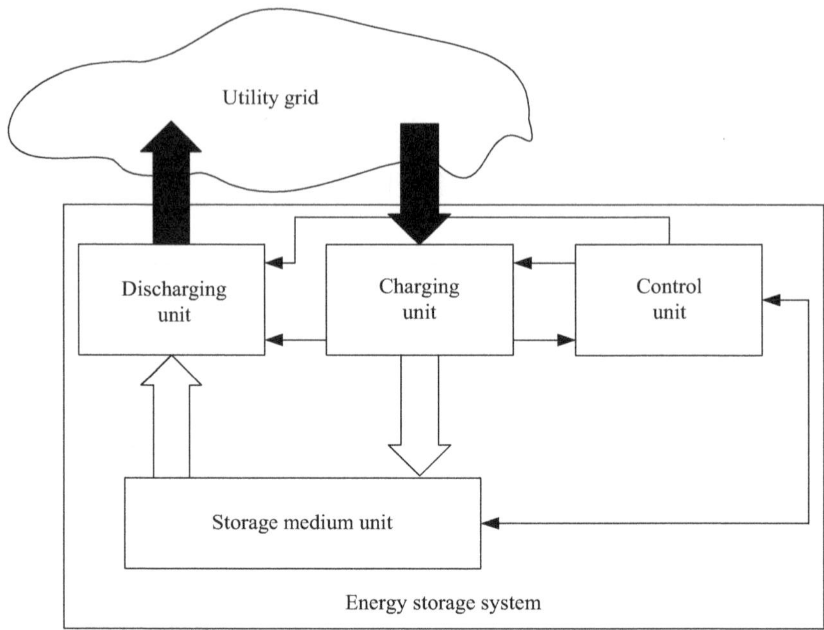

Figure 15.1 General configuration of energy storage system

General configuration of an energy storage system is shown in Figure 15.1. It mainly comprises four components: (i) the energy storage medium, which sets the basic system storage capability limits, (ii) the charging system, which takes power from the utility system (Grid) and converts it into the form that can be stored in the storage medium, (iii) the discharging system, which takes energy from the storage and converts it back to electricity and then delivers to the utility grid and (iv) the control system, which is used to monitor the performance of the unit as well as to control the how and when of flow of electrical energy between the storage system and the grid.

A limited amount of bulk energy storage, mainly in the form of pumped hydroelectric storage (PHES), has long played a role in the electric power grid, and storage continues to grow in importance as a component of the electric power infrastructure [3]. Advances in storage technologies and the needs of the electric power grid enable energy storage to become a more substantial component of the electric power grid of the future. Several primary drivers described below have increased interest in energy storage:

- The increase in peak demand and the need to respond quickly and efficiently to changes in demand given constraints on generation and transmission capacity.
- The need to integrate distributed and intermittent renewable energy resources into the electricity supply system.
- The need for investments in transmission and distribution systems that are experiencing increasing congestion.

- The need to provide grid ancillary services critical to the efficient and reliable operation of the grid.
- The increase in the need for high-quality, reliable power as a result of increased use of consumer power electronics and information and communication systems that are highly sensitive to power fluctuations.

15.1.1 Chemical energy storage systems (batteries)

Batteries have the potential to span a broad range of energy storage applications due in part to their portability, ease of use and variable storage capacity. In particular, they can stabilise electrical systems by rapidly providing extra power and by smoothing out ripples in voltage and frequency. Currently, numerous batteries including lead-acid, flow, sodium-sulphur and lithium-ion have commercial applications. However, many battery types have only limited market penetration, are expensive or have short lifetime in terms of charge/discharge cycles [4]. Efforts to develop battery technologies to improve their power and energy density characteristics, life cycle and costs are in progress on various fronts. These efforts may result in better storage options in the future.

15.1.2 Flywheel energy storage

Flywheel energy storage (FES) converts electricity to rotational kinetic energy in the form of the momentum of a spinning mass. Simply, a flywheel is a disc with a certain amount of mass that spins, holding kinetic energy. The disc is attached to a rotor in an upright position to prevent the influence of gravity. The spinning mass, rotor and disc, rests on bearings that facilitate its rotation and altogether are contained in a sealed housing designed to reduce friction between them and the surrounding environment as well as to provide a safeguard against hazardous failure modes. Reducing the friction increases the efficiency. Therefore, the spin mass spins in a vacuum, i.e. no air friction, and has electromagnetic bearings [3, 5].

FES has several advantages: low maintenance cost, fast access to the stored energy, no need for toxic resources and no carbon emissions. On the other hand, FES has the disadvantages of high cost and low capacity compared to systems such as the pumped hydro-storage.

As flywheels can be charged and discharged quickly and frequently, they can be used to maintain power quality and reliability of power systems by regulating frequency and providing protection against transient interruptions in the power supply.

The flywheel model was initially developed and supplied by Beacon Power Corporation [6]. The model incorporates charging and discharging losses, floating losses and auxiliary power as shown in Figure 15.2.

15.1.3 Compressed air energy storage

Compressed air energy storage (CAES) system is a hybrid generation/storage technology in which electricity is used to inject air at high pressure into underground geologic formations ‘cavern’. When demand for electricity is high, the

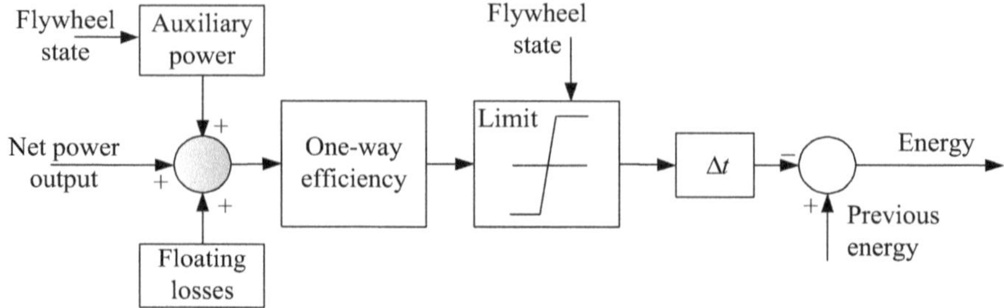

Figure 15.2 Flywheel model [6]

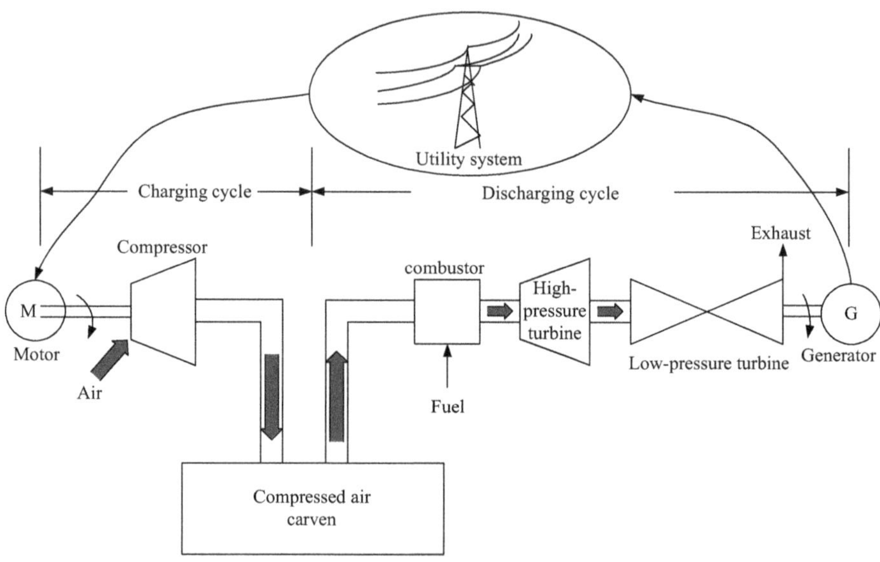

Figure 15.3 Schematic diagram of compressed air energy storage system

high-pressure air is released from the underground cavern and used to help power natural gas-fired turbines (Figure 15.3). The pressurised air allows the turbines to provide a generator with the kinetic energy necessary to generate electricity using significantly less natural gas. CAES system is appropriate for load levelling as it can be constructed in capacities of a few hundred megawatts and can be discharged over long (4–24 hours) period of time.

15.1.4 Pumped hydroelectric energy storage

PHES uses low-cost electricity generated during periods of low demand to pump water from a lower level reservoir to a higher elevation reservoir. During periods of high electricity demand, the water is released to flow back down to the lower reservoir while turning turbines to generate electricity, similar to conventional

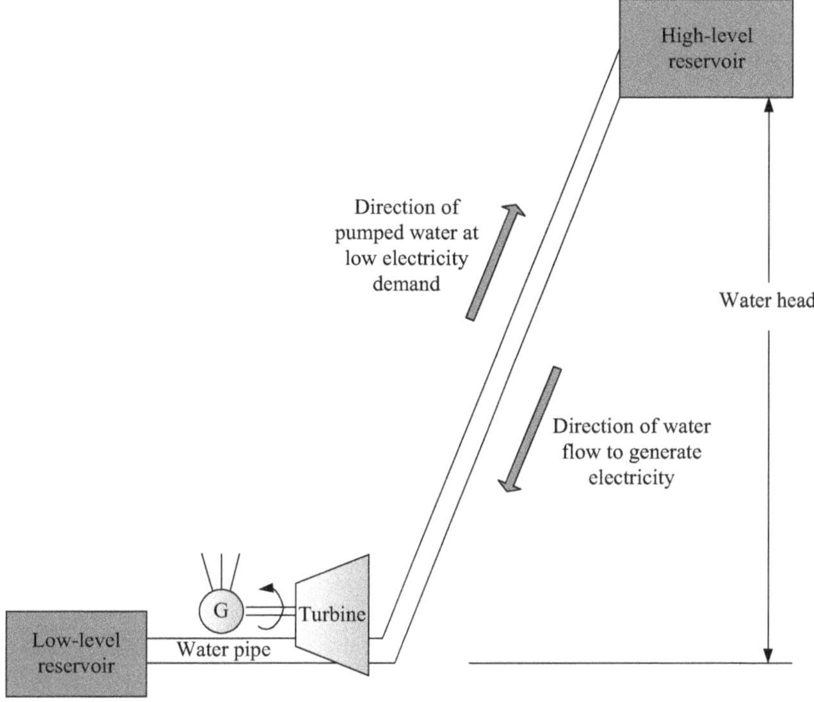

Figure 15.4 Schematic diagram of a pumped hydroelectric storage

hydropower plants. PHES is appropriate for load levelling as it can be constructed for large capacities of hundreds to thousands of megawatts and discharge over long periods of time (4–10 hours).

PHES is considered to be one of the possible ways to store energy in a large amount while maintaining high efficiency, being economical and having fast response. A schematic illustration of a pumped storage plant is shown in Figure 15.4. It basically contains two water reservoirs at different elevations: one at low level and the other at high level. The water is pumped from the lower reservoir into the higher reservoir storing electricity in the form of potential energy of water. When needed, e.g. on peak demand or transmission congestion, water can be released flowing down the pipes again and back through the turbine, which then generates electricity. The output power, P, is given as

$$P = Q \times h \times \eta \times g \times \rho \tag{15.1}$$

where

h is the water head
Q is the volume flow rate passing the turbine
η is the turbine efficiency
ρ is the water density
g is the gravitational acceleration

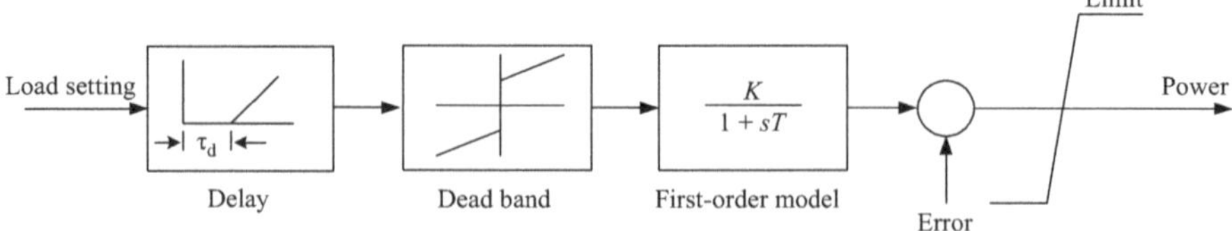

Figure 15.5 Hydro power plant model [6]

Kaplan or Francis turbines are commonly used to maximise efficiency. They are reversible and capable of handling both the pumping and generating process. Similarly, synchronous machine can be operated as motor during pumping and as a generator to generate electricity.

The developed hydro power plant model is shown in Figure 15.5. It includes delay block simulating the delay in the plant's response to the changing regulation signal, dead band element, first-order plant response model, error range simulating deviations of the actual plant response from the load setting and limiting element restricting the maximum and minimum regulation output provided by the plant [6].

15.1.5 Super capacitors

Super capacitors, like traditional dielectric capacitors, store energy by increasing the electric charge accumulation on the metal plates and discharge energy when the electric charges are released by the metal plates. The energy density of super capacitor is, however, much higher than that of traditional capacitors and could be used to improve power quality as they can rapidly provide short bursts of energy (less than a second) and store energy for a few minutes. The current applications typically take place in combination with batteries or other storage or power supplies, in situations where a low average, but high pulse, power is needed.

15.1.6 Superconducting energy storage

First, before dealing with superconducting energy storage, it is of crucial importance to understand what superconductors (SCs) are and how superconductivity works? It can be said that superconductivity is a phenomenon that occurs in certain materials and is characterised by the absence of electrical resistivity. Superconductivity is the property of complete disappearance of electrical resistance in substances when they are cooled below a characteristic temperature (Figure 15.6) [7]. This temperature is called transition temperature or critical temperature (T_C). So, for a substance to be superconductive it must be cooled to below its T_C. T_C varies with the substance used: this means that for a superconductive substance, once a current is set up in a closed circuit comprising only superconductive wires, a current will flow forever. Superconductivity is essentially a macroscopic quantum phenomenon.

15.1.6.1 Types of SCs

SCs are divided into two types depending on their characteristic behaviour in the presence of a magnetic field. Type I SCs comprise pure metals, whereas Type II

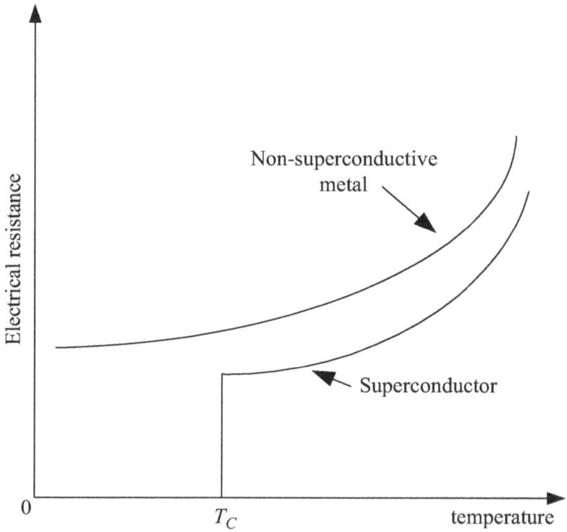

Figure 15.6 Electrical resistance versus temperature

SCs comprise primarily alloys or intermetallic compounds. Both, however, have one common feature: below a T_C their resistance vanishes.

Type I SCs: They were the first to be discovered and generally are pure metals or metal alloys. These are often referred to as conventional SCs and are also known as low-temperature SCs (LTS) due to the fact that the highest T_C of a type I SC is only 23.2 K. They are operated in liquid helium (LH_2). Type I SCs have been explained using BCS theory by Leon Cooper, John Bardeen and Robert Schrieffer [7]. The theory states that electrons pair up in what is known as (Cooper pairs). In a typical metal at room temperature, electrons are able to move throughout the lattice structure of metals, giving metals their conductive properties. However, due to the temperature, vibrations occur inside the lattice and this causes collisions between electrons and the lattice causing resistance and a loss of energy. However, when a metal is super cooled, the lattice gets to a point (T_C) where the lattice effectively stops vibrating and the Cooper pairs of electrons work together to overcome any remaining obstacles and avoid collisions. These two electrons work together to create a slipstream in much the same way that a car will be dragged along a highway by a semi-trailer in front. Table 15.1 gives the transition temperature of some elements of type 1 superconductive metals.

Type II SCs: They are known as high-temperature SCs (HTS) and are commonly made from ceramic compounds. The first and most common type of high-temperature SC is the YBCO ($YBa_2Cu_3O_7$) SC and has a T_C of around 92 K. This type is not a metal as type I, and then it does not contain a lattice structure that would allow the Cooper pairs to flow. HTS SCs provide new impetus for pursuing superconducting applications in different fields because of the prospect for higher temperature operation at liquid nitrogen (LN_2) (77 K) temperatures or above. A list

Table 15.1 Values of T_C for different LTS materials [7]

Lead (Pb)	7.196 K
Lanthanum (La)	4.88 K
Tantalum (Ta)	4.47 K
Mercury (Hg)	4.15 K
Tin (Sn)	3.72 K
Indium (In)	3.41 K
Thallium (Tl)	2.38 K
Rhenium (Re)	1.697 K
Protactinium (Pa)	1.40 K
Thorium (Th)	1.38 K
Aluminium (Al)	1.175 K
Gallium (Ga)	1.083 K
Molybdenum (Mo)	0.915 K
Zinc (Zn)	0.85 K
Osmium (Os)	0.66 K
Zirconium (Zr)	0.61 K
Americium (Am)	0.60 K
Cadmium (Cd)	0.517 K
Ruthenium (Ru)	0.49 K
Titanium (Ti)	0.40 K
Uranium (U)	0.20 K
Hafnium (Hf)	0.128 K
Iridium (Ir)	0.1125 K
Beryllium (Be)	0.023 K (SRM 768)
Tungsten (W)	0.0154 K
Lithium (Li)	0.0004 K
Rhodium (Rh)	0.000325 K

of some high-T_C SCs with their respective transition temperature and the number of Cu–O layers present in unit cell, *n*, is summarised in Table 15.2. Transition temperature has been found to increase as the number of Cu–O layers increases to three in Bi–Sr–Ca–Cu–O, Tl–Ba–Ca–Cu–O and Hg–Ba–Ca–Cu–O compounds [8].

15.1.6.2 Magnetic properties of superconductive materials

The experiment described below can be implemented to recognise the magnetic properties of SCs [9]. The procedures are:

1. Preparing a ring made of superconductive material.
2. Applying an external magnetic field to the ring in its normal state, i.e. at $T > T_C$, it is found that the magnetic field penetrates the ring (Figure 15.7(a)).
3. Reducing the temperature to be $T < T_C$, then removing the external magnetic field.
4. It is seen that the magnetic field applied in Step 2 remains there although it is turned off in Step 3. Thus, the magnetic flux remains trapped in the ring opening (Figure 15.7(b)).

Table 15.2 Critical temperature of some high-T_C SCs [8]

High-T_C SCs		**T_C (K)**	***n***
Formula	**Notation**		
$La_{1.6}Ba_{0.4}CuO_4$	214	30	1
$La_{2-x}Sr_xCuO_4$	214	38	1
$YBa_2Cu_3O_7$	123	92	2
$YBa_2Cu_4O_8$	124	80	2
$Y_2Ba_2Cu_7O_{14}$	247	80	2
$Bi_2Sr_2CuO_6$	Bi-2201	20	1
$Bi_2Sr_2CaCu_2O_8$	Bi-2212	85	2
$Bi_2Sr_2Ca_2Cu_3O_{10}$	Bi-2223	110	3
$TIBa_2CuO_5$	TI-1201	25	1
$TIBa_2CaCu_2O_7$	TI-1212	90	2
$TIBa_2Ca_2Cu_3O_9$	TI-1223	110	3
$TIBa_2Ca_3Cu_4O_{11}$	TI-1234	122	4
$TI_2Ba_2CuO_6$	TI-2201	80	1
$TI_2Ba_2CaCu_2O_8$	TI-2212	108	2
$TI_2Ba_2Ca_2Cu_3O_{10}$	TI-2223	125	3
$HgBa_2CuO_4$	Hg-1201	94	1
$HgBa_2CaCu_2O_6$	Hg-1212	128	2
$HgBa_2Ca_2Cu_3O_8$	HG-1223	134	3
$(Nd_{2-x}Ce_x)CuO_4$	T	30	1

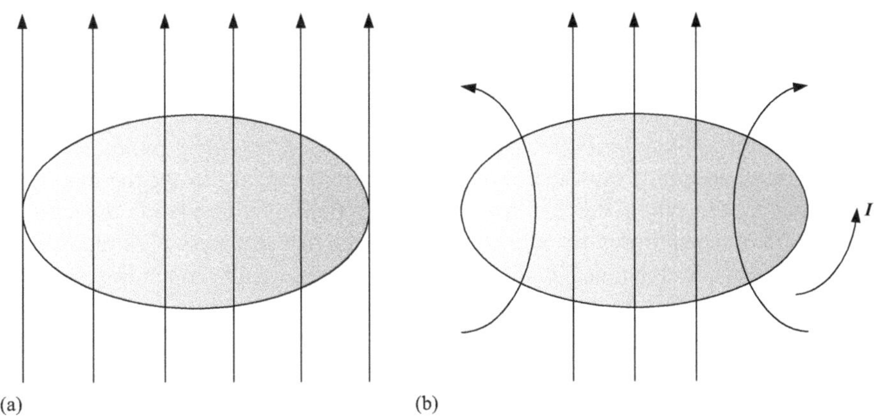

Figure 15.7 Ring of superconductive material (a) at $T > T_C$ under external magnetic field and (b) at $T < T_C$ and removed external magnetic field

Referring to Faraday's law of induction:

$$\oint \boldsymbol{E}.\mathrm{d}\boldsymbol{l} = -\frac{\mathrm{d}\phi}{\mathrm{d}t} \tag{15.2}$$

where $\boldsymbol{E}$ is the electric field along the closed loop, ϕ is the magnetic flux through the opening of the ring = **B***(area); it is found that before turning the external magnetic field off there was a magnetic flux through the ring. When the temperature of the SC is decreased to be below T_C its resistivity becomes zero and consequently the electric field inside the SC is zero as well. Thus,

$$\oint \boldsymbol{E}.\mathrm{d}\boldsymbol{l} = 0 \tag{15.3}$$

Hence, (15.2) gives

$$\frac{\mathrm{d}\phi}{\mathrm{d}\boldsymbol{t}} = 0, \text{ i.e. } \phi = \mathbf{B}^*(\text{area}) = \text{constant} \tag{15.4}$$

Therefore, the magnetic flux, ϕ, through the ring must remain constant. So, it remains trapped in the opening of the ring after turning the external magnetic field off. In addition, when the external magnetic field is turned off a current is induced in the ring resulting in trapped magnetic field passing through the ring (internal field). This induced current is called the persistent current, and it does not decay as the resistance of the ring is zero.

15.1.6.3 Meissner effect

It is a phenomenon of magnetic flux expulsion during the transition from normal conductor to a superconductor. The Meissner effect is an effect whereby the magnetic field created in a superconductor will repel all other magnetic fields, regardless of whether they change. This is because the creation of magnetic field in a superconductor results in the creation of poles to repel all fields.

Expulsion of magnetic field from the interior of the superconductor: Considering a sphere made out of superconductive material, at $T > T_C$ the material is in normal state. When the external magnetic field is turned on, the external magnetic field penetrates through the material. On the basis of Faraday's law (15.2), it is expected that at $T < T_C$ the magnetic field would remain trapped in the material after the external field has been turned off. Trapping of the magnetic field actually does not happen because of the Meissner effect as shown in Figure 15.8.

The magnetic field is expelled from the interior of the superconductor by setting up electric current at the surface. The surface current creates a magnetic field that exactly cancels the external magnetic field. It appears at $T < T_C$ in order that $\mathbf{B} = 0$ inside the superconductor and is distributed in the surface layer. As the layer carrying the electric current has a finite thickness, λ, the external magnetic field partially penetrates into the interior of the superconductor with a value given by (15.5) and as depicted in Figure 15.9(a).

$$\mathbf{B}(x) = \mathbf{B}_{\text{external}} \mathrm{e}^{-\frac{x}{\lambda}} \tag{15.5}$$

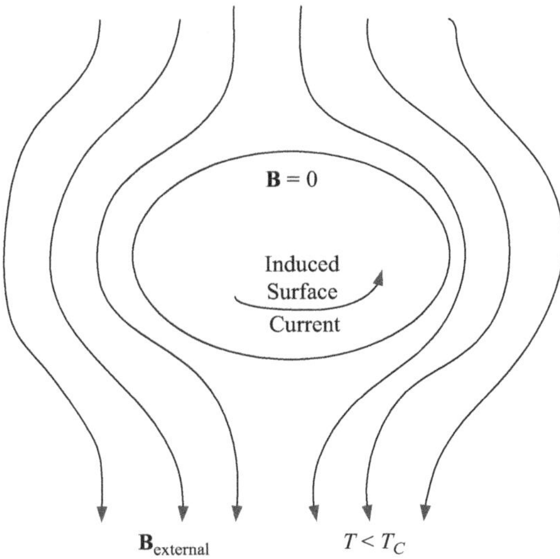

Figure 15.8 Superconductor in magnetic field

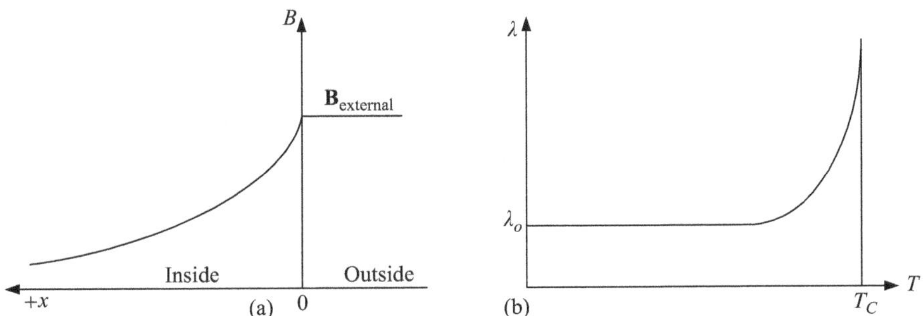

Figure 15.9 (a) magnetic field versus distance and (b) penetration distance versus temperature

where

x is the distance from the surface inside the superconductor
λ is given and defined as the penetration distance at temperature T

The value of λ can be calculated in terms of the penetration distance at temperature $T = 0$, λ_o, by

$$\lambda = \frac{\lambda_o}{\sqrt{1 - \left(\frac{T}{T_C}\right)^4}} \tag{15.6}$$

λ_o ranges from 30 to 130 nm, depending on the superconductor material. Accordingly, λ varies with time as shown in Figure 15.9(b).

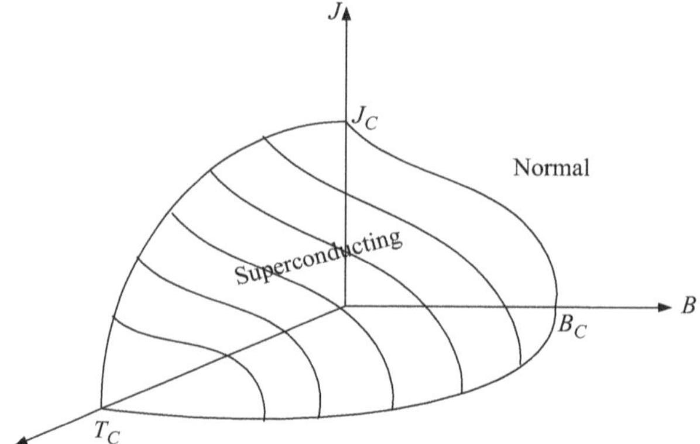

Figure 15.10 Three-dimensional manifold comprising superconducting space

From what is mentioned above, it is seen that in addition to the loss of resistance, superconductors prevent external magnetic field from penetrating the interior of the superconductor. This expulsion of external magnetic fields takes place for magnetic fields that are less than the critical field, B_C. External magnetic fields greater than B_C destroy the superconductive state. Therefore, the critical field can be defined as the maximum field that can be applied to a superconductor at a particular temperature and still maintain superconductivity. It can be concluded that the ability of superconductivity depends on meeting three conditions: (i) temperature, T_C; (ii) magnetic field strength, B_C; and (iii) current density, J_C, where [10]

T_C is the critical temperature in the absence of external magnetic field and with no current flowing in the sample
B_C is the critical magnetic field strength with no current flowing at $T=0$
J_C is the critical current density at $T=0$ with no external magnetic field strength

The variation of T, B and J represents a manifold in three-dimensional space separating the superconductivity and normal state as shown in Figure 15.10.

The critical magnetic field as a function of temperature is given by

$$B_C(T) = B_{Co}\left[1 - \left(\frac{T}{T_C}\right)^2\right] \qquad (15.7)$$

where B_{Co} is the critical magnetic field at $T=0$

The variation of B_{Co} against temperature change (15.7) variation of resistivity ρ versus external magnetic field and internal magnetic field versus external field are shown in Figure 15.11(a–c), respectively.

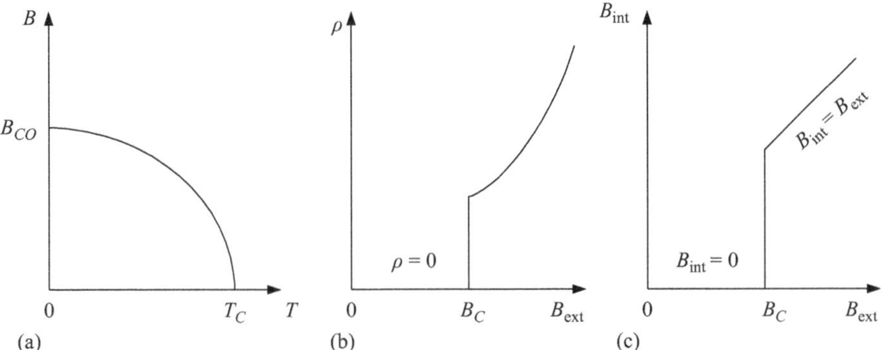

Figure 15.11 (a) B versus T; (b) ρ versus B_{ext}; (c) B_{int} versus B_{ext}

The maximum superconductive current, I_C, carried by the superconductor corresponds to the critical magnetic field, B_C, at its surface where I_C in terms of B_C is obtained as

$$I_C = \frac{2\pi * R * B_C}{\mu_0} \tag{15.8}$$

Consequently, the current density, magnetic field and T_C are all interdependent. By increasing any of these parameters to a sufficiently high value, superconductivity can be destroyed and the conductor will revert to a normal, non-superconducting state.

Type I and Type II superconductors exhibit different response to external magnetic field. A type I superconductor completely excludes the magnetic field from the interior for $\boldsymbol{B} < \boldsymbol{B}_C$. Type II superconductors, however, permit the field to partially penetrate through the material in quantised amounts of flux for $B_{C1} < B < B_{C2}$, whereas for $B < B_C$ the magnetic field is completely expelled, the same as type I superconductor, and for $B > B_{C2}$ the superconductor reverts to its normal state, i.e. full penetration by external magnetic field (Figure 15.12).

Values of critical current, I_C, and current density, J_C, for some HTS materials provided in Table 15.2 are summarised in Table 15.3.

Based on the study of superconducting and magnetic properties explained above, electricity can be stored in the form of d.c. magnetic field as described in Section 15.1.6.4.

15.1.6.4 Superconducting magnetic energy storage

Superconducting magnetic energy storage (SMES) is an energy storage device that stores energy in the form of direct current that is the source of a d.c. magnetic field. The conductor operates at cryogenic temperatures where it is a superconductor and thus has virtually no resistive losses as it produces the magnetic field. Consequently, the energy can be stored in a persistent mode until required.

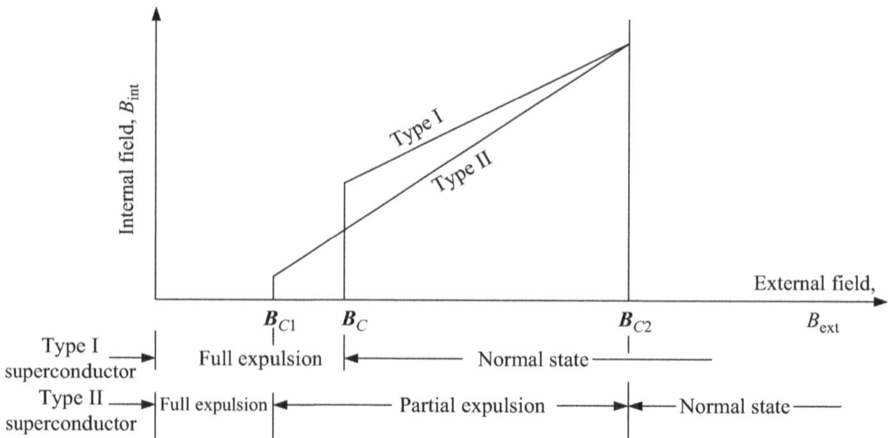

Figure 15.12 B_{ext} versus B_{int} for type I and type II superconductors

Table 15.3 Values of I_C and J_C for some HTS materials [11]

Material	**I_C (A)**	**J_C (kA/cm^2)**	**Condition**
Bi-2223/Ag sheath	70	27.8	77 K, 114 (m)-length
Bi-2212/Ag sheath	500	490	4.2 K, 50 (m)-length
Bi-2212 coated	130	100	4.2 K, 450 (m)-length
TI-1223 coated	18	8	77 K, 0.02 (m)-length

The core element of an SMES unit is a superconducting coil of high inductance (L_{coil} in Henrys). It stores energy in the magnetic field generated by a d.c. current (I_{coil} in amperes) flowing through the coil. Therefore, the inductively stored energy (E in joules) and the rated power (P in watts) are the common specifications for SMES devices. They are expressed as

$$E = \frac{1}{2} L I_{coil}^2 \tag{15.9}$$

$$P = \frac{dE}{dt} = L I_{coil} \frac{dI_{coil}}{dt} = V_{coil} I_{coil} \tag{15.10}$$

As the energy is stored as circulating current, energy can be discharged from, or stored in, a SMES unit with almost instantaneous response over periods ranging from a fraction of a second to several hours [12].

The entire SMES unit consists of four parts: (i) large superconducting coil with the magnet (SCM) at the desired cold temperature; (ii) power conditioning system (PCS) to interface the AC utility and SCM. Through the PCS the power is converted from a.c. to d.c. or inversely; (iii) the cryogenic system (CS) that is required to cool the SCM and keep it at the operating temperature; and (iv) the controller

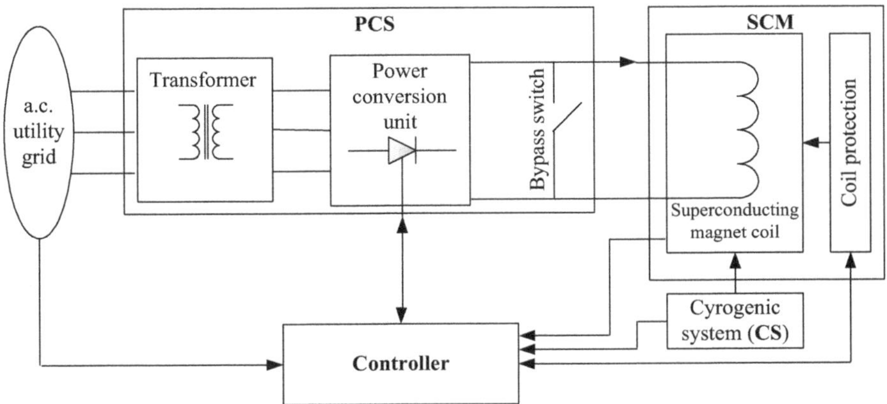

Figure 15.13 Typical SMES unit components

that is the essential part of SMES system as it performs various functions: controlling the modes of charge/discharge/standby by controlling the voltage across the magnet coil, controlling CS and controlling the power conversion unit. A typical system is shown in Figure 15.13.

Because of refrigeration needs, SMES technology is costly compared with other energy storage technologies, but with the development of HTS, SMES has become a bit more cost effective. Electric utilities pay more attention towards SMES because of its fast response capability, flexibility, reliability and high efficiency.

15.2 Superconductivity applications

15.2.1 Superconducting synchronous generators

The use of superconductors, LTS or HTS, in a.c. machines is primarily motivated by achieving higher current densities that allow overall reduction in cross-section and field winding volume compared to ordinary copper-wound rotors. Reduction of winding volumes leads to a reduction in the size and weight of the entire machine. During the 1970s a stationary room-temperature armature with a rotating SC field winding was used for the design of an ac machine. This design faced problems concerning transferring cryogen into a rotating vacuum-insulated container [11].

In a synchronous machine, a.c. current is supplied to the armature to provide a flux component that rotates in synchronism with the flux component produced by the rotating field winding. The rotor is thus phase-locked at the synchronous speed and under balanced load conditions will see essentially a d.c. field from the armature. The armature, however, is connected to the electric power system, which experiences load-related electrical disturbances under steady-state and transient conditions. These electrical disturbances are reflected back into the armature and produce non-synchronous AC effects that impact the rotor. Rapid changes in the DC excitation also occur as a result of load changes. The primary armature disturbance is caused by unbalanced loads, which gives rise to negative sequence currents, proportional to the load and degree of unbalance, that counter-rotate at twice the synchronous speed.

Transient events caused by system faults provide the other major source for non-synchronous a.c. effects on the rotor. These time-varying fields in the armature, at frequencies different from the synchronous frequency, induce compensating currents to flow in the rotor and produce heating in the d.c. superconducting winding and structural support. This a.c. influence can effectively be minimised by incorporating two concentric electromagnetic shields, called as inner cold damper shield and outer warmer cold damper shield, between the two windings that attenuate these non-synchronous a.c. fields. The outer damper shield acts as a damper and provides damping for the mechanical oscillations of the rotor related to the phase of the system. In addition to preventing the magnetic flux with high frequencies, it also helps to shield the entire cryogenic zone. The inner cold damper shield helps in shielding the superconducting field winding from time-varying magnetic field. Therefore, the shielding must be carefully designed, taking into account the limitation of the thermal margin for the superconductors, to minimise AC heating of the field winding, to prevent degradation of the superconductor J_C and magnetic field capability, and possible normalisation during extreme transient conditions [13, 14].

The important aspect of the superconducting synchronous generator (SSG) is to keep the field winding at low temperature that is enhanced by the refrigerator. The refrigeration unit has four parts: cryostat, cryogenic pump, heat exchanger and liquid coupling junction. Depending on the type of superconductors (LTS or HTS) liquid cryogen such as helium or nitrogen is chosen and the most used cryo-cooler is Gifford–McMahon cryo-cooler. At present all known superconductors have to be operated at cryogenic temperature between 4 K and 80 K [15, 16].

15.2.1.1 Benefits of SSGs

- *Steady-state stability improvement*: The SSG is characterised by low synchronous reactance that helps to enhance the power transfer capacity of a transmission line. It is represented as a voltage source, E, connected to its terminal bus. When the machine is connected to an infinite bus through an external reactance X_e the maximum power transmitted is given, as known, by

$$P_{\max} = \frac{EV}{X_g + X_e} \tag{15.11}$$

where V is the infinite bus voltage

Thus, as the machine reactance is low the maximum power is high. Compared with a conventional synchronous generator of the same rating, the per unit synchronous reactance of the SSG ranges from one-third to one-quarter of that of the conventional one. This means that the maximum power transmitted is getting higher and the system has more marginal stability.

- *Improved voltage regulation (VR)*: The field current of SSG between no load and full load is much smaller than the conventional generator. Consequently, the SSG can be operated at any power factor within its MVA rating. In addition, the short circuit ratio of SSG is much higher than the conventional generator due to the low synchronous impedance [17]. This results in providing stable operation by SSG for any type of load. The VR of SSG is calculated as

$$VR_{SSG} = \frac{V_{NL} - V_{FL}}{V_{FL}} \times 100 \qquad (15.12)$$

where V_{NL} and V_{FL} are the no-load and full-load terminal voltages, respectively.
- *Higher current density*: This permits higher magnetic fields and allows a reduction in weight and size. In addition, the machine efficiency is increased because of the elimination of I^2R heating in the field winding.

15.2.2 Superconducting transmission cables

One of the fields that will benefit most from superconductor technology is transmission. When it comes to the transmission of electricity through the power grid, this is where superconductors can have arguably the biggest impact. If superconducting technology is implemented into the power grid right now, the LN_2-cooled cables could be placed underground in place of copper cables. These superconductor cables are 7000 per cent more space efficient than their copper rivals. These new power lines effectively have negligible energy losses, reducing the need for boosting of voltage at substations. By using superconducting electrical cables as opposed to copper, the cost of the transmission is reduced and very large current densities are able to be transmitted with three to five times the current of regular wires. The only concern with this, as mentioned previously, is the practicality of cooling kilometres of underground cables [11].

Sumitomo Cable has been working with TEPCO in Japan, since 1995, for developing HTS power transmission cables. The best performance has been demonstrated in both fundamental materials development and cable construction. A typical design is configured as shown schematically in Figure 15.14 for a cross-section liquid hydrogen (LH_2) transmission pipeline with HTS electrical power cable [18]. The pipeline has the multi-layer insulator and LN_2 channel as thermal shield. HTS tapes are used for both current transmission and shielding the external pipe from the magnetic fields generated by the tapes transmitting the power.

15.2.3 Superconducting transformers

In a conventional power transformer, under load condition: Joule heating (I^2R losses) of the copper coil adds considerably to the amount of lost energy. Almost 80 per cent of load losses represent I^2R losses and the remaining 20 per cent consist of stray and eddy current losses. Therefore, power engineers pay more attention towards the reduction of load losses. Unlike copper and aluminium, superconductors present no resistance to the flow of electricity, with the consequence that I^2R losses become essentially zero, yielding a dramatic reduction in overall losses. Previously developed

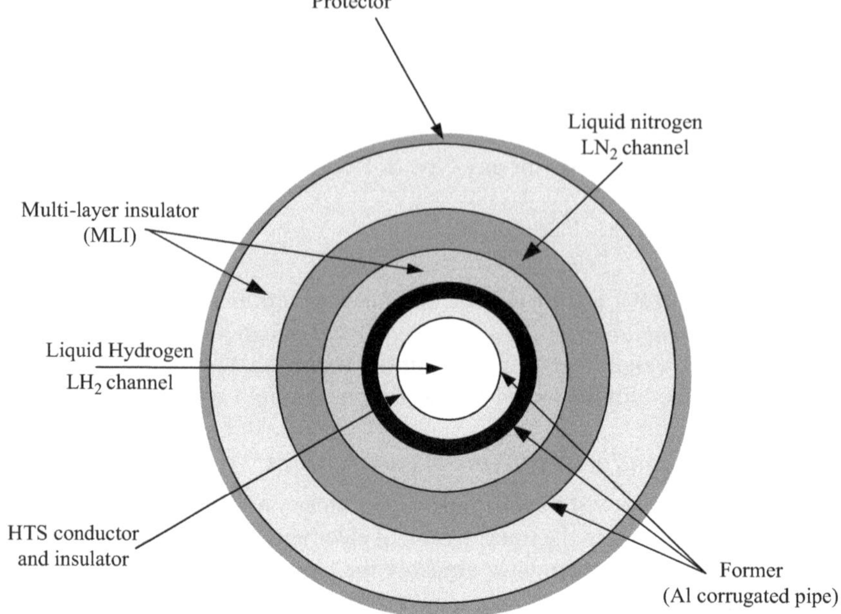

Figure 15.14 Configuration of 7-m HTS cable

© 2009 IEEE. Reprinted with permission from Nakayama T., Yagai T., Tsuda M., Hamajima T. 'Micro power grid system with SMES and superconducting cable modules cooled by liquid hydrogen'. IEEE Transactions on Applied Superconductivity, Vol. 19(3), June 2009, pp. 2062–5

LTS required cooling by LH_2 to about 4.2 K, with advanced cryogenic technology that is expensive in terms of both cost and refrigeration power expended per unit of heat power removed from the cryostat. The HTS, based on LN_2 at temperature up to 78 K, is simpler, cheaper and a reduced ratio of refrigeration power is used to remove heat. Even with the added cost of refrigeration, HTS-based transformers of rating 10 MVA and higher range are projected to be substantially more efficient and less expensive than their conventional counterparts. Furthermore, due to the increased capacity of HTS transformers, it is easy to replace the existing oil transformers in the grid by HTS transformers of the same size to meet the growth in power demand [19].

15.2.4 Superconducting fault current limiters

The electric power demand has been continuously growing and, as a consequence, the power system must be expanded to meet this growth in demand. System expansion may need replacement of transformers and addition of new generators to increase the power capacity of the system. This results in an increase of fault current, but the existing buses and switchgear are not rated for the new fault current. Other sources can increase the fault current such as addition of distributed generation to the system and using parallel feeder to increase system reliability. Therefore, the trade-off between fault current control and bus capacity is a problem facing power system engineers.

The fault current must be controlled in a manner that limits the investments required for equipment replacement or addition. Different solutions can be applied, for instance, (i) using high impedance transformers but degradation of VR arises for the loads on the bus; (ii) expanding the bus by adding new sections through bus-tie circuit breakers and supplying each section by a small transformer; and (iii) adding reactors in the path of fault current.

An alternative solution that overcomes the effects resulting from the classic solutions of fault current control is using SFCL. The SFCL has the advantage of inserting high resistance in the system in the event of a fault while at the normal operation state its resistance vanishes. Thus, the SFCL acts as a non-linear resistance that varies according to the operating conditions. The earliest designs used LTS materials cooled with LH_2, but with the development of superconductive technology, HTS materials are used with LN_2 for cooling to decrease the expenses of SFCLs.

Two approaches of applying SFCLs can be presented in power systems: series-resistive limiter and inductor limiter. In the first approach, *series-resistive limiter*, the SFCL is inserted in the circuit with a critical current of two or three times the full load current. The fault current, during the fault, causes the SFCL to be in a resistive state and resistance R appears in the circuit. To limit the energy absorbed by the SFCL, a shunt coil called 'trigger coil' can be used allowing the bulk of fault current to flow through the paralleled resistor and inductor (Figure 15.15(a)). It is noted that the SFCL at the normal state is a short circuit across the copper-inductive element. In addition, limiting the energy absorbed by the current limiter allows the power system designers to use it for transmission line applications.

The second approach, *inductive limiter*, uses a transformer. Its primary copper winding is connected in series with the circuit and the secondary is connected with a resistive HTS limiter as shown in Figure 15.15(b). At normal operation the limiter is in its steady state with impedance nearly zero, which in turn leads to reflection of zero impedance of secondary HTS winding to the primary. During the fault, a large current is induced in the secondary because of the large current flow in the circuit and then the SFCL loses its superconductivity. Consequently, a resistance is developed in the HTS winding and reflected to the primary limiting the fault

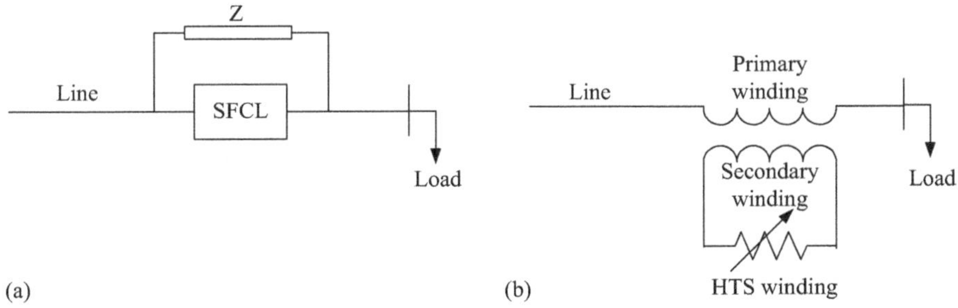

Figure 15.15 (a) Resistive current limiter approach and (b) inductive current limiter

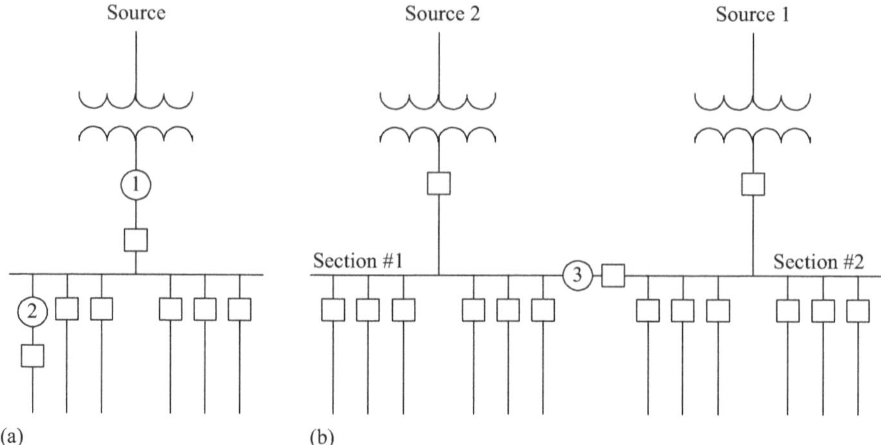

Figure 15.16 Locations of SFLC in a substation at (a) outgoing or incoming feeder and (b) bus tie

current. This approach is appropriate for high current circuits. Whatever the approach used, it can be seen that with SFCLs, the utilities can provide a low impedance system with a low fault current level.

As an illustration of the benefits of applying SFCLs to the power system, the possible locations of inserting the limiters in substations are investigated. When the limiter is inserted in the main feeder feeding the substation (e.g. location #1 (marked inside a circle), Figure 15.16(a)), the fault current in case of faulted bus is limited. So, a larger low impedance transformer can be used to meet increased demand and maintain VR at the new power level without replacing or upgrading the switchgear. In addition, as the fault current is limited, damage of transformer due to I^2R is avoided and the voltage dip on the upstream high-voltage bus during the fault is minimised as well. For limiter located at the outgoing feeder (location #2, Figure 15.16(a)), less-expensive limiters can be used to protect the existing over-stressed equipment without replacement. Another possible location is at bus-tie position (location #3, and Figure 15.16(b)). In this case the limiter requires a small load current rating. During a fault on one of the two sections of the bus, a large voltage drop across the limiter helps to maintain the voltage level on the un-faulted section. In contingencies, the two sections of the bus can be tied together without a large increase in the fault duty on both sections.

Use of SFCL in power systems not only allows more capacity of generation and transformation equipment but also improves the system transient stability [20]. As explained in Chapter 9, Section 9.2, and considering the system shown in Figure 15.17 where each circuit breaker is provided with FSCL, when a three-phase short-circuit occurs at the beginning of one of the two transmission lines and is cleared by disconnecting the faulty line, it is found that the P versus δ curves (pre, during and post fault) are as given in Figure 15.18. Neglecting the resistance of

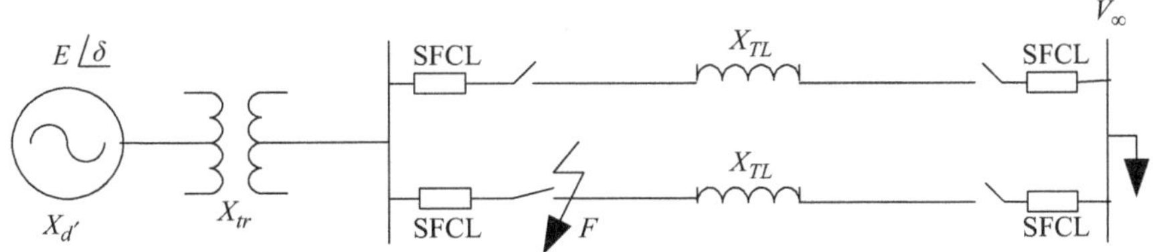

Figure 15.17 One machine to an infinite bus system (circuit breakers are provided with SFCL)

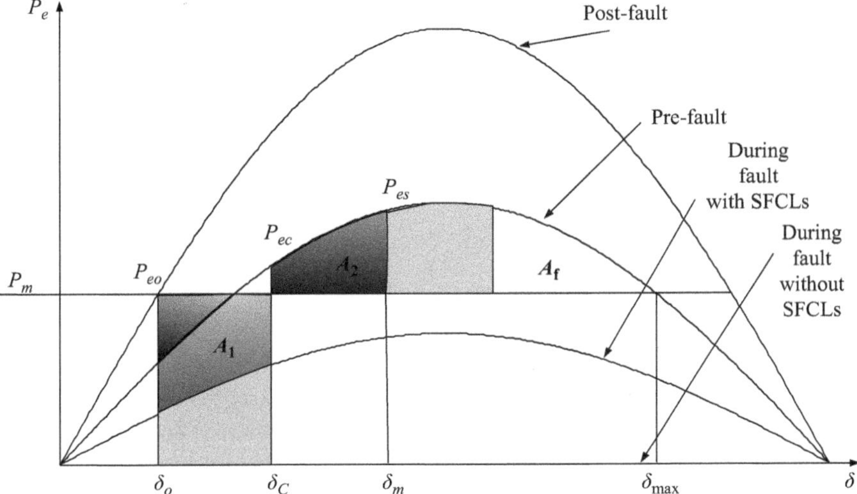

Figure 15.18 Power–angle curves for pre, during and post fault: the shadowed areas and the dark-shadowed areas represent the energy balance without and with SMES, respectively

system elements, the power delivered from the generator to the infinite bus in terms of the power angle, is expressed as

$$P_e = \frac{EV_\infty}{X_{eq}} \sin\delta \tag{15.13}$$

where E is the internal machine voltage, X'_d is the transient direct axis reactance and X_{eq} is the equivalent transfer reactance between the generator and the infinite bus.

Before the fault and after the fault, the power–angle curves of the system with and without SFCL are the same as the system has not any added impedance due to SFCLs. The value X_{eq} is derived as below:

$$X_{eq(pre\text{-}fault)} = X'_d + X_{tr} + \frac{X_{TL}}{2} \quad \text{and} \quad X_{eq(post\text{-}fault)} = X'_d + X_{tr} + X_{TL} \tag{15.14}$$

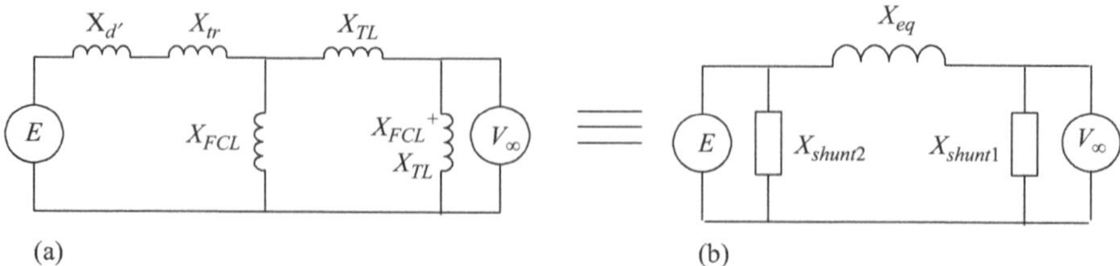

Figure 15.19 (a) Reactance diagram and (b) equivalent reactance diagram

During the fault and without using SFCLs, the output power of the generator is zero. On the other hand, with SFCLs the electric power has a specific value corresponding to the equivalent reactance that can be deduced as below.

The reactance diagram of the system is shown in Figure 15.19 from which the equivalent reactance, X_{eq}, can be obtained by Y–Δ transformation as below:

$$\begin{aligned} X_{eq} &= \frac{1}{X_{TL}}[X_{TL}X_{FCL} + X_{FCL}(X'_d + X_{tr}) + (X'_d + X_{tr})X_{TL}] \\ &= X_F + X'_d + X_{tr} + \frac{X_F}{X_{TL}}(X'_d + X_{tr}) \neq \infty \end{aligned} \tag{15.15}$$

Accordingly, and as depicted in Figure 15.18, the accelerating area of machine rotor that is the area below the input power line is decreased when using SFCLs and becomes A_1 (dark shadowed). Consequently, the required decelerating energy represented by the area above the input power line is also decreased to achieve the energy balance and the system in this case has more marginal stability. Therefore, it is concluded that the use of SFCLs in power system substations improves the system stability in addition to the other mentioned benefits.

15.2.5 SMES applications

Power systems during the transient period following a disturbance such as component failure, line switching, load changes and fault clearance should be equipped with the devices that provide adequate damping of oscillations in the system. So, countermeasures such as power system stabilisers (PSSs), turbine governor control system and phase shifters have been used to prevent system collapse due to loss of synchronism or voltage instability.

SMES offers a feasible application to improve transmission capacity by damping inter-area modal oscillations. It can actively dampen these system oscillations through modulation of both active and reactive power. The first full commercial application of superconducting power grid is SMES in Bonneville Power Administration 1981. It is American superconductor's SMES system for power quality and grid stability and was located along 500 kV Pacific Intertie that interconnects California and the Northwest.

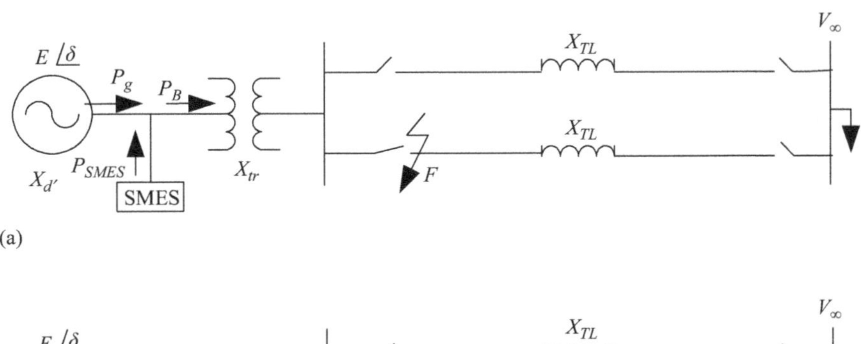

(a)

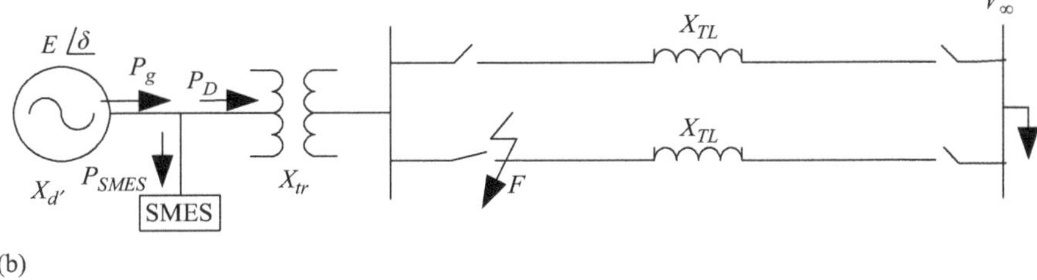

(b)

Figure 15.20 Power system with SMES connection: (a) SMES in energy output mode and (b) SMES in energy input mode

Transient stability improvement by using SMES can be demonstrated by considering the power system shown in Figure 15.20 [21, 22]. The system is subjected to a three-phase short circuit at the beginning of one of the transmission lines. The SMES is shunt connected at the generator terminal node. At the normal state 'before the fault event' the SMES operates in the energy output mode (Figure 15.20(a)). Thus, the power balance relation is

$$P_B = P_g + P_{SMES} \tag{15.16}$$

At the instant of fault occurrence the control system of the SMES detects the fault and switches the SMES to go into the energy input mode after reverse switching time t_{rev}. The SMES acts as an additional load to the generator, which is necessary to keep the system stability and prevent the loss of synchronism. Thus, the power balance at generator terminal node (Figure 15.20(b)) is:

$$P_D = P_g - P_{SMES} \tag{15.17}$$

The SMES power versus time can be expressed by

$$P_{SMES}(t) = P_{SMES}e^{-(t+t_{rev}/T_{SMES})} \tag{15.18}$$

where P_{SMES} is the power of SMES at $t=0$ when released instantly to the power system and T_{SMES} is the time constant of SMES model.

The SMES operates in energy input mode during the accelerating power to the rotor and is disconnected from the system when the generator rotor starts deceleration. Therefore, the system stability condition can be determined by getting the accelerating power (absorbed by the rotor) and decelerating power (delivered from the rotor) in balance, i.e.

$$\int_{\delta_o}^{\delta_{rev}} P_B \mathrm{d}\delta + \int_{\delta_{rev}}^{\delta_{dsc}} P_D \mathrm{d}\delta = \int_{\delta_{dsc}}^{\delta_m} \left(P_{\max(A)} \sin\delta - P_D\right) \mathrm{d}\delta \tag{15.19}$$

where $P_{\max(A)}$ is the maximum power delivered to the transmission system after the fault clearance and can be calculated by (15.11). δ_o, δ_{rev}, δ_{dsc} and δ_m are the rotor's relative angles in normal operating mode, at the instant of SMES connection, at the instant of clearing the fault and maximum desired angle of rotor oscillation, respectively.

Equation (15.19) can be solved to give

$$P_D = \frac{P_{\max(A)}(\cos\delta_{dsc} - \cos\delta_m) - P_B(\delta_{rev} - \delta_o)}{(\delta_m - \delta_{rev})} \tag{15.20}$$

Thus, the angle at which the SMES is disconnected and the fault is cleared can be obtained by

$$\cos\delta_{dsc} = \frac{P_D(\delta_m - \delta_{rev}) + P_B(\delta_{rev} - \delta_o) + P_{\max(A)} \cos\delta_m}{P_{\max(A)}} \tag{15.21}$$

Subtracting (15.17) from (15.16) to give

$$P_{SMES} = 0.5(P_B - P_D) \tag{15.22}$$

The SMES parameters (t_{rev}, δ_{dsc}, P_{SMES}, etc.) can be decided by using the aforementioned relations to keep the system stability when the system is subjected to transient disturbances. SMES systems have some prominent performances such as rapid response (millisecond), high power, high efficiency and four-quadrant control due to the advantages in both superconducting technologies and power electronics. Thus, SMES systems offer flexible, reliable and fast-acting power compensation. Superconducting technology is a promising technology in the future. Developing this technology is very attractive for power engineers and researchers to work on it. Paying more attention and more research to developing the superconducting technology will lead to more applications in real practices.

15.2.6 *Features of storage systems*

Some features of aforementioned storage systems are listed in Table 15.4 [6].

Table 15.4 Some features of storage systems [6]

Storage system	Advantages	Disadvantages	Time-to-release power	Duration	Technology maturity
Batteries (lead-acid)	High power capacity; low energy density; low capital cost; long life time	Low efficiency; potential adverse environmental impact	Instantaneous discharge	Rated power for a few seconds to a couple of hours	Limited commercial
Flywheel	High power capacity; short access time; long life time, low maintenance effort; small environmental impact	Low energy density	Milliseconds	Full power for 15 minutes	Limited commercial
CAES	Very high energy and power capacity; long life time	Low efficiency; adverse environmental impact	9 minutes full or 6 minutes Emergency start	Rated power for long time	Commercial
PHES	Very high energy and power capacity; long life time; moderate access time	Special site requirements; adverse environmental impact; moderate efficiency	Seconds to 1–3 minutes	Rated power for long time	Commercial
Super-capacitors	High efficiency, long life cycle	Low energy density; few power system applications	<1 minute	Rated power from seconds up to several minutes	Commercial
SMES	High power capacity; short access time; long life time; high efficiency	Low energy density; high production cost; potential adverse health impact	Milliseconds	High power for several seconds	Commercial

15.3 Phasor measurement units

The continuous load growth necessitates an increase in transmission networks capacities. Otherwise, the transmission lines are congested and the power system is stressed as well as it may be forced to operate closer to its stability limit. Furthermore, penetration of renewable energy sources into power systems is inevitable and adds more uncertainty that requires more severe operation. To achieve flexible operation, power system monitoring should be considered [23]. It needs installation of excessive number of measuring instruments to monitor active power, reactive power, bus voltage and frequency at different points in a wide-area power system. Wide-area means different adjacent areas belonging to different utilities. Supervisory control and data acquisition (SCADA) system, two-to-four second measurements, has been applied to locally monitor and control these areas separately. Tremendous efforts have been made to design controllers such as PSSs for damping electromechanical oscillations as explained in Chapter 11. They provide supplementary control action through the generator excitation systems to improve the power system stability [24].

The wide-area measurement system (WAMS) based on rms values of the measured parameters can be associated with a state estimator (SE) that is responsible for estimating the phase angles of voltages and currents. Of course, it will be more advanced to measure both the voltage/current magnitudes and their angles as estimation is less than measurement in accuracy. In addition, the measured voltages and their angles at all measuring points must be determined by time synchronisation of a fraction of milliseconds accuracy [25]. Fortunately, the recent global positioning system (GPS) and development of communication systems for fast and large data transmission satisfy the time synchronisation of measured parameters at different remote locations and help the controllers for decision making on real time. With the aid of GPS and communication systems, the phasor measurement units (PMUs) that are able not only to measure the phasor voltages but also to determine the phase reference effectively help the WAMS to be a promising technique for power system monitoring.

15.3.1 Structure of WAMS

PMUs are located at different measurement points in the wide-area power system. By offline analysis the optimal placement and number of PMUs can be determined [26–30]. The data are measured by each PMU and then collected by phasor data concentrators (PDCs) through narrow-band channel communication network. Each PDC receives data from multiple PMUs and sorts its frames. The concentrated data are locally stored and can be exchanged with other PDCs belonging to other utilities by using standard format including the time stamp of the synchronised GPS time. An efficient server is used to receive all concentrated data through wide-band communication network for processing, analysis and applications such as monitoring, state estimation, protection, real time control, contingency analysis, oscillation detection and stability estimation [23, 31].

As shown in Figure 15.21, the wide-area comprises adjacent areas in the power system belonging to different utilities (area #1, 2, . . . , n). Each area has a number of

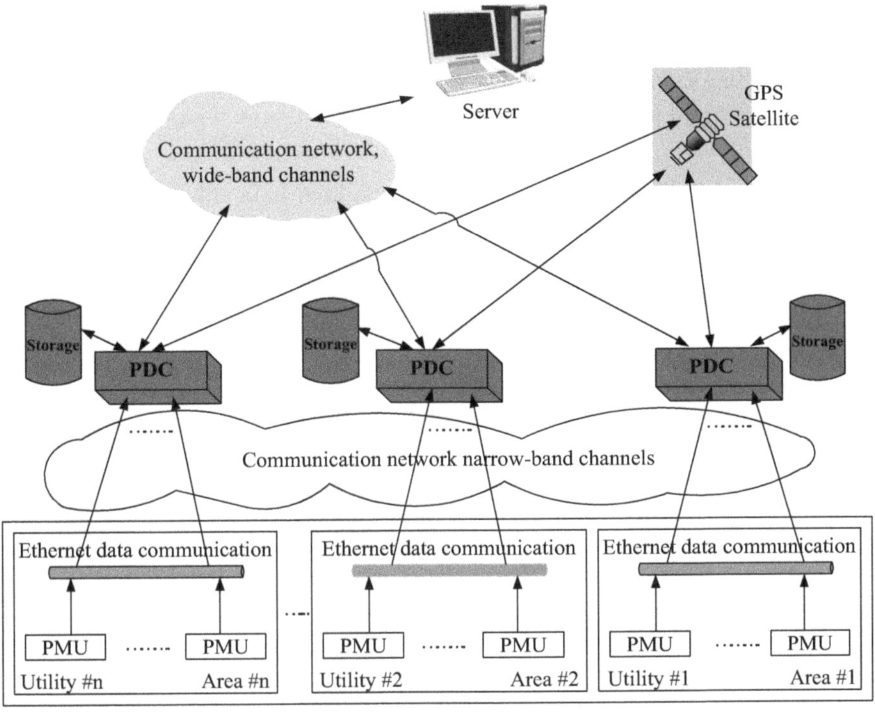

Figure 15.21 Schematic diagram of PMUs and wide-area measurement structure

PMUs inserted at predetermined measurement points. The phasor measurements may be sampled at rates of 30 times a second or more. They can also be time-stamped and synchronised with very high accuracy and millisecond resolution across a wide geographic area [32]. All the data measured by PMUs in that area are gathered and sent to a PDC in compliance with the standard data format. These huge data sorted by the PDCs of the different areas are collected at a server through wide-band channels communication network for processing to give the information required for system performance improvement. Exchange of information between the PDCs of the different areas can be implemented. GPS satellite is used for time synchronisation between PDCs and determining the time reference. For instance, measurement of frequency, voltage and current magnitude and angles, and voltage/current phase difference by PMUs pre, during and post-fault can be processed and sent to the real-time controller for online decision-making to keep the system stability. Exchange of information between the PDCs helps to decide which areas are controllable and which areas are observable.

15.3.2 Benefits of WAMS

The traditional security assessment of power systems based on offline analysis and SCADA data becomes increasingly unreliable for real-time operation, in particular, for wide-area power systems. WAMS and PMUs allow monitoring, assessing and taking the control action in real time to prevent or alleviate power system problems. The capability of synchronised measurements of system parameters in magnitude and angle at different locations in a wide-area provides the benefits below:

- The possibility of directly measuring the system state rather than estimating it by state estimation techniques.
- Real-time monitoring gives the system operators real-time information that help increasing their operational efficiency under normal conditions and enabling them to detect problems during abnormal conditions.
- Avoiding large area disturbances.
- Maximising exploitation of present networks.
- Increasing power transmission capability.
- Damping of electromechanical oscillations when the system is subjected to disturbances.
- Improving transient and voltage stability since adding and networking MPUs gives the utilities the capability to more closely monitor the stability of the bulk electric system and control remedial actions with their neighbours [33, 34].
- Avoiding system congestion.
- Increasing the accuracy of SEs, which in turn provides a valid best estimate of a consistent network model that can be used as a starting point for real-time applications, e.g. VAr optimisation, constrained re-dispatch and contingency analysis [35].
- Enhancing relays performance and protection schemes, because use of synchronised phasor measurement, certain relays (adaptive relays) could be made to adapt to the prevailing system conditions [36].

15.3.3 Case studies

In Finland, a WAMS project was launched in 2006 by Fingrid, the Finish transmission system owner and operator (TSO), to obtain real-time information about damping of 0.3 Hz electromechanical inter-area oscillations, which typically is the main factor limiting power transfer capability from South Finland to North and further down to South Scandinavia. Then the project has been developed and in 2011 the system deployed 12 PMUs and a PDC. The measurements of PMUs are also streamed from and to wide-area measurements in Norway and Denmark. PMUs in addition to WAMS have been successfully applied for other power system planning, analysis and control purposes. For instance, two PMUs are installed as an integral part of SVC POD controls to provide local frequency measurement and positive sequence voltage signals for the POD controls. More details about this project can be found in [31].

In Brazil, a large-scale blackout of the Brazilian interconnected power system (BIPS) with a 40 per cent load loss occurred due to a fault at the AC Itaipu transmission system. This event was recorded at the low voltage level by a synchronised phasor measurement prototype, the LVPMS, with PMUs installed in nine universities throughout Brazil. The recorded data contain relevant information on the events leading to the blackout and the BIPS restoration [37]. Many other installations of PMU prototype are presented in different countries such as Austria, China, Canada, USA, Sweden and Switzerland.

References

1. Ham W.K., Hwang S.W., Kim J.H. 'Active and reactive power control model of superconducting magnetic energy storage (SMES) for the improvement of power system stability'. *Journal of Electrical Engineering & Technology*. 2008;**3**(1):1–7
2. Elamana S., Rathinam A. 'Interarea oscillation damping by unified power flow controller-superconducting magnetic energy storage integrated system'. *International Journal of Engineering and Advanced Technology (IJEAT)*. 2013;**2**(3):221–5
3. Carnegie R., Gotham D., Nderitu D., Preckel P.V. *Utility scale energy storage systems: benefits, applications, and technologies*. Report of State Utility Forecasting Group, Jun 2013
4. APS Panel on Public Affairs, Committee on Energy and Environment (US). *Challenges of electricity storage technologies* [online]. Report, May 2007. Available from www.aps.org/policy/reports/popa-reports/upload/Energy_2007_Report_ElectricityStorageReport.pdf [Accessed Oct 2014]
5. Oberhofer A., Meisen P. *Energy storage technologies & their role in renewable integration*. Report, Global Energy Network Institute (GENI), Jul 2012
6. Makarov Y.V., Nyeng P., Yang B., DeSteese J.G., Ma J., Hammerstorm D.J., Lu S., *et al. Wide-area energy storage and management system to balance intermittent resources in the Bonneville power administration and California ISO control areas*. Report, prepared by Pacific Northwest National Laboratory for U.S. Department of Energy, Jun 2008
7. Science Educators. *Type 1 Superconductors* [online]. May 2007. Available from http://superconductors.org/Type1.htm [Accessed 12 Nov 2014]
8. Science Educators. *Type 2 Superconductors* [online]. Oct 2014. Available from http://superconductors.org/Type2.htm [Accessed 12 Nov 2014]
9. Khare N. *Handbook of High-Temperature Superconductor Electronics*. New York, NY, US: Marcel Dekker, Inc.; 2003
10. *Superconductor Terminology and the Naming Scheme* [online]. Available from http://superconductors.org/terms.htm [Accessed 23 Sept 2014]
11. World Technology Evaluation Center (WTEC) Panel. *Power applications of superconductivity in Japan and Germany*. Report, International Technology Research Institute, Loyola College in Maryland, Sept 1997
12. Ribeiro P., Johnson B., Crow M., Arsoy A., Steurer M., Liu Y. 'Energy Storage Systems'. *Encyclopedia of Life Support Systems (EOLSS), Electrical Engineering*. 2012;**3**:1–11

13. Lawrenson P.J., Miller T.J.E., Stephenson J.M., Ula A.H.M.S. 'Damping and screening in the synchronous superconducting generator'. *Electrical Engineers, Proceedings of the Institution of.* 1976;**123**(8):787–94
14. Takao T., Tsukamoto O., Hirao T., Morita M., Ikeda B. 'Quench characteristics of rotor winding of superconducting generator in static and rotating conditions'. *IEEE Transactions on Magnetics*. 1996;**32**(4):2365–8
15. Singh K.S. *Applications of High Temperature Superconductors to Electric Equipment*. NJ, US: Wiley-IEEE Press; 2011
16. Lynn R.W. *High Temperature Superconductors: Materials, Properties, and Applications*. Dordrecht, Netherlands: Kluwer Academic Publisher; 1998
17. Suryanarayana T., Bhattacharya J.L., Raju K.S.N., Durga Prasad K.A. 'Development and performance testing of a 200 kVA damperless superconducting generator'. *IEEE Transactions on Energy Conversion*. 1997; **12**(4):330–6
18. Nakayama T., Yagai T., Tsuda M., Hamajima T. 'Micro power grid system with SMES and superconducting cable modules cooled by liquid hydrogen'. *IEEE Transactions on Applied Superconductivity*. 2009;**19**(3):2062–5
19. Sykulski J. (eds.). 'Superconducting transformers'. *Advanced Research Workshop on Modern Transformers*; Vigo, Spain, Oct 2004. pp. 1–44
20. Kopylov S.I., Palashov N.N., Ivanov S.S., Veselovsky A.S., Zhemerikin V.D. (eds.). 'Joint operation of the superconducting fault current limiter and magnetic energy storage system in an electric power network'. *9th European Conference on Applied Superconductivity (EUCAS 09)*; Dresden, Germany, Sept 2009. pp. 912–17
21. Xue X.D., Cheng K.W.E., Sutanto D. (eds.). 'Power system applications of superconducting magnetic energy storage systems'. *Industrial Applications Conference 2005, Fortieth IAS Annual Meeting*; Kowloon, Hong Kong, Oct 2005, vol. 2. pp. 1524–9
22. Torre W.V., Eckroad S. (eds.). 'Improving power delivery through the application of superconducting magnetic energy storage (SMES)'. *IEEE Power Engineering Society Winter Meeting Conference*; Colombus, OH, US, Jan/Feb 2001, vol. 1. pp. 81–7
23. Bevrani H., Watanabe M., Mitani Y. *Power System Monitoring and Control*. Hoboken, NJ, US: John-Wiley Press; 2014
24. Ma J., Wang T., Wu J., Thorp J.S. (eds.). 'Design of global power systems stabilizer to damp inter-area oscillations based on wide-area collocated control technique'. *IEEE Power and Energy Society General Meeting*; Detroit, MI, US, Jul 2011
25. Giri J. (eds.). 'Enhanced power grid operations with a wide-area synchrophasor measurement & communication network'. *IEEE Power and Energy Society General Meeting*; San Diego, CA, US, Jul 2012
26. Kulkarni S., Allen A., Santoso S., Grady W.M. (eds.). 'Phasor measurement unit placement algorithm'. *IEEE Power and Energy Society General Meeting*; Calgary, AB, Canada, Jul 2009

27. Li Q., Negi R., LLić M.D. (eds.). 'Phasor measurement units placement for power system state estimation: a greedy approach'. *IEEE Power and Energy Society General Meeting*; Calgary, AB, Canada, Jul 2009
28. Bahabadi H.B., Mirzaei A., Moallem M. (eds.). 'Optimal placement of phasor measurement units for harmonic state estimation I unbalanced distribution system using genetic algorithms'. *21st International Technical Conference on Industrial & Commercial Power Systems (ICPS) IEEE*; Newport Beach, CA, US, May 2011. pp. 100–05
29. Zadeh A.K., Masshadi H.R., Abadi M.E.H. (eds.). 'Optimal placement of a defined number of phasor measurement units in power systems'. *2nd Iranian Conference on Smart Grids (ICSG) 2012*; Iran, May 2012
30. Gao Y., Hu Z., He X., Liu D. (eds.). 'Optimal placement of PMUs in power systems based on improved PSO algorithm'. *3rd IEEE Conference on Industrial Electronics and Applications (ICIEA) 2008*; Harbin, China, Jun 2008. pp. 2464–69
31. Rauhala T., Saarinen K., Latvala M., Laasonen M., Uusitalo M. (eds.). 'Applications of phasor measurement units and wide-area measurement system in Finland'. *Power Tech. 2011 IEEE Trondheim*, 2011. pp. 1–8
32. Carty D., Atanacio M. (eds.). 'PMUs and their potential impact on real-time control center operations'. *IEEE Power and Energy Society General Meeting*; Minneapolis, MN, US, Jul 2010
33. Alsafih H.A., Dunn R. (eds.). 'Determination of coherent clusters in a multi-machine power system based on wide-area signal measurements'. *IEEE Power and Energy Society General Meeting*; Minneapolis, MN, US, Jul 2010
34. Glavic M., Custem T.V. (eds.). 'Detecting with PHUs the onset of voltage instability caused by a large disturbance'. *IEEE Power and Energy Society General Meeting, Conversion and Delivery of Electrical Energy in the 21st century 2008 IEEE*; Pittsburgh, PA, US, Jul 2008. pp. 1–8
35. Liu Z, Llić D. (eds.). 'Toward PMU-based robust automatic voltage control (AVC) and automatic flow control (AFC)'. *IEEE Power and Energy Society General Meeting*; Minneapolis, MN, US, Jul 2010. pp. 1–8
36. Skok S., Ivankovic I., Cerina Z. (eds.). 'Applications based on PMU technology for improved power system utilization'. *IEEE Power and Energy Society General Meeting*; Tampa, FL, US, Jun, 2007. pp. 1–8
37. Decker I.C., Agostini M.N., e Silva A.S., Dotta D. (eds.). 'Monitoring of a large scale event in the Brazilian power system by WAMS'. *2010 IREP Symposium-Bulk Power System Dynamics and Control-VIII (IREP)*; Buzios, RJ, Brazil, Aug 2010

Appendix I
Calculation of synchronous machine parameters in per unit/normalised form

I.1 Per unit values

I.1.1 Base quantities for stator

It is common to choose the following three base quantities for armature (stator) variables:

- $S_{\rm B} \triangleq$ Base power = three-phase stator rated power (VA rms)
- $V_{\rm B} \triangleq$ Base voltage = stator rated line-to-line voltage, $V_{\rm L-L}$ (V rms)
- $t_{\rm B} \triangleq$ Base time (s)

The other base quantities can be accordingly determined as

- $I_{\rm B} \triangleq$ Base current $= \frac{S_{\rm B}}{V_{\rm B}} = \sqrt{3} \times$ rated line current, $I_{\rm L}$
- $\omega_{\rm B} \triangleq$ Generator rated speed $(\omega_{\rm o}) = \frac{1}{t_{\rm B}}$ (elec. rad/s)
- $Z_{\rm B} \triangleq$ Base impedance $= \frac{V_{\rm B}}{I_{\rm B}} = \frac{V_{L-L}}{\sqrt{3}I_L}$
- $\Psi_{\rm B} \triangleq$ Base flux linkage $= L_{\rm B}I_{\rm B} = V_{\rm B}t_{\rm B} = \frac{V_{\rm B}}{\omega_B}$
- $L_{\rm B} \triangleq$ Base inductance $= \frac{\Psi_{\rm B}}{I_{\rm B}} = \frac{Z_{\rm B}}{\omega_{\rm B}}$

To ensure the validity of base quantities identified above, the total power in the three stator phases, P_{abc}, must be the same as the power in the *d–q* circuits. This can be proved as below:

$$P_{abc} = v_a i_a + v_b i_b + v_c i_c = v^t_{abc} i_{abc} \tag{I.1}$$

Applying Park's transformation: $v^t_{abc} = v^t_{dqo}\mathbf{P}^t$ and $i_{abc} = \mathbf{P}i_{dqo}$

As $\mathbf{P}$ is an orthogonal matrix (power invariant), i.e. $\mathbf{P}^t = \mathbf{P}^{-1}$ and assuming the zero sequence power is zero, (I.1) becomes

$$P_{abc} = v^t_{dq} i_{dq} = v_d\, i_d + v_q i_q \tag{I.2}$$

The *d–q* voltages can be written as

$v_d = V \sin\delta$ and $v_q = V\cos\delta$, where V is the line-to-line voltage, and in pu values $v_{du} = V_u \sin\delta$ and $v_{qu} = V_u \cos\delta$, where the subscript u indicates the pu values

Thus,

$$v_{du}^2 + v_{qu}^2 = V_u^2 \tag{I.3}$$

Equation (I.3) shows that the *d–q* axis voltages are numerically equal to the line-to-line voltages.

Similarly, the *d–q* axis currents are

$i_d = I \sin\gamma$ and $i_q = I \cos\gamma$ (where I is the line current),

and in pu values:

$$i_{du} = I_u \sin\gamma \quad \text{and} \quad i_{qu} = I_u \cos\gamma$$

Substituting the pu values of *d–q* axis voltages and currents into (I.2), the three-phase stator power is given by

$$\begin{aligned} P_{abc} &= V_u I_u (\sin\delta \sin\gamma + \cos\delta \cos\gamma) \\ &= V_u I_u \cos(\delta - \gamma) \end{aligned}$$

In terms of the base quantities, the *d–q* stator voltages (2.30) can be written in pu values as

$$\left.\begin{aligned} v_{du} &= -\frac{1}{\omega_B}\frac{d\Psi_{du}}{dt} - \frac{\omega}{\omega_B}\Psi_{qu} - R_{au} i_{du} \\ v_{qu} &= -\frac{1}{\omega_B}\frac{d\Psi_{qu}}{dt} + \frac{\omega}{\omega_B}\Psi_{du} - R_{au} i_{qu} \end{aligned}\right\} \tag{I.4}$$

To normalise any quantity, it is divided by the base quantity of the same dimension, e.g.

$$i_{qu} = \frac{i_q}{I_B}, \quad i_{du} = \frac{i_d}{I_B}, \quad v_{du} = \frac{v_d}{V_B}, \quad \Psi_{du} = \frac{\Psi_d}{\Psi_B}$$

More details about normalisation of equations are given in Section I.2.

I.1.2 Base quantities for rotor

The base power, S_B, of the armature is based on its rating and the time base is fixed by the rated radian frequency. These base quantities must be the same for rotor circuits as the rotor and stator circuits are coupled electromagnetically. To satisfy this condition, the numeric value of the rotor quantities in per unit is small, because the stator base power is much larger than the rated power of the rotor circuits.

Therefore, it is necessary to decide on a suitable base quantity in the rotor that gives the correct base quantity in the stator. Equality of mutual flux linkages is the main concept on which the choice of base quantity in the rotor is based. This implies that the base currents in rotor circuits in the *d*-axis (base field current or base damper current) are chosen in such a way that they produce the same space fundamental of air gap flux as produced by the base stator current flowing in the *d*-axis stator windings.

Again, (2.20) is rewritten below in an expanded form as

$$\begin{bmatrix} \Psi_d \\ \Psi_q \\ \Psi_o \\ \Psi_f \\ \Psi_{kd} \\ \Psi_{kq} \end{bmatrix} = \begin{bmatrix} L_d & 0 & 0 & kM_f & kM_{kd} & 0 \\ 0 & L_q & 0 & 0 & 0 & kM_{kq} \\ 0 & 0 & L_o & 0 & 0 & 0 \\ kM_f & 0 & 0 & L_f & L_{fkd} & 0 \\ kM_{kd} & 0 & 0 & L_{fkd} & L_{kd} & 0 \\ 0 & kM_{kq} & 0 & 0 & 0 & L_{kq} \end{bmatrix} = \begin{bmatrix} i_d \\ i_q \\ i_o \\ i_f \\ i_{kd} \\ i_{kq} \end{bmatrix} \tag{I.5}$$

In (I.5), assume that the currents $i_d = I_{\mathrm{B}}$, $i_f = I_{f\mathrm{B}}$ and $i_{kd} = I_{kd\mathrm{B}}$ are applied one at a time with other currents set to zero. Then equating the mutual flux linkages in each d-axis winding (Ψ_{md}, Ψ_{mf}, Ψ_{mkd}), the equations below are obtained.

$$(L_d - \ell_d)I_{\mathrm{B}} = kM_f I_{f\mathrm{B}} = kM_{kd} I_{kd\mathrm{B}} \tag{I.6}$$

where

ℓ_d is the leakage inductance of the d-axis armature winding

$I_{f\mathrm{B}}$ and $I_{kd\mathrm{B}}$ are the base currents in the rotor field and damper windings, respectively

Hence,

$$I_{f\mathrm{B}} = \frac{L_{md}}{kM_f} I_{\mathrm{B}} \quad \text{and} \quad I_{kd\mathrm{B}} = \frac{L_{md}}{kM_{kd}} I_{\mathrm{B}} \tag{I.7}$$

where $L_{md} = L_d - \ell_d$

The base flux linkages for rotor circuits are chosen such that

$$\Psi_{\mathrm{B}} I_{\mathrm{B}} = \Psi_{f\mathrm{B}} I_{f\mathrm{B}} = \Psi_{kd\mathrm{B}} I_{kd\mathrm{B}}, \quad \Psi_{f\mathrm{B}} = \frac{I_{\mathrm{B}}}{I_{f\mathrm{B}}} \Psi_{\mathrm{B}} \quad \text{and} \quad \Psi_{kd\mathrm{B}} = \frac{I_{\mathrm{B}}}{I_{kd\mathrm{B}}} \Psi_{\mathrm{B}} \tag{I.8}$$

Similarly, for the q-axis rotor circuits (KQ coil), the base current and flux linkages are given by

$$I_{kq\mathrm{B}} = \frac{L_{mq}}{kM_{kq}} I_{\mathrm{B}} \quad \text{and} \quad \Psi_{kq\mathrm{B}} = \frac{I_{\mathrm{B}}}{I_{kq\mathrm{B}}} \Psi_{\mathrm{B}} \tag{I.9}$$

where $L_{mq} = L_q - \ell_q$ and $\ell_q \triangleq$ the leakage inductance of the q-axis armature winding.

Commonly, $\ell_d = \ell_q$ and can be written as ℓ_a.

As the base quantity S_{B} for the stator must be equal to S_{B} for the rotor, the relations below can be computed.

$$\left.\begin{aligned} \frac{V_{f\mathrm{B}}}{V_{\mathrm{B}}} &= \frac{I_{\mathrm{B}}}{I_{f\mathrm{B}}} = \frac{kM_f}{L_{md}} \\ \frac{V_{kd\mathrm{B}}}{V_{\mathrm{B}}} &= \frac{I_{\mathrm{B}}}{I_{f\mathrm{B}}} = \frac{kM_{kd}}{L_{md}} \\ \frac{V_{kq\mathrm{B}}}{V_{\mathrm{B}}} &= \frac{I_{\mathrm{B}}}{I_{f\mathrm{B}}} = \frac{kM_{kq}}{L_{mq}} \end{aligned}\right\} \tag{I.10}$$

I.1.3 Conversion of rotor quantities to equivalent stator EMF

In synchronous machine equations, it is preferable to convert the rotor current, flux linkage and voltage to an equivalent stator EMF as below.

In steady state and at open circuit conditions, the field current i_f corresponds to a peak stator EMF of $(\omega_o M_f i_f)$. Then, $\omega_o M_f i_f = \sqrt{2}E$, where E is the rms of stator EMF as a line-to-neutral value. As the coupling between the d-axis rotor and stator windings involves the factor $k = \sqrt{(3/2)}$ as formerly explained, this relation can be written as

$$\omega_o k M_f i_f = \sqrt{3}E \quad \text{or} \quad \omega_o k M_f i_f = E_I, \quad \text{where } E_I \text{ is the line-to-line rms value} \tag{I.11}$$

in compliance with the American National Standard Institute (ANSI) or $\omega_o k M_f i_f = E_q$ in compliance with the International Electro-technical Commission (IEC) notations. It is to be noted that the field current i_f corresponds to a given EMF by a scaling factor where ω_o and M_f are constants for a given machine. Therefore, E_I in pu corresponds to i_f in pu.

The flux linkage Ψ_f can also be converted to a corresponding stator EMF. In steady-state and at open circuit conditions $i_f = \Psi_f / L_f$. Multiplying the field current by $\omega_o k M_f$ to give the d-axis stator EMF, E_q' (as line-to-line rms value) corresponding to the flux linkage Ψ_f is

$$\omega_o k M_f \Psi_f / L_f = E_q' \tag{I.12}$$

where E_q' is represented by the quadrature component of stator voltage behind the transient reactance.

Similarly, at steady state, the field voltage v_f corresponds to a field current $i_f = v_f / R_f$. Consequently, it corresponds to a peak stator EMF as $i_f \omega_o M_f = (v_f / R_f) \omega_o M_f$. If its line-to-line rms value is denoted by E_{fd}, the d-axis stator EMF corresponds to a field voltage v_f given by

$$(v_f / R_f) \omega_o k M_f = E_{fd} \tag{I.13}$$

It is to be noted that the values of stator EMF in (I.11)–(I.13) are considered as line-to-line rms values as the base value of the voltage V_B is taken as line-to-line rms value (Section I.1.1). In some literature, V_B is taken as line-to neutral rms voltage and that leads to different equations using a factor of $\sqrt{3}$ as described in Section I.3.

I.2 Normalising the synchronous machine voltage equations

Based on a choice of appropriate base values, the voltage equations can be normalised so that they are easier to deal with, as the numerical values of voltage and current in the normalised form will be of the same order of magnitude. The subscript u is added to all pu quantities and is omitted later when all values are normalised. As the actual value of any quantity = the pu value × base value, (2.32) is rewritten as

$$
\begin{bmatrix} v_{du}\,V_{\mathrm{B}} \\ v_{qu}\,V_{\mathrm{B}} \\ v_{ou}\,V_{\mathrm{B}} \\ -v_{fu}\,V_{f\mathrm{B}} \\ 0 \\ 0 \end{bmatrix}
= -\left[\begin{array}{ccc:ccc}
R_a & \omega L_a & 0 & 0 & 0 & \omega kM_{ka} \\
-\omega L_d & R_a & 0 & -\omega kM_f & -\omega kM_{kd} & 0 \\
0 & 0 & R_a + 3R_n & 0 & 0 & 0 \\
\hdashline
0 & 0 & 0 & R_f & 0 & 0 \\
0 & 0 & 0 & 0 & R_{kd} & 0 \\
0 & 0 & 0 & 0 & 0 & R_{ka}
\end{array}\right]
\begin{bmatrix} i_{du}\,I_{\mathrm{B}} \\ i_{qu}\,I_{\mathrm{B}} \\ i_{ou}\,I_{\mathrm{B}} \\ i_{fu}\,I_{f\mathrm{B}} \\ i_{kdu}\,I_{kd\mathrm{B}} \\ i_{kqu}\,I_{kq\mathrm{B}} \end{bmatrix}
$$

$$
-\left[\begin{array}{ccc:ccc}
L_d & 0 & 0 & kM_f & kM_{kd} & 0 \\
0 & L_q & 0 & 0 & 0 & kM_{kq} \\
0 & 0 & L_o + 3L_n & 0 & 0 & 0 \\
\hdashline
kM_f & 0 & 0 & L_f & L_{fkd} & 0 \\
kM_{kd} & 0 & 0 & L_{fkd} & L_{kd} & 0 \\
0 & kM_{kq} & 0 & 0 & 0 & L_{kq}
\end{array}\right]
\begin{bmatrix} (\mathrm{p}i_{qu})I_{\mathrm{B}} \\ (\mathrm{p}i_{qu})I_{\mathrm{B}} \\ (\mathrm{p}i_{ou})I_{\mathrm{B}} \\ (\mathrm{p}i_{fu})I_{f\mathrm{B}} \\ (\mathrm{p}i_{kdu})I_{kd\mathrm{B}} \\ (\mathrm{p}i_{kqu})I_{kq\mathrm{B}} \end{bmatrix} \tag{I.14}
$$

This system of equations can be written in expanded form by substituting $\omega = \omega_u \omega_o$ (the rated angular speed ω_o is taken as the base value ω_{B}) as below:

$$
v_{du} = -R_a \frac{I_{\mathrm{B}}}{V_{\mathrm{B}}} i_{du} - \omega_u \omega_o L_q \frac{I_{\mathrm{B}}}{V_{\mathrm{B}}} i_{qu} - \omega_u \omega_o kM_{kq} \frac{I_{kq\mathrm{B}}}{V_{\mathrm{B}}} i_{kqu} - L_d \frac{I_{\mathrm{B}}}{V_{\mathrm{B}}} \mathrm{p}i_{du}
$$

$$
- kM_f \frac{I_{f\mathrm{B}}}{V_{\mathrm{B}}} \mathrm{p}i_{fu} - kM_{kd} \frac{I_{kd\mathrm{B}}}{V_{\mathrm{B}}} \mathrm{p}i_{kdu} \ \mathrm{pu} \tag{I.15}
$$

Hence,

$$
v_{du} = -\frac{R_a}{R_{\mathrm{B}}} i_{du} - \omega_u \frac{L_q}{L_{\mathrm{B}}} i_{qu} - \omega_u \frac{\omega_o I_{kq\mathrm{B}}}{V_{\mathrm{B}}} kM_{kq} i_{kqu} - \frac{L_d}{\omega_o L_{\mathrm{B}}} \mathrm{p}i_{du} - \frac{kM_f}{\omega_o} \frac{\omega_o I_{f\mathrm{B}}}{V_{\mathrm{B}}} \mathrm{p}i_{fu}
$$

$$
- \frac{kM_{kd}}{\omega_o} \frac{\omega_o I_{kd\mathrm{B}}}{V_{\mathrm{B}}} \mathrm{p}i_{kdu} \tag{I.16}
$$

By definition

$$
R_u = R_a/R_{\mathrm{B}}, \quad L_{du} = L_d/L_{\mathrm{B}}, \quad M_{fu} = M_f \omega_o I_{f\mathrm{B}}/V_{\mathrm{B}}, \quad L_{kdu} = L_{kd}/L_{\mathrm{B}},
$$
$$
M_{kdu} = M_{kd} \omega_o I_{kd\mathrm{B}}/V_{\mathrm{B}}, \quad M_{kqu} = M_{kq} \omega_o I_{kq\mathrm{B}}/V_{\mathrm{B}}
$$

Then, by substituting into (I.16), it gives

$$v_{du} = -R_u i_{du} - \omega_u L_{qu} i_{qu} - \omega_u k M_{kqu} i_{kqu} - \frac{L_{du}}{\omega_o} \mathrm{p} i_{du} - k\frac{M_{fu}}{\omega_o} \mathrm{p} i_{fu} - k\frac{M_{kdu}}{\omega_o} \mathrm{p} i_{kdu} \tag{I.17}$$

Similarly, applying similar analysis, the q-axis voltage equation in pu is

$$v_{qu} = -R_u i_{qu} + \omega_u L_{du} i_{du} + \omega_u k M_{fu} i_{fu} + \omega_u k M_{kdu} i_{kdu} - \frac{L_{qu}}{\omega_o} \mathrm{p} i_{qu} - k\frac{M_{kqu}}{\omega_o} \mathrm{p} i_{kqu} \text{ pu} \tag{I.18}$$

The equation of v_{ou} can be written as below, and it is noted that it vanishes under balanced conditions:

$$v_{ou} = -\frac{R_a + 3R_n}{R_B} i_{ou} - \frac{L_o + 3L_n}{\omega_o L_B} \mathrm{p} i_{ou}$$

Thus,

$$v_{ou} = -(R_a + 3R_n)_u i_{ou} - \frac{1}{\omega_o}(L_o + 3L_n)_u \mathrm{p} i_{ou} \text{ pu} \tag{I.19}$$

The rotor equations are normalised on the rotor base values. The pu field voltage is

$$v_{fu} = R_f \frac{I_{fB}}{V_{fB}} i_{fu} + k\frac{M_f}{\omega_o}\frac{\omega_o I_B}{V_{fB}} \mathrm{p} i_{du} + \frac{L_f}{\omega_o}\frac{\omega_o I_{fB}}{V_{fB}} \mathrm{p} i_{fu} + \frac{L_{fkd}}{\omega_o}\frac{\omega_o I_{kdB}}{V_{fB}} \mathrm{p} i_{kdu} \text{pu} \tag{I.20}$$

The last two terms are normalised by incorporating the base rotor inductance as

$$L_{fu} = L_f / L_{fB} \quad \text{and} \quad L_{fkdu} = L_{fkd} / L_{fkdB}$$

Thus, the normalised field voltage equation is

$$v_{fu} = R_{fu} i_{fu} + \frac{k M_{fu}}{\omega_o} \mathrm{p} i_{\mathrm{du}} + \frac{L_{fu}}{\omega_o} \mathrm{p} i_{fu} + \frac{L_{fkdu}}{\omega_o} \mathrm{p} i_{kdu} \tag{I.21}$$

Applying the same procedure to damper winding equations for circuits KD and KQ, the following normalised equations are obtained:

$$v_{kdu} = 0 = R_{kdu} i_{kdu} + \frac{k M_{kdu}}{\omega_o} \mathrm{p} i_{du} + \frac{L_{fkdu}}{\omega_o} \mathrm{p} i_{fu} + \frac{L_{kdu}}{\omega_o} \mathrm{p} i_{kdu} \tag{I.22}$$

$$v_{kqu} = 0 = R_{kqu} i_{kqu} + \frac{k M_{kqu}}{\omega_o} \mathrm{p} i_{qu} + \frac{L_{kqu}}{\omega_o} \mathrm{p} i_{kqu} \tag{I.23}$$

Under balanced conditions and incorporating the normalised equations in a matrix form, where the first three rows express the voltage relations in the d-axis, the fourth and fifth rows express the voltage relations in the q-axis; the following form can be written as

$$\begin{bmatrix} v_d \\ -v_f \\ 0 \\ v_q \\ 0 \end{bmatrix} = -\begin{bmatrix} R_a & 0 & 0 & \omega L_q & \omega k M_{kq} \\ 0 & R_f & 0 & 0 & 0 \\ 0 & 0 & R_{kd} & 0 & 0 \\ -\omega L_d & -\omega k M_f & -\omega k M_{kd} & R_a & 0 \\ 0 & 0 & 0 & 0 & R_{kq} \end{bmatrix} \begin{bmatrix} i_d \\ i_f \\ i_{kd} \\ i_q \\ i_{kq} \end{bmatrix}$$
$$- \begin{bmatrix} L_d & kM_f & kM_{kd} & 0 & 0 \\ kM_f & L_f & L_{fkd} & 0 & 0 \\ kM_{kd} & L_{fkd} & L_{kd} & 0 & 0 \\ 0 & 0 & 0 & L_q & kM_{kq} \\ 0 & 0 & 0 & kM_{kq} & L_{kq} \end{bmatrix} \begin{bmatrix} \mathrm{p}i_d \\ \mathrm{p}i_f \\ \mathrm{p}i_{kd} \\ \mathrm{p}i_q \\ \mathrm{p}i_{kq} \end{bmatrix} \qquad \text{(I.24)}$$

In (I.24), the subscript u is dropped as all values are in pu as well as this form is adequate to analyse the system in time domain (time in seconds).

I.3 Alternative per unit/normalising systems

In some literature phase quantities are used as base quantities. Therefore, the stator base quantities are chosen as below.

I.3.1 Base quantities for stator

- $S_B \triangleq$ Base power = stator rated power/phase (VA rms)
- $V_B \triangleq$ Base voltage = stator rated line-to-neutral voltage, $V_{L\text{-}N}$ (V rms)
- $t_B \triangleq$ Base time (s)

The other base quantities can accordingly be determined as

- $I_B \triangleq$ Base current $= \frac{S_B}{V_B}$ = rated phase current, or in star connection line current, I_L
- $\omega_B \triangleq$ Generator rated speed $(\omega_o) = \frac{1}{t_B}$ (elec. rad/s)
- $Z_B \triangleq$ Base impedance $= \frac{V_B}{I_B} = \frac{V_{L\text{-}N}}{I_L}$
- $\Psi_B \triangleq$ Base flux linkage $= L_B I_B = V_B t_B = V_{L\text{-}N} t_B = \frac{V_{L\text{-}N}}{\omega_B}$
- $L_B \triangleq$ Base inductance $= \frac{\Psi_B}{I_B} = \frac{Z_B}{\omega_B}$

Under balanced conditions, the d–q axis pu quantities, such as v_{dqo}, i_{dqo} and the total power in the three stator phases, P_{abc}, can be obtained as below.

Assuming the stator voltages in the form

$$\left.\begin{aligned} v_a &= V_{\max}\sin(\delta+\alpha) = \sqrt{2}V\sin(\delta+\alpha) \\ v_b &= \sqrt{2}V\sin\left(\delta+\alpha-\frac{2\pi}{3}\right) \\ v_c &= \sqrt{2}V\sin\left(\delta+\alpha+\frac{2\pi}{3}\right) \end{aligned}\right\} \tag{I.25}$$

where $V\angle\alpha$ is the rms phase voltage.

Applying Park's transformation to give v_{dqo} as

$$\begin{bmatrix} v_d \\ v_q \\ v_o \end{bmatrix} = \begin{bmatrix} \sqrt{3}V\sin\alpha \\ \sqrt{3}V\cos\alpha \\ 0 \end{bmatrix} \tag{I.26}$$

Thus, the pu voltages in d–q frame of reference are

$$\left.\begin{aligned} v_{du} &= v_d/V_{\mathrm{B}} = \sqrt{3}(V/V_{\mathrm{B}})\sin\alpha = \sqrt{3}V_u\sin\alpha \\ v_{qu} &= v_q/V_{\mathrm{B}} = \sqrt{3}(V/V_{\mathrm{B}})\cos\alpha = \sqrt{3}V_u\cos\alpha \end{aligned}\right\} \tag{I.27}$$

Hence,

$$v_{du}^2 + v_{qu}^2 = 3V_u^2 \tag{I.28}$$

Equation (I.27) illustrates that the d- and q-axis voltages are numerically equal to $\sqrt{3}$ times the pu voltages.

Similarly, assuming the rms phase current is $I\angle\gamma$. The stator currents in d–q frame of reference are

$$\begin{bmatrix} i_d \\ i_q \\ i_o \end{bmatrix} = \begin{bmatrix} \sqrt{3}I\sin\gamma \\ \sqrt{3}I\cos\gamma \\ 0 \end{bmatrix} \tag{I.29}$$

Thus, the pu currents can be obtained by

$$i_{du} = \sqrt{3}I_u\sin\gamma, \quad i_{qu} = \sqrt{3}I_u\cos\gamma \tag{I.30}$$

Using (I.27) and (I.30), the total power in the three stator phases, P_{abc}, is given by

$$P_{abc} = i_{du}v_{du} + i_{qu}v_{qu} = 3I_uV_u(\sin\alpha\sin\gamma + \cos\alpha\cos\gamma) = 3I_uV_u\cos(\alpha-\gamma)\mathrm{pu} \tag{I.31}$$

It is seen that (I.31) validates the equality of the power in d–q circuits and the power in the three phases of the stator.

Table I.1 Base quantities for per unit systems I.1 and I.3

Base quantity	**Per unit system**	
	System I.1	**System I.3**
Base quantities for stator		
S_B	$\sqrt{3}V_{L\text{–}L}I_L$	$V_{L\text{–}N}I_L$
V_B	$V_{L\text{–}L}$	$V_{L\text{–}N}$
I_B	$(S_B/V_B)=\sqrt{3}I_L$	I_L
Z_B	$(V_B/I_B)=V_{L\text{–}L}/(\sqrt{3}I_L)$	$(V_B/I_B)=V_{L\text{–}N}/(I_L)$
L_B	Z_Bt_B	Z_Bt_B
Ψ_B	V_Bt_B	V_Bt_B
Base quantities for rotor		
I_{fB}	$(L_{md}/kM_f)I_B$	$(L_{md}/kM_f)I_B$
I_{kdB}	$(L_{md}/kM_{kd})I_B$	$(L_{md}/kM_{kd})I_B$
I_{kqB}	$(L_{mq}/kM_{kq})I_B$	$(L_{mq}/kM_{kq})I_B$
Ψ_{fB}	$(I_B/I_{fB})\Psi_B$	$(I_B/I_{fB})\Psi_B$
Ψ_{kdB}	$(I_B/I_{kdB})\Psi_B$	$(I_B/I_{kdB})\Psi_B$
Ψ_{kqB}	$(I_B/I_{kqB})\Psi_B$	$(I_B/I_{kqB})\Psi_B$
V_{fB}	$(kM_f/L_{md})V_B$	$(kM_f/L_{md})V_B$
V_{kdB}	$(kM_{kd}/L_{md})V_B$	$(kM_{kd}/L_{md})V_B$
V_{kqB}	$(kM_{kq}/L_{md})V_B$	$(kM_{kq}/L_{md})V_B$

I.3.2 Base quantities for rotor

Based on the concept of equating the mutual flux linkages in each *d*-axis winding (Ψ_{md}, Ψ_{mf}, Ψ_{mkd}), the same relations obtained in Section I.2.2 to calculate the base rotor quantities I_{fB}, I_{kdB} and I_{kqB}, (I.7), (I.9) and (I.10) can be applied.

It has been shown that in the per unit system explained in Section I.1, the stator base quantities are the three-phase rated power and rated line-to-line voltage, whereas in the alternative per unit system, Section I.3, the rated power/phase and rated line-to-neutral voltage are used as stator base quantities. The other base quantities, calculated for each system accordingly, are summarized in Table I.1.

As summarised in Table I.1, it is noted that:

- The base rating in system I.1 is three times its value in system I.3.
- The stator base voltage and base current in system I.1 is $\sqrt{3}$ times its value in system I.3.
- The stator base impedance and inductance are the same for the two pu systems.
- The base value of stator flux linkage is $\sqrt{3}$ times its value in system I.3.
- The *d–q* base currents, flux linkages and voltages of rotor circuits are $\sqrt{3}$ times their values in system I.3.
- The pu value of $v_{du}^2 + v_{qu}^2$ equals V_u^2 in system I.1 while it equals 3 V_u^2 in system I.3.
- The pu three-phase stator power in system I.1 equals V_uI_u cos$(\delta-\gamma)$ and in system I.3 equals $3I_uV_u\cos(\alpha-\gamma)$.

Appendix II
Nine-bus test system

Single-line diagram

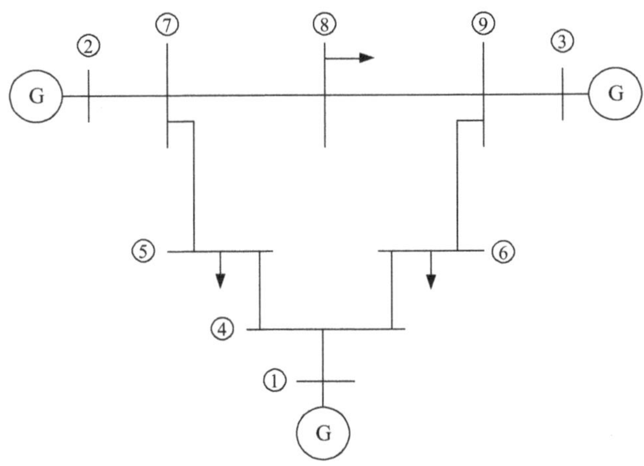

System data

Table II.1 Transmission line data on 100 MVA base

From bus number	To bus number	Series resistance (R_s) pu	Series resistance (X_s) pu	Shunt susceptance (B) pu
1	4	0	0.0576	0
4	6	0.0170	0.0920	0.1580
6	9	0.0390	0.17	0.3580
9	3	0	0.0586	0
9	8	0.0119	0.1008	0.2090
8	7	0.0085	0.0720	0.1490
7	2	0	0.0625	0
7	5	0.0329	0.1610	0.3060
5	4	0.0100	0.0850	0.1760

Table II.2 Bus data of the system

Bus no.	Bus type	Generation (pu)		Load (pu)		Voltage magnitude
		P_G	Q_G	P_L	Q_L	
1	Swing	–	–	0	0	1.04
2	PV	1.63	–	0	0	1.025
3	PV	0.85	–	0	0	1.025
4	PQ	0	0	0	0	–
5	PQ	0	0	1.25	0	–
6	PQ	0	0	0.9	0	–
7	PQ	0	0	0	0	–
8	PQ	0	0	1	0	–
9	PQ	0	0	0	0	–

Table II.3 Generator data

Generator	1	2	3
Rated MVA	247.5	192	128
KV	16.5	18	13.8
Power factor	1	0.85	0.85
Type	Hydro	Steam	Steam
Speed	180 r/min	3600 r/min	3600 r/min
X_d	0.1460	0.8958	1.3125
$X_{d'}$	0.0608	0.1198	0.1813
X_q	0.0969	0.8645	1.2578
$X_{q'}$	0.0969	0.1969	0.25
X_l (leakage)	0.0336	0.0521	0.0742
T'_{d0}	8.96	6	5.89
T'_{q0}	0	0.535	0.6
Stored energy at rated speed	2364 MW·s	640 MW·s	301 MW·s
H (MW·s/MVA)	9.55	3.33	2.35

Appendix III
Numerical integration techniques

Consider a first-order non-linear ordinary differential equation (ODE), $y' = f(x,y)$, $y(x_o) = y_o$. It has a unique solution $y = u(x)$ on the interval $I = [x_o, b]$. The solution $u(x)$ is a function at each point of I. The task of finding an approximate solution to the ODE is, therefore, one of approximating the (usually unknown) function $y = u(x)$ on I.

In general, the approximation of $u(x)$ at any one point will involve a number of arithmetic operations. Because there are an infinite number of points in I, it is not proposed to calculate an approximate value of $u(x)$ at each individual point. Therefore, the task is to find approximate values of $u(x)$ on a certain finite subset of I. The points of this subset will be denoted by $x_o, x_1, \ldots, x_m$. While it is not necessary that the points be equally spaced, it is more convenient computationally to have them so. Thus, it is assumed to approximate $u(x)$ at points $x_i = x_o + ih (i = 0, 1, \ldots, m)$. The quantity h is called the step size. The integer m is such that $x_m \leq b$, while $x_m + h > b$.

In the literature on differential equations, the exact solution at a point x_i is usually denoted by $y(x_i)$, whereas an approximation to this is denoted by y_i. Thus, the objective is to look for $y_1, y_2, \ldots, y_m$, which approximate $y(x_1), y(x_2), \ldots, y(x_m)$. Various methods described in the following sections can be used to obtain the numerical solution of ODE.

III.1 Euler's method

This method is perhaps the simplest of all numerical methods. Studying its application helps in understanding the basic ideas involved in the numerical solution of ODE.

Assuming that $f(x, y)$, x_o, y_o, h, and m are given, the numbers $x_1, x_2, \ldots, x_m$ and $y_1, y_2, \ldots, y_m$ are formed by the rules:

$$\begin{aligned} x_{i+1} &= x_i + h \\ y_{i+1} &= y_i + hf(x_i, y_i) \quad \text{for } i = 1, 2, \ldots, m - 1 \end{aligned}$$

A geometric picture of Euler's method is depicted in Figure III.1. An initial point (x_o, y_o) on the solution curve is given. The slope of the solution curve at this point is given by $f(x_o, y_o)$. Thus, the tangent line to the solution curve at the initial

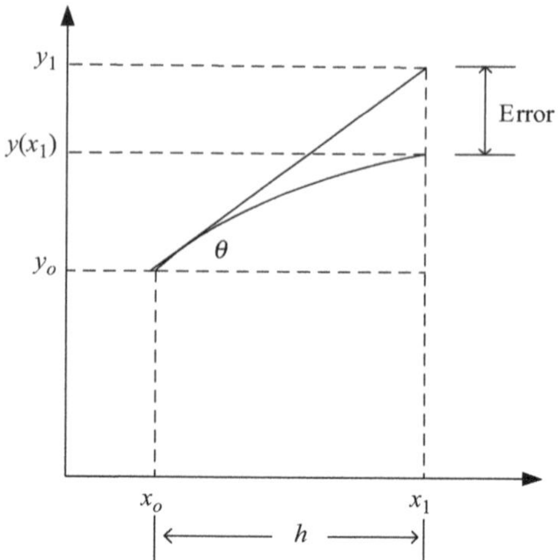

Figure III.1 One step of Euler's method

point can be determined. Euler's method consists in approximating the solution function by this tangent line.

$$\tan\theta = f(x_o, y_o) = \frac{y_1 - y_o}{h}$$

This is easily solved for y_1 giving:

$$y_1 = y_o + hf(x_o, y_o)$$

The error equals $y_1 - y(x_1)$.

This is one step of the Euler algorithm. If (x_1, y_1) is considered the initial point and the entire process is repeated, the result will be (x_2, y_2) by the rules. An error, i.e. the difference between the calculated value y_1 and the actual value $y(x_1)$, arises from the first step. Then, each subsequent step is made using an incorrect value of the slope and moving from an incorrect point under the incorrect assumption that the solution curve is a linear function.

Euler's algorithm can be extended to a system of simultaneous first-order ODEs. Suppose that the equations are given in the form

$$y_1' = f_1(x, y_1, y_2, \ldots, y_n)\; y_1(x_o) = y_{1o}$$
$$y_2' = f_2(x, y_1, y_2, \ldots, y_n)\; y_2(x_o) = y_{2o}$$
$$\vdots$$
$$y_n' = f_n(x, y_1, y_2, \ldots, y_n)\; y_n(x_o) = y_{no}$$

The problem is to find approximate values for the unknown functions $y_1(x)$, $y_2(x), \ldots, y_n(x)$ at the points $x_1, x_2, \ldots, x_m$. For a particular function, say $y_j(x)$, the exact solution is denoted by $y_j(x_o)$, $y_j(x_1), \ldots,$ $y_j(x_m)$ whereas an approximate solution is denoted by $y_{jo}, y_{j1}, y_{j2}, \ldots, y_{jm}$.

Assuming that $x_o, y_{1o}, y_{2o}, \ldots, y_{no}$, h and m are given, the numbers y_{ji} ($j = 1, 2, \ldots, m$) can be formed by the rules: $x_{i+1} = x_i + h$

$$\begin{aligned} y_{1,i+1} &= y_{1i} + hf_1(x_i, y_{1i}, y_{2i}, \ldots, y_{ni}) \\ y_{2,i+1} &= y_{2i} + hf_2(x_i, y_{1i}, y_{2i}, \ldots, y_{ni}) \\ &\vdots \\ y_{n,i+1} &= y_{ni} + hf_n(x_i, y_{1i}, y_{2i}, \ldots, y_{ni}) \quad \text{for } i = 0, 1, \ldots, m-1 \end{aligned}$$

This algorithm has been modified to what is called *modified Euler–Cauchy method* by using the rule:

$$y_{i+1} = y_i + hf\left(x_i + \frac{h}{2}, y_i + \frac{h}{2}f(x_i, y_i)\right)$$

In general, if the solution to an ODE is known to possess many derivatives, then the more involved methods will provide more accurate approximations to the solution as explained in the next sections.

III.2 Trapezoidal method

This method is attributed to Heun and sometimes called 'Heun's method'. It satisfies more accurate approximation to the solution of an ODE than that obtained by Euler's method. A differential equation in the form $y' = f(x, y), y(x_o) = y_o$ is given. If $x_{i+1} = x_i + h$, by integrating each side of the equation

$$\int_{x_o}^{x_1} y'(x)\mathrm{d}x = \int_{x_o}^{x_1} f(x, y(x))\mathrm{d}x$$

The left side may be simplified to obtain

$$y(x_1) = y(x_o) + \int_{x_o}^{x_1} f(x, y(x))\mathrm{d}x$$

If the integral on the right side is approximated by the trapezoidal rule, then

$$y(x_1) = y(x_o) + \frac{h}{2}[f(x_o, y(x_o)) + f(x_1, y(x_1))] + \textit{remainder}$$

Finally, if the quantity $y(x_1)$ in the right side is approximated by the use of the Euler's method and all remainder terms are ignored, the result is

$$y_1 = y_o + \frac{h}{2}[f(x_o, y_o) + f(x_1, y_o + hf(x_o, y_o))]$$

The result can be stated as an algorithm as below.

III.2.1 The algorithm

Assuming that $f(x, y)$, x_o, y_o, h and m are given, the numbers $x_1, x_2, \ldots, x_m$ and $y_1, y_2, \ldots, y_m$ are formed by the rules

$$x_{i+1} = x_i + h$$
$$y_{i+1} = y_i + \frac{h}{2}[f(x_i, y_i) + f(x_{i+1}, y_i + hf(x_i, y_i))] \quad \text{for } i = 0, 1, \ldots, m-1$$

It is noted that each step involves two evaluations of the function $f(x,y)$. To get y_{i+1}, it is necessary to evaluate $f(x,y)$ at (x_i, y_i) and at $f(x_{i+1}, y_i + hf(x_i, y_i))$. It is assumed that the value computed in the first evaluation will be stored so that it need not be recomputed in the second evaluation.

III.3 Runge–Kutta Methods

Different algorithms are used to numerically solve the ODE:

$$y' = f(x, y), \quad y(x_o) = y_o \quad \text{for } x \in [x_o, b]$$

to obtain the points y_{i+1} as an approximation to $y(x_{i+1})$ for $i = 0, 1, 2, \ldots, m-1$ by applying the form:

$$y_{i+1} = y_i + h\emptyset(x_i, y_i; h)$$

III.3.1 Second-Order Runge–Kutta Method

The following form is used:

$$y_{i+1} = y_i + \frac{h}{2}[K_1 + K_2]$$

where

$$K_1 = f(x_i, y_i) \quad \text{and} \quad K_2 = f\left(x_i + \frac{h}{2}, y_i + \frac{h}{2}K_1\right)$$

A geometric picture of second-order Runge–Kutta method is depicted in Figure III.2. An initial point (x_o, y_o) on the solution curve is given. The slope of the solution curve at this point is given by $K_1 = f(x_o, y_o)$. Then, a tangent line at the point $(x_o + h/2, y_o + K_1h/2)$ is drawn to determine the slope of the solution curve K_2. Thus, the tangent line to the solution curve at the initial point can be determined as $(K_1 + K_2)/2$.

The Runge–Kutta methods of higher orders have the same basic idea, but they differ in calculating the slope of the tangent line at the initial point of each step as stated below.

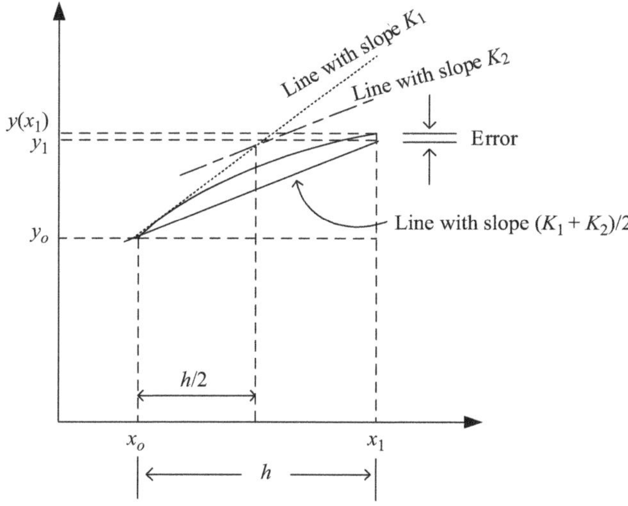

Figure 3.2 A geometric picture of one step in second-order Runge–Kutta method

III.3.2 Third-Order Runge–Kutta Method

The form used is

$$y_{i+1} = y_i + \frac{h}{6}[K_1 + 4K_2 + K_3]$$

where

$$K_1 = f(x_i, y_i), \quad K_2 = f\left(x_i + \frac{h}{2}, y_i + \frac{h}{2}K_1\right) \quad \text{and} \quad K_3 = f(x_i + h, y_i + hK_1)$$

III.3.3 Fourth-Order Runge–Kutta Method

The form used is

$$y_{i+1} = y_i + \frac{h}{6}[K_1 + 2K_2 + 2K_3 + K_4]$$

where

$$K_1 = f(x_i, y_i), \quad K_2 = f\left(x_i + \frac{h}{2}, y_i + \frac{h}{2}K_1\right),$$

$$K_3 = f\left(x_i + \frac{h}{2}, y_i + \frac{h}{2}K_2\right), \quad K_4 = f(x_i + h, y_i + hK_3)$$

This method can be extended to solve a system of simultaneous first-order ODEs. For instance, assume the two equations below are given.

$$y_1' = f_1(x, y_1, y_2) y_1(x_o) = y_{1o}$$
$$y_2' = f_2(x, y_1, y_2) y_2(x_o) = y_{2o}$$

for $x \in [x_o, b]$. Using the notation $x_i = x_o + ih$, and $y_{1,i}$, $y_{2,i}$ as the numerical approximations to $y_1(x_i)$, $y_2(x_i)$ gives

$$y_{1,i+1} = y_{1,i} + h\phi_1(x_i, y_{1,i}, y_{2,i}; h)$$
$$y_{2,i+1} = y_{2,i} + h\phi_2(x_i, y_{1,i}, y_{2,i}; h)$$

where

$$\phi_1(x_i, y_{1,i}, y_{2,i}; h) = \frac{1}{6}(K_{11} + 2K_{12} + 2K_{13} + K_{14})$$
$$\phi_2(x_i, y_{1,i}, y_{2,i}; h) = \frac{1}{6}(K_{21} + 2K_{22} + 2K_{23} + K_{24})$$

and

$$K_{11} = f(x_i, y_{1,i}, y_{2,i})$$
$$K_{21} = f(x_i, y_{1,i}, y_{2,i})$$
$$K_{12} = f_1\left(x_i + \frac{h}{2}, y_{1,i} + \frac{h}{2}K_{11}, y_{2,i} + \frac{h}{2}K_{21}\right)$$
$$K_{22} = f_1\left(x_i + \frac{h}{2}, y_{1,i} + \frac{h}{2}K_{11}, y_{2,i} + \frac{h}{2}K_{21}\right)$$
$$K_{13} = f_1\left(x_i + \frac{h}{2}, y_{1,i} + \frac{h}{2}K_{12}, y_{2,i} + \frac{h}{2}K_{22}\right)$$
$$K_{23} = f_1\left(x_i + \frac{h}{2}, y_{1,i} + \frac{h}{2}K_{12}, y_{2,i} + \frac{h}{2}K_{22}\right)$$
$$K_{14} = f_1(x_i + h, y_{1,i} + hK_{13}, y_{2,i} + hK_{23})$$
$$K_{24} = f_2(x_i + h, y_{1,i} + hK_{13}, y_{2,i} + hK_{23})$$

Appendix IV

15-bus, 4-generator system data

Single-line diagram

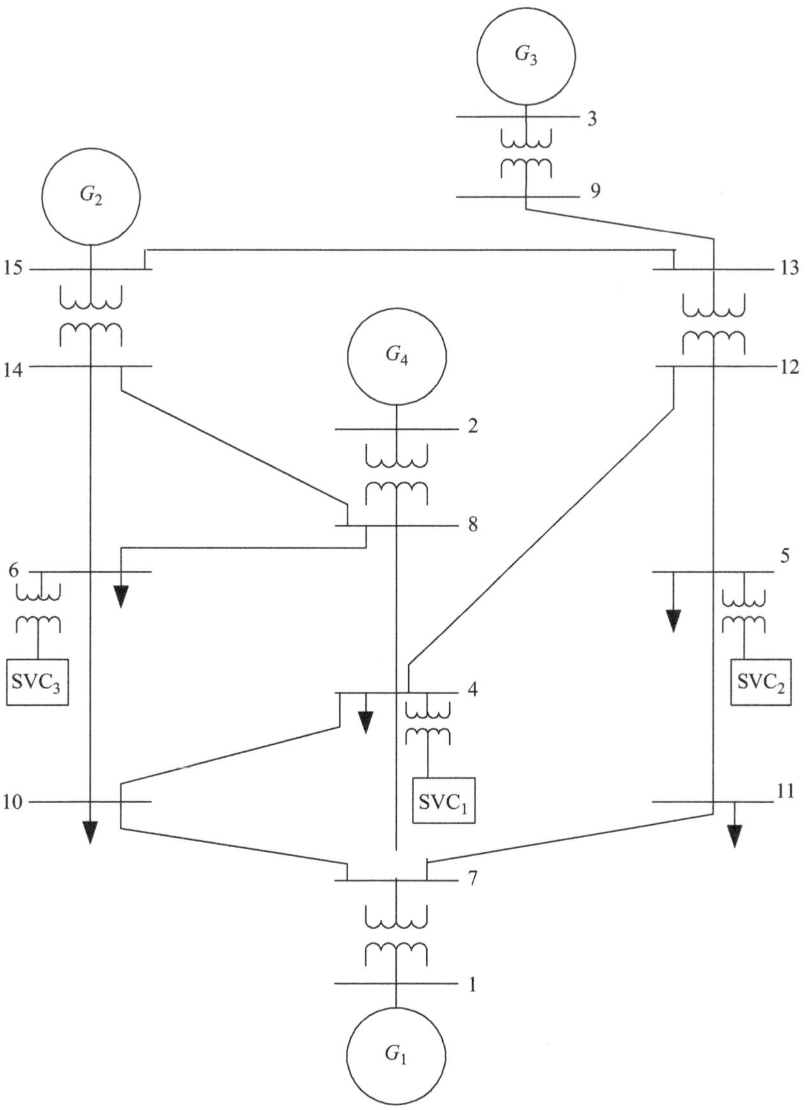

System data

Table IV.1 Impedance and line-charging data (400 MVA base)

Line designation	*R* (pu)	*X* (pu)	Line-charging* (pu)
1-7	0	0.085	0
2-8	0	0.110	0
3-9	0	0.095	0
4-7	0.0364	0.3925	0.0265
4-8	0.0276	0.2983	0.0201
4-10	0.0334	0.3611	0.0243
4-12	0.0290	0.3140	0.0212
5-11	0.0262	0.2826	0.0190
5-12	0.0378	0.4082	0.0275
6-8	0.0233	0.2512	0.0169
6-10	0.0116	0.1256	0
6-14	0.0349	0.3768	0.0254
7-10	0.0029	0.0314	0
7-11	0.0035	0.0377	0
8-14	0.0262	0.2826	0.0190
9-13	0.0026	0.0283	0
12-13	0	0.1000	0
13-15	0.0116	0.1256	0
14-15	0	0.1000	0

*Line-charging one-half of total charging of line.

Table IV.2 Static capacitor data (400 MVA base)

Bus no.	Susceptance (pu)
4	0.463
5	0.295
6	0.419

Table IV.3 Operating conditions (400 MVA base)

Bus no.	Generation (pu)		Impedance load (pu)	
	Real power	Reactive power	Real power	Reactive power
1	0.850	0.080	0	0
2	0.720	0.050	0	0
3	0.680	0.039	0	0
4	0	0	0.950	0.400
5	0	0	0.700	0.200
6	0	0	0.800	0.320
10	0	0	0.180	0.090
11	0	0	0.210	0.110
15	0.800	0.011	0	0

Table IV.4 Generator data (400 MVA base)

Generator	H (s)	D pu (MWs/rad)	X_d pu	X_q pu	X_{md} pu	X'_d pu	T_{do} pu
1	3	0.0121	1.73	1.73	1.61	0.26	7.0
2	3.3	0.0110	1.82	1.82	1.71	0.27	6.4
3	2.95	0.0117	1.80	1.80	1.68	0.31	6.0
4	3.10	0.0113	1.75	1.75	1.70	0.29	6.6

Index